THE OXFORD ENGINEERING SCIENCE SERIES

GENERAL EDITORS

A. ACRIVOS, J. M. BRADY, F. W. CRAWFORD,
A. L. CULLEN, L. C. WOODS, C. P. WROTH

THE OXFORD ENGINEERING SCIENCE SERIES

1 D. R. RHODES: *Synthesis of planar antenna sources*
2 L. C. WOODS: *Thermodynamics of fluid systems*
3 R. N. FRANKLIN: *Plasma phenomena in gas discharges*
4 G. A. BIRD: *Molecular gas dynamics*
5 H.-G. UNGER: *Optical wave guides and fibres*
6 J. HEYMAN: *Equilibrium of shell structures*
7 K. H. HUNT: *Kinematic geometry of mechanisms*
8 D. S. JONES: *Methods in electromagnetic wave propagation (two volumes)*
9 W. D. MARSH: *Economics of electric utility power generation*
11 R. HILL: *Mathematical theory of plasticity*
12 D. J. DAWE: *Matrix and finite element displacement analysis of structures*
13 N. W. MURRAY: *Introduction to the theory of thin-walled structures*
14 R. I. TANNER: *Engineering rheology*
15 M. F. KANNINEN and C. H. POPELAR: *Advanced fracture mechanics*
16 R. H. T. BATES and M. J. McDONNELL: *Image restoration and reconstruction*
18 K. HUSEYIN: *Multiple-parameter stability theory and its applications*
19 R. N. BRACEWELL: *The Hartley transform*
20 J. WESSON: *Tokamaks*

METHODS IN ELECTROMAGNETIC WAVE PROPAGATION

VOLUME 2

Radiating Waves

D. S. JONES
PROFESSOR OF MATHEMATICS,
UNIVERSITY OF DUNDEE

CLARENDON PRESS · OXFORD
1987

Oxford University Press, Walton Street, Oxford OX2 6DP
Oxford New York Toronto
Delhi Bombay Calcutta Madras Karachi
Petaling Jaya Singapore Hong Kong Tokyo
Nairobi Dar es Salaam Cape Town
Melbourne Auckland
and associated companies in
Beirut Berlin Ibadan Nicosia

Oxford is a trade mark of Oxford University Press

Published in the United States
by Oxford University Press, New York

First published (as a single volume), 1979
First issued in paperback,
(as two volumes), 1987

British Library Cataloguing in Publication Data
Jones, D. S. (Douglas Samuel)
Methods in electromagnetic wave propagation
—(Oxford engineering science series; 8)
1. Electromagnetic waves
I. Title
530.1'41 QC670
ISBN 0-19-856189-X v.1
ISBN 0-19-856190-3 v.2

Library of Congress Cataloging in Publication Data
Jones, D. S. (Douglas Samuel)
Methods in electromagnetic wave propagation
Includes bibliographies and indexes
Contents: v.1. Theory and guided waves—v.2
Radiating waves.
1. Electromagnetic waves—Transmission. I. Title
QC665.T7J66 1987 621.38 87-5472
ISBN 0-19-856189-X (pbk.: v.1)
ISBN 0-19-856190-3 (pbk.: v.2)

Printed in Great Britain
at the University Printing House, Oxford
by David Stanford
Printer to the University

Preface to Volume 2

Modern methods of tackling problems associated with electromagnetic waves involve a judicious mixture of analysis and computation. The analysis occurs in the mathematical formulation and in establishing that it has the requisite properties. Conversion to a form suitable for the computer entails numerical analysis, whose justification may also rest on a considerable body of analysis. Therefore, the aim of these two volumes is to develop a suitable framework of theory and numerical analysis with applications to various aspects of the propagation of electromagnetic waves. An attempt has been made to couch the explanation in as comprehensible a language as possible and to assume a starting point as early as commensurate with the size of the text. To assist with the understanding numerous exercises have been inserted at convenient points and some of these are open-ended so that any instructor has plenty of freedom in determining the mode of treatment. Complementary material will be found in D. S. Jones, *Acoustic and electromagnetic waves*, Oxford University Press (1986).

The subject of this volume is radiating waves whether produced directly from a transmitter or indirectly by scattering from an irradiated obstacle. Antennas are discussed in Chapter 6 with separate sections for wires, solids, and dielectrics. The analysis in Chapter 6 is concerned with the frequency domain. The changes necessary in the time domain are examined in Chapter 7, including singularity expansion method. The well-known geometric theory of diffraction receives an extensive review in Chapter 8. Finally, Chapter 9 investigates inverse scattering, embracing holography and adaptive arrays as well as other applications.

Again my thanks are due to my wife and Mrs Ross for their constant help and encouragement.

Dundee
December 1986

D. S. J.

This book is dedicated with deep affection to the Streather family and, in particular, Bessie, Kittie, Nell, Flo, Alice, Peg, and Frank.

Contents: Volume 2

CONTENTS: VOLUME 1

6

ANTENNAS AND INTEGRAL EQUATIONS

WIRE ANTENNAS

6.1. Introduction

The concept of an antenna as a piece of wire or portion of dielectric which radiates electromagnetic energy is simple enough in principle, but the derivation of quantitative results of value for design purposes is fraught with difficulties. Even when the isolated antenna can be described as a straightforward boundary-value problem, it can rarely be solved with any ease. In fact, the antenna, to be of any use as an element of a communication system, must be coupled with a transmission line or waveguide, and that coupling forms an important but complicated part of any real system. For these reasons a substantial amount of analysis has been devoted to antennas, not always with great success. The advent of large computers has made it possible to generate numerical answers to problems which had hitherto defied solution.

It must be confessed, however, that the mathematical detail has often obscured the physical principles involved leaving the engineer up in the air when both analysis and computer fail. For example, to keep computer requirements reasonable, some type of symmetry is often assumed but the symmetry is usually lost as soon as a transmission line is connected. While it is our purpose to enumerate some of the analytical and numerical techniques that have been tried, it is hoped not to lose sight completely of physical principles which may be helpful.

6.2. The perfectly conducting wire

The first antenna to be considered is a metallic wire whose cross-section is much smaller than a wavelength. The cross-section is assumed to be circular of radius a and the centre of the circle is taken to lie on a curve on which the arc length is s (Fig. 6.1). The basic approximation is to regard

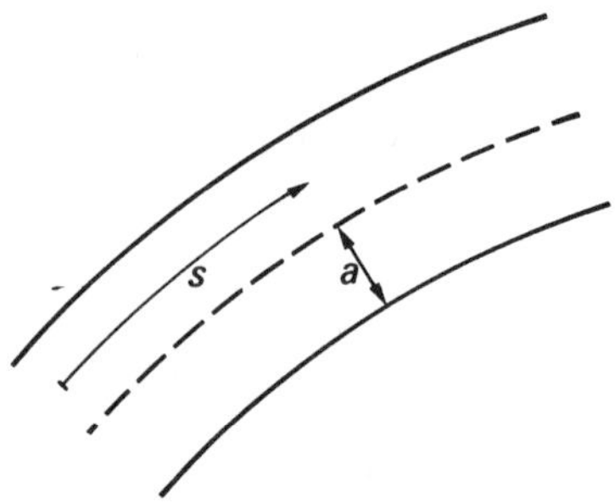

FIG. 6.1. The thin-wire antenna.

the field due to the wire as produced by a current filament along the curve of centres. Then, if $\mathbf{E}_0(\mathbf{x})$ represents the electric intensity at the point $\mathbf{x}$ from any sources present and ignoring the wire, the total electric intensity at $\mathbf{x}$ due to both sources and wire is given by

$$\mathrm{E}(\mathbf{x}) = \mathbf{E}_0(\mathbf{x}) + \frac{1}{i\omega\varepsilon_0}(\text{grad div} + k^2)\int_0^l \mathbf{I}(\xi)\psi(\mathbf{x},\xi)ds \tag{6.1}$$

for a time dependence $e^{i\omega t}$, k being the wave number and the region outside the wire being free space. The arc length of the axis of the wire has been taken as l and $\mathbf{I}$ is the current vector tangential to the axis, ξ being the point identified by the arc length s. The function ψ is defined by

$$\psi(\mathbf{x},\xi) = \frac{\exp(-ik|\mathbf{x}-\xi|)}{4\pi|\mathbf{x}-\xi|}\ . \tag{6.2}$$

The boundary conditions on the metallic wire are that the tangential components of the electric intensity vanish on the surface of the wire. Only the longitudinal component parallel to the axis of the wire will be considered here. Then, if E_s is the longitudinal component of $\mathbf{E}_0$ and $\mathbf{x}_p$ is a point on the boundary of the wire,

$$-i\omega\varepsilon_0 E_s(\mathbf{x}_p) = (\text{grad div} + k^2)_s \int_0^l \mathbf{I}(\boldsymbol{\xi})\psi(\mathbf{x}_p,\boldsymbol{\xi})\,ds \tag{6.3}$$

follows from (6.1). The current **I** is to be determined from the integral equation (6.3).

If the wire is straight so that the current filament lies along the z axis, (6.3) simplifies to

$$-i\omega\varepsilon_0 E_s(z) = \left(\frac{\partial^2}{\partial z^2}+k^2\right) \int_0^l I(s)\psi(z,s)\,ds \tag{6.4}$$

where now

$$\psi(z,s) = \frac{\exp[-ik\{(z-s)^2+a^2\}^{\frac{1}{2}}]}{4\pi\{(z-s)^2+a^2\}^{\frac{1}{2}}} . \tag{6.5}$$

It will be noticed that the right-hand side of (6.4) does not vary round the perimeter of the cross-section and therefore (6.4) can be consistent only if E_s enjoys the same property. In other words, the representation of the current distribution by an axial current filament may be accepted as a satisfactory approximation only when E_s is effectively constant on a cross-section, i.e. for sufficiently thin wires.

Since $\partial\psi(z,s)/\partial z = -\partial\psi(z,s)/\partial s$, an alternative way of writing (6.4) is

$$-i\omega\varepsilon_0 E_s(z) = \int_0^l \{k^2 I(s)\psi(z,s) + \frac{dI(s)}{ds}\frac{\partial}{\partial z}\psi(z,s)\}ds$$
$$- \frac{\partial}{\partial z}\{I(l)\psi(z,l)-I(0)\psi(z,0)\} . \tag{6.6}$$

If it be assumed that the current vanishes at the ends of the wire both $I(l)$ and $I(0)$ are zero with the consequence that the last term in (6.6) disappears.

The derivative in (6.4) can be removed by introducing two

arbitrary constants A and B. There is then the equivalent form

$$\int_0^l I(s)\psi(z,s)ds = A\cos kz + B\sin kz$$

$$- i\omega\varepsilon_0 \int_0^z E_s(t)\sin k(z-t)dt . \qquad (6.7)$$

Impositions of conditions such as the vanishing of the current at the ends of the wire will supply sufficient information to determine A and B. If the wire is in the form of a hollow tube it is reasonable to take the current at the ends as zero but, for a solid rod, currents will flow across the ends. Neglect of these currents will introduce an error of the order of ka.

Integral equations for wires were first derived by Pocklington (1897) and they were subsequently used to find the current distribution on a semi-infinite wire by MacDonald (1902).

In the absence of an external field $E_s = 0$ and an approximate solution of (6.4) is sought. Let $\delta \gg a$ be such that $k\delta \ll 1$, which is possible when a is small enough. Then, for $|s-z| > \delta$, the contribution of the integral in (6.4) is $O(1/\delta)$. In $|s-z| \leqslant \delta$, $I(s)$ may be approximated by $I(z)$ and the exponential replaced by unity. Then (6.4) becomes

$$O = \ln\left(\frac{2\delta}{a}\right)\left(\frac{\partial^2}{\partial z^2}+k^2\right)\frac{I(z)}{2\pi} + O\left(\frac{1}{\delta}\right) .$$

Thus, as a diminishes, we are forced to conclude that

$$I(z) = A_1\exp(-ikz) + B_1\exp(ikz)$$

where A_1 and B_1 are constants. Consequently, the current distribution on a thin wire is, to a first approximation, sinusoidal and propagates, without loss of amplitude, at the speed of light. In practice, this will not be quite true because of corrections allowing for a being non-zero but it is sufficiently close to the truth to be a valuable concept.

When $I(z) = I_0\exp(-ikz)$ it is possible to express the integral in (6.1) in terms of tabulated functions. Select

$$\{x^2+y^2+(z-s)^2\}^{\frac{1}{2}} + s - z$$

as a new variable of integration. Then

$$\int_0^l I_0\exp(-iks)\psi ds = I_0\exp(-ikz)\{\mathrm{Ei}(-iku_2)-\mathrm{Ei}(-iku_1)\} \tag{6.7}$$

where

$$\mathrm{Ei}(\pm ix) = -\int_x^\infty \frac{\exp(\pm iu)}{u}du = \mathrm{Ci}x \pm i \,\mathrm{si}\, x, \tag{6.8}$$

$$\mathrm{si}\, x = -\int_x^\infty \frac{\sin t}{t}dt$$

x being positive. The parameters u_1 and u_2 are defined by $u_2 = R_2 + l - z$, $u_1 = R_1 - z$ where R_1 and R_2 are the distances to the point of observation from the ends of the wire (Fig. 6.2). The electric field may now be calculated from (6.1) and has the form (the subscript w denoting the contribution of the wire)

$$(E_x)_\mathrm{w} = \frac{I_0 x}{4\pi i\omega\varepsilon_0}\left[\frac{\exp\{-ik(R_2+l)\}}{R_2^3}\left\{1-\frac{ikR_2}{u_2}(z-l)\right\}-\frac{\exp(-ikR_1)}{R_1^3}\left(1-\frac{ikR_1 z}{u_1}\right)\right] \tag{6.9}$$

with a corresponding expression for E_y while

$$(E_z)_\mathrm{w} = \frac{I_0}{4\pi i\omega\varepsilon_0}\left(ik - \frac{\partial}{\partial z}\right)\left[\frac{\exp\{-ik(R_2+l)\}}{R_2} - \frac{\exp(-ikR_1)}{R_1}\right]. \tag{6.10}$$

The evaluation of (6.9) and (6.10) entails only elementary functions which makes currents of the form $\exp(-ikz)$ very

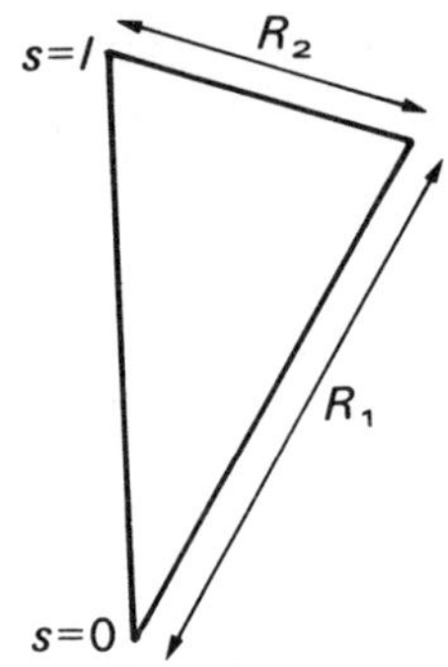

FIG. 6.2. Parameters for field due to current $\exp(-iks)$.

convenient to handle. It will also be noticed that the field involves only distances from the ends of the wire and becomes negligibly small as the length of the wire increases to infinity. Therefore, in this mode of operation, the radiation of a thin wire may be conceived of as originating from the ends of the wire. Of course, the excitation or otherwise of this mode depends upon $\mathbf{E}_0$ and so more general possibilities are examined in the next section.

If $I(z) = I_0\exp(ikz)$ the only changes to (6.9) and (6.10) are that (i) $\exp(-ikl)$ is replaced by $\exp(ikl)$, (ii) the operator $ik - \partial/\partial z$ of the right-hand side of (6.10) is altered to $-ik - \partial/\partial z$, and (iii) $u_2 = R_2 - l + z$, $u_1 = R_1 + z$.

6.3. General excitation of the infinite wire

As a start to the problem of general excitation for a straight wire the case in which $E_s(z) = \exp(-i\alpha z)$ will be considered. The constant α will be assumed to be real. Suppose firstly that the wire is infinitely long. Then it is plausible to try $I(z) = I_0\exp(-i\alpha z)$. Substitution in (6.4) gives, after the change of variable $s - z = t$,

$$-4\pi i\omega\varepsilon_0 = (k^2-\alpha^2)I_0\int_{-\infty}^{\infty}\exp\{-i\alpha t-ik(t^2+a^2)^{\frac{1}{2}}\}\,\frac{dt}{(t^2+a^2)^{\frac{1}{2}}}\,.$$

Now

$$\int_{-\infty}^{\infty} \exp\{-i\alpha t - ik(t^2+a^2)^{\frac{1}{2}}\}\frac{dt}{(t^2+a^2)^{\frac{1}{2}}} = -\pi i H_0^{(2)}\{a(k^2-\alpha^2)^{\frac{1}{2}}\} \tag{6.11}$$

where $(k^2-\alpha^2)^{\frac{1}{2}}$ is positive when $k^2 > \alpha^2$ and equals $-i(\alpha^2-k^2)^{\frac{1}{2}}$ when $k^2 < \alpha^2$. The function $H_0^{(2)}$ is the standard Hankel function of the second kind. Hence in this case

$$I(z) = \frac{4\omega\varepsilon_0 \exp(-i\alpha z)}{(k^2-\alpha^2)H_0^{(2)}\{a(k^2-\alpha^2)^{\frac{1}{2}}\}} \,. \tag{6.12}$$

The solution for the infinite straight wire under general excitation may now be deduced by Fourier transforms. For instance, if

$$E_s(z) = \int_{-\infty}^{\infty} f(\alpha)\exp(-i\alpha z)\,d\alpha \,, \tag{6.13}$$

$$I(z) = 4\omega\varepsilon_0 \int_{-\infty}^{\infty} \frac{f(\alpha)\exp(-i\alpha z)\,d\alpha}{(k^2-\alpha^2)H_0^{(2)}\{a(k^2-\alpha^2)^{\frac{1}{2}}\}} \,. \tag{6.14}$$

As an illustration, consider the model when the source is a *magnetic frill* or *ring current* of radius b concentric with the wire (Fig. 6.3). Such a model corresponds approximately to the physical situation of an infinite straight perfectly conducting cylinder excited by a coaxial transmission line, the inner conductor having the same radius as the antenna, surrounded by an infinite ground plane (Fig. 6.4). The opening in the ground plane can be regarded as creating a ring of magnetic current with no ground plane, the radius of the ring being the mean of the radii of the inner and outer conductors of the coaxial line.

Assuming symmetrical illumination, a solution is required of the equations

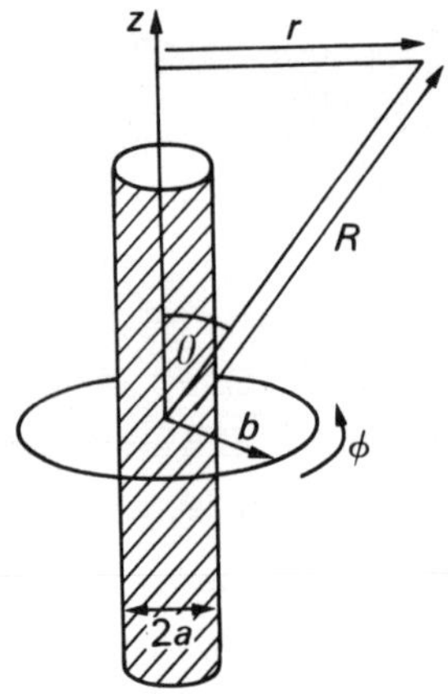

FIG. 6.3. Magnetic ring current of radius b.

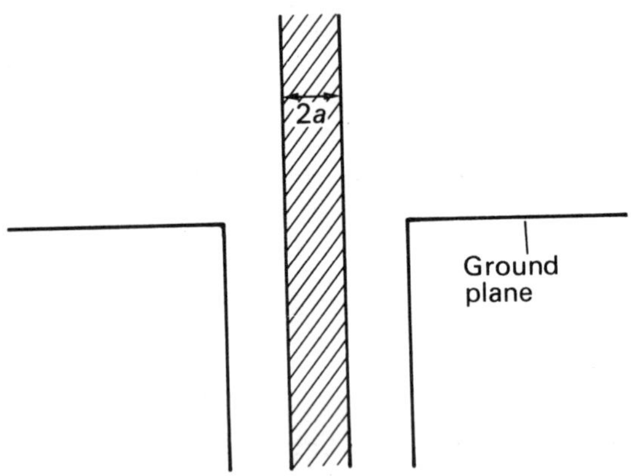

FIG. 6.4. A coaxial line feeding a semi-infinite monopole.

$$\operatorname{curl} \mathbf{E}_0 + i\omega\mu_0\mathbf{H}_0 = -\delta(r-b)\delta(z)\hat{\boldsymbol{\phi}}, \tag{6.15}$$

$$\operatorname{curl} \mathrm{H}_0 - i\omega\varepsilon_0\mathbf{E}_0 = 0 \tag{6.16}$$

where r,ϕ,z are cylindrical polar coordinates and $\hat{\boldsymbol{\phi}}$ is a unit vector in the direction of ϕ increasing. It may be verified that the solution obtained by Fourier transform in the absence of the wire is for $r > b$

$$E_{0z} = -\tfrac{1}{4}\mathrm{i}b \int_{-\infty}^{\infty} \kappa H_0^{(2)}(\kappa r)J_0'(\kappa b)\exp(-\mathrm{i}\alpha z)\,\mathrm{d}\alpha\ , \tag{6.17}$$

$$E_{0r} = -\tfrac{1}{4}b \int_{-\infty}^{\infty} \alpha H_0^{(2)\prime}(\kappa r)J_0'(\kappa b)\exp(-\mathrm{i}\alpha z)\,\mathrm{d}\alpha\ , \tag{6.18}$$

$$H_{0\phi} = -\tfrac{1}{4}\omega\varepsilon_0 b \int_{-\infty}^{\infty} H_0^{(2)\prime}(\kappa r)J_0'(\kappa b)\exp(-\mathrm{i}\alpha z)\,\mathrm{d}\alpha \tag{6.19}$$

where $\kappa^2 = k^2 - \alpha^2 = \omega^2\mu_0\varepsilon_0 - \alpha^2$. The other components of the field are zero. When $r < b$, the non-zero components are

$$E_{0z} = -\tfrac{1}{4}\mathrm{i}b \int_{-\infty}^{\infty} \kappa J_0(\kappa r)H_0^{(2)\prime}(\kappa b)\exp(-\mathrm{i}\alpha z)\,\mathrm{d}\alpha\ , \tag{6.20}$$

$$E_{0r} = -\tfrac{1}{4}b \int_{-\infty}^{\infty} \alpha J_0'(\kappa r)H_0^{(2)\prime}(\kappa b)\exp(-\mathrm{i}\alpha z)\,\mathrm{d}\alpha\ , \tag{6.21}$$

$$H_{0\phi} = -\tfrac{1}{4}\omega\varepsilon_0 b \int_{-\infty}^{\infty} J_0'(\kappa r)H_0^{(2)\prime}(\kappa b)\exp(-\mathrm{i}\alpha z)\,\mathrm{d}\alpha\ . \tag{6.22}$$

Formula (6.20) is the one for the representation analogous to (6.13). Therefore, from (6.14), the current induced in the wire by the field $\mathbf{E}_0$ is given by

$$I(z) = -\mathrm{i}\omega\varepsilon_0 b \int_{-\infty}^{\infty} \frac{J_0(\kappa a)H_0^{(2)\prime}(\kappa b)}{\kappa H_0^{(2)}(\kappa a)} \exp(-\mathrm{i}\alpha z)\,\mathrm{d}\alpha\ . \tag{6.23}$$

The contour in (6.23) may be deemed to pass below the branch point $\alpha = -k$ and above the branch point $\alpha = k$ (Fig. 6.5). For positive values of z the contour may be deformed into the path

FIG. 6.5. Contour of integration for the infinite wire.

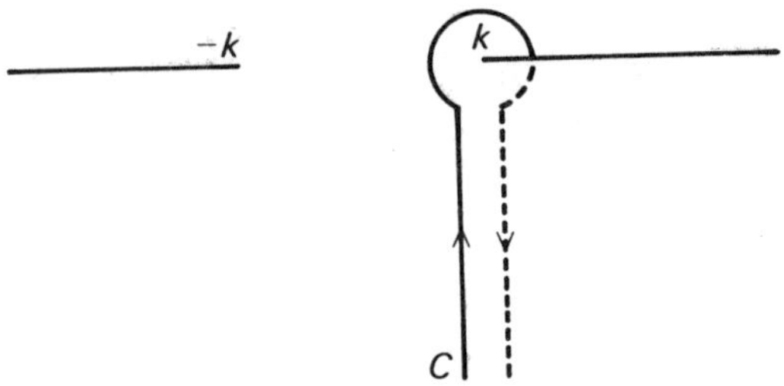

FIG. 6.6. Deformed contour of integration.

C (Fig. 6.6) which goes parallel to the imaginary axis from $-i\infty$, circles $\alpha = k$, crosses the branch line, and returns to $-i\infty$. Thus

$$I(z) = -i\omega\varepsilon_0 b \int_C \frac{J_0(\kappa a)H_0^{(2)\prime}(\kappa b)}{\kappa H_0^{(2)}(\kappa a)} \exp(-i\alpha z)\mathrm{d}\alpha \; . \tag{6.24}$$

In the deformation contributions due to the residues at the poles (if any) might be picked up. However, $H_0^{(2)}(z)$ is free of zeros in $0 \geqslant \mathrm{ph}\, z \geqslant -\pi$ and so there are no poles to be accounted for.

When z is large, the integral in (6.24) can be evaluated asymptotically. By the general theory for such integrals (see, for example, Jones 1972) the dominant part comes from a neighbourhood of the branch point $\alpha = k$. Therefore, as $z \to \infty$,

$$I(z) \sim -i\omega\varepsilon_0 \int_C \frac{\exp(-i\alpha z)\mathrm{d}\alpha}{\kappa^2(\ln\frac{1}{2}\kappa a+\gamma+\frac{1}{2}\pi i)}$$

where $\gamma = 0.5772\ldots$ is Euler's constant. Replacing $k + \alpha$ by $2k$ we may simplify this to

$$I(z) \sim -\frac{2\pi\omega\varepsilon_0}{k} \exp(-ikz)f_1(z) \tag{6.25}$$

where

$$f_1(z) = \int_0^\infty \frac{\exp(-t)dt}{t[\{\gamma+\ln(ka^2t\ \exp(\frac{1}{2}\pi i)/2z)\}^2+\pi^2]} . \tag{6.26}$$

It will be noticed that (6.25) does not involve b so that the radial position of the magnetic frill is of no consequence as far as the current far along the wire is concerned. Therefore, the same distant current would be obtained if the rod had been excited by a narrow gap in its length.

The function f_1 depends only on the single parameter $2z/ka^2$ in which $\frac{1}{2}ka^2$ is the cross-sectional area divided by the wavelength. It may therefore be tabulated conveniently (Kunz 1963, Bach Andersen 1968). This is best done by numerical integration but a rough idea of the behaviour of f_1 can be obtained by considering what happens when $2z/ka^2 \gg 1$. In that case

$$f_1(z) \sim \int_0^d \frac{e^{-t}dt}{t\{\ln(t/d^2)\}^2} \tag{6.27}$$

where $d^2 = 2z/ka^2\exp(\gamma+\frac{1}{2}\pi i)$. The error in (6.27) is $O(e^{-d})$. An integration by parts gives

$$\begin{aligned} f_1(z) &\sim -\int_0^d \frac{e^{-t}dt}{\ln(t/d^2)} \\ &\sim \int_0^d e^{-t} \int_0^\infty \exp\{s\ln(t/d^2)\}ds\,dt \\ &\sim \int_0^\infty \exp(-2s\ln d) \int_0^d \exp(-t)t^s dt\,ds \\ &\sim \frac{1}{2\ln d}\left\{\int_0^d \exp(-t)dt + \int_0^\infty \exp(-2s\ln d) \int_0^d \exp(-t)t^s \ln t\,dt\,ds\right\} \end{aligned}$$

by a further integration by parts. It is evident that

$$f_1(z) \sim \frac{1}{2\ln d} = \frac{1}{\ln\{2z/ka^2\exp(\gamma+\frac{1}{2}\pi i)\}} \tag{6.28}$$

when $|d| \gg 1$.

According to (6.28), $|f_1|$ decreases steadily as $|d|$ increases whereas ph f_1 increases steadily. Similar behaviour is exhibited by (6.26). The approximation (6.28) should be accurate to within 10 per cent for the modulus when $|d| > 10$ and for the phase when $|d| > 100$.

On the basis of (6.28) and (6.25)

$$I(z) = \frac{4\pi V\exp(-ikz)f_1(z)}{Z_0} = Y(z)V\exp(-ikz) \tag{6.29}$$

if the strength of the magnetic ring is multiplied by $-2V$ and $Z_0 = (\mu_0/\varepsilon_0)^{\frac{1}{2}}$ is the impedance of free space. The admittance $Y(z)$ is given approximately by

$$Y(z) \sim 4\pi/2Z_0\ln d\ . \tag{6.30}$$

The current wave decays slowly with distance along the rod. It travels nearly at the speed predicted by Pocklington for an infinitesimally thin wire. The slow diminution in amplitude means that the wave is guided by the wire over long distances, gradually radiating energy.

The radiation field can be calculated from (6.1) and

$$\mathbf{H}(\mathbf{x}) = \mathbf{H}_0(\mathbf{x}) + \operatorname{curl}\int_{-\infty}^{\infty}\mathbf{I}(\xi)\psi(\mathbf{x},\xi)\mathrm{d}s\ . \tag{6.31}$$

Thus, when $r > b$, (6.19), (6.23), and (6.11) supply

$$Z_0H_\phi = -\tfrac{1}{4}kb\int_{-\infty}^{\infty}H_0^{(2)\prime}(\kappa r)\left\{J_0'(\kappa b)-J_0(\kappa a)\frac{H_0^{(2)\prime}(\kappa b)}{H_0^{(2)}(\kappa a)}\right\}\exp(-\mathrm{i}\alpha z)\mathrm{d}\alpha\ . \tag{6.32}$$

If $kr \gg 1$, the derivative of the Hankel function can be replaced by the first term in its asymptotic development so long as α is not near $\pm k$, which can always be arranged by deforming the contour in Fig. 6.5 so that it does not pass too close to these points. Hence

$$Z_0H_\phi \sim \tfrac{1}{4}ikb \int_{-\infty}^{\infty} \exp\{-i(\kappa r - \tfrac{1}{4}\pi)\}\left(\frac{2}{\pi\kappa r}\right)^{\frac{1}{2}}\left\{J_0'(\kappa b) - J_0(\kappa a)\frac{H_0^{(2)\prime}(\kappa b)}{H_0^{(2)}(\kappa a)}\right\}\exp(-i\alpha z)\,d\alpha \, .$$

The exponent $\kappa r + \alpha z$ has a stationary point at $\alpha = kz/R$ or $\alpha = k\cos\theta$ where R and θ are spherical polar coordinates. Hence the method of stationary phase gives

$$Z_0H_\phi \sim \tfrac{1}{2}kb\{J_0(ka\sin\theta)H_0^{(2)\prime}(kb\sin\theta) - J_0'(kb\sin\theta)H_0^{(2)}(ka\sin\theta)\}\frac{\exp(-ikR)}{RH_0^{(2)}(ka\sin\theta)} \, . \tag{6.33}$$

If $kb \ll 1$, this formula simplifies to

$$Z_0H_\phi \sim \left[2\sin\theta\{\gamma + \ln(\tfrac{1}{2}ka\exp(\tfrac{1}{2}\pi i)\sin\theta)\}\right]^{-1}\frac{\exp(-ikR)}{R} \tag{6.34}$$

and to this order of approximation the radial position of the magnetic ring is of no significance. It will be observed that the radiation is largest when θ is near 0 or π, consistent with the earlier conclusion that the rod guides the field over long distances. However, it must be borne in mind that near the wire kr is not large and so the approximation made in going from (6.32) to (6.34) is invalid.

Other components of the distant field can be written down from the radiation conditions. Thus $E_\theta = Z_0H_\phi$ and

$E_r = -E_\theta \cos\theta$, $E_z = E_\theta \sin\theta$. The total power radiated can be calculated by integrating the Poynting vector over a surface at infinity. This may be expressed as $\frac{1}{8}G$ where G is the *radiation conductance*, the reciprocal of the *radiation resistance*. Hence

$$G = \frac{4\pi}{Z_0}(kb)^2 \int_0^{\frac{1}{2}\pi} |J_0(ka \sin\theta) H_0^{(2)\prime}(kb \sin\theta) - J_0'(kb \sin\theta) H_0^{(2)}(ka \sin\theta)|^2$$

$$\times \frac{\sin\theta d\theta}{|H_0^{(2)}(ka \sin\theta)|^2} \tag{6.35}$$

if (6.33) is used even for $\theta \approx 0$.

The integral in (6.35) has been investigated numerically (Duncan 1962). If $kb \ll 1$, the simplification (6.34) may be employed. In the resulting integral, $\sin\theta$ may be replaced by θ in the interval $(0,\varepsilon)$ where $\ln\varepsilon \gg \ln(\frac{1}{2}ka e^\gamma)$ and $\ln \sin\theta$ may be neglected in the interval $(\varepsilon, \frac{1}{2}\pi)$. Hence

$$G \approx -4\pi/Z_0 \ln(\tfrac{1}{2}ka e^\gamma) \tag{6.36}$$

for $kb \ll 1$. Formula (6.36) is fairly accurate for $ka < 1/20$ so long as kb does not exceed moderate values because (6.35) is relatively insensitive to variations in kb.

If one wishes to regard the effect of the antenna in Fig. 6.4 on the coaxial line as the same as a circuit element placed appropriately on the line, the conductance will be $\frac{1}{2}G$ because the radiation takes place only above the ground plane. There will also be a susceptance to take account of the energy stored in the capacitor composed of the ground plane and rod. Good theoretical estimates of this susceptance do not seem to be available (Bach Andersen 1968) so no details will be given.

It is worth emphasizing that the formulae of this section also hold for thick antennas providing that the initial requirement of E_s not varying round the cross-section is met. In particular, this will be true whenever the rod is illuminated symmetrically.

There is one further point to be elucidated and that is the admittance observed at the ring current. The admittance well away from the source has been derived in (6.29). Letting $z \to 0$, $I(0) = Y(0)V$ where, however, $Y(0)$ is no longer expressed in terms of f_1 but is obtained from (6.24) by putting $z = 0$. If $kb \ll 1$, only integration near k is relevant and if an analysis similar to that in the derivation of (6.36) is carried out it is found that

$$Y(0) = -\frac{2\pi\{\ln ka+\gamma+\frac{1}{2}\pi i\}^{-1}}{Z_0} \qquad (6.37)$$

for very thin wires. The resemblance of the real part of (6.37) to $\frac{1}{2}G$ will be noted; it is an indication of the connection between the admittance at the ring and the radiated energy.

6.4. The finite travelling-wave antenna

Before examining the finite wire it is worth spending a moment on the semi-infinite wire supporting a travelling wave. Such an antenna can be achieved in practice by adding a resistor one-quarter wavelength from the end (Altshuler 1961).

If the rod extends from $-\infty$ to $z = l$ ($l>0$) the current distribution can be regarded as $I(z)$ from $z = -\infty$ to $z = \infty$ together with $-I(z)$ from $z = l$ to $z = \infty$. The radiated pattern will therefore consist of (6.33) plus a correction factor. The correction may be evaluated by taking $I(z)$ to have the form (6.25) which is legitimate so long as the source is not near $z = l$. Thus the correction term in Z_0H_ϕ is

$$\frac{1}{2}ik \sin\theta \frac{\exp(-ikR)}{R} \int_l^\infty f_1(s) \exp\{-iks(1-\cos\theta)\}ds \;. \qquad (6.38)$$

If $kl(1-\cos\theta) \gg 1$, this may be approximated by

$$\frac{1}{2l}\frac{\sin\theta}{1-\cos\theta} f_1(l) \frac{\exp\{-ikR-ikl(1-\cos\theta)\}}{R} \;. \qquad (6.39)$$

For general purposes, one of the integrations in (6.38) may be carried out with the result that

$$\int_{l}^{\infty} f_1(s)\exp(-\mathrm{i}\beta s)\mathrm{d}s$$

$$= \exp(-\mathrm{i}\beta l)\int_0^{\infty} \frac{\mathrm{e}^{-t}\mathrm{d}t}{(t+\mathrm{i}\beta)t[\{\gamma+\ln(ka^2 t\exp(\frac{1}{2}\pi\mathrm{i})/l\}^2+\pi^2]} . \tag{6.40}$$

For a finite travelling-wave antenna, extending from $z = -l_1$ to $z = l$, with a magnetic ring source at $z = 0$ between the two ends the correction to the pattern for the infinite wire is

$$\tfrac{1}{2}\mathrm{i}k\sin\theta\,\frac{\exp(-\mathrm{i}kR)}{R}\left\{\int_{l}^{\infty} f_1(s)\exp\{-\mathrm{i}ks(1-\cos\theta)\}\mathrm{d}s\, + \right.$$

$$\left. + \int_{l_1}^{\infty} f_1(s)\exp\{-\mathrm{i}ks(1+\cos\theta)\}\mathrm{d}s\right\} \tag{6.41}$$

and both terms may be expressed in the form (6.40).

In one limiting case integration may be dispensed with. Suppose $l = l_1$ and that kl is sufficiently large for (6.40) to be applicable. Suppose also that kb is small enough for (6.28) and (6.34) to hold. Then the combination of (6.34) and (6.41) gives

$$Z_0H_\phi \sim \left(\frac{1}{2\sin\theta\,\ln\{\frac{1}{2}ka\exp(\gamma+\frac{1}{2}\pi\mathrm{i})\sin\theta\}} - \right.$$

$$- \frac{\sin\theta}{2\ln\{\frac{1}{2}ka^2\exp(\gamma+\frac{1}{2}\pi\mathrm{i})/l\}}\left\{\frac{\exp\{-\mathrm{i}kl(1-\cos\theta)\}}{1-\cos\theta} + \right.$$

$$\left.\left. + \frac{\exp\{-\mathrm{i}kl(1+\cos\theta)\}}{1+\cos\theta}\right\}\right)\frac{\exp(-\mathrm{i}kR)}{R} . \tag{6.42}$$

6.5. The finite wire

When the wire is not specially terminated so that only a travelling-wave current develops, any wave set up will be reflected at the end and a standing-wave pattern may be expected on the wire. The problem is then much more complicated. Several analytical devices have been employed to tackle the solution of the integral equation (see, for example, Jones 1964) and numerical methods will be described later. Here we shall content ourselves with an approach based on splitting the standing wave into travelling waves.

The main difficulty is to account for the reflection at an end. Obviously, this will depend to some extent on the exact shape of the end and whether it is flat or rounded. Nevertheless some insight can be gained by considering the particular case in which the antenna is a hollow tube with very thin walls. It is then plausible to visualize the reflection as creating a magnetic ring current at the end.

Suppose this ring current is of magnitude $-2V_1$ (that in (6.15) being of unit magnitude). Then, if a travelling wave due to a ring current of magnitude $-2V$ at the origin encounters the end at $z = l$, the current (including the reflected wave) is

$$I(z) = VY(z)\exp(-ikz) + V_1Y(l-z)\exp\{ik(z-l)\} .$$

The current I must, however, vanish at $z = l$ and so

$$VY(l)\exp(-ikl) + V_1Y(0) = 0$$

which determines V_1. Hence

$$I(z) = V[Y(z)\exp(-ikz)-Y(l)Y(l-z)\exp\{ik(z-l)\}/Y(0)] .$$

An estimate for $Y(0)$ is provided by (6.37) for wires of very small radius. A check on it may be derived by solving the canonical problem of a semi-infinite tube exactly. It is found (Hallén 1956) that there is agreement with (6.37). Thus $Y(0)$ is known for thin wires. For larger radii recourse to the original defining integral is necessary; empirical determination from experiment is also a possibility.

The wave emanating from $z = l$ now travels down the wire to $z = -l$ and is reflected there, the reflected wave being $-V_1 Y(2l)\exp\{-2ikl-ik(z+l)\}Y(z+l)/Y(0)\}$. This wave is reflected at $z = l$ and the process continues. The net contribution to the current in $z > 0$ is

$$VY(z)\exp(-ikz) - \frac{VY(l)\exp(-ikl)Y(0)}{\{Y(0)\}^2-\{Y(2l)\}^2\exp(-4ikl)}\left[Y(l-z)\exp\{ik(z-l)\}-\right.$$

$$\left. - \frac{Y(2l)}{Y(0)}\exp(-2ikl)Y(z+l)\exp\{-ik(z+l)\}\right] .$$

In addition, there is a set of waves initiated by the wave going from the source to the end $z = -l$. When the two sets are combined it is discovered that

$$I(z) = VY(z)\exp(-ikz) -$$

$$- \frac{VY(l)\exp(-ikl)}{Y(0) + Y(2l)\exp(-2ikl)}\left[Y(l-z)\exp\{ik(z-l)\} + \right.$$

$$\left. + Y(z+l)\exp\{-ik(z+l)\}\right] \tag{6.43}$$

on $z > 0$.

If we write $I(0) = VY$, (6.43) gives

$$Y = Y(0) - \frac{2\{Y(l)\}^2\exp(-2ikl)}{Y(0) + Y(2l)\exp(-2ikl)} . \tag{6.44}$$

For thin rods where (6.37) and (6.30) may be employed, (6.44) reduces to

$$Y = -\frac{2\pi}{Z_0 \ln\{\frac{1}{2}ka \exp(\gamma+\frac{1}{2}\pi i)\}} +$$

$$+\frac{16\pi \ln\{\frac{1}{2}ka \exp(\gamma+\frac{1}{2}\pi i)\}}{Z_0\left\{\ln \frac{ka^2\exp(\gamma+\frac{1}{2}\pi i)}{2l}\right\}^2\left[\exp(2ikl)+1+\ln\{4kl \exp(\gamma+\frac{1}{2}\pi i)/\ln\} \frac{ka^2\exp(\gamma+\frac{1}{2}\pi i)}{4l}\right.}.$$

If $kl = \frac{1}{2}\pi$ so that the antenna is a half wavelength long, (6.45) becomes

$$Y = -\frac{2\pi}{Z_0 \ln\{\frac{1}{2}ka \exp(\gamma+\frac{1}{2}\pi i)\}} + \\ +\frac{16\pi \ln\{\frac{1}{2}ka \exp(\gamma+\frac{1}{2}\pi i)\}\ln\{k^2a^2\exp(\gamma+\frac{1}{2}\pi i)/2\pi\}}{Z_0[\ln\{k^2a^2\exp(\gamma+\frac{1}{2}\pi i)/\pi\}]^2\ln\{2\pi \exp(\gamma+\frac{1}{2}\pi i)\}}.$$

In the limit as $ka \to 0$, this simplifies to

$$Y = \frac{8\pi}{Z_0 \ln\{2\pi \exp(\gamma+\frac{1}{2}\pi i)\}}.$$

Since $Z_0 = 120\pi$ ohms, this gives an impedance of 36.3 + 23.6i ohms in agreement with the other theories mentioned earlier.

The admittance Y is what will be observed by the feed in Fig. 6.4 when the antenna projects a distance l if the equivalent circuit element of the antenna is placed in the coaxial line at a half wavelength from the opening in the ground plane (assuming that the ground plane is perfectly conducting). In general, it will be best to determine $Y(0)$ from measurements with one or two different values of l since (6.37) holds only for tubes of small cross-section. The experiments of Hartig (see King 1956) provide a convenient reference point for antennas with hemispherical ends.

The influence of a load inserted in the antenna can also be calculated by the above technique provided that it can be treated as a lumped circuit.

Exercises

1. The current $I_0\exp(-ikz)$ flows in a thin wire along the positive z axis. Show that

$$E_z = \frac{I_0}{4\pi i\omega\varepsilon_0}\left(\frac{\partial}{\partial z} - ik\right)\frac{\exp(-ikR)}{R}$$

where R is the distance from the origin.

2. A travelling wave of current $I_0\exp(-ikz)$ flows along a thin straight wire extending from $z = -\frac{1}{2}l$ to $z = \frac{1}{2}l$. Show that the radiation resistance is

$$60\left\{\ln\,2kl - \mathrm{Ci}\,2kl + \gamma - 1 + \frac{\sin\,2kl}{2kl}\right\}.$$

3. If the *gain* of an antenna is defined as 4π times the maximum value of the polar diagram divided by the total power radiated show that the gain in dB in Exercise 2 when $kl \gg 1$ is

$$5.97 - 10\log_{10}\frac{0.915 + \log_{10}(l/\lambda)}{l/\lambda}$$

assuming that $\frac{1}{2}\tan 1.16 = 1.16$.

4. In Fig. 6.4 a resistor R is inserted in the antenna at a distance h from the ground plane. Find the admittance Y observed at the reference plane in the coaxial line.

If $ka = 1/25$ find the value of R which gives a travelling wave up to the resistor when $l-h = \frac{1}{4}\lambda$ (experimentally R is about 240 ohms). If $R = \infty$ is there a value of h for which there is a travelling wave?

5. By combining (6.42) with the patterns from the magnetic ring currents at the ends show that the radiation pattern of an infinitely thin antenna of length $2l$ is given by

$$Z_0H_\phi \sim \frac{\cos kl - \cos(kl\cos\theta)}{\sin\theta\cos kl}\,\frac{\exp(-ikR)}{R}.$$

6.6. The receiving antenna

Most of the preceding theory has been concerned with the radiation characteristics of a straight wire when excited by a magnetic ring current surrounding the wire and therefore refers, in essence, to the situation in which the antenna is used as a transmitting device. There is no real loss of generality in concentrating on the transmitting mode because properties of the antenna in a receiving state can be deduced from it. The connection between transmission and reception can be elucidated under very general circumstances and therefore specific reference to the straight wire can be omitted.

The basic feature is to represent the antenna as a network placed in some suitable position. Usually this is somewhere in the feed to the antenna where it is feasible to talk in terms of currents and voltages. Here currents and voltages are used in a generalized sense. For instance, if a single-mode waveguide is conveying the energy to or from the antenna, they can be transverse components of the dominant mode at a convenient reference plane (see §2.2). A place where this representation in currents and voltages is adopted will be called a *port*.

The antenna will be supposed to possess two surfaces S_o, S_i, of which it presents S_o to the outside world while S_i is coupled directly to ports (multiple feeds being permitted). The surfaces S_o and S_i are bounded, with V the volume between, and parts of them may coincide (Fig. 6.7). The only requirement on V is that it be linear and passive; the tangential components of **E** and **H**

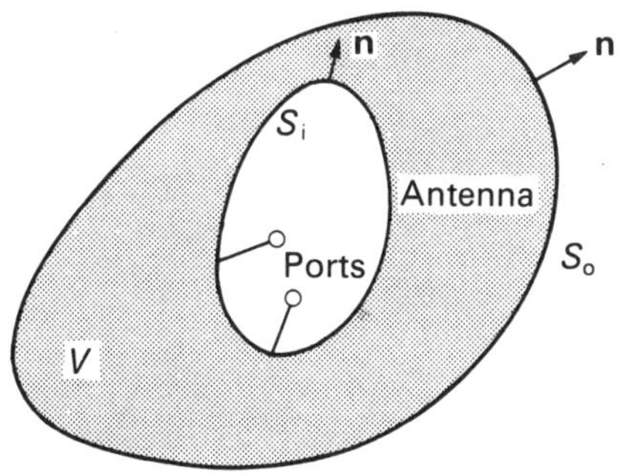

FIG. 6.7. Configuration of an antenna.

are continuous across any discontinuities in the medium. Outside S_0 is free space though it could be replaced by any linear, homogeneous, isotropic, and lossless medium.

When the antenna is transmitting let the current I_n^T and the voltage V_n^T be supplied at the nth port. Then, owing to the linearity of the governing equations,

$$V_m^T = \sum_{n=1}^{N} Z_{mn} I_n^T \qquad (m = 1, \ldots, N) \tag{6.46}$$

if there are N ports. The impedances Z_{mn} define a network characterization of the antenna in the transmitting mode. The average power supplied to the system is

$$P^T = \tfrac{1}{2}\mathcal{R} \sum_{m=1}^{N} V_m^T I_m^{T*} = \tfrac{1}{2}\mathcal{R} \sum_{m=1}^{N} \sum_{n=1}^{N} Z_{mn} I_n^T I_m^{T*} \tag{6.47}$$

from (6.46), the asterisk indicating a complex conjugate. Owing to the linearity of the system

$$P^T = \tfrac{1}{2}\mathcal{R} \int_{S_i} \mathbf{E}^T \wedge \mathbf{H}^{T*}.\mathbf{n}\,\mathrm{d}S = \tfrac{1}{2}\mathcal{R} \int_{S_o} \mathbf{E}^T \wedge \mathbf{H}^{T*}.\mathbf{n}\,\mathrm{d}S$$

when the antenna between S_i and S_o is lossless.

The surface S_o may be replaced by a large sphere at infinity without altering P^T. The radiation conditions (§2.1) imply that on this sphere

$$\mathbf{E}^T \sim \mathbf{e}^T \frac{\exp(-ikR)}{R}, \qquad \mathbf{H}^T \sim \mathbf{h}^T \frac{\exp(-ikR)}{R}$$

where the fields $\mathbf{e}^T$, $\mathbf{h}^T$ depend only upon angle and are transverse to the radius vector. Thus, if Ω is the surface of the unit sphere

$$P^{\mathrm{T}} = \tfrac{1}{2}\mathcal{R}\int_{\Omega} \mathbf{e}^{\mathrm{T}} \wedge \mathbf{h}^{\mathrm{T}*}.\mathbf{n}\mathrm{d}\Omega = \frac{1}{2Z_0}\int_{\Omega} \mathbf{e}^{\mathrm{T}}.\mathbf{e}^{\mathrm{T}*}\mathrm{d}\Omega$$

since $\mathbf{h}^{\mathrm{T}} \sim \hat{\mathbf{R}} \wedge \mathbf{e}^{\mathrm{T}}/Z_0$. The dependence of $\mathbf{e}^{\mathrm{T}}$ on I_n^{T} may be denoted by writing

$$\mathbf{e}^{\mathrm{T}} = \sum_{n=1}^{N} I_n^{\mathrm{T}}\mathbf{e}_n^{\mathrm{T}}. \tag{6.48}$$

Then (6.47) gives the equation

$$\mathcal{R}\sum_{m=1}^{N}\sum_{n=1}^{N} Z_{mn}I_n^{\mathrm{T}}I_m^{\mathrm{T}*} = \frac{1}{Z_0}\sum_{m=1}^{N}\sum_{n=1}^{N} I_n^{\mathrm{T}}I_m^{\mathrm{T}*}\int_{\Omega} \mathbf{e}_n^{\mathrm{T}}.\mathbf{e}_m^{\mathrm{T}*}\mathrm{d}\Omega . \tag{6.49}$$

Put $I_m^{\mathrm{T}} = 1$ and all other currents zero in (6.49). Then

$$\mathcal{R}Z_{mm} = \frac{1}{Z_0}\int_{\Omega} |\mathbf{e}_m^{\mathrm{T}}|^2\mathrm{d}\Omega . \tag{6.50}$$

More generally, it is clear that

$$Z_{mn} + Z_{nm}^{*} = \frac{2}{Z_0}\int_{\Omega} \mathbf{e}_n^{\mathrm{T}}.\mathbf{e}_m^{\mathrm{T}*}\mathrm{d}\Omega \tag{6.51}$$

which includes (6.50). Eqn (6.51) relates the impedance elements of the network to the radiated far field (de Hoop 1975, van Bladel 1966).

The derivation of (6.51) is based on the constitutive equations relating the flux densities **D**, **B** linearly to **E**, **H**. A further result can be obtained if **D** depends only on **E** while **B** depends only on **H**. In that case let $\mathbf{E}^{\mathrm{T}(1)}$ and $\mathbf{E}^{\mathrm{T}(2)}$ be two possible fields produced by different excitations of the transmitting ports. Then

$$\int_{S_o} \{\mathbf{E}^{T(1)} \wedge \mathbf{H}^{T(2)} - \mathbf{E}^{T(2)} \wedge \mathbf{H}^{T(1)}\} \cdot \mathbf{n} dS = 0$$

as can be seen by pushing the surface of integration out to infinity and applying the radiation conditions. Because of the particular constitutive relations under consideration the surface S_o can be changed to S_i. But, on S_i the description in terms of the ports is valid. Hence

$$\sum_{m=1}^{N} V_m^{T(1)} I_m^{T(2)} = \sum_{m=1}^{N} V_m^{T(2)} I_m^{T(1)}.$$

The insertion of (6.46) then enables the deduction

$$Z_{mn} = Z_{nm} \tag{6.52}$$

for any pair of subscripts m and n. The symmetry displayed in (6.52) cannot be expected in the general case.

When the antenna is acting as a receptor the excitation will be assumed to be a plane wave travelling in the direction of the unit vector $\mathbf{n}_0$. Consequently the incident field can be expressed as

$$\mathbf{E}^i = \mathbf{E}_0 \exp(-ik\mathbf{n}_0 \cdot \mathbf{x}) ,$$

$$Z_0 \mathbf{H}^i = \mathbf{n}_0 \wedge \mathbf{E}_0 \exp(-ik\mathbf{n}_0 \cdot \mathbf{x})$$

where the complex constant vector $\mathbf{E}_0$ satisfies $\mathbf{n}_0 \cdot \mathbf{E}_0 = 0$ in order that the wave be plane. Let the total field generated by this incident wave be $\mathbf{E}^R$, $\mathbf{H}^R$. Now

$$\int_{S_i} (\mathbf{E}^T \wedge \mathbf{H}^R - \mathbf{E}^R \wedge \mathbf{H}^T) \cdot \mathbf{n} dS = \int_{S_o} (\mathbf{E}^T \wedge \mathbf{H}^R - \mathbf{E}^R \wedge \mathbf{H}^T) \cdot \mathbf{n} dS \tag{6.53}$$

so long as

$$\mathbf{H}^R \cdot \mathbf{B}^T + \mathbf{E}^T \cdot \mathbf{D}^R - \mathbf{H}^T \cdot \mathbf{B}^R - \mathbf{E}^R \cdot \mathbf{D}^T = 0 \tag{6.54}$$

holds throughout V. The relation (6.54) is a requirement on the electromagnetic properties of the antenna in the transmitting and receiving modes. It will certainly be valid when the medium is reciprocal and may also be true for non-reciprocal media if the constitutive equations have the appropriate form.

The integral on the left-hand side of (6.53) can be rewritten in terms of the currents and voltages on the ports. In the receiving situation let V_n^R be the voltage at the nth port and I_n^R the current flowing *into* the port (i.e. in the opposite direction to I_n^T). Then

$$\sum_{m=1}^{N} (V_m^T I_m^R + V_m^R I_m^T) = -\int_{S_0} (\mathbf{E}^T \wedge \mathbf{H}^R - \mathbf{E}^R \wedge \mathbf{H}^T) \cdot \mathbf{n}\, dS\ .$$

The surface of integration may be shifted out to infinity since (6.54) is certainly satisfied outside S_0. Since the difference between $\mathbf{E}^R, \mathbf{H}^R$ and $\mathbf{E}^i, \mathbf{H}^i$ is a radiating field we deduce that

$$\sum_{m=1}^{N} (V_m^T I_m^R + V_m^R I_m^T) = -\int_{S_R} (\mathbf{E}^T \wedge \mathbf{H}^i - \mathbf{E}^i \wedge \mathbf{H}^T) \cdot \mathbf{n}\, dS$$

where S_R is the surface of the sphere of radius R. In view of the special form of $\mathbf{E}^i$ the integral can be evaluated by the method of stationary phase with the result

$$\sum_{m=1}^{N} (V_m^T I_m^R + V_m^R I_m^T) = -\frac{4\pi}{i\omega\mu_0} \mathbf{e}^T(-\mathbf{n}_0) \cdot \mathbf{E}_0 \tag{6.55}$$

where $\mathbf{e}^T(-\mathbf{n}_0)$ is the value of $\mathbf{e}^T$ in the direction of $-\mathbf{n}_0$. The substitution of (6.46) and (6.48) into (6.55) leads to an equation which must be true for arbitrary I_m^T. Hence

$$\sum_{n=1}^{N} Z_{nm} I_n^R + V_m^R = -\frac{4\pi}{i\omega\mu_0} \mathbf{e}_m^T(-\mathbf{n}_0) \cdot \mathbf{E}_0 \tag{6.56}$$

for $m = 1, \ldots, N$.

Eqns (6.56) provide a network description of the receiving situation. The right-hand side represents the driving force due to the incident plane wave while V_m^R is due to an internal voltage source. The impedance matrix is the transpose of the matrix which occurs when the antenna is transmitting.

The internal voltage depends upon how the ports are loaded. If, for example, there is a load impedance Z_{mn}^L then

$$V_m^R = \sum_{n=1}^{N} Z_{mn}^L I_n^R \qquad (m = 1, \ldots, N).$$

If all the ports are open $I_n^R = 0$ for all n and measurement of V_m^R determines the right-hand side of (6.56). If the measurements are repeated for two distinct polarizations of the plane wave $\mathbf{e}_m^T(-\mathbf{n}_0)$ is determined. By varying $\mathbf{n}_0$ the variations of $\mathbf{e}^T$ over the unit sphere can be found. Thus the radiation pattern when transmitting can be derived from the characteristics of the antenna when it is subject to an incident plane wave, provided that the condition (6.54) is complied with.

It is of some interest to examine under what circumstances the power dissipated in the load is a maximum for a given incident plane wave, i.e. when the receiving antenna is *matched* for maximum power transfer. Let Z and Z^L be the matrices with elements Z_{mn} and Z_{mn}^L respectively. Then (6.56) may be abbreviated to

$$(Z'+Z^L)\mathbf{I} = \mathbf{V}$$

where the prime indicates a transpose (to avoid using T which has just been employed to identify the transmitting situation). The power dissipated in the load is

$$\tfrac{1}{2}\mathcal{R}\mathbf{I}^* Z^L \mathbf{I} = \tfrac{1}{2}\mathcal{R}\mathbf{V}^* Y^{*\prime} Z^L Y\mathbf{V}$$

where Y is the inverse of $Z' + Z^L$. Thus the dissipated power is

$$\tfrac{1}{2}\mathcal{R}\mathbf{V}^*(Y^{*\prime}-Y^{*\prime}Z'Y)\mathbf{V}.$$

If changes in Z^L are made so that Y becomes $Y + \varepsilon Y_1$, the alteration in dissipated power is

$$\tfrac{1}{2}\mathcal{R}\mathbf{V}^*(\varepsilon Y_1^{*\prime}-\varepsilon Y_1^{*\prime}Z'Y-\varepsilon Y^{*\prime}Z'Y_1-\varepsilon^2 Y_1^{*\prime}Z'Y_1)\mathbf{V}\ . \tag{6.57}$$

Since $\mathcal{R}\mathbf{V}^*Z\mathbf{V} = \mathcal{R}\mathbf{V}^*Z^{*\prime}\mathbf{V}$, this is stationary if

$$\tfrac{1}{2}\mathcal{R}\mathbf{V}^*(Y_1^{*\prime}-Y_1^{*\prime}Z'Y-Y_1^{*\prime}Z^*Y)\mathbf{V} = 0\,.$$

But Y_1 is an arbitrary complex matrix so this is possible only if $(Z'+Z^*)Y$ is the unit matrix, i.e.

$$Z^L = Z^*\ . \tag{6.58}$$

Granted the truth of (6.58), maximum power transfer must occur because the only remaining term in (6.57) is negative since the real part of Z' is forced to be positive semi-definite in a passive system. The condition for matching therefore is that the receiver load impedance should be the complex conjugate of the impedance matrix of the antenna when transmitting. In this optimal regime the powers absorbed by the load and internally by the network are obviously the same. Hence the maximum power in the load is half that delivered by the sources. The condition of optimality (6.58) is clearly independent of the incoming plane wave.

Exercises

6. If $\mathbf{E}^R - \mathbf{E}^i \sim \mathbf{e}^s\exp(-ikR)/R$ as $R \to \infty$ show that

$$\sum_{m=1}^{N}\sum_{n=1}^{N}(Z^L_{mn}+Z^L_{nm}{}^*)I^R_m{}^*I^R_n + \frac{2}{Z_0}\int_\Omega \mathbf{e}^s.\mathbf{e}^s{}^*d\Omega = -\mathcal{R}\frac{\mathbf{E}_0^*\cdot\mathbf{e}^s(\mathbf{n}_0)}{i\omega\mu_0}\ .$$

7. If there is only a single port, define the efficiency η of the load by

$$\eta = \frac{16P^{\mathrm{T}}P^{\mathrm{L}}}{|V_1^{\mathrm{T}}I_1^{\mathrm{R}}+V_1^{\mathrm{R}}I_1^{\mathrm{T}}|^2}$$

where P^{L} is the power dissipated in the load. Prove that

$$\eta = 1 - \frac{|Z_{11}^*-Z_{11}^{\mathrm{L}}|^2}{|Z_{11}+Z_{11}^{\mathrm{L}}|^2}$$

and deduce that the maximum efficiency of unity is obtained when the load is matched.

6.7. Numerical methods

The analysis so far presented is admirable in bringing out the physical features of the behaviour but it has its limitations. Much of it relies upon the antenna being straight and of very small radius, as well as being illuminated by a relatively simple field. When one or more of these conditions are not met, numerical techniques need to be employed. For the moment, it will be assumed that the representation (6.1) is still a valid approximation, i.e. attention will be confined to antennas of small cross-section.

The governing integral equation is then (6.3) and its form will now be considered in more detail. The point $\mathbf{x}_{\mathrm{p}}$ may be taken as a point $\mathbf{x}$ of the filament to a first approximation. Let $\boldsymbol{\tau}(\mathbf{x})$ be a unit vector tangential to the filament at the point $\mathbf{x}$. Then, if $\boldsymbol{\xi}$ is the point where the arc length is s, the integral equation may be written as

$$-i\omega\varepsilon_0\boldsymbol{\tau}(\mathbf{x}) \cdot \mathbf{E}_0(\mathbf{x}) = \int_0^l I(s)[k^2\boldsymbol{\tau}(\mathbf{x}).\boldsymbol{\tau}(\boldsymbol{\xi}) + \{\boldsymbol{\tau}(\mathbf{x}).\mathrm{grad}\}\{\boldsymbol{\tau}(\boldsymbol{\xi}).\mathrm{grad}\}]\psi_1(\mathbf{x},\boldsymbol{\xi})ds \qquad (6.59)$$

where

$$\psi_1(\mathbf{x},\boldsymbol{\xi}) = \frac{\exp(-ik|\mathbf{x}-\boldsymbol{\xi}+\mathbf{a}|)}{4\pi|\mathbf{x}-\boldsymbol{\xi}+\mathbf{a}|} .$$

The vector $\mathbf{a}$ has magnitude equal to the radius of the cross-section at $\boldsymbol{\xi}$. Its direction is parallel to $\tau(\boldsymbol{\xi}) \wedge (\mathbf{x}-\boldsymbol{\xi})$; if τ and $\mathbf{x} - \boldsymbol{\xi}$ are parallel, any convenient orientation perpendicular to τ may be chosen so long as the direction varies continuously along the wire.

Eqn (6.59) is of the first kind, and efforts to solve such equations numerically go back as far as 1879 when Maxwell was attempting to determine the capacitance of a square metal plate. In essence, his method was that of collocation (§5.6(b)) to find the parameters in a finite-element analysis (§5.8). Nowadays, Galerkin (§5.1) and point-matching (§5.6(a)) techniques are available as well as Fourier methods (§5.6(c)). Nevertheless, it is still by no means a trivial business to deduce the best way of tackling (6.59). One general observation should be made. It has been explained in §5.7 that it is advantageous for certain numerical aspects if the kernel of an integral equation of the first kind is as singular as possible. Therefore, results for tiny radii of cross-section are likely to be more reliable than those for thicker wires.

The fundamental idea in all approaches is to select a set of known basis functions $\{\phi_m\}$ and to expand the unknown I as a linear combination of them, i.e.

$$I = \sum_{m=1}^{M} a_m \phi_m . \qquad (6.60)$$

The values of the coefficients a_m are to be discovered and any of the techniques in Chapter 5 may be adopted, but it must not be supposed that they will all lead to the same answers.

The first matter to be examined is the choice of ϕ_m. The evidence of §5.4, although applying to positive operators and thereby excluding (6.59), suggests that at the least they should be strongly minimal for a stable numerical performance. Since an orthonormal set is strongly minimal early analytical work based on Fourier series was in the right direction (though, strictly, orthonomality with respect to the operator is preferable). There has been some numerical investigation (Richmond 1965) along these lines including also expansion in Chebyshev, Hermite, and Legendre (§1.5) polynomials. Generally speaking, though, the trend has been away from such full-domain representations and towards expansions based on subdomains. The reason is partly that full-domain representations are less flexible when dealing with wires of arbitrary shape and partly because subdomain expansions give comparable accuracy with less computer effort. Primarily this is because integration has to be carried out over the whole wire for each term in a full-domain representation whereas integration over short segments is all that is necessary when subdomains are handled.

Let the axis be subdivided at arc lengths $s_0, s_1, \ldots, s_M$ with s_0 and s_M corresponding to the ends of the wire so that $s_0 = 0$ and $s_M = l$. There are many possible choices for basis functions for these subdivisions. A list of some that have been employed is as follows (only the interval where the basis function is non-zero is shown).

Pulse or *piecewise constant* functions:

$$\phi_m(s) = 1 \quad (s_{m-1} \leqslant s \leqslant s_m).$$

Piecewise linear or *triangle* functions:

$$\phi_m(s) = \frac{h - |s-s_{m-1}|}{h} \quad (s_{m-2} \leqslant s \leqslant s_m).$$

Piecewise sinusoidal:

$$\phi_m(s) = \frac{\sin k(h-|s-s_{m-1}|)}{\sin kh} \quad (s_{m-2} \leqslant s \leqslant s_m).$$

Trigonometric:

$$\phi_m(s) = 1 + B_m \cos k\{s-\tfrac{1}{2}(s_m+s_{m-1})\} + C_m \sin k\{s-\tfrac{1}{2}(s_m+s_{m-1})\}$$

$$(s_{m-1} \leqslant s \leqslant s_m).$$

Quadratic:

$$\phi_m(s) = 1 + B_m(s-s_{m-1}) + C_m(s-s_{m-1})^2 \qquad (s_{m-1} \leqslant s \leqslant s_m).$$

In the triangle and piecewise sinusoidal approximations the points of subdivision are equally spaced with $s_m - s_{m-1} = h$ for all m. It will also be noticed that these two require an extra segment of length h on the ends so that there are $M + 1$ basis functions rather than M. The trigonometric and quadratic representations contain two constants B_m and C_m which are at our disposal; they may be adjusted so as to satisfy extra conditions such as the continuity of I.

To apply the boundary conditions that $I(0) = 0$, $I(l) = 0$ it is merely necessary to set $a_1 = 0$ and $a_M = 0$ for the pulse basis functions. For the piecewise linear and piecewise sinusoidal put a_1 and a_{M+1} both zero. In the case of the trigonometric and quadratic basis functions the constants B_1, C_1, B_{M+1}, and C_{M+1} of the end-most intervals are adjusted to make $I(0)$ and $I(l)$ vanish.

Both the trigonometric and piecewise sinusoidal expansions have the advantage that a numerical integration can be evaded for a straight antenna by drawing benefit from (6.10). For example,

$$\left(\frac{\partial^2}{\partial z^2} + k^2\right) \int_{s_{m-2}}^{s_m} \sin k(h-|s-s_{m-1}|)\psi ds$$

$$= \frac{k}{4\pi}\{\psi_m(z)+\psi_{m-2}(z)-2\psi_{m-1}(z)\cos kh\} \qquad (6.61)$$

where

$$\psi_m(z) = \frac{\exp[-ik\{(z-s_m)^2+a^2\}^{\frac{1}{2}}]}{\{(z-s_m)^2+a^2\}^{\frac{1}{2}}}.$$

Therefore if an expansion of I is made in terms of the piecewise sinusoidal basis function and subsectional collocation is employed with matching at $s = s_m$, the system of equations

$$\frac{k}{\sin kh}\sum_{m=2}^{M} a_m\{\psi_m(s_n)+\psi_{m-2}(s_n)-2\psi_{m-1}(s_n)\cos kh\}$$

$$= - 4\pi i\omega\varepsilon_0 E_{0z}(s_n) \qquad (6.62)$$

is deduced from (6.61) and (6.4). The coefficients a_1 and a_{M+1} are missing because of the boundary conditions at the end of the wire and (6.62) is implemented at $n = 1, \ldots, M - 1$. There are therefore exactly the right number of equations to determine the unknown coefficients a_m and their solution may be dealt with by the techniques of §§1.12 and 1.13. Apparently, as M increases from 3 or 4, the numerical answer converges rapidly to the correct solution, but this is illusory because further increase in M makes the linear system (6.62) ill conditioned (§1.12) (Taylor and Wilton 1972). This is related to the fact that $\psi_m(s_m)$ is very large for small radii so that the combination $a_n + a_{n+2} - 2a_{n+1}\cos kh$ dominates in (6.62). This combination is a multiple of the discontinuity of the derivative of the representation (6.60) as s goes from just above s_n to just below. Thus the main effort of the process is expended in trying to secure the continuity of the first derivative of the representation. Because of this emphasis with its concentration on the points s_n the coefficients a_m found may well be such that the approximate current gives an E_z which is badly in error when s is not a matching point, as may be verified easily in practice (Pearson and Butler 1975). Only when the discontinuity in the derivative can be expected to be naturally small is (6.62) likely to furnish good results. Such an eventuality will arise when the current distribution on the antenna is sinusoidal, i.e. when $kl = n\pi$ and the antenna is resonant. Although (6.62) dispenses with the need for numerical integration, the procedure cannot be recommended except possibly for the resonant antenna whose length is an integral number of half wavelengths or almost so.

One may attempt to improve the smoothness by averaging the equations instead of enforcing them at the solitary points s_n. In other words, a Galerkin procedure (§5.1) is adopted in which integration is performed after multiplication by a suitable weight function. Thus, in (6.62), $\psi_m(s_n)$ is replaced by $\int_0^l \psi_m(z) w_n(z) dz$. If the weight functions are chosen as piecewise sinusoidal, numerical integration is again escaped. In practice, this choice yields good results (Butler and Wilton 1975) when the antenna is split into 30 or 40 segments. A typical output is shown in Fig. 6.8. However, the selection of weight functions is crucial because it is known that some options deliver erroneous answers. For example (Klein and Mittra 1975), if the weight functions are pulse functions whose width need not be h, wildly different approximations arise for different pulse widths and the differences are not due to the numerical integration involved. While the influence of weight functions is incompletely understood it is always wise to check how well the approximate current arrived at reproduces E_{0z} over the whole length of the wire. If the boundary condition is satisfied everywhere with tolerable accuracy one can hope to have attained a reasonable approximation, but wide discrepancies in various intervals will, in general, imply a poor approximation to the current.

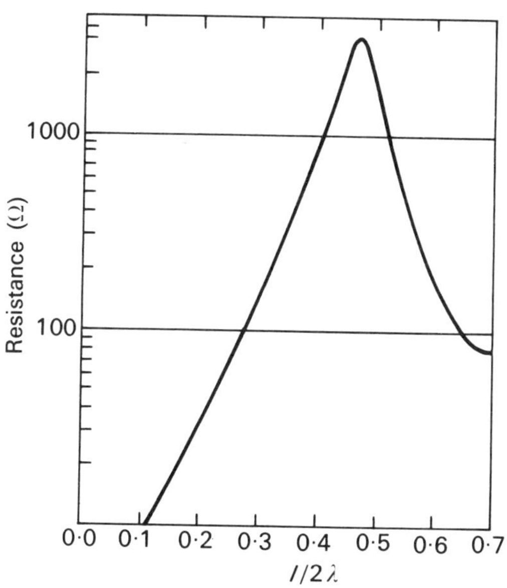

FIG. 6.8. Input resistance as a function of frequency when $l/2a$ = 2000.

All of these remarks apply with equal strength if the triangle functions are adopted for the basis since, when $h \ll \lambda$ (which is advisable to keep proper track of phase variations), there is not much difference between the piecewise sinusoidal and triangle functions.

A trigonometric basis offers also the opportunity of avoiding numerical integration. The constants B_m and C_m are selected so that the current vanishes at the ends and the current, together with its first derivative, is continuous across $s = s_m$ $(m = 1, \ldots, M - 1)$. There are exactly the right number of conditions to fix the B_m and C_m. Then the field is matched at M points so that the a_m are educed. This method, which overcomes the embarrassment of the discontinuous first derivative described above, has accuracy comparable with the Galerkin method (with piecewise sinusoidal weight functions) already mentioned.

Another possibility for the straight antenna is to work with the integral equation (6.7). Because of the integration on the right-hand side the equations are less critically dependent on the values of the applied electric field at particular points. Furthermore, the absence of a derivative on the left-hand side means that it is less sensitive to discontinuities in the approximating currents. Accordingly, the use of triangle functions will not cause trouble. By matching at $s = s_0, \ldots, s_M$ we obtain $M + 1$ equations for the $M - 1$ current coefficients and the two unknown constants A and B. The accuracy gained by this method is of the same order as that for the trigonometric basis except, perhaps, at the lower values of M.

There is one topic which has not yet been alluded to and that is the feeding arrangement. The case of the antenna as a transmitting device may be handled by propounding the existence of a gap $l_1 \leqslant s \leqslant l_2$ in which the feed is introduced (Fig. 6.9). Then, if

$$\int_{l_1}^{l_2} E_z(s)\,ds = -V \quad ,$$

V may be regarded as the voltage of the source. The input impedance Z and input admittance Y of the antenna can then be

FIG. 6.9. The gap where the feed is inserted.

defined by

$$Z = V/I(l_2) = 1/Y \ . \tag{6.63}$$

However, things are not quite straightforward because now it is required that the current in the wire produce zero E_z on the wire and the correct E_z in the gap. However, the integral equation holds only on the wire and so, if only points on the wire are considered, we are led to a homogeneous system of linear equations whose solution is, consequently, non-unique. Two ways out of this impasse have been suggested.

The first is to take E_z in the gap as $-V/(l_2-l_1)$ and then to require the current to reproduce this value. The non-uniqueness disappears but it is found, in practice, that no reliance can be placed on the impedance (Miller and Deadrick 1973) calculated by this means. Some improvement occurs if the end segments of the wires adjacent to the gap are made the same length as the gap. This is because, as pointed out in §6.3, the input susceptance is very responsive to small changes near the gap. The input conductance is relatively insensitive, so any numerical method that entails fluctuations in the conductance for minor variations in the gap length may be safely discarded as unreliable.

The second method is to specify the current at the gap, say

$I(l_2) = I(l_1) = 1$. The system of equations from the wire then becomes determinate. With $I(s)$ known, E_z can be evaluated in the gap and thus V acquired. In fact, the integral can be taken from 0 to l since E_z is supposed to be zero on the wire. Owing to the approximations carried out this will not be exactly true and the alterations in V with integration length provide a measure of reliability. Usually V settles down once the integration extends beyond two or three central segments. It is, of course, vital to employ a numerical method which eliminates the spikiness in E_z already referred to in connection with piecewise sinusoidal basis functions. On the whole, this method leads to more satisfactory values than the first.

As soon as a satisfactory approximant to the current has been achieved, the far-field pattern can be computed from

$$\mathbf{E}(\mathbf{x}) \sim \mathbf{E}_0(\mathbf{x}) + \\ + \frac{k^2}{4\pi i\omega\varepsilon_0}\frac{\exp(-ikR)}{R}\sum_{m=1}^{M} a_m \int_0^l \left[\boldsymbol{\tau}(s) - \left\{\boldsymbol{\tau}(s)\cdot\frac{\mathbf{x}}{R}\right\}\frac{\mathbf{x}}{R}\right]\phi_m(s)\exp\left(\frac{ik\mathbf{x}\cdot\boldsymbol{\xi}}{R}\right)ds \tag{6.64}$$

where the fast Fourier transform (§2.14) may be expedient. The first term is absent when the antenna is transmitting. The *power gain* of the transmitter may be taken as

$$\frac{\{|E_\theta|^2 + |E_\phi|^2\}R^2}{30|I|^2 \mathcal{R}Z}$$

where I is the current at the feed. The total power radiated also follows in a straightforward way and, if it be written as $\frac{1}{2}|I|^2 R_r$, the *radiation resistance* R_r is known. The input power is $\frac{1}{2}|I|^2 \mathcal{R}Z$ so the radiation efficiency is $R_r/\mathcal{R}Z$.

A measure of the back-scattering by a receiving antenna is the *radar cross-section* σ_r. This is defined for an incident plane wave by means of the intensity scattered back towards the origins of the plane wave. Thus

$$\sigma_r = \lim_{R\to\infty} \frac{4\pi R^2 |\mathbf{E}_s|^2}{|\mathbf{E}_0|^2}$$

where $\mathbf{E}_s$ is the field scattered in the specified direction. The radar cross-section depends not only on the geometry and materials of the antenna but also on the wavelength, polarization, and direction of arrival of the incident wave. The problem of determining the reflection characteristics of an antenna of complex shape via the radar cross-section is therefore a formidable task (cf. §7.7).

If the incident wave is not truly plane the radar cross-section may misrepresent the back-scattering because of the influence of the curvature of the wavefront.

Exercises

8. A plane wave falls on a straight receiving antenna of length l and radius a. Compare the current obtained at the centre of the wire by the various numerical methods described in this section as M increases from 5. How large does M have to be for you to have confidence in the results? Take $a = 0.01\lambda$ and give l a number of values between 0.3λ and 1.5λ.

Would your conclusions be modified if (i) $a = 0.001\lambda$, (ii) $a = 0.1\lambda$?

9. Examine how the input admittance of a centre-fed antenna varies as the width of the gap is narrowed. Use both methods above and decide which you prefer. Consider the same values of a and l as in Exercise 8.

10. Calculate the radiation pattern and radiation efficiency for the antennas in Exercise 9. Do the variations in the pattern surprise you?

6.8. Curved antennas

It is pretty futile to try to circumvent numerical integration by working with piecewise sinusoidal basis functions when the antenna is not straight unless perhaps it is polygonal in shape. Even then extreme care would be necessary to mitigate the bother caused by discontinuities in derivatives already encountered. One may also anticipate that approximating the antenna by a polygon will lead to numerical difficulties because of the discontinuities in slope so embodied. Therefore, it is fitting to investigate schemes which incorporate an appropriate amount of continuity right from the beginning.

Apposite basis functions are the B-splines discussed in §1.1 and now commonly utilized for computer-aided design in mechanical engineering, but it will be convenient to consider them more fully than was done there. Let $t_1, t_2, \ldots, t_m$ be m real numbers such that $t_1 < t_2 < \ldots < t_m$; these are to be the knots of the spline. With x_+^n meaning x^n if $x > 0$ and 0 if $x \leqslant 0$, the B-spline of degree $n - 1$ is defined by

$$M_{ni}(t) = \sum_{r=i-n}^{i} \frac{(t_r - t)_+^{n-1}}{\omega'(t_r)} \tag{6.65}$$

where $\omega(t) = (t-t_{i-n})(t-t_{i-n+1}) \ldots (t-t_i)$ and the prime is a derivative with respect to the argument. The definition (6.65) is more general than that of §1.1, allowing the knots to be arbitrarily spaced and the degree of the B-spline to be chosen freely. On taking $n = 4$ and the knots equispaced at integer intervals, we can confirm without difficulty that (6.65) gives $\frac{1}{6}B_i^s$ as defined in (1.14).

The knots of M_{ni} are at $t_{i-n}, t_{i-n+1}, \ldots, t_i$ and M_{ni} vanishes identically for $t \leqslant t_{i-n}$ and $t \geqslant t_i$. For $t_{i-n} < t < t_i$, the B-spline is strictly positive (Schoenberg 1946, Curry and Schoenberg 1947, 1966). In addition, M_{ni} and its derivatives of order $1, 2, \ldots, n-2$ are continuous for all t.

A general spline for the finite interval $a \leqslant t \leqslant b$ can be constructed by taking $t_1 > a$, $t_m < b$ and adding n knots at each end so that

$$t_{1-n} < t_{2-n} < \ldots < t_0 \leqslant a, \qquad b \leqslant t_{m+1} < t_{m+2} < \ldots < t_{m+n} .$$

Then

$$S(t) = \sum_{i=1}^{m+n} c_i M_{ni}(t).$$

An arbitrary function may be likewise expanded and the coefficients determined by reproducing the function values at $m + n$

nodes which may be distributed conveniently apart from some mild restrictions on their relation to knots. The equations for the coefficients have a simple structure because every row of the matrix has at most four non-zero elements (three if the nodes and knots coincide).

It is tempting to substitute the definition (6.65) and write the expansion as

$$\sum_{i=1}^{n} a_i t^{i-1} + \sum_{j=1}^{m} b_j (t-t_j)_+^{n-1}$$

but this is unprofitable for numerical purposes, being especially bad for interpolation. The equations to determine a_i and b_j are extremely ill conditioned even if n is small. Moreover, even when a_i and b_j are known accurately, the evaluation of the sum engenders many arithmetic operations and can experience severe loss of accuracy due to cancellation (Carasso 1966). Nor is it really satisfactory to compute M_{ni} direct from (6.65) on account of instability which arises. Instead, it is better to use the recurrent relation

$$M_{ni}(t) = \frac{(t-t_{i-n})M_{n-1,i-1}(t) + (t_i-t)M_{n-1,i}(t)}{t_i - t_{i-n}} . \qquad (6.66)$$

This process is stable, accurate, and fast (Cox 1972) as well as permitting one to take advantage of the fact that $M_{1j}(t)$ is zero except in $[t_{j-1}, t_j)$ where it is $1/(t_j-t_{j-1})$.

Stable algorithms for derivatives and integrals of B-splines are also available (de Boor 1972, 1973).

The B-splines of (6.65) are normalized so that

$$\int_{-\infty}^{\infty} M_{ni}(t)\,dt = 1/n.$$

For our purposes, renormalization possesses some convenience. Define

$$N_{ni}(t) = (t_i - t_{i-n})M_{ni}(t) \; . \tag{6.67}$$

Then

$$\sum_{i=1}^{m+1} N_{ni}(t) = 1 \tag{6.68}$$

for $a \leqslant t \leqslant b$, as may be established quickly by induction from (6.66) and the fact that (6.68) is true when $n = 1$.

From now on only cubic B-splines will be discussed, so that $n = 4$. For simplicity, N_i will be written instead of N_{4i}.

In view of the particular simple form which B-splines assume when the interval between the knots is unity, a parameter will be introduced to convert the knots to integer values. In other words, a class of *parametric* B-splines will be the subject of investigation. Such parametric B-splines are used extensively in the aircraft industry in planning the geometry of an aeroplane (Roberts and Rundle 1972).

Let $N_i(\sigma)$ be the normalized cubic B-spline which has knots at the integers $i - 4$, $i - 3$, $i - 2$, $i - 1$, i. Then $N_{i+1}(\sigma)$ has the same shape as N_i but is shifted along the σ axis by a unit length. It is intended to regard σ as a parameter so that a given curve is described as σ goes from 1 to T. Only the integer values of σ will be made to correspond exactly to specific points on the curve so that in between the curve is being approximated. Therefore, the approximant for the curve is

$$\left.\begin{aligned} x(\sigma) &= \sum_{i=2}^{T+3} X_i N_i(\sigma) \; , \\ y(\sigma) &= \sum_{i=2}^{T+3} Y_i N_i(\sigma) \; . \end{aligned}\right\} \tag{6.69}$$

Only two-dimensional curves are treated in detail but the extension to three dimensions is immediate. The coefficients X_i and Y_i are fixed, first by requiring that, when $\sigma = j$ ($j = 1, \ldots, T$), $x = x_j$ and $y = y_j$ where (x_j, y_j) are specified points on the given curve. This gives $2T$ conditions for $2T + 4$ unknowns. If the curve is closed, periodicity supplies the missing information. Otherwise, two extra points can be designated for matching, say $\sigma = \frac{1}{2}$ and $\sigma = T - \frac{1}{2}$. Since, from (1.14),

$$N_i(\sigma) = \tfrac{1}{6}\{(\sigma-i)_+^3-4(\sigma-i+1)_+^3+6(\sigma-i+2)_+^3-4(\sigma-i+3)_+^3+(\sigma-i+4)_+^3\} \quad (6.70)$$

$N_i(i) = 0 = N_i(i-4)$, $N_i(i-1) = \frac{1}{6}$, $N_i(i-2) = \frac{4}{6}$, $N_i(i-3) = \frac{1}{6}$. Thus the matrix associated with finding X_i and Y_i will be tri-diagonal with the only non-zero elements on each row being $\frac{1}{6}$, $\frac{4}{6}$, $\frac{1}{6}$ except possibly for those rows relating to the two extra points or special properties of the curve. Computation of the X_i and Y_i should therefore be rapid.

The arc length between successive data points becomes unity when the σ variable is employed. It is therefore desirable to aim at making the spacing between data points roughly uniform so that the spline curve is reasonably representative of the original. At any rate, the arc lengths between successive data points should not fluctuate wildly. It is, of course, vital to choose T large enough to ensure that the approximant is a tolerable replica of the given curve.

The B-spline representation guarantees continuity of the first two derivatives. If the antenna exhibits discontinuities, say in slope, it may be undesirable for the approximating curve to display higher continuity there. In that case, the antenna should be split into a number of macro-elements, in each of which the B-spline approximation is adopted but without undue continuity enforced at their joins. For simplicity of presentation, it will be assumed in the following that there is only one macro-element.

As soon as eqn (6.69) of the wire has been settled, the evaluation of the integrals in (6.59) has to be dealt with. In these both $\mathbf{x}$ and $\boldsymbol{\xi}$ are replaced by the same formula (6.69) but σ and σ' are employed as their respective parameters to distinguish between them. Then

$$\boldsymbol{\tau}(\boldsymbol{\xi}) = \frac{d\boldsymbol{\xi}}{ds} = \frac{d\boldsymbol{\xi}}{d\sigma'}\frac{d\sigma'}{ds}$$

and

$$\frac{ds}{d\sigma'} = \left\{\left(\frac{d\xi}{d\sigma'}\right)^2 + \left(\frac{d\eta}{d\sigma'}\right)^2\right\}^{\frac{1}{2}}$$

where $\boldsymbol{\xi} \equiv (\xi,\eta)$, again restricting attention to two-dimensional wires. Consequently, τ is readily available. The derivatives are evaluated analytically before the substitution (6.69).

By this process the integral equation goes over to the form

$$f(\sigma) = \int_1^T I(\sigma')K(\sigma,\sigma')d\sigma'$$

where f and K are known but I has to be found. It is now desired to insert a suitable approximation for I so that a numerical solution is revealed.

Since the typical distance between knots for σ is unity a canonical integral to be computed is

$$\int_m^{m+1} I(\sigma')K(\sigma,\sigma')d\sigma'$$

where m is a positive integer. The first stage is to approximate the integral by a quadrature formula. It need not be of high order and a choice appropriate to our method of approximation by cubic splines such as a Gaussian method of order 3 (§5.12) should be sufficient since the kernel is non-singular. (If a is very small it may be helpful to consider the methods for singular kernels discussed later in §6.19.) Hence

$$\int_m^{m+1} I(\sigma')K(\sigma,\sigma')d\sigma' \approx \sum_{r=1}^{n} w_r I(m+\sigma_r)K(\sigma,m+\sigma_r)$$

where w_r and σ_r are the weights and points respectively required by the quadrature rule applied to the interval (0,1). Next we assume that I can be approximated by cubic B-splines with the same knots as those of the representation of the wire, i.e.

$$I(\sigma') = \sum_{i=2}^{T+3} I_i N_i(\sigma')$$

where the complex constants I_i are to be determined from the integral equation and any end conditions imposed. For the interval under consideration only four of the B-splines will be non-zero. Also $N_{i+1}(\sigma) = N_i(\sigma-1)$ and hence

$$\int_m^{m+1} I(\sigma')K(\sigma,\sigma')\mathrm{d}\sigma' \approx \sum_{r=1}^{n} w_r K(\sigma,m+\sigma_r) \sum_{j=1}^{4} I_{m+j}N_j(\sigma_r)$$

$$\approx \sum_{j=1}^{4} I_{m+j} \sum_{r=1}^{n} w_r N_j(\sigma_r)K(\sigma,m+\sigma_r) \; .$$

The constant $w_r N_j(\sigma_r)$ is independent of m and so, once computed for one interval, is available for all intervals.

By point matching at the integer values of σ the values of I_i solve the linear system of equations resulting from the integral equation. The antenna characteristics and radiation pattern may then be determined as in previous sections. A numerical procedure for bent antennas has therefore been devised.

Exercises

11. A linearly polarized plane wave with its electric vector parallel to the z axis is incident on a circular antenna of radius b in the plane $y = 0$. The direction of propagation of the incident wave makes an angle χ with the x axis and the origin is at the centre of the antenna. If $kb \ll 1$, show that the current in the wire is approximately the same at all points of the wire and decreases steadily from its value at $\chi = 0$ to zero at $\chi = \frac{1}{2}\pi$.

By considering the cases $\lambda/4a = 500$, 1000, and 2000 with kb round about 1, demonstrate that the greatest current at $(b,0,0)$ occurs when the length of the loop is about 4 per cent greater than the wavelength.

12. Calculate the radiation pattern in $z = 0$ and $y = 0$ of the scattered field for the cases of Exercise 11.

13. The incident field of Exercise 11 with $\chi = 0$ falls on a helical antenna wound on a cylinder of radius b with pitch angle β, the axis of the helix being the z axis. If the helix has only three or four turns examine the radiation pattern of the scattered field and decide whether it is elliptically polarized in general. Would it be circularly polarized if $kb = 2 \tan \beta$?

14. If in Exercise 13, kb is about 1 and the direction of propagation of the incident wave is not perpendicular to the axis of the helix, find the current induced at the central point of the antenna. For what range of lengths of wire (β being fixed) is the current a maximum when the direction of propagation is along the axis of the helix?

6.9. Log-periodic antennas

The log-periodic antenna was introduced as a mechanism for producing frequency-independent patterns over moderate bandwidths (Rumsey 1957, DuHamel and Isbell 1957). The aim is to provide a structure whose electrical properties vary periodically with the logarithm of the frequency. How this is achieved may be seen by approaching first the problem of the log-periodic transmission line.

Suppose that it is asked that the voltage and current at the position x and complex frequency ω shall be equal to those at the position τx and frequency ω/τ, where the real constant τ is known as the *log-periodic expansion parameter*. Then

$$\begin{aligned} V(x,\omega) &= V(x',\omega'), \\ I(x,\omega) &= I(x',\omega') \end{aligned} \tag{6.71}$$

where $x' = \tau x$, $\omega' = \omega/\tau$. Now

$$\frac{\mathrm{d}}{\mathrm{d}x} V(x,\omega) + Z(x,\omega)I(x,\omega) = 0,$$

$$\frac{\mathrm{d}}{\mathrm{d}x} I(x,\omega) + Y(x,\omega)V(x,\omega) = 0,$$

$$\frac{d}{dx'} V(x',\omega') + Z(x',\omega')I(x',\omega') = 0,$$

$$\frac{d}{dx'} I(x',\omega') + Y(x',\omega')V(x',\omega') = 0$$

where Z and Y are the series impedance and shunt admittance respectively. Sources could be included but are omitted to simplify the analysis. Dividing the first set of equations by τ and using (6.71) we deduce that

$$Z(x,\omega) = \tau Z(x',\omega') ,$$

$$Y(x,\omega) = \tau Y(x',\omega').$$

These are the conditions to be satisfied if the transmission line is to be log-periodic.

Put $x = \tau^p$, $\omega = \tau^q$. Then the equation for Z becomes

$$Z(\tau^p,\tau^q) = \tau Z(\tau^{p+1},\tau^{q-1}) .$$

The general solution of this difference equation is

$$Z(\tau^p,\tau^q) = \tau^{-p}F_1(p,q)G_1(p+q)$$

where F_1 is periodic with period 1 in both its arguments and G_1 is an arbitrary function. Hence.

$$Z(x,\omega) = \frac{F(\ln x,\ln\omega)G(\omega x)}{x}$$

where the periods of F are both $\ln\tau$ and G is arbitrary. On account of (6.71), the structure of V will be similar without the denominator x; the actual functions occurring may, however, be different.

If (6.71) holds for all τ, then $F \equiv 1$ in order to satisfy its periodicity condition. Consequently, if G is independent of frequency a transmission line of frequency-independent structure has been obtained. Notice that (6.71) specifies phase only to within a multiple of 2π and this flexibility may be important in practical circumstances.

For two-dimensional electromagnetic fields in a medium with permittivity ε, permeability μ, and conductivity σ (assumed independent of ω) the analogue of (6.71) is expressed in terms of the polar coordinates ρ,ϕ when the fields are independent of z. For example,

$$E(\rho,\phi,\omega) = \tau^{\frac{1}{2}}E(\tau\rho,\phi-\phi_0,\omega/\tau) \tag{6.72}$$

and a similar equation for $\mathbf{H}$ are conditions for log-periodicity. By operating with Maxwell's equations in the same way as for the transmission line we find

$$\varepsilon = F(\ln\rho,\eta\phi)G\{\rho\exp(-\eta\phi)\}$$

with parallel formulae for μ and $\sigma\rho$. Here $\eta = -(\ln\tau)/\phi_0$ and F and G have the same properties as in the formula for Z. Evidently G will be constant on a logarithmic spiral

$$\rho = C\exp(\eta\phi).$$

If $\phi_0 = 0$, the equation for ε reduces to

$$\varepsilon = F(\ln\rho)G(\phi).$$

Thus if the material constants are arranged to spiral logarithmically, a log-periodic structure is produced. In practice, it has to be of finite size and only experiment can confirm that the lack of infinite periodicity does not destroy the desired properties. In fact, antennas of moderate size operate satisfactorily and are usually sufficiently compact for the tapering of the conductivity of σ with ρ to be ignored. Indeed, the elements are often deformed into straight lines (the spiral is almost straight for angular deviations which are not too large) especially when τ is not far from unity. For example, the monopole and zigzag are exhibited in Figs. 6.10 and 6.11, where the *relative* distance from the origin of the points

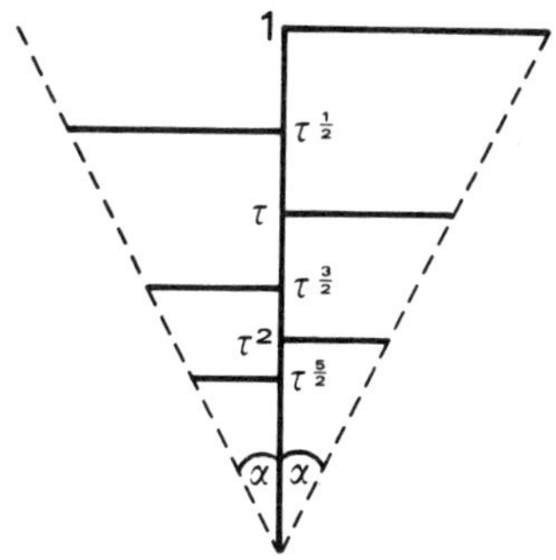

FIG. 6.10. A log-periodic monopole.

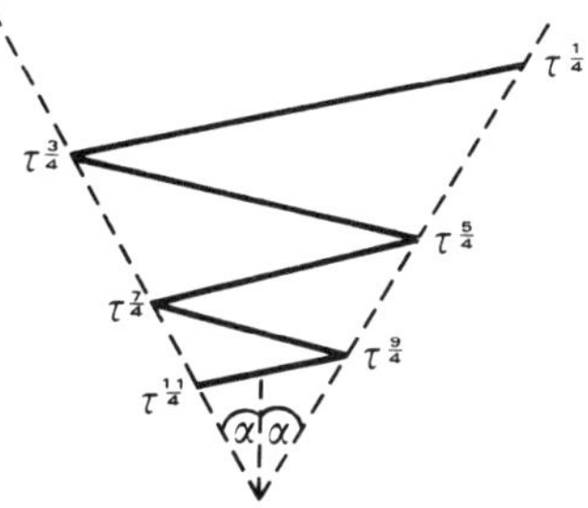

FIG. 6.11. A log-periodic zigzag.

marked is also displayed (Isbell 1960, Greiser and Mayes 1961). The sought-for frequency independence is attained and some useful polarization effects are also available as a bonus.

Exercises

15. In three dimensions the relation corresponding to (6.72) is

$$E(R,\theta,\phi,\omega) = \tau E(\tau R,\theta,\phi-\phi_0,\omega/\tau)$$

in spherical polar coordinates R, θ, ϕ. Show that the analogue of the logarithmic spiral in two dimensions is a logarithmic spiral wrapped round a cone.

16. Why are the multiplying factors $\tau^{\frac{1}{2}}$ and τ inserted in (6.71) and Exercise 15? Hint: consider the power flow through a circle and sphere respectively.

17. A zigzag antenna is constructed with five elements as in Fig. 6.11 with $\alpha = 5°$ and $\tau = 0.7$. A plane wave is incident vertically upwards with the electric vector in the plane of the antenna. If the wire radius is 0.005λ compute the radar cross-section as a function of the ratio of the distance of the end of the first element from the origin to the wavelength. Is there a region in which the response is reasonably flat?

6.10. Loads and arrays

Once a satisfactory technique has been established for a single wire there is no quandary about what to do for arrays in principle. Simultaneous integral equations arise, but since the numerical procedure replaces them by linear algebraic equations there is no fresh feature. Admittedly, the size of the matrices soon becomes huge as the number of elements shoots up. Nevertheless, with a sound numerical technique and taking what advantage one can of any sparseness, the main limitation will be the size of the computer. Further remarks about junctions are contained in the next section.

No awkwardness should be encountered when an antenna is loaded provided that the load may be treated as a lumped circuit. Then the additional equation $V = Z_l I$ has to be satisfied at the load, V being an appropriately defined voltage. Such an equation can be incorporated by following a similar route to that described in the second method for input impedance in §6.7.

Exercises

18. Compare your predictions for the radiation pattern of a Yagi array with the experimental results given by Fishenden and Wiblin (1949).

19. Find the radar cross-section for two concentric coplanar circular wires and compare your results with the measurements of Gans (1965).

20. Repeat Exercise 17 for the monopole of Fig. 6.10 and compare the radar cross-sections for the two antennas.

21. Repeat Exercise 20 when the side elements are swept down at an angle of 60° to the central radius in herring-bone fashion.

22. Two parallel wires of lengths 0.5λ and 4λ are placed so that a wire joining their centres is perpendicular to both. Examine the radar cross-section in the central plane perpendicular to the parallel wires and show

that for certain separations the main scattering occurs in the direction from the larger to smaller element.

In effect, the wave travelling along the central wire is reflected by the longer wire in such a way as to reinforce the radiation from the shorter at some angles. The structure is often known as a *back-fire antenna*.

23. Six circular wires are arranged so that their planes are parallel and their centres lie on a straight line. The dimensions are such that the antenna is log-periodic. If $\tau = 0.8$ find the radar cross-section for a plane wave incident along the line of centres as a function of the diameter of the smallest circle.

SOLID ANTENNAS

6.11. Wire grid models

It is well known experimentally that a wire grid is equivalent to a conducting sheet provided that the mesh spacing is small compared with the wavelength. On this basis the idea that a solid perfectly conducting antenna might be modelled by replacing its surface by an equivalent wire grid was proposed (Richmond 1966). Only the integral equations for thin wires need to be solved. Typical possibilities for a sphere plus monopole and an aircraft are shown in Figs. 6.12 and 6.13. Owing to the simplicity of this approach and its malleability in adapting to complex structures, it has been widely used. At least two computer programs are to hand for its implementation (Burke and Selden 1973, Forgan 1974).

The application of such programs to problems like the sphere of Fig. 6.12 yields promising results, especially when collated with experimental measurements (Albertsen, Hansen, and Eilskov Jensen 1972) (see Fig. 6.14). Other elementary shapes such as discs and squares also offer encouragement. However, no theory has been developed to back up the method, so that assessment of how to select the wire radius and positions of the grid wires rests on numerical experimentation and comparison with laboratory measurements. Unfortunately, this leaves open the contingency that in a new situation the numerical predictions cannot be relied on until proved in the laboratory, in which

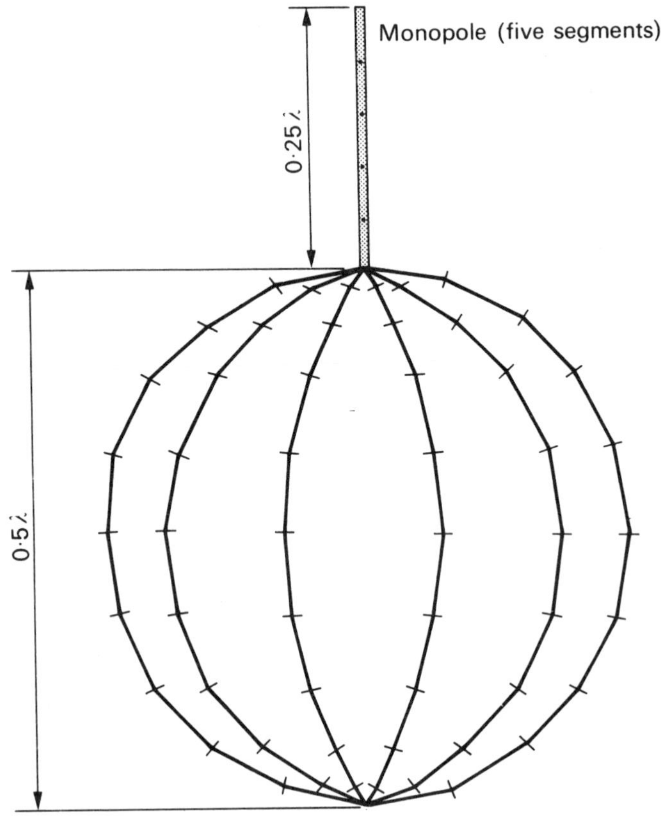

FIG. 6.12. Wire-grid model for a sphere with monopole

case it may be more cost effective not to undertake the computation. It is therefore worthwhile indicating some of the points where experience has shown that particular care is called for.

The programs are based on the trigonometric expansions of §6.7. Generally speaking, these give satisfactory results when segments of equal length are employed but their performance when there are junctions between several segments of different length has room for considerable improvement. In this respect, the cubic B-splines of §6.8 with their automatic conversion to equispacing should be superior. One way of overcoming the difficulty is always to use segments of equal length, but this may involve many more segments than are necessary to model the antenna. An irregular structure may need very short segments

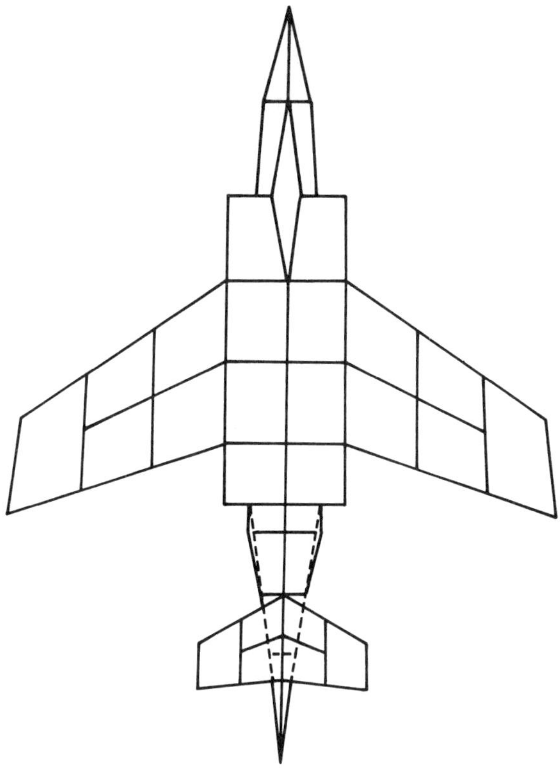

FIG. 6.13. Wire-grid model of an aircraft

in some directions to specify its outline but long segments may be perfectly adequate in other directions. Such variations may, of course, entail the use of higher-order quadrature formulae such as the Gauss eight-point rule.

Another problem is the choice of wire radius. A close mesh of very thin wire should be ideal as a prototype of the body but it incurs a vast computational effort. It therefore seems preferable to make the radius as large as is consistent with the radius being less than one-hundredth of a wavelength and adjacent wire segments not overlapping. This may force different radii in different parts of the grid which can cause difficulty at junctions because of the introduction of capacitance thereby. The way round this in the RAE program (Forgan 1974) is to taper the wires near a junction so that there is a smooth transition; the wire radius in the integrals is then a function

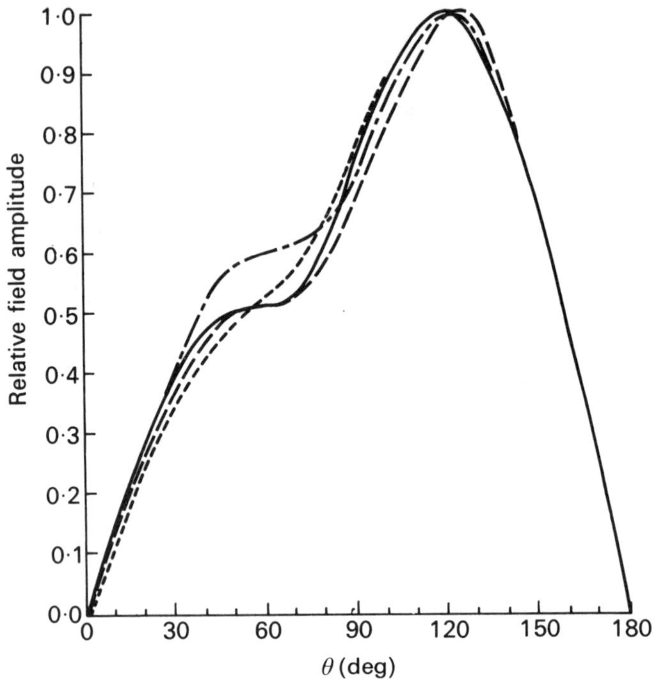

FIG. 6.14. Radiation pattern of a ½λ sphere plus monopole: solid curve, measured (Lyngby); chain curve, computed (Lyngby); broken curve, computed (AMP); dotted curve, computed (RAE).

of position.

Again, with large systems of segments small errors may easily accumulate so that it is wise to prevent them doing so if possible. For instance, the radiated power should be calculated by integration of the radiation pattern because this procedure is relatively insensitive to small errors.

Exercises

24. A plane wave is incident along the axis of a circular sheet of perfectly conducting metal. Find the radar cross-section as the radius varies and compare your results with those of Richmond (1966).

25. A square antenna contains a central rectangular slot parallel to the sides. Find the radar cross-section for a plane wave incident perpendicular to the plane of the antenna and linearly polarized (*a*) parallel to the longer side of the rectangle, (*b*) in the perpendicular direction (see

also Miller and Morton 1970).

26. Determine the radiation pattern for Fig. 6.12 when the length of the monopole is $\frac{1}{8}\lambda$ (compare also Tesche and Neureuther 1970).

27. Find the radar cross-section of a spheroid for incidence along a principal axis and check against the results of Oshiro, Torres, and Heath (1966).

28. Model a cone with a flat base by radial wires on the base and their continuations along generators of the slant sides. Determine the scattered pattern for a plane wave incident along the axis towards (*a*) the apex, (*b*) the base.

29. A quarter-wave monopole is added to the base of the cone of Exercise 28 and is along the axis away from the apex. It is fed by a small magnetic frill surrounding it in the base. Calculate the radiation pattern produced by this transmitting antenna (Thiel, Travieso-Davies, and Jones 1969).

30. Model a paraboloidal antenna by a wire grid and calculate its transmitting characteristics for excitation from a small dipole at its focus (Poggio and Miller 1970).

31. A cone–sphere antenna is manufactured by placing a hemisphere on the base of a cone so that the perimeters of the base and hemisphere coincide. Compute the scattering characteristics for a wave incident along the axis (Mautz and Harrington 1969).

6.12. The electric-field integral equation

The derivation of integral equations for antennas of arbitrary shape commences from a standard representation for electromagnetic fields in terms of surface integrals. Let S be a closed surface which can be the boundary of the antenna but may be some other conveniently disposed surface. The infinite region outside S will be denoted by S_+ while the bounded interior will be called S_- (Fig. 6.15). Points in S_+ will be distinguished by capital letters such as P and Q; moreover, the notation will be simplified so that, for example,

$$\psi(P,Q) = \psi(\mathbf{x}_P,\mathbf{x}_Q)$$

where the right-hand side is defined by (6.2). Lower-case letters such as p and q will signify points of S.

The radiation conditions to be imposed on scattered fields,

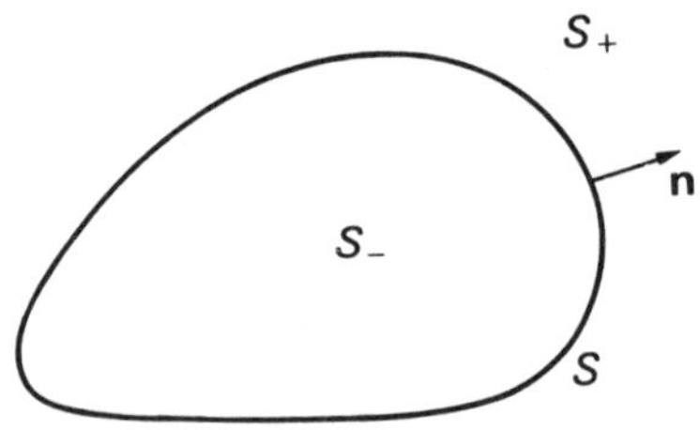

FIG. 6.15. The antenna configuration.

assuming the medium outside S to have constant μ and ε, are

$$\left.\begin{aligned} &R\mathbf{E}, R\mathbf{H} \text{ bounded,} \\ &R(\mathbf{E} - Z\mathbf{H}\wedge\hat{\mathbf{R}}) \to 0, \\ &R(\mathbf{H} - \hat{\mathbf{R}}\wedge\mathbf{E}/Z) \to 0 \end{aligned}\right\} \tag{6.73}$$

as $R \to \infty$, $\hat{\mathbf{R}}$ being a unit vector along the radius from the origin and $Z = (\mu/\varepsilon)^{\frac{1}{2}}$ is the impedance of the medium.

The problem for which a solution is sought is to satisfy Maxwell's equations in S_+, the radiation conditions (6.73) at infinity, and to have a prescribed tangential electric intensity on S, i.e. if $\mathbf{n}_p$ is a unit normal from the point p of S into S_+ it is required that

$$\mathbf{n}_p \wedge \mathbf{E}(p) = \mathbf{n}_p \wedge \mathbf{E}_0(p) \tag{6.74}$$

on S where $\mathbf{n} \wedge \mathbf{E}_0$ is a known field. The field $\mathbf{E}_0$ might originate from sources inside S, as when S encloses a waveguide with radiating slots. In that case (6.74) is effectively the condition for a transmitting antenna. On the other hand, $\mathbf{E}_0$ might be due to a signal whose origins lie outside S. Then (6.74) corresponds to S being perfectly conducting and $\mathbf{E}$ is the field scattered by such a receiving antenna. It may be desirable to specify alternative boundary conditions, e.g. of impedance type, on S but that will not alter the essence of the principles involved and so (6.74) will be concentrated on. Observe that the field $\mathbf{E}$, $\mathbf{H}$ does not have any sources in S_+; they must all be on S or in S_-.

Having formulated the fundamental antenna or exterior problem, one has to consider what techniques are to be adopted to solve it. A natural thought is to apply finite differences which have proved so successful in waveguide problems (Chapter 2) or the variational methods of Chapter 4. There are several objections. First the bounds and positive-definite nature of variational methods for waveguides disappear for the exterior problem. Next, the generality of the surface makes finite differences very cumbersome. The most important reason, though, is that S_+ is of infinite extent. One way round this is to place a fictitious boundary surrounding S at a large, but finite, distance away (McDonald and Wexler 1972, Silvester and Hsieh 1971, James 1973). On the artificial boundary (6.73) is supposed to hold exactly. No great confidence can be placed in the accuracy of a solution to such a modified problem even if the solution has been repeated several times with outer boundaries of increasing size. The solution of the modified problem will always be different from that of the original problem posed. Finite elements (§5.8), used to minimize energy, will not obviate this difficulty.

Combinations of integral equations with finite differences or finite elements for a finite part of the region have been proposed (Mei 1974, Shaw 1974), but in view of the difficulties that are encountered with pure integral equations these methods must be treated with caution (§6.28). The approach will therefore be that of integral equations solely and some of the considerations in solving them, including finite-element methods. For a method based on an infinite system of state-space differential equations see Hizal (1974).

The basic field representation can be written, on the assumption that $\mathbf{E}$ and $\mathbf{H}$ satisfy the radiation conditions (6.73) and Maxwell's equations

$$\operatorname{curl} \mathbf{E} + ikZ\mathbf{H} = 0, \quad \operatorname{div} \mathbf{E} = 0\ ,$$

$$\operatorname{curl} Z\mathbf{H} - ik\mathbf{E} = 0, \quad \operatorname{div} \mathbf{H} = 0$$

as (see for example Jones 1964)

$$\int_S [\{\mathbf{n}_q \wedge \mathbf{E}(q)\} \wedge \operatorname{grad}_q \psi(P,q) + \{\mathbf{n}_q . \mathbf{E}(q)\} \operatorname{grad}_q \psi(P,q) -$$

$$- \mathrm{i}\omega\mu \{\mathbf{n}_q \wedge \mathbf{H}(q)\}\psi(P,q)]\,\mathrm{d}S_q = \mathbf{E}(P) \quad (P \in S_+),$$

$$= \mathbf{0} \quad (P \in S_-) \qquad (6.75)$$

$$\int_S [\{\mathbf{n}_q \wedge \mathbf{H}(q)\} \wedge \operatorname{grad}_q \psi(P,q) + \{\mathbf{n}_q . \mathbf{H}(q)\} \operatorname{grad}_q \psi(P,q)$$

$$+ \mathrm{i}\omega\varepsilon \{\mathbf{n}_q \wedge \mathbf{E}(q)\}\psi(P,q)]\,\mathrm{d}S_q = \mathbf{H}(P) \quad (P \in S_+)$$

$$= \mathbf{0} \quad (P \in S_-) \qquad (6.76)$$

where, now, $k^2 = \omega^2\mu\varepsilon$ and the time dependence $\mathrm{e}^{\mathrm{i}\omega t}$ has been suppressed. The symbol $\mathbf{n}_q$ indicates the unit normal at q of S into S_+ and the subscript on a vector operator shows the variable to which it is applied, i.e.

$$\operatorname{grad}_q \psi(P,q) = \left(\mathbf{i}\frac{\partial}{\partial x_q} + \mathbf{j}\frac{\partial}{\partial y_q} + \mathbf{k}\frac{\partial}{\partial z_q}\right)\psi(P,q)$$

where **i**, **j**, and **k** are unit vectors along the Cartesian axes.

The integrals in (6.75) and (6.76) contain the normal components **n.E** and **n.H**, but the boundary condition (6.74) involves only tangential components. It is therefore advantageous to express the normal components in terms of tangential ones. This may be accomplished by introducing the *surface divergence* of a tangential vector **u** on S.

Draw a small curve C on S and let $\boldsymbol{\nu}$ be the unit vector which is perpendicular to both **n** and C, as well as being out of the portion of S containing **n** (Fig. 6.16). If the area of this portion of S enclosed by C is A, the surface divergence $\operatorname{div}_0 \mathbf{u}$ is defined by

$$\mathrm{div}_0 \mathbf{u} = \lim \frac{1}{A} \int_C \mathbf{u}.\boldsymbol{\nu} \, \mathrm{d}s$$

where the integration is round C and the limit is taken as C contracts to the point where $\mathbf{n}$ is drawn. If C is arbitrary it follows, by covering the enclosed part of S by a grid of curves each surrounding a small area, that

$$\int_{S_C} \mathrm{div}_0 \mathbf{u} \, \mathrm{d}S = \int_C \mathbf{u}.\boldsymbol{\nu} \, \mathrm{d}s \qquad (6.77)$$

where S_C is the portion of S with C as the rim out of which $\boldsymbol{\nu}$ is pointing. If S is closed then C can be allowed to contract to a point in such a way that S_C becomes S. Hence

$$\int_S \mathrm{div}_0 \mathbf{u} \, \mathrm{d}S = 0 \qquad (6.78)$$

when S is closed.

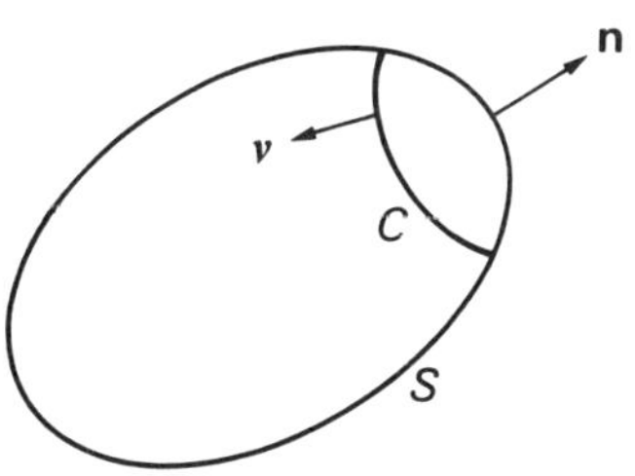

FIG. 6.16. The calculation of surface divergence.

The element $\mathrm{d}\mathbf{s}$ of arc of C satisfies $\mathrm{d}\mathbf{s} = \mathbf{n} \wedge \boldsymbol{\nu} \, \mathrm{d}s$. Hence

$$\mathrm{div}_0(\mathbf{n} \wedge \mathbf{E}) = \lim \frac{1}{A} \int_C \mathbf{n} \wedge \mathbf{E} \, . \, \boldsymbol{\nu} \, \mathrm{d}s = - \lim \frac{1}{A} \int_C \mathbf{E} \, . \, \mathrm{d}\mathbf{s}$$

$$= - \lim \frac{1}{A} \int_{S_C} (\mathbf{n} \, . \, \mathrm{curl} \, \mathbf{E}) \mathrm{d}S$$

by Stokes's theorem.

By Maxwell's equations, $-i\omega\mu\mathbf{H}$ may be substituted for curl $\mathbf{E}$. The integrand may be regarded as constant if C is sufficiently small and hence

$$\mathrm{div}_0(\mathbf{n}\wedge\mathbf{E}) = i\omega\mu\mathbf{n}\cdot\mathbf{H}. \tag{6.79}$$

It may be shown similarly that

$$\mathrm{div}_0(\mathbf{n}\wedge\mathbf{H}) = -i\omega\varepsilon\mathbf{n}\cdot\mathbf{E}. \tag{6.80}$$

The surface divergence requires only derivatives along the surface and so (6.79) and (6.80) permit the determination of normal components from tangential ones.

Write

$$\mathbf{n}\wedge\mathbf{H} = -\mathbf{j}, \quad -\varepsilon\mathbf{n}\cdot\mathbf{E} = \rho\,. \tag{6.81}$$

Then, from (6.80),

$$\mathrm{div}_0\mathbf{j} + i\omega\rho = 0. \tag{6.82}$$

The vector $\mathbf{j}$ may be thought of as a surface electric current while ρ is a surface electric charge. Thus (6.82) is a continuity equation on the surface.

Surface magnetic currents and charges may also be injected via

$$\mathbf{n}\wedge\mathbf{E} = \mathbf{j}_m, \quad -\mu\mathbf{n}\cdot\mathbf{H} = \rho_m \tag{6.83}$$

with the surface continuity equation

$$\mathrm{div}_0\mathbf{j}_m + i\omega\rho_m = 0 \tag{6.84}$$

from (6.79).

For our immediate purpose, the substitution (6.83) will not be made since $\mathbf{n}\wedge\mathbf{E}$ is known on S from (6.74). Therefore the following electromagnetic field will be considered:

$$\mathbf{E}(P) = \int_S [\{\mathbf{n}_q \wedge \mathbf{E}_0(q)\} \wedge \operatorname{grad}_q \psi(P,q) - \frac{\rho(q)}{\varepsilon} \operatorname{grad}_q \psi(P,q) + $$

$$+ i\omega\mu \mathbf{j}(q)\psi(P,q)] \mathrm{d}S_q \ , \tag{6.85}$$

$$\mathbf{H}(P) = \int_S [-\mathbf{j}(q) \wedge \operatorname{grad}_q \psi(P,q) + \operatorname{div}_0\{\mathbf{n}_q \wedge \mathbf{E}_0(q)\} \frac{\operatorname{grad}_q \psi(P,q)}{i\omega\mu} +$$

$$+ i\omega\varepsilon \mathbf{n}_q \wedge \mathbf{E}_0(q)\psi(P,q)] \mathrm{d}S_q \ . \tag{6.86}$$

In (6.85) and (6.86) it is assumed that (6.82) is valid so that ρ is known once $\mathbf{j}$ has been specified. Thus the integrals of (6.85) and (6.86) involve only the single unknown tangential surface current $\mathbf{j}$.

No matter how $\mathbf{j}$ is chosen, so long as (6.82) is satisfied, (6.85) and (6.86) are solutions of Maxwell's equations in S_+. The verification is straightforward so that only the hardest step need be considered in detail. Put $\mathbf{n}_q \wedge \mathrm{E}_0(q) = \mathbf{a}$. Then

$$\operatorname{curl}_P \int_S \mathbf{a} \wedge \operatorname{grad}_q \psi(P,q) \mathrm{d}S_q$$

$$= \int \{\mathbf{a}.\operatorname{div}_P \operatorname{grad}_q \psi(P,q) - (\mathbf{a}.\operatorname{grad}_P)\operatorname{grad}_q \psi(P,q)\} \mathrm{d}S_q$$

$$= \int \{(\mathbf{a}.\operatorname{grad}_q)\operatorname{grad}_q \psi(P,q) - \mathbf{a}.\nabla_P^2 \psi(P,q)\} \mathrm{d}S_q$$

because ψ depends only on $|\mathbf{x}_P - \mathbf{x}_q|$. Since ψ satisfies Helmholtz's equation in S_+ the last term reproduces the final term of (6.86). Further

$$\int_S (\mathbf{a}.\mathrm{grad}_q)\mathrm{grad}_q\psi(P,q)\mathrm{d}S_q = -\,\mathrm{grad}_P \int_S \mathbf{a}\;.\;\mathrm{grad}_q\psi\mathrm{d}S_q$$

$$= -\,\mathrm{grad}_P \int_S \{\mathrm{div}_0(\mathbf{a}\psi)-\psi\;\mathrm{div}_0\mathbf{a}\}\mathrm{d}S_q$$

$$= -\int_S \mathrm{div}_0\mathbf{a}\;\mathrm{grad}_q\psi(P,q)\mathrm{d}S_q \qquad (6.87)$$

from (6.78) because $\psi\mathbf{a}$ is a tangential vector. The second term of (6.86) has consequently been recovered. A similar procedure is necessary in evaluating curl $\mathbf{H}$ but no other terms give any difficulty.

It will be remarked that the argument applies equally if P is in S_- i.e. (6.85) and (6.86) are also solutions of Maxwell's equations inside S. In contrast to (6.75) and (6.76), it cannot be guaranteed that a free choice of $\mathbf{j}$ will lead to an identically zero field in S_-. Of course, if the right choice is made it is expected that the field will be zero in the interior.

Again, (6.85) and (6.86) comply with the radiation conditions (6.73) for arbitrary choices of $\mathbf{j}$. This may be confirmed by incorporating the asymptotic behaviour of ψ as $P \to \infty$ (which is permissible because S is bounded) and by noting that

$$\int_S (\mathrm{div}_0\mathbf{j}+ik\mathbf{j}.\hat{\mathbf{x}}_P)\exp(ik\mathbf{x}_q.\hat{\mathbf{x}}_P)\mathrm{d}S_q = \int_S \mathrm{div}_0\{\mathbf{j}\;\exp(ik\mathbf{x}_q.\hat{\mathbf{x}}_P)\}\mathrm{d}S_q$$

$$= 0$$

from (6.78) since S is closed.

Consequently, the representation (6.85) and (6.86) satisfies all the conditions which have been demanded of the solution with the sole exception of (6.74). The next aim is to rectify this deficiency by letting P tend to a point of S.

Some care has to be exercised in this process because ψ becomes singular as P approaches S. While a rigorous analysis can be undertaken (Kellogg 1929) we shall be content with a plausible investigation which gives the same results. The singularity of $\psi(p,q)$ is integrable so that, if f is reasonably continuous on S,

$$\lim_{P\to p} \int_S f(q)\psi(P,q)\,dS_q = \int_S f(q)\psi(p,q)\,dS_q \,. \tag{6.88}$$

Integrals with derivatives of ψ, however, display discontinuities. Split S into two sections, S_1 and S_2, of which S_2 is a small region surrounding p and S_1 the remainder of S (Fig. 6.17). If S has a continuously turning tangent plane in a neighbourhood of p it may be supposed that S_2 is effectively a circular disc of radius δ with centre p. The z axis may also be taken along the outward normal at p.

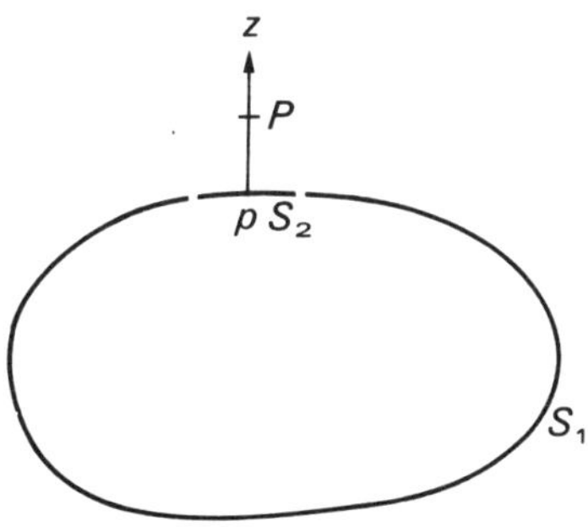

FIG. 6.17. Determination of the discontinuities of surface integrals.

An integral over S_1 is continuous as $P \to p$ and causes no trouble. In the limit as $\delta \to 0$ it will lead to a principal value. With regard to S_2, assume firstly that P is on the z axis, a small distance z from p. Then the phase of ψ can be ignored and

$$\int_{S_2} f(q)\psi(P,q)\,dS_q \approx \frac{f(p)}{4\pi}\int_0^{2\pi}\int_0^{\delta} \frac{t\,dt\,d\phi}{(t^2+z^2)^{\frac{1}{2}}}$$

$$= \tfrac{1}{2}f(p)\{(\delta^2+z^2)^{\frac{1}{2}}-z\} \,. \tag{6.89}$$

As $z \to 0$ this may be approximated by $\frac{1}{2}f(p)\{\delta+\frac{1}{2}z^2/\delta-z\}$. P may now be moved a small distance r off the z axis so long as a solution of Laplace's equation is retained. Thus, in general,

$$\int_{S_2} f(q)\psi(P,q)\mathrm{d}S_q \approx \frac{1}{2}f(p)\left\{\delta-z+\frac{\frac{1}{2}\left(z^2-\frac{1}{2}r^2\right)}{\delta}\right\} .$$

A derivative with respect to z along the normal gives a non-zero value as $P \to p$, whereas tangential derivatives supply no contribution in the limit. Hence

$$-\lim_{P\to p} \operatorname{grad}_P \int_S f(q)\psi(P,q)\mathrm{d}S_q = \lim_{P\to p} \int_S f(q)\operatorname{grad}_q\psi(P,q)\mathrm{d}S_q$$

$$= \frac{1}{2}f(p)\mathbf{n}_p + ⨍_S f(q)\operatorname{grad}_q\psi(p,q)\mathrm{d}S_q \tag{6.90}$$

where the bar on the integral sign signifies a principal value.

If P is inside S, at P_i say, the only difference is that $-z$ in (6.89) is changed to z. Therefore

$$\lim_{P_i\to p} \int_S f(q)\operatorname{grad}_q\psi(P,q)\mathrm{d}S_q = -\frac{1}{2}f(p)\mathbf{n}_p + ⨍_S f(q)\operatorname{grad}_q\psi(p,q)\mathrm{d}S_q . \tag{6.91}$$

From (6.88) and (6.90) it follows that

$$\lim_{P\to p} \mathbf{E}(p) = \frac{1}{2}\{\mathbf{n}_p\wedge\mathbf{E}_0(p)\}\wedge\mathbf{n}_p - \frac{1}{2}\frac{\rho(p)\mathbf{n}_p}{\varepsilon} +$$

$$+ ⨍_S [\{\mathbf{n}_q\wedge\mathbf{E}_0(q)\}\wedge\operatorname{grad}_q\psi(p,q) - \frac{\rho(q)}{\varepsilon}\operatorname{grad}_q\psi(p,q) +$$

$$+ \mathrm{i}\omega\mu\mathbf{j}(q)\psi(p,q)]\mathrm{d}S_q , \tag{6.92}$$

$$\lim_{P\to p} \mathbf{H}(P) = -\tfrac{1}{2}\mathbf{j}(p) \wedge \mathbf{n}_p + \frac{1}{2}\frac{\mathrm{div}_0\{\mathbf{n}_p\wedge\mathbf{E}_0(p)\}\mathbf{n}_p}{\mathrm{i}\omega\mu} +$$

$$+ \int_S [-\mathbf{j}(q)\wedge\mathrm{grad}_q\psi(p,q) + \frac{\mathrm{div}_0\{\mathbf{n}_q\wedge\mathbf{E}_0(q)\}\mathrm{grad}_q\psi(p,q)}{\mathrm{i}\omega\mu} +$$

$$+ \mathrm{i}\omega\varepsilon\mathbf{n}_q \wedge \mathbf{E}_0(q)\psi(p,q)]\mathrm{d}S_q . \tag{6.93}$$

Now apply the boundary conditions (6.74) to (6.92). Then

$$\mathbf{n}_p \wedge \int_S \left\{\mathrm{i}\omega\mu\mathbf{j}(q)\psi(p,q) - \frac{\rho(q)}{\varepsilon}\,\mathrm{grad}_q\psi(p,q)\right\}\mathrm{d}S_q$$

$$= \tfrac{1}{2}\mathbf{n}_p \wedge \mathrm{E}_0(p) - \mathbf{n}_p \wedge \int_S \{\mathbf{n}_q\wedge\mathbf{E}_0(q)\} \wedge \mathrm{grad}_q\psi(p,q)\mathrm{d}S_q . \tag{6.94}$$

The right-hand side of (6.94) is known in principle and so (6.94) constitutes an integral equation to determine the unknown **j**. It is known as the *electric-field integral equation* (EFIE). For an elongated antenna it effectively degenerates to one of the forms discussed in connection with wires.

The integral equation (6.94) only applies at points where the tangent plane is continuously turning. At edges or conical points the formulae (6.90) and (6.91) are no longer valid and neither is (6.94). However, (6.74) cannot be strictly applied at such points since **n** is not well defined there. Therefore, although (6.94) can be imposed arbitrarily close to an edge, it must be avoided actually on an edge — an aspect to be watched when employing point matching in numerical techniques.

When $\mathbf{E}_0$ is due to an electromagnetic field arriving from outside S, the integration on the right-hand side of (6.94) can be escaped. The incident field has no sources inside S and so

$$\int_S [\{\mathbf{n}_q\wedge\mathbf{E}_0(q)\} \wedge \mathrm{grad}_q\psi(P_i,q) + \{\mathbf{n}_q\cdot\mathbf{E}_0(q)\}\mathrm{grad}_q\psi(P_i,q) -$$

$$- \mathrm{i}\omega\mu\{\mathbf{n}_q\wedge\mathbf{H}_0(q)\}\psi(P_i,q)]\mathrm{d}S_q = -\mathbf{E}_0(P_i) \tag{6.95}$$

from (6.75). Let $P_i \to p$ and then (6.95) becomes, on account of (6.91),

$$\oint [i\omega\mu\{\mathbf{n}_q\wedge\mathbf{H}_0(q)\}\psi(p,q) - \{\mathbf{n}_q\cdot\mathbf{E}_0(q)\}\mathrm{grad}_q\psi(p,q)]\mathrm{d}S_q$$

$$= \tfrac{1}{2}\mathbf{E}_0(p) + \oint_S \{\mathbf{n}_q\wedge\mathbf{E}_0(q)\} \wedge \mathrm{grad}_q\psi(p,q)\mathrm{d}S_q \ . \qquad (6.96)$$

Multiply (6.96) vectionally by $\mathbf{n}_p$ and add to (6.94). Relabel $\mathbf{j} + \mathbf{n}_q \wedge \mathbf{H}_0$ and $\rho/\varepsilon + \mathbf{n}.\mathbf{E}_0$ as $\mathbf{j}$ and ρ respectively; no violation of (6.82) occurs by virtue of (6.80). Hence the electric-field integral equation

$$\mathbf{n}_p \wedge \oint_S \{i\omega\mu\mathbf{j}(q)\psi(p,q) - \frac{\rho(q)}{\varepsilon}\,\mathrm{grad}_q\psi(p,q)\}\mathrm{d}S_q = \mathbf{n}_p \wedge \mathbf{E}_0(p) \qquad (6.97)$$

is obtained, when the sources of excitation are in S_+.

6.13. Uniqueness

Despite the fact that it has been demonstrated that a solution to our problem satisfies (6.94) (or (6.97)) it has not been shown that solving (6.94) necessarily provides the answer to our problem. If (6.94) possesses only a single solution the difficulty is resolved, but if (6.94) has several solutions the identification of the desired one has to be settled. Now, if (6.94) has two solutions, their difference must satisfy

$$\mathbf{n}_p \wedge \oint_S \{i\omega\mu\mathbf{j}(q)\psi(p,q) - \frac{\rho(q)}{\varepsilon}\,\mathrm{grad}_q\psi(p,q)\}\mathrm{d}S_q = 0 \qquad (6.98)$$

and the question may be reformulated to ask whether (6.98) holds for any $\mathbf{j} \not\equiv 0$. It will be shown now that there are such solutions.

Consider the electromagnetic field $\mathbf{E}^0$, $\mathbf{H}^0$ defined by

$$\mathbf{E}^0(P) = \int_S \{i\omega\mu\mathbf{j}(q)\psi(P,q) - \frac{\rho(q)}{\varepsilon}\operatorname{grad}_q\psi(P,q)\}dS_q ,$$

$$\mathbf{H}^0(P) = -\int_S \mathbf{j}(q) \wedge \operatorname{grad}_q\psi(P,q)dS_q .$$

Owing to (6.90) and (6.98), $\lim_{P\to p} \mathbf{n}_p \wedge \mathbf{E}^0(P) = 0$. Accordingly, $\mathbf{E}^0$, $\mathbf{H}^0$ is an electromagnetic field in S_+ which satisfies the radiation conditions and has vanishing tangential components on S. By the standard uniqueness theorem (Jones 1964) for radiating fields $\mathbf{E}^0 \equiv 0$ and $\mathbf{H}^0 \equiv 0$ in S_+. By a similar argument based on (6.91), we can assert that the tangential component of $\mathbf{E}^0(P_i)$ vanishes as P_i tends to S. It does not follow that $\mathbf{E}^0$ is identically zero in S_- because there exist values of k for which electric modes of oscillation take place inside S which is then acting as a cavity resonator (§3.5). If $\mathbf{E}^0(P_i)$ is identically zero, and thereby $\mathbf{H}^0(P_i)$ is also zero, then from (6.92) and (6.93)

$$\mathbf{E}^0_+(p) - \mathbf{E}^0_-(p) = -\frac{\rho(p)\mathbf{n}_p}{\varepsilon} , \tag{6.99}$$

$$\mathbf{H}^0_+(p) - \mathbf{H}^0_-(p) = -\mathbf{j}(p) \wedge \mathbf{n}_p \tag{6.100}$$

where the subscripts + and - signify the values obtained as S is approached from S_+ and S_- respectively. Evidently, ρ and $\mathbf{j}$ must be identically zero. Consequently, (6.94) has a unique solution unless k has a value at which the interior of S resonates in an electric mode. When k does have a resonant value there is more than one solution of (6.94). For suppose $\mathbf{E}_i$, $\mathbf{H}_i$ is an interior electric model. Choose $\mathbf{j}(q) = -\mathbf{n}_q \wedge \mathbf{H}_i(q)$ and define ρ from (6.82). Then, by letting $P_i \to p$, we can check the validity of (6.98) and confirm the existence of a non-trivial solution.

While (6.94) fails only for isolated values of k the shortcoming can be disastrous for the application of numerical

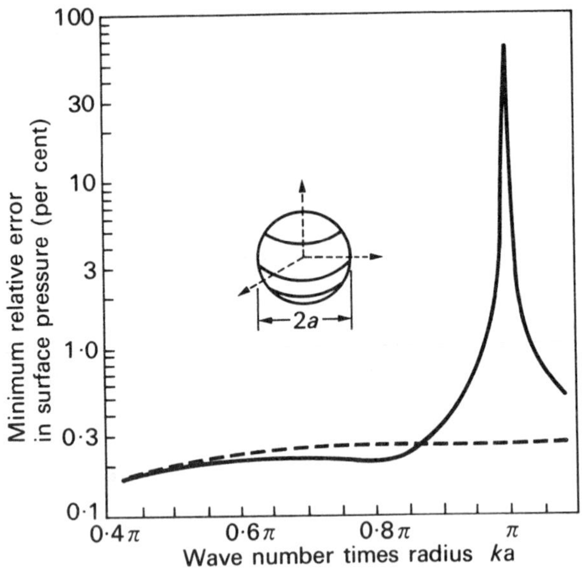

FIG. 6.18. The relative error in surface pressure on a vibrating sphere: solid curve, no correction for interior resonance; broken curve, exact theory.

methods. The matrix of the algebraic system which replaces the integral equation must be singular or nearly so when k is in a neighbourhood of a point of non-uniqueness. How far away k can be to eliminate the uncertainty is less easy to explain.

Fig. 6.18 shows the relative error as the wavenumber varies for a sphere in an acoustic field where a similar phenomenon rears its ugly head. Clearly, errors start to make themselves manifest at wavenumbers as little as half the first resonant wavenumber. The situation is worse at the higher wavenumbers because the interior eigenvalues tend to cluster with increasing frequency.

It may therefore be concluded that numerical results founded on the EFIE must be viewed with suspicion unless the dimensions of the body do not exceed half a wavelength when they are probably acceptable provided that errors have not crept in from sources other than interior resonance.

The EFIE has one virtue which does not seem to have been exploited to any extent. It is transparent from Fig. 6.18 that it exposes in a distinct fashion where resonance occurs and so it could supply a means of finding the wavenumbers of electric modes of oscillation.

Exercise

32. A perfectly conducting sphere of radius a has its centre at the origin. In terms of the spherical polar co-ordinates R, θ, ϕ an electromagnetic field is required outside the sphere with

$$(\mathbf{E}_0)_\theta = 0, \quad (\mathbf{E}_0)_\phi = h_1^{(2)}(ka)\sin\theta$$

where $h_1^{(2)}$ is the spherical Hankel function defined by

$$h_1^{(2)}(z) = (\pi/2z)^{\frac{1}{2}} H_{3/2}^{(2)}(z).$$

Assume that $j_\theta = 0$ and $j_\phi = C \sin\theta$ where C is a constant. Show that the EFIE (6.94) is satisfied independent of C whenever k satisfies $j_1(ka) = 0$ where j_1 is the spherical Bessel function.

6.14. The magnetic field integral equation

The downfall of the EFIE cannot be attributed to posing the original problem incorrectly because that does have a unique solution. The fault must lie with the representation of $\mathbf{E}$. Perhaps that for $\mathbf{H}$ is better.

Suppose we let $P \to p$ in (6.86). We are immediately faced with the problem that we have no boundary condition on $\mathbf{n} \wedge \mathrm{H}$ and j is unknown. However, from (6.93) (*cf*. (6.100))

$$\mathbf{n}_p \wedge \{\mathrm{H}_+(p) - \mathrm{H}_-(p)\} = -\,\mathbf{j}(p) \; .$$

It is desirable that the representation should have the same property as (6.76) since that is exact. Therefore $\mathbf{j}$ ought to be such that $\mathbf{n}_p \wedge \mathrm{H}_-(p) = \mathbf{0}$. Accepting this condition leads to the integral equation

$$-\tfrac{1}{2}\mathbf{j}(p) + \mathbf{n}_p \wedge \int_S \mathbf{j}(q) \wedge \operatorname{grad}_q \psi(p,q)\,dS_q$$

$$= \mathbf{n}_p \wedge \int_S \left[\frac{\operatorname{div}_0\{\mathbf{n}_q \wedge \mathbf{E}_0(q)\}\operatorname{grad}_q \psi(p,q)}{i\omega\mu} + \right.$$

$$\left. + \; i\omega\varepsilon\mathbf{n}_q \wedge \mathbf{E}_0(q)\psi(p,q)\right] dS_q \qquad (6.101)$$

which is called the *magnetic field integral equation* (MFIE).

If $\mathbf{E}_0$ is caused by a field incident from outside S the right-hand side of (6.101) may be replaced by $-\mathbf{n}_p \wedge \mathbf{H}_0(p)$ in a similar way to (6.97) provided that $\mathbf{j} + \mathbf{n}_q \wedge \mathbf{H}_0$ is again relabelled as $\mathbf{j}$.

Whether (6.101) has a unique solution turns on whether

$$-\tfrac{1}{2}\mathbf{j}(p) + \mathbf{n}_p \wedge \int_S \mathbf{j}(q) \wedge \operatorname{grad}_q \psi(p,q)\,dS_q = 0 \qquad (6.102)$$

is satisfied by non-trivial $\mathbf{j}$. Define an inner product (§3.2) for tangential vectors $\mathbf{j}$ and $\mathbf{h}$ by

$$\langle \mathbf{j},\mathbf{h}\rangle = \int_S \mathbf{j} \cdot \mathbf{h}^* dS_q \,. \qquad (6.103)$$

Then the adjoint (§3.2) of (6.102) is

$$\tfrac{1}{2}\mathbf{h}(p) + \int_S \{\mathbf{n}_q \wedge \mathbf{h}(q)\} \wedge \operatorname{grad}_q \psi^*(p,q)\,dS_q = 0. \qquad (6.104)$$

By the *Fredholm alternative* (to be discussed in the following section), whenever (6.102) possesses a non-trivial solution so does (6.104) and vice versa. Therefore construction of a non-trivial solution of (6.104) will imply non-uniqueness of (6.102).

Take the complex conjugate of (6.104) and then multiply vectorially by $\mathbf{n}_p$. With $\mathbf{j}_m(p) = \mathbf{n}_p \wedge \mathbf{h}^*(p)$ there results

$$\tfrac{1}{2}\mathbf{j}_m(p) + \mathbf{n}_p \wedge \fint_S \mathbf{j}_m \wedge \mathrm{grad}_q\psi(p,q)\mathrm{d}S_q = 0. \qquad (6.105)$$

Thus to every non-trivial solution of (6.104) corresponds a non-trivial solution of (6.105). Conversely, when (6.105) is satisfied non-trivially so is (6.104) as may be confirmed by defining $\mathbf{h}^* = \fint_S \mathbf{j}_m \wedge \mathrm{grad}_q\psi(p,q)\mathrm{d}S_q$ whence $\tfrac{1}{2}\mathbf{j}_m = -\mathbf{n} \wedge \mathbf{h}^*$ and the result follows by taking the complex conjugate. Hence, it is sufficient for our purposes to demonstrate the non-uniqueness of (6.105).

Consider the electromagnetic field

$$\mathbf{E}'(P) = \int_S \mathbf{j}_m(q) \wedge \mathrm{grad}_q\psi(P,q)\mathrm{d}S_q\,, \qquad (6.106)$$

$$\mathbf{H}'(P) = \int_S \{\mathrm{i}\omega\varepsilon\mathbf{j}_m(q)\psi(P,q) - \frac{\rho_m}{\mu}\,\mathrm{grad}_q\psi(P,q)\}\mathrm{d}S_q \qquad (6.107)$$

where ρ_m is given by (6.84). It satisfies Maxwell's equations in S_+, the radiation conditions, and $\mathbf{n} \wedge \mathbf{E}'_+(p) = 0$ on account of (6.90) and (6.105). Hence it is identically zero in S_+ and $\mathbf{n} \wedge \mathbf{H}'_+(p) = 0$. An immediate consequence, by virtue of (6.91), is that $\mathbf{n} \wedge \mathbf{H}'_-(p) = 0$. Thus (6.105) can have non-trivial solutions only when there are interior magnetic modes of oscillations. On the other hand, if $\mathbf{E}_m$, $\mathbf{H}_m$ is a magnetic mode of resonance choose $\mathbf{j}_m = \mathbf{n} \wedge \mathbf{E}_m$ and then the application of (6.90) and (6.91) to (6.106) implies (6.105).

The deduction is that the uniqueness of (6.101) fails precisely for those values of k for which there are interior modes of magnetic resonance. Thus the MFIE suffers from the same disadvantage as the EFIE, and accepting numerical results based on the MFIE should be tempered with prudence. Notwithstanding, if a choice has to be made between the MFIE and the EFIE, the better gambit is the MFIE. One reason is that the EFIE is a singular integral equation of the first kind, whereas the MFIE is one of the second kind, for which the theory has sounder foundations. The second motive for the selection is numerical in origin. Often the algebraic system, substituting for an integral equation, is increased in size by subdividing a mesh on the surface with the aim of improving the accuracy. However, as the size of a mesh element diminishes, the value of the integral over it tends to decrease. Thus the elements of the matrix for the EFIE will be becoming smaller as its order is becoming larger. There will therefore be a tendency for the matrix to become ill conditioned. In contrast, the MFIE contains a **j** which does not involve integration. Its matrix is therefore likely to exhibit more and more diagonal dominance with steadily decreasing mesh size. Hence, reduction in mesh size should not cause the deterioration of accuracy for the MFIE that would be expected with the EFIE.

Before concluding this section there is one further implication of the Fredholm alternative that needs checking. This is the assertion that at a resonant value of k, (6.101) possesses a solution if and only if the right-hand side is orthogonal (with respect to the inner product) to every non-trivial solution of (6.104). For the receiving antenna the right-hand side is a multiple of $\mathbf{n} \wedge \mathbf{H}_0$ and the requirement is

$$\langle \mathbf{n}\wedge\mathbf{H}_0, \mathbf{h}\rangle = 0. \qquad (6.108)$$

Now

$$\int_S \mathbf{n}_q \wedge \mathbf{H}_0(q) \cdot \mathbf{h}^*(q)\, dS_q = -\int_S \mathbf{H}_0(q) \cdot \mathbf{j}_m(q) dS_q$$

$$= \int_S \{\mathbf{H}_m(q)\cdot\mathbf{n}_q\wedge\mathbf{E}_0(q)-\mathbf{H}_0(q)\cdot\mathbf{n}_q\wedge\mathbf{E}_m(q)\}dS_q$$

since $\mathbf{j}_m$ stems from a magnetic mode of oscillation. The last integral is zero because the divergence theorem can be applied in S_- since neither field has any singularity there. Hence (6.108) is verified for a receiving antenna with no sources inside S.

6.15. The Fredholm alternative

This section is a digression from the theory of antennas because its aim is to establish the Fredholm alternative for operators sufficiently general to cover the needs of the preceding section. The notation of Chapter 3 will be used but since (6.102) and (6.104) are plainly not self-adjoint a generalized version of the theory of that chapter will be necessary. The proof will be restricted to operators defined on Hilbert spaces but, in fact, the results hold for normed linear spaces though the proofs have to be changed.

Let T be a compact operator on a complex Hilbert space H into H. By §3.3 T is both bounded and continuous. Let N_1 be the set of x such that $Tx = x$. It is obvious that N_1 is a linear subspace of H. Indeed, N_1 is of finite dimension and thereby closed (§3.1). For, if N_1 contained an infinite basis, an infinite orthonormal set $\{x_n\}$ could be constructed by the Schmidt process (§1.5). Then $Tx_n = x_n$ and so the definition of a compact operator demands that $\{x_n\}$ has a convergent subsequence. However, $\|x_m-x_n\|^2 = 2$ prevents the existence of any such convergent subsequence. Hence N_1 is of finite dimension.

Since T is into H, T^2 can be defined by $T^2x = T(Tx)$ and higher powers of T follow in a similar manner. Now if $\{x_j\}$ is a bounded sequence so is $\{Tx_j\}$ since T is bounded. Hence $T(Tx_j)$ has a convergent subsequence because T is compact. Consequently, T^2 is compact and, in general, so is T^m.

Let N_m be the set of x such that $(T-I)^m x = 0$, I being the identity operator. Since T^p is compact it follows from the result for N_1 that N_m is of finite dimension. Clearly, if $x \in N_m$, then $x \in N_{m+1}$ and the question arises whether the dimension of N_m increases without limit as m grows.

Suppose there is $x \in N_{m+1}$ which is not in N_m. Then $(T-I)^{m-n}x \in N_{n+1}$ for $n \leqslant m$ but does not belong to N_n. Therefore, either N_{n+1} and N_n are different spaces for all n or they coincide for some n and do not change for any further increase in n. To show that the first possibility cannot occur choose in each N_n an element x_n which is orthogonal to all members of

N_{n-1} and such that $\|x_n\| = 1$. Then, for $n > m$,

$$Tx_n - Tx_m = \{(T-I)x_n-(T-I)x_m-x_m\} + x_n = y + x_n$$

where $y \in N_{n-1}$. Since x_n is orthogonal to y

$$\|Tx_n-Tx_m\|^2 = \|y\|^2 + \|x_n\|^2 \geqslant 1.$$

Therefore $\{Tx_n\}$ does not possess a convergent subsequence contradicting T being compact. We deduce that there is a finite integer ν such that $N_\nu = N_{\nu+1} = N_{\nu+2} = \ldots.$

An associated space is M_n which consists of all those y which can be written as $y = (T-I)^n x$ for some $x \in H$, i.e. M_n is the range of the operator $(T-I)^n$. Evidently, M_n is linear and it is, in fact, a closed linear subspace of H. This will be proved for M_1 and the general demonstration left to the reader.

Let $\{y_m\}$ be a convergent sequence in M_1. If $y_m = (T-I)x_m$ put $x_m = u_m + v_m$ where $u_m \in N_1$ and v_m is orthogonal to N_1. Then $y_m = (T-I)v_m$. Suppose that there is a subsequence on which $\|v_m\| \to \infty$. Then $(T-I)v_m/\|v_m\| = y_m/\|v_m\| \to 0$ since $\{y_m\}$ is convergent and hence bounded. Therefore the sequence $\{Tv_m/\|v_m\|\}$ contains a subsequence $\{Tv_{m_j}/\|v_{m_j}\|\}$ which converges. Therefore $\{Tv_{m_j}-(T-I)v_{m_j}/\|v_{m_j}\|\}$ or $\{v_{m_j}/\|v_{m_j}\|\}$ is also convergent. If its limit is v then

$$(T-I)v = \lim_{j\to\infty} \frac{(T-I)v_{m_j}}{\|v_{m_j}\|} = 0$$

i.e. $v \in N_1$. However, v_{m_j} is orthogonal to N_1 and so

$$\left\|\frac{v-v_{m_j}}{\|v_{m_j}\|}\right\|^2 = \|v\|^2 + 1$$

which is a contradiction. Hence $\|v_m\|$ is bounded.

Consequently, $\{Tv_m\}$ possesses a convergent subsequence $\{Tv_{m_k}\}$. The sequence $\{Tv_{m_k}-(T-I)v_{m_k}\}$ or $\{v_{m_k}\}$ converges also.

Let its limit be v_0. Then

$$y_{m_k} = (T-I)v_{m_k} \to (T-I)v_0$$

and so the limit of $\{y_m\}$ is in M_1, i.e. M_1 is closed.

Obviously an element of M_{n+1} is an element of M_n, but the spaces cannot all be different. If they could, pick for each M_n a y_n which is orthogonal to M_{n+1} and of unit norm. A similar sort of argument to that applied to N_m demonstrates that this is impossible. Accordingly, there is a finite integer μ such that $M_\mu = M_{\mu+1} = M_{\mu+2} = \ldots$.

The remarkable fact is that $\mu = \nu$. In order to prove this the spaces are first supplemented by M_0, which is to be the same as H, and N_0, which contains only the zero element. These spaces are consistent with the earlier definitions of M_n and N_m since $(T-I)^0$ can be regarded as the identity operator.

Next, the aim is to show that any $x \in H$ can be expressed as $x = y + z$ where $y \in M_\mu$ and $z \in N_\mu$. Moreover, the only element M_μ and N_μ have in common is the zero element. Briefly, $H = M_\mu \oplus N_\mu$.

Now $(T-I)^\mu x$ is in M_μ and therefore in $M_{2\mu}$ since $M_{2\mu} = M_\mu$. Hence there is some $w \in H$ such that $(T-I)^\mu x = (T-I)^{2\mu} w$. Therefore if we write

$$x = \{x-(T-I)^\mu w\} + (T-I)^\mu w$$

the first member is an element of N_μ whereas the second lies in M_μ. The desired decomposition has therefore been achieved if zero is the only common element of M_μ and N_μ. Suppose that there is a $z \neq 0$ which is in both M_μ and N_m $(m \geqslant 0)$. For $n \geqslant \mu$, $z \in M_n$ and so, for each n, there is a z_n such that $z = (T-I)^n z_n$. Since $z \in N_m$ it follows that $z_n \in N_{m+n}$ but is not in N_n, contrary to the properties of N_ν as soon as $n \geqslant \nu$. Hence the only possible common element of N_m and M_μ is zero, which is a rather stronger result than was stated. At any rate $H = M_\mu \oplus N_\mu$.

One consequence of this result is that the solution of $(T-I)x = y$ is unique when $x, y \in M_\mu$. There is a solution $x \in M_\mu$ since $M_{\mu+1} = M_\mu$ and it will be unique if $(T-I)x = 0$

implies that $x = 0$. However, this is true by the preceding paragraphs because x is in both M_μ and N_1, whose only common element is zero.

Let now $x \in N_{\mu+1}$ so that $(T-I)^{\mu+1}x = 0$. The equation $(T-I)y = 0$ has solution $y = (T-I)^\mu x$ which is in M_μ and so $y = 0$. Therefore $x \in N_\mu$, i.e. $N_{\mu+1} = N_\mu$ whence $\mu \geqslant \nu$. An instant deduction is that $\nu = 0$ when $\mu = 0$.

If $\mu \geqslant 1$, let $y = (T-I)^{\mu-1}x \in M_{\mu-1}$ which is not in M_μ. Since $z = (T-I)^\mu x \in M_\mu$ the equation $(T-I)^\mu w = z$ has the unique solution w_0 (say) in M_μ. Hence $x - w \in N_\mu$. On the other hand $(T-I)^{\mu-1}(x-w_0) = y - (T-I)^{\mu-1}w_0 \neq 0$ since $(T-I)^{\mu-1}w_0 \in M_\mu$ but y is not in M_μ. Hence $x - w_0$ is not in $N_{\mu-1}$ and so $N_{\mu-1} \neq N_\mu$. Therefore $\mu \leqslant \nu$.

Combining the last two paragraphs we conclude that $\mu = \nu$. There is no longer any necessity to distinguish between μ and ν.

Suppose that $\mu = 0$. Since $M_0 = H$, the equation $(T-I)x = y$ has a unique solution for any $y \in H$. The correspondence could be expressed as $x = (T-I)^{-1}y$ and the inverse is, in fact, bounded. For, if this were not so, there would be a sequence $\{y_n\}$ such that $\|x_n\|/\|y_n\|$ was unbounded. Put $z_n = x_n/\|x_n\|$; then $(T-I)z_n = y_n/\|x_n\| \to 0$. On account of $\|z_n\| = 1$, the sequence $\{Tz_n\}$ contains a convergent subsequence $\{Tz_{n_k}\}$ with limit, say z_0. Then

$$z_{n_k} = Tz_{n_k} - (T-I)z_{n_k} \to z_0.$$

By virtue of the continuity of $T - I$, $(T-I)z_{n_k} \to (T-I)z_0$. The limit of the left-hand side is zero and hence $z_0 = 0$ by uniqueness. This contradicts $\|z_0\| = \lim\|z_{n_k}\| = 1$ and so there is a finite C such that $\|(T-I)^{-1}\| < C$.

Further, if $T^A y = y$,

$$0 = \langle x, T^A y - y\rangle = \langle Tx - x, y\rangle$$

for all $x \in H$. This shows that y cannot be in $M_1 = M_0 = H$ since $\mu = 0$ unless $y = 0$. Accordingly, when $\mu = 0$ the homogeneous adjoint equation possesses only a trivial solution and the inverse $(T^A-I)^{-1}$ exists. It is indeed bounded by the same constant as $(T-I)^{-1}$ because, in general,

$$\|T^{A}y\|^{2} = \langle T^{A}y, T^{A}y\rangle = \langle TT^{A}y, y\rangle \leqslant \|TT^{A}y\| \; \|y\| \leqslant \|T\| \; \|T^{A}y\| \; \|y\| .$$

If $\mu \geqslant 1$ the situation is different because the equation $Tx = x$ always admits the solution $x = (T-I)^{\mu-1}z$ where $z \in N_{\mu}$ but is not in $N_{\mu-1}$. Thus $T - I$ has no inverse. However, it has been proved above that $Tx - x = y$ has a unique solution when both x and y are in M_{μ}. Arguing as in the case $\mu = 0$ we deduce that there is a C such that for $x \in M_{\mu}$, $\|x\| \leqslant C\|(T-I)x\|$. Since $(T-I)^{n}x \in M_{\mu}$, it follows that $\|x\| \leqslant C^{\mu}\|(T-I)^{\mu}x\|$ for $x \in M_{\mu}$.

It has already been shown that any $x \in H$ can be expressed uniquely in the form $x = u + v$ where $u \in M_{\mu}$ and $v \in N_{\mu}$. Define operators P and Q such that $Px = u$, $Qx = v$. Obviously P and Q are linear. Also $Pu = u$ so that $P^{2} = P$. Similarly $Q^{2} = Q$ and so both P and Q can be considered as *projection operators*. Moreover, $Qu = 0$ and $Pv = 0$ whence $PQ = QP = O$; clearly, $P + Q = I$. That P and Q are bounded is evident from

$$\|u\| \leqslant C^{\mu}\|(T-I)^{\mu}u\| \leqslant C^{\mu}\|(T-I)^{\mu}x\|$$

$$\leqslant C^{\mu}\|(T-I)^{\mu}\| \; \|x\|$$

and $\|v\| \leqslant \|x\| + \|u\|$. Another property is $PT = TP$, $QT = TQ$ because $(T-I)u \in M_{\mu}$, $(T-I)v \in N_{\mu}$ with the consequence

$$P(T-I)x = P(T-I)u = (T-I)u = (T-I)Px .$$

Put $R = PT$ and $S = QT$. Then R and S are compact because, for example,

$$\|Rx_{n}\| \leqslant \|P\| \; \|Tx_{n}\|$$

implying that the left-hand side has a convergent subsequence when the right-hand side does. Furthermore

$$R + S = (P+Q)T = T ,$$

$$RS = PTQT = TPQT = O, \qquad SR = O$$

so that

$$I - T = (I-R)(I-S) = (I-S)(I-R) .$$

It will now be demonstrated that $(I-R)^{-1}$ exists. If this were not so there would be $u \in M_\mu$ such that $(I-R)u = 0$. But then $(I-T)u = 0$ contrary to the fact that such an equation has no solution in M_μ other than the trivial one.

Consequently, any solution of $(I-T)x = 0$ must satisfy $(I-S)x = 0$ and conversely. More generally

$$(I-T)^n = (I-R)^n(I-S)^n = (I-S)^n(I-R)^n ,$$

$$(I-S)^n = (I-R)^{-n}(I-T)^n = (I-T)^n(I-R)^{-n}$$

showing that the spaces M_n and N_n are the same for both T and S. Since N_μ is of finite dimension each of its elements can be represented in terms of a finite orthonormal set $\{\phi_m\}$ whence

$$Sx = \sum_{m=1}^{r} a_m\phi_m$$

where $a_m = \langle Sx,\phi_m\rangle$. Since a_m is a linear functional of x, there are $\psi_j \in H$ such that $a_m = \langle x,\psi_m\rangle$. An equation such as $(I-S)x = y$ will therefore take the form

$$x - \sum_{m=1}^{r} \langle x,\psi_m\rangle\phi_m = y \qquad (6.109)$$

whence

$$x_n - \sum_{m=1}^{r} c_{nm}x_m = \langle y,\psi_n\rangle \quad (n = 1, \ldots, r) \qquad (6.110)$$

where $x_n = \langle x,\psi_n\rangle$ and $c_{nm} = \langle\phi_m,\psi_n\rangle$. Once x_m has been found from (6.110), x can be written down from (6.109).

With regard to the adjoint we have

$$T^{\mathrm{A}} = R^{\mathrm{A}} + S^{\mathrm{A}}, \qquad R^{\mathrm{A}}S^{\mathrm{A}} = S^{\mathrm{A}}R^{\mathrm{A}} = O$$

so that the adjoint enjoys similar properties to the original. In addition

$$<x, S^{\mathrm{A}}y> = <Sx, y> = \sum_{m=1}^{r} <x, \psi_m> <\phi_m, y>$$

for any $x, y \in H$ so we see, recalling eqn (1.21), that

$$S^{\mathrm{A}}y = \sum_{m=1}^{r} <y, \phi_m> \psi_m .$$

The equation $(I - S^{\mathrm{A}})w = z$ therefore gives

$$w_n - \sum_{m=1}^{r} c^*_{mn} w_m = z_n , \qquad (6.111)$$

where $w_n = <w, \phi_n>$ and $z_n = <z, \phi_n>$, as the analogue of (6.110).

The matrix on the left-hand side of (6.110) may be non-singular. In that case, the left-hand side of (6.110), being the Hermitian (§1.10) of (6.110), is also non-singular. Both equations and the corresponding operator equations have a unique solution.

Per contra the matrix on the left of (6.110) may be singular and of rank p, say. Then the matrix on the left of (6.111) is also of rank p and the homogeneous equation possesses $r - p$ linearly independent solutions $w^{(1)}, w^{(2)}, \ldots, w^{(n-p)}$. Then

$$\sum_{n=1}^{r} w_n^{(j)} <y, \psi_n>^* = \sum_{n=1}^{r} w_n^{(j)} \left\{ x^*_n - \sum_{m=1}^{r} c^*_{nm} x^*_m \right\}$$

$$= \sum_{n=1}^{r} w_n^{(j)} x^*_n - \sum_{m=1}^{r} w_m^{(j)} x^*_m = 0.$$

Since $w^{(j)} = S^{A}w^{(j)} = \Sigma_{m=1}^{r} w_m^{(j)}\psi_m$, this may be expressed as $\langle w^{(j)}, y\rangle = 0$. Thus (6.109) cannot be soluble unless y is orthogonal to each of $w^{(1)}, \ldots, w^{(r-p)}$, i.e. y must be orthogonal to all solutions of $(I-S^{A})w = 0$ and thereby of $(I-T^{A})w = 0$. If y is orthogonal in this way the right-hand side of (6.110) is in a space of p dimensions. Since this space can always be generated by giving x_m all possible values (6.111) will have a solution though it will not be unique.

These results may now be summarized in a theorem. If λ is a complex number, λT is compact, so that T and T^{A} in the preceding may be replaced by λT and $\lambda^* T^{A}$ respectively without affecting the conclusions. Hence, we enunciate

THEOREM 6.15 (THE FREDHOLM ALTERNATIVE). *If T is a compact operator from H into H EITHER the equations $x - \lambda Tx = 0$ and $w - \lambda^* T^{A} w = 0$ have only the solutions $x = 0$ and $w = 0$, and $x - \lambda Tx = y$ possesses a unique solution OR the equations $x - \lambda Tx = 0$ and $w - \lambda^* T^{A} w = 0$ each have the same number of linear independent solutions $x^{(1)}, \ldots, x^{(r-p)}$ and $w^{(1)}, \ldots, w^{(r-p)}$. Then $x - \lambda Tx = y$ has a solution if and only if $\langle w^{(j)}, y\rangle = 0$ for $j = 1, \ldots, r - p$.*

When the second alternative is relevant and y has the necessary orthogonality the solution of $x - \lambda Tx = y$ can be written as

$$x = z + \sum_{m=1}^{r-p} b_m x^{(m)} \tag{6.112}$$

where z is a particular solution and the b_m are arbitrary constants.

It will be observed that Theorem 6.15 also resolves an eigenvalue problem. In particular, any eigenvalue λ (which may be complex) is of finite multiplicity by the second part of the theorem. It may be deduced that the eigenvalues λ_n are countable and have no limit point in the finite part of the complex plane.

Often applications provide an operator T which is bounded but not compact, though some positive integer power T^n is compact. In that case, Theorem 6.15 still stands, the necessary modifications to the proof being based on the fact that there is a bounded operator B such that $I - T^n = B(I-T) = (I-T)B$.

6.16. Compactness and other properties of the MFIE

To apply the Fredholm alternative to (6.102) it is necessary to show that the operator T, defined by

$$T\mathbf{j} = \mathbf{n}_p \wedge \int_S \mathbf{j}(q) \wedge \operatorname{grad}_q \psi(p,q) \mathrm{d}S_q ,$$

is compact. The theory at the beginning of §3.4 cannot be cited because the kernel is not square integrable. Instead, advantage is taken of the device described after eqn (3.17) of that section.

Define $\boldsymbol{\Psi}_1$ and $\boldsymbol{\Psi}_2$ by

$$\begin{aligned} \boldsymbol{\Psi}_1 &= \operatorname{grad}_q \psi(p,q) && (|\mathbf{x}_p - \mathbf{x}_q| \geqslant \delta) \\ &= 0 && (|\mathbf{x}_p - \mathbf{x}_q| < \delta), \\ \boldsymbol{\Psi}_2 &= 0 && (|\mathbf{x}_p - \mathbf{x}_q| \geqslant \delta) \\ &= \operatorname{grad}_q \psi(p,q) && (|\mathbf{x}_p - \mathbf{x}_q| < \delta). \end{aligned}$$

Then $T = T_1 + T_2$ where

$$T_1\mathbf{j} = \mathbf{n}_p \wedge \int_S \mathbf{j}(q) \wedge \boldsymbol{\Psi}_1 \mathrm{d}S_q ,$$

$$T_2\mathbf{j} = \mathbf{n}_p \wedge \int_S \mathbf{j}(q) \wedge \boldsymbol{\Psi}_2 \mathrm{d}S_q .$$

The kernel for T_1 has no singularity and is square integrable. Therefore, by §3.4, T_1 is compact. Hence T will be compact, by the first criterion at the end of §3.3, if it can be shown that the norm of T_2 can be made arbitrarily small.

For given $\mathbf{x}_p$ select temporary axes with the origin at $\mathbf{x}_p$

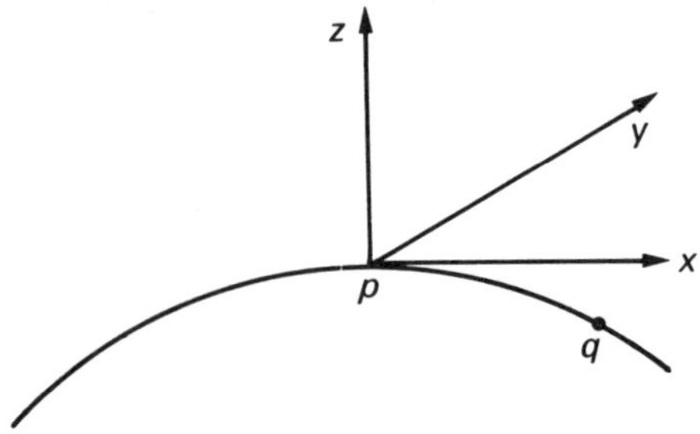

FIG. 6.19. The geometrical configuration for the norm of T_2.

and the z axis along $\mathbf{n}_p$, assuming that S has a continuously turning tangent plane near $\mathbf{x}_p$. The integral in T_2 will then be effectively limited to $x^2 + y^2 \leqslant \delta^2$ and the equation of the surface will be approximately $2z = ax^2 + by^2$ if the x and y axes are arranged to lie in the planes of principal curvature. If $\mathbf{j}$ has components j_1, j_2, and j_3 parallel to the coordinate axes

$$j_3 = axj_1 + byj_2 \tag{6.113}$$

because $\mathbf{j}$ is tangential to the surface. Also

$$T_2\mathbf{j} = \oint_S \{(\mathbf{k}.\boldsymbol{\Psi}_2)\mathbf{j}-(\mathbf{k}.\mathbf{j})\boldsymbol{\Psi}_2\}\mathrm{d}S_q$$

and now $\psi = \mathrm{e}^{-ikr}/4\pi r$ where $r^2 = x^2 + y^2 + z^2$. Hence

$$|T_2\mathbf{j}| \leqslant \oint_{r\leqslant\delta}\left\{\left|\frac{\partial\psi}{\partial z}\right|(|j_1|+|j_2|)+|j_3|\left(\left|\frac{\partial\psi}{\partial x}\right|+\left|\frac{\partial\psi}{\partial y}\right|\right)\right\}\mathrm{d}S_q\ .$$

Now $|z|/r^2 \leqslant |a| + |b|$ and so, if (6.113) is substituted,

$$|T_2\mathbf{j}| \leqslant B\int_{r\leqslant\delta}\frac{|\mathbf{j}|}{r}\,\mathrm{d}S_q$$

for some finite constant B. Hence, in general,

$$|T_2\mathbf{j}|^2 \leqslant B^2\left\{\int_{|\mathbf{x}_p-\mathbf{x}_q|\leqslant\delta} \frac{|\mathbf{j}|}{|\mathbf{x}_p-\mathbf{x}_q|}\,\mathrm{d}S_q\right\}^2$$

$$\leqslant B^2 \int_{|\mathbf{x}_p-\mathbf{x}_q|\leqslant\delta} \frac{|\mathbf{j}|^2}{|\mathbf{x}_p-\mathbf{x}_q|}\,\mathrm{d}S_q \int_{|\mathbf{x}_q|\leqslant\delta} \frac{\mathrm{d}S_q}{|\mathbf{x}_q|}$$

by the Schwarz inequality. Consequently

$$\|T_2\mathbf{j}\|^2 = \int_S |T_2\mathbf{j}|^2\mathrm{d}S \leqslant B^2\left\{\int_S\int_S \frac{|\mathbf{j}|^2}{|\mathbf{x}_p-\mathbf{x}_q|}\mathrm{d}S_p\,\mathrm{d}S_q\right\} \int_{|\mathbf{x}_q|\leqslant\delta} \frac{\mathrm{d}S_q}{|\mathbf{x}_q|}$$

$$\leqslant B^2 \int_S |\mathbf{j}|^2\mathrm{d}S_q \int_{|\mathbf{x}_p|<d} \frac{\mathrm{d}S_p}{|\mathbf{x}_p|} \int_{|\mathbf{x}_q|\leqslant\delta} \frac{\mathrm{d}S_q}{|\mathbf{x}_q|}$$

$$\leqslant B^2\|\mathbf{j}\|^2 \int_{|\mathbf{x}_p|<d} \frac{\mathrm{d}S_p}{|\mathbf{x}_p|} \int_{|\mathbf{x}_q|\leqslant\delta} \frac{\mathrm{d}S_q}{|\mathbf{x}_q|}$$

where d is the maximum separation of two points of S. Since S is bounded, d is finite and the first integral is finite. The second integral can be made as small as we like by making δ sufficiently tiny. Hence it can be arranged that $\|T_2\mathbf{j}\|^2 \leqslant \varepsilon\|\mathbf{j}\|^2$ for any $\varepsilon > 0$, i.e. $\|T_2\| \leqslant \varepsilon$ and therefore T is compact.

Although compactness has been proved for a surface which has a local quadratic approximation[†] a check of the proof reveals that it is still valid if $|z|/r^2 < 1/r^\alpha$ where $0 \leqslant \alpha < 1$ which may be otherwise stated as the angle between two normals to S must not exceed a finite multiple of $|\mathbf{x}_p-\mathbf{x}_q|^{1-\alpha}$.

The argument that our integral operator is compact is legitimate for any complex value of k. In particular, the static

---

[†] For a more precise argument, including a proof that T is a bounded operator, see Müller 1969 (see also §7.7).

operator in which $k = 0$ is compact. However, the solution of the static integral equation is unique subject to the static surface continuity condition $\text{div}_0\mathbf{j} = 0$. To prove this define

$$\mathbf{J} = \int_S \mathbf{j}(q) \wedge \text{grad}_q \psi_0(P,q)\text{d}S_q$$

where ψ_0 is ψ with $k = 0$. Then $\text{div}_P\mathbf{J} = 0$ and $\text{curl}_P\mathbf{J} = \mathbf{0}$ (from (6.87) and the continuity condition) in both S_+ and S_-. Thus $\mathbf{J} = \text{grad}\, V$ where V satisfies Laplace's equation in both regions. The static analogue of (6.102) implies that $\mathbf{n}_p \wedge \mathbf{J}_- = \mathbf{0}$ and so V_- is constant on S. Hence V is constant throughout S_- which entails $\mathbf{J} = \mathbf{0}$ in S_-. Therefore $\mathbf{n}_p \cdot \mathbf{J}_- = 0$. But (6.90) and (6.91) require the normal component of $\mathbf{J}$ to be continuous whence $\mathbf{n}_p \cdot \mathbf{J}_+ = 0$. Consequently, the normal derivative of V vanishes on S and, since V tends to zero at infinity, V must be zero in S_+. Therefore $\mathbf{n}_p \wedge \mathbf{J}_+ = \mathbf{0}$ with the implication from (6.90) and (6.91) that $\mathbf{j} = 0$ and the statement about uniqueness for the static problem is proved.

Having established this fact, we can drop the surface continuity condition and still retain uniqueness, for, although $\text{curl}_P\mathbf{J} = \mathbf{0}$ cannot now be asserted, it is true that $\text{curl}_P\text{curl}_P\mathbf{J} = \mathbf{0}$ by virtue of (6.87). Accordingly,

$$\int_{S_-} (\text{curl}\ \mathbf{J})^2\text{d}\tau = \int_S \{\mathbf{n}_q.\mathbf{J}_-\wedge\text{curl}_q\mathbf{J}_-\}\text{d}S_q = 0 \tag{6.115}$$

by means of the static analogue of (6.102). Therefore $\text{curl}_P\mathbf{J} = 0$ in S_- and we conclude, as before, that $\mathbf{J} = \mathbf{0}$ in S_-. In particular, $\mathbf{n}_p \wedge \text{curl}\ \mathbf{J}_- = \mathbf{0}$ on S. It follows from (6.87), (6.90), and (6.91) that $\mathbf{n}_p \wedge \text{curl}\ \mathbf{J}_+ = \mathbf{0}$. A repetition of (6.115) for S_+, bearing in mind the behaviour of $\mathbf{J}$ and curl $\mathbf{J}$ at infinity, then leads to $\text{curl}_P\mathbf{J} = \mathbf{0}$ in S_+. A return to the previous case has therefore been arrived at and uniqueness of the static integral equation has been demonstrated without imposition of the surface continuity condition.

6.17. Other integral equations

To avoid the failure of the MFIE the representation of the field in terms of a magnetic current instead of an electric current might be contemplated, for example

$$\mathrm{E}(P) = \int_S \mathbf{J}(q) \wedge \operatorname{grad}_q \psi(P,q)\mathrm{d}S_q \ .$$

The integral equation

$$\tfrac{1}{2}\mathbf{J}(p) + \mathbf{n}_p \wedge \int_S \mathbf{J} \wedge \operatorname{grad}_q \psi(p,q)\mathrm{d}S_q = \mathbf{n}_p \wedge \mathbf{E}_0(p) \qquad (6.116)$$

then arises. Since the left-hand side of (6.116) coincides with that of (6.105) the discussion of §6.14 reveals that uniqueness of (6.116) will fail when interior magnetic resonance can occur. Whether (6.116) is soluble at such frequencies depends upon the adjoint equation

$$\tfrac{1}{2}\mathbf{k}(p) - \int_S \{\mathbf{n}_q \wedge \mathbf{k}(q)\} \wedge \operatorname{grad}_q \psi^*(p,q)\mathrm{d}S_q = 0. \qquad (6.117)$$

It will be noted that $\mathbf{j} = \mathbf{n} \wedge \mathbf{k}^*$ supplies a solution of (6.102) so the conditions for the solubility of (6.116) can be expressed as

$$\int_S \mathbf{E}_0 \cdot \mathbf{j}\ \mathrm{d}S_q = 0\ .$$

Because $\mathbf{j}$ is connected to an exterior magnetic field there is no guarantee that these equations can be complied with even by a receiving antenna. In that sense (6.116) is worse than the MFIE. Nevertheless it is possible to show how to solve the electromagnetic problem by modifying the representation.

A relationship between the solutions of (6.102) and (6.116) has first to be established. If $\mathbf{j}$ is any solution of (6.102) let $\mathbf{E}^2$, $\mathbf{H}^2$ be the electromagnetic field defined by

$$\mathbf{E}^2(P) = \int_S \{i\omega\mu\mathbf{j}\psi(P,q) - \frac{\rho(q)}{\varepsilon}\,\mathrm{grad}_q\psi(P,q)\}\mathrm{d}S_q \;.$$

From (6.102), $\mathbf{n} \wedge \mathbf{H}_-^2 = 0$ and hence, by applying the analogue of (6.75) for the interior of S,

$$\mathbf{n}_p \wedge \int_S \{\mathbf{n}_q \wedge \mathbf{E}^2(q)\} \wedge \mathrm{grad}_q\psi(p,q)\mathrm{d}S_q + \tfrac{1}{2}\mathbf{n}_p \wedge \mathbf{E}^2(p) = 0$$

i.e. $\mathbf{n}_p \wedge \mathbf{E}^2(p)$ is a solution of (6.105). If, with all choices of $\mathbf{j}$, the space spanned by the $\mathbf{n}_p \wedge \mathbf{E}^2(p)$ were smaller than the space of solutions of (6.105) there would be some $\mathbf{j}$ for which $\mathbf{n}_p \wedge \mathbf{E}^2(p) = 0$. Since $\mathbf{n}_p \wedge \mathbf{E}_+^2(p) = \mathbf{n}_p \wedge \mathbf{E}_-^2(p) = \mathbf{n}_p \wedge \mathbf{E}^2(p)$ the uniqueness theorem for the exterior problem would imply that the field was identically zero in S_+ and so $\mathbf{n} \wedge \mathbf{H}_+^2(p) = 0$. But that would force $\mathbf{j} = 0$ and a zero field. Accordingly, every solution of (6.105) is of the form $\mathbf{n}_p \wedge \mathbf{E}^2(p)$.

Now, if $\mathbf{j}_m$ is a solution of (6.105), it generates an electromagnetic field $\mathbf{E}_m$, $\mathbf{H}_m$ inside S such that $\mathbf{n}_p \wedge \mathbf{H}_{m-} = 0$ and $\mathbf{j}_m = \mathbf{n}_p \wedge \mathbf{E}_{m-}$. However, $\mathbf{E}_m^*$, $-\mathbf{H}_m^*$ is also a solution of Maxwell's equations in S_- with zero tangential magnetic component on S. Therefore $\mathbf{j}^*$ also satisfies (6.105). Thus it can be concluded that to each solution $\mathbf{j}$ of (6.102) there is related a unique solution $\mathbf{j}_m$ of (6.105) such that

$$\mathbf{j}_m^*(p) = \mathbf{n}_p \wedge \mathbf{E}^2(p).$$

It follows obviously that solutions of the adjoint (6.104) satisfy

$$\mathbf{n}_p \wedge \mathbf{h} = \mathbf{n}_p \wedge \mathbf{E}^2 \;.$$

With this information it is possible to check that no solution of (6.105) is orthogonal to all the solutions of its adjoint (6.117) or, what is equivalent, that no solution of the homogeneous MFIE is orthogonal to all the solutions of its adjoint. For such orthogonality would require

$$\langle \mathbf{j},\mathbf{h}\rangle = 0$$

or

$$\int_S \mathbf{n} \wedge \mathbf{H}_+^2 \cdot \mathbf{h}^* \mathrm{d}S_q = 0$$

i.e.

$$\int_S \mathbf{n} \wedge \mathbf{H}_+^2 \cdot \mathbf{E}^{2*} \mathrm{d}S_q = 0$$

from what has just been proved. However, the exterior uniqueness theorem would then entail $\mathbf{n} \wedge \mathbf{H}_+^2 = 0$ on S or $\mathbf{j} = 0$ and this is not permitted.

In view of this property, linear independent functions $\mathbf{J}_r$, $\mathbf{k}_s$ can be found spanning the spaces of solutions of (6.105) and (6.117) and such that

$$\langle \mathbf{J}_r,\mathbf{k}_s\rangle = \delta_{rs}$$

where δ_{rs} is the Kronecker symbol.

Define the surface field $\mathbf{F}$ by

$$\mathbf{F}(p) = \mathbf{n}_p \wedge \mathbf{E}_0(p) - \sum_{r=1}^{n} a_r \mathbf{J}_r$$

where

$$a_r = \langle \mathbf{n}\wedge\mathbf{E}_0,\mathbf{k}_r\rangle .$$

Then $\langle \mathbf{F},\mathbf{k}_s \rangle = 0$ for $s = 1, \ldots, n$ and the equation

$$\tfrac{1}{2}\mathbf{J}'(p) + \mathbf{n}_p \wedge \int_S \mathbf{J}'(q) \wedge \mathrm{grad}_q \psi(p,q) \mathrm{d}S_q = \mathbf{F}(p) \qquad (6.118)$$

is soluble by the Fredholm alternative.

Clearly $\mathbf{J}'$ is not determined uniquely because any linear combination of the $\mathbf{J}_r$ can be added to it. However, the exterior electromagnetic field E^3, H^3 generated by it is unique, for any $\mathbf{J}_r$, being a solution of (6.105), will give rise to an electromagnetic field in the exterior whose tangential electric vector is zero on S and so is identically zero in S_+. Indeed $\mathbf{n}_p \wedge \mathrm{E}_+^3(p) = \mathrm{F}(p)$.

The interconnection between $\mathbf{j}_m$, $\mathbf{E}^2$, $\mathbf{j}$, and $\mathbf{k}$ means that constants a_{rs} can be found such that

$$\mathrm{J}_r(p) = \mathbf{n}_p \wedge \sum_{s=1}^{N} a_{rs} \int_S \Big\{ \mathrm{i}\omega\mu \mathbf{n}_q \wedge \mathbf{k}_s^* \psi(p,q) + \frac{\mathrm{div}_0(\mathbf{n}_q \wedge \mathbf{k}_s^*) \mathrm{grad}_q \psi(p,q)}{\mathrm{i}\omega\varepsilon} \Big\} \mathrm{d}S_q \, .$$

Consider now the electromagnetic field $\mathbf{E}^4$, $\mathbf{H}^4$ in which

$$\mathbf{E}^4 = \mathrm{E}^3 + \sum_{r=1}^{n} a_r \sum_{s=1}^{n} a_{rs} \int_S \Big\{ \mathrm{i}\omega\mu \mathbf{n}_q \wedge \mathbf{k}_s^* \psi(P,q) + \frac{\mathrm{div}_0(\mathbf{n}_q \wedge \mathbf{k}_s^*) \mathrm{grad}_q \psi(P,q)}{\mathrm{i}\omega\varepsilon} \Big\} \mathrm{d}S_q \, . \qquad (6.119)$$

Since the tangential component of the second term on the right-hand side of (6.119) is continuous across S

$$\mathbf{n}_p \wedge \mathrm{E}_+^4 = \mathbf{F} + \sum_{r=1}^{n} a_r \mathbf{J}_r = \mathbf{n}_p \wedge \mathrm{E}_0 \, .$$

Thus the representation (6.119) provides an exterior electromagnetic field which satisfies the required boundary condition even though interior resonance is taking place.

Observe that (6.119) involves both electric and magnetic currents. Moreover, it is subject to the practical disadvantage that not only must the integral equation (6.119) be solved but also the eigenfunctions of (6.105) and its adjoint must be constructed. It is therefore eminently desirable to devise integral equations which do not need such special treatment for particular values of k.

One device is to draw benefit from (6.75) and impose the requirement from (6.85) that

$$\int_S \{\mathrm{i}\omega\mu \mathbf{j}(q)\psi(P_i,q) - \frac{\rho(q)}{\varepsilon}\,\mathrm{grad}_q\psi(P_i,q)\}\mathrm{d}S_q$$

$$= -\int_S \{\mathbf{n}_q \wedge \mathbf{E}_0(q)\} \wedge \mathrm{grad}_q\psi(P_i,q)\mathrm{d}S_q \qquad (6.120)$$

for P_i in S_-. The integral equation (6.120), which holds at all interior points, is known as the *interior electric integral equation* although it is a relation rather than an equation. It can be shown that (6.120) does not suffer from any uniqueness difficulties but it has enormous drawbacks in practice, for, in any numerical scheme, it can be implemented at only a finite number of points inside S. There is no known method of choosing that finite number of points so that satisfaction of (6.120) at them will ensure that (6.120) holds at all interior points. One consequence is that it cannot be certified that the uniqueness of the analytical formulation will carry over to a finite numerical scheme.

There is an analogous *interior magnetic integral equation* derived from (6.86) and (6.76), namely

$$\int_S \mathbf{j}(q) \wedge \mathrm{grad}_q\psi(P_i,q)\mathrm{d}S_q = \text{known} \qquad (6.121)$$

for $P_i \in S_-$. Its deficiencies are the same as those of the interior electric integral equation.

The *Schenck method* is to employ the MFIE and supplement it by requiring that (6.121) be satisfied at a number of interior points as well. In some cases it is found that a few interior points are sufficient to give good results but in other cases several interior points fail to prevent poor answers. The lack of knowledge on how to select the interior points for reliable calculations is the stumbling block.

There is also the alternative of supplementing the EFIE by satisfying (6.120) at a number of interior points. In general, the results may be expected to be poorer than those based on the MFIE. The reason for this stems from the fact that the usual numerical methods split S into sub-areas by a mesh and assume some approximating formula for the unknown on each sub-area. In the absence of error estimates accuracy is checked by reducing the size of the mesh until two successive combinations agree to a prescribed number of figures. As the size of the sub-area diminishes, the value of the integral over it will tend to decrease. Thus the coefficients in the algebraic system derived from the integrals will become smaller and smaller at the same time as the order of the system becomes larger and larger. Consequently, there is a tendency for the determinant of the matrix to approach zero and the system becomes ill conditioned — for very fine meshes some of the numbers may even be too small to be retained by the machine. These comments apply in particular to the EFIE and the interior integral equations when used by themselves. On the other hand, the MFIE contains a term $\mathbf{j}$ which does not involve integration. Therefore the MFIE escapes the above defect and becomes more and more diagonally dominant as the mesh size decreases. Methods based on the MFIE are hence to be preferred since a reduction in mesh size may cause a deterioration of accuracy in other methods.

There is another way of implementing (6.113). Choose some interior point as origin and introduce spherical polar coordinates r, θ, ϕ. Let

$$\psi_{mn}(P) = h_m^{(2)}(kr_P)P_m^n(\cos\theta_P)\cos n\phi_P ,$$

$$\Psi_{mn}(P) = h_m^{(2)}(kr_P)P_m^n(\cos\theta_P)\sin n\phi_P$$

where P_m^n is the standard associated Legendre polynomial and $h_\nu^{(2)}$ is the spherical Hankel function defined by

$$h_\nu^{(2)}(x) = \left(\frac{\pi}{2x}\right)^{\frac{1}{2}} H_{\nu+\frac{1}{2}}^{(2)}(x) .$$

Then, for $r_P < r_Q$,

$$\psi(P,Q) = -\frac{ik}{4\pi}\sum_{m=0}^{\infty}(2m+1)\psi_{m0}(Q)\alpha_{m0}(P) -$$

$$-\frac{ik}{2\pi}\sum_{m=1}^{\infty}(2m+1)\sum_{n=1}^{m}\frac{(m-n)!}{(m+n)!}\{\psi_{mn}(Q)\alpha_{mn}(P)+\Psi_{mn}(Q)A_{mn}(P)\}$$

where α_{mn} and A_{mn} are the real parts of ψ_{mn} and Ψ_{mn} respectively.

If this result is substituted in (6.121) and P is allowed to vary near the origin, the right-hand side may be expanded in terms of the α_{mn} and A_{mn}. We deduce that

$$\int_S \mathbf{j}(q) \wedge \operatorname{grad}_q\psi_{mn}(q)\,dS_q = \text{known} \qquad (6.122)$$

for $m = 0, 1, \ldots, n = 1, \ldots, m$. There is a corresponding equation with Ψ_{mn} in place of ψ_{mn}. If all these equations, which are imposed at a single point, are satisfied, it follows from the analytic nature of the field that (6.121) holds throughout S_-. Therefore the proposal is that the MFIE should be solved subject to the enforcement of a finite number of (6.122) and its correspondent for Ψ_{mn}. Such a system can be expected to be overdetermined but a solution can be specified by least squares.

The Schenck method requires the selection of a number of interior points with no governing criterion other than that they should avoid the nodal surfaces of the interior modes of resonance. Unfortunately, the location of the nodal surfaces is not known, so that the placing of the interior points has to be based on experience and intuition. The implementation of (6.122) removes the arbitrariness in the selection of interior points, replacing the selection by a demand that certain partial derivatives of the left-hand side of (6.121) should vanish at the origin. On the other hand the computation of (6.121) needs only the computation of ψ whereas each of (6.122) involves a different ψ_{mn}. It may be possible to have the best of both worlds by choosing the points of the Schenck method at positions which are used for the calculation of derivatives by difference formulae. Providing that the choice of positions were close enough for the differences to approximate the derivatives and sufficiently separated for numerical distinguishability one would hope that the Schenck method would achieve the theoretical advantages of (6.122).

Yet another integral equation may be formulated by combining the EFIE and MFIE. Since the EFIE and MFIE fail for different kinds of interior resonances their combination might have the desired uniqueness property. Multiply (6.94) vectorially by $\mathbf{n}_p(\varepsilon/\mu)^{\frac{1}{2}}$ and add it to the scalar multiple by β of (6.101). Then

$$\left(\frac{\varepsilon}{\mu}\right)^{\frac{1}{2}} \mathbf{n}_p \wedge [\mathbf{n}_p \wedge \int_S \{i\omega\mu\mathbf{j}(q)\psi(p,q) - \frac{\rho(q)}{\varepsilon}\operatorname{grad}_q\psi(p,q)\}dS_q] +$$

$$+ \beta\{-\frac{1}{2}\mathbf{j}(p) + \mathbf{n}_p \wedge \int \mathbf{j}(q) \wedge \operatorname{grad}_q\psi(p,q)\,dS_q\} = \text{known}. \qquad (6.123)$$

When the left-hand side of (6.123) vanishes there is an electromagnetic field without sources inside S_- such that

$$\left(\frac{\varepsilon}{\mu}\right)^{\frac{1}{2}} \mathbf{n}_p \wedge [\mathbf{n}_p \wedge \mathbf{E}_-(p)] - \beta\mathbf{n}_p \wedge \mathbf{H}_-(p) = \mathbf{0}\,. \qquad (6.124)$$

Now, by the divergence theorem and Maxwell's equations,

$$\int_S \mathrm{n}_q \cdot \{\mathbf{E}_-(q)\wedge\mathbf{H}_-^*(q)+\mathbf{E}_-^*(q)\wedge\mathbf{H}_-(q)\}\mathrm{d}S_q = 0.$$

Hence, substitution for $\mathbf{n} \wedge \mathbf{H}$ from (6.124) leads to

$$\left(\frac{1}{\beta} + \frac{1}{\beta^*}\right) \int_S |\mathbf{n}_q\wedge\mathbf{E}_-(q)|^2\mathrm{d}S_q = 0. \tag{6.125}$$

Thus, so long as $1/\beta + 1/\beta^*$ is non-zero, $\mathbf{n}_p \wedge \mathbf{E}_-(p) = \mathbf{0}$ and then the field representation (6.85) implies that $\mathbf{n}_p \wedge \mathbf{E}_+(p) = \mathbf{0}$. From the exterior uniqueness theorem follows $\mathbf{n}_p \wedge \mathbf{H}_+(p) = \mathbf{0}$. Moreover, when $\mathbf{n}_p \wedge \mathbf{E}_-(p) = \mathbf{0}$, $\mathbf{n}_p \wedge \mathbf{H}_-(p) = \mathbf{0}$ on account of (6.124) provided that $\beta \neq 0$. The consequence is that $\mathbf{j} \equiv \mathbf{0}$. Therefore the solution of (6.123) is unique so long as β is finite and neither zero nor pure imaginary. That a solution to (6.123) exists can be deduced immediately from the fact that the sum of two compact operators is compact.

The integral equation (6.123) enjoys the same superiority as the MFIE over the EFIE in that it becomes diagonally dominant as the mesh size diminishes. Furthermore, it has supremacy over the MFIE in possessing no uniqueness difficulties. The price to be paid for these advantages is a substantial increase in computer requirements.

A wide choice is available for β — to make it a positive real number is an obvious possibility. Theoretically, the results should be independent of β but no numerical approximation can be expected to display this property. The size of the discrepancies which occur with variations in β supplies a measure of the numerical accuracy. Actually, β must not be chosen too small otherwise (6.123) is too close to the EFIE for numerical convenience. Similarly, if β is too large (6.123) is not sufficiently far enough away from the MFIE. A value around 5 for β would seem to be about right.

Exercises

33. It may be shown that, when $\mathbf{j}$ is continuous, the integral in the MFIE has continuous first partial derivatives. Deduce that, if the right-hand side of the MFIE is continuous with continuous first partial derivatives, the solution $\mathbf{j}$ has the same properties when k does not correspond to an interior resonance.

34. An electromagnetic field is required in S_- such that $\mathbf{n} \wedge \mathbf{E}_- = \mathbf{g}$ on S where $\mathbf{g}$ is continuously differentiable. Show that it is necessary that

$$\int_S \mathbf{g} \cdot \mathbf{h}^* \mathrm{d}S = 0$$

for every solution $\mathbf{h}$ of (6.104). If these conditions are satisfied show that the electromagnetic field can be found, it being expressed in terms of a magnetic current only which has the same continuity properties as in Exercise 33.

35. Let the set $\{\mathbf{j}_r\}$ be a basis spanning the space of solutions of (6.102) and let $\{\mathbf{J}_r\}$ be formed from it by the rule $\mathbf{J}(p) = \mathbf{n}_p \wedge \mathbf{E}^2(p)$. Prove that the matrix with elements $\int_S \mathbf{n}_q \wedge \mathbf{j}_r(q) \cdot \mathbf{J}_s(q)\mathrm{d}S_q$ is symmetric and has non-zero determinant. Deduce that, for each solution $\mathbf{k}'$ of (6.117), there is a corresponding $\mathbf{J}'$ such that

$$\int_S (\mathbf{k}'-\mathbf{J}') \cdot \mathbf{k}^* \mathrm{d}S = 0$$

for all solutions $\mathbf{k}$ of (6.117).

36. For the exterior problem in which $\mathbf{n} \wedge \mathbf{E}_+ = \mathbf{g}$ on S choose $\mathbf{k}'$ (in the notation of Exercise 35) so that

$$\int_S (\mathbf{g}-\mathbf{k}') \cdot \mathbf{k}^* \mathrm{d}S = 0$$

for all solutions $\mathbf{k}$ of (6.117). Find a solution in terms solely of an electric current such that $\mathbf{n} \wedge \mathbf{E}' = \mathbf{J}'$ on S. Deduce that a solution of $\mathbf{n} \wedge (\mathbf{E}_+ - \mathbf{E}') = \mathbf{g} - \mathbf{J}'$ can be found involving only magnetic current and hence that a solution of the exterior problem can be determined though the solution may not be unique.

37. Do you think that there would be trouble from non-uniqueness in the wire-grid model?

38. If in (6.123) β is taken as a function of p instead of being constant prove that uniqueness still holds provided that β never vanishes nor assumes purely imaginary values.

6.18. Numerical considerations for surfaces

Whatever integral equation an investigator decides is most suited to his needs, the problem of the numerical approximation for surface integrals will have to be faced. Usually, the surface will be divided into sub-areas or *patches* and the surface integral expressed as a sum of integrals over the patches. No approximation is involved in this process so that local approximations can be introduced without worrying about the global properties of the integrand.

There are two different aspects to be considered. One is where the surface is given and the other is where the designer is attempting to draw the surface subject to certain criteria about its shape and boundaries. The latter problem lies more in the realm of computer-aided design and here techniques such as *Coons' patches* (Coons 1967, Forrest 1968) in which a patch is defined in terms of four boundary curves and the slopes across these boundary curves; the corner points and twists there may well be relevant. However, we shall concentrate on the situation in which the surface is known and the main concern is to arrive at its scattering properties.

The splitting of the surface into patches permits the transfer of ideas that originated with finite elements. One of the first approaches (Wait 1973, Wilton 1973) used plane triangular finite elements (subsequently extended to quadratic triangular elements) with either piecewise constant, linear, or quadratic approximation to the unknown function. Every triangular surface element is mapped on to a plane right-angled triangle (cf. §5.8); the same quadrature nodes can then be employed to compute all integrals over patches of the same type.

More complicated representations for functions are possible, though in finite elements it is usual for the approximant to be an interpolation polynomial with parameters which are the actual function values (or maybe derivatives) at specified nodes. If the function being interpolated is unknown every extra node adds an equation to the algebraic system to be solved eventually. To set against that is the fact that, with more nodes, higher-order approximations with a reduction in the number of elements are feasible. There is some evidence that higher-order methods pay off (Hess 1973). It is advantageous to have as many nodes on the boundary as possible so as to improve efficiency; also including derivatives by means of Hermite elements (Strang and Fix 1973) increases the order but becomes unwieldy when the surface is represented parametrically.

Continuity of the interpolant and maybe some derivatives is often required if convergence of the numerical approximation to the analytical solution is to be secured. Elements which comply with theoretical requirements are often said to be *conforming*. Elements which do not have the requisite degree of continuity are frequently simpler to work with; they are known as *non-conforming*. With non-conforming elements there will be convergence at some places but not at others and *Irons' patch test* (Irons and Razzaque 1972, Ciarlet 1975) provides a means of deciding which elements are well behaved. If the surface is not closed special consideration of the behaviour at its boundaries will be necessary, and here it is important to note that replacement of the boundary by a polygon may lead to the *Babuska paradox* in which the numerical approximation converges but to the solution of the wrong problem (Birkhoff 1969). For this reason methods have been developed for treating a boundary exactly (Marshall and Mitchell 1973) but they would generally need a prohibitive labour for integral equations. Representation of the surface by portions of quadrics also has been tried (Van Buren 1970).

These general points, together with the observations of §6.7 in the simpler case of the wire antenna, prompt one to continue the pattern of §6.8 and avail oneself of cubic B-splines. If ξ and η are surface parameters, the natural generalization is provided by the *bicubic* (because it is a cubic in both ξ and η) *B-spline basis* $N_i(\xi)N_j(\eta)$. The analogue of (6.69) for a three-dimensional surface is then

$$x(\xi,\eta) = \sum_{i=2}^{T+3} \sum_{j=2}^{T+3} X_{ij}N_i(\xi)N_i(\eta) ,$$

$$y(\xi,\eta) = \sum_{i=2}^{T+3} \sum_{j=2}^{T+3} Y_{ij} N_i(\xi) N_j(\eta) ,$$

$$z(\xi,\eta) = \sum_{i=2}^{T+3} \sum_{j=2}^{T+3} Z_{ij} N_i(\xi) N_i(\eta)$$

thereby mapping the surface onto the square $1 \leqslant \xi \leqslant T$, $1 \leqslant \eta \leqslant T$ in the (ξ,η) plane. (The image can be made a rectangle if different ranges for ξ and η are utilized.) As in §6.8 the coefficients X_{ij}, Y_{ij}, Z_{ij} are fixed by specifying the points corresponding to integer values of ξ and η. Again special treatment at any perimeter is necessary but the associated matrix is still tridiagonal.

In order that the mapping be invertible it is necessary that the three determinants

$$D_1 = \frac{\partial(y,z)}{\partial(\xi,\eta)}, \qquad D_2 = \frac{\partial(z,x)}{\partial(\xi,\eta)}, \qquad D_3 = \frac{\partial(x,y)}{\partial(\xi,\eta)}$$

do not vanish simultaneously. When that is so a typical surface integral is converted according to

$$\int_S f\mathrm{d}S = \sum_m \int_{Q_m} fJ\mathrm{d}\xi\mathrm{d}\eta$$

where Q_m is a unit square and

$$J^2 = D_1^2 + D_2^2 + D_3^2 .$$

The procedure of §6.8 may now be followed, the principal differences being the extra variable and the singular integrands which will be discussed in the next section.

Exercises

39. Prove that $J\mathbf{n} = D_1\mathbf{i} + D_2\mathbf{j} + D_3\mathbf{k}$ where **i**, **j**, and **k** are unit vectors parallel to the Cartesian coordinates.

40. If $r^2 = (x_q-x_p)^2 + (y_q-y_p)^2 + (z_q-z_p)^2$ show that

$$\frac{\partial r}{\partial n_q} = \frac{(x_q-x_p)D_1(q) + (y_q-y_p)D_2(q) + (z_q-z_p)D_3(q)}{rJ(q)} .$$

41. Use Exercise 40 to find r and its normal derivatives at p and q for (a) a sphere, (b) an ellipsoid, and (c) a truncated hyperboloid by adopting the following strategy (for a suitable parametric representation): (i) assume p is fixed and $\mathbf{x}_p$, $\partial\mathbf{x}/\partial n_p$ are known; (ii) determine $\mathbf{x}_q$ from the values of ξ and η for q; (iii) calculate the partial derivatives of x, y, and z with respect to ξ and η; (iv) evaluate D_1, D_2, and D_3.

6.19. Singular integrals

The kernel of the MFIE contains a weak singularity of the form of $1/r$ as the analysis of §6.16 reveals and the same is true of part of the kernel of the EFIE at any rate. Therefore, although 3 × 3 Gaussian quadrature is suitable for patches not containing the singularity, a different process is essential for elements in the immediate neighbourhood of the singularity. Because of the weakness of the singularity it is probably sufficient to withhold special treatment except for the nearest four elements.

Of the one-dimensional methods that have been invented to deal with singularities the one which seems to be most adaptable to surface integrals is the following (Takahasi and Mori 1973). Let $f(t)$ be a function which may have singularities at ±1 but not elsewhere. By making the transformation

$$t = \frac{2}{\pi^{\frac{1}{2}}} \int_0^u \exp(-v^2)dv = \text{erf } u$$

we obtain

$$\int_{-1}^{1} f(t)dt = \frac{2}{\pi^{\frac{1}{2}}} \int_{-\infty}^{\infty} f(\text{erf } u)\exp(-u^2)du .$$

The integral on the right is approximated by the trapezoidal rule with constant interval h. There results

$$\int_{-1}^{1} f(t)\,\mathrm{d}t \approx \frac{2h}{\pi^{\frac{1}{2}}} \sum_{n=-N}^{N} f(\operatorname{erf} nh)\exp(-n^2h^2) \tag{6.126}$$

which is a quadrature rule with the weights $(2h/\pi^{\frac{1}{2}})\exp(-n^2h^2)$ attached to the points erf nh. Owing to the rapid convergence of (6.126) it is rarely necessary for N to exceed 5 when h is of the order of unity.

The switch to integrals over a square is immediate by changing both variables of integration to erf. Although (6.126) applies even if f has no singularities it tends to be less efficient and that is why its application should be limited.

For singularities of higher order such as occur in a Cauchy principal value the following device can be helpful. If f has no singularities

$$\begin{aligned}
\int_{-1}^{1} \frac{f(t)}{t}\mathrm{d}t &= \lim_{\varepsilon\to+0}\left(\int_{-1}^{-\varepsilon} + \int_{\varepsilon}^{1}\right)\frac{f(t)}{t}\mathrm{d}t \\
&= \lim_{\varepsilon\to+0}\int_{\varepsilon}^{1}\{f(t)-f(-t)\}\frac{\mathrm{d}t}{t} \\
&= \int_{0}^{1}\{f(t)-f(-t)\}\frac{\mathrm{d}t}{t}
\end{aligned}$$

by changing t to $-t$ in the first integral on the right of the first line. Now the singularity has disappeared and straightforward quadrature can be accepted.

Exercises

42. Evaluate $\int_{-1}^{1}(1+t)^p(1-t)^q\mathrm{d}t$ numerically for various values of p and q between -1 and 0. Check how close your answer is to π when $p = q = -\frac{1}{2}$.

43. The integrals (*a*) $\int_0^T t^{-\frac{1}{2}}(1+t)^{-1}\mathrm{d}t$, (*b*) $\int_0^T t^{-\frac{3}{4}}\mathrm{d}t/(1-t)$ with $T \geqslant 2$ are supposed to be approximations to π. By evaluating them numerically for increasing values of T decide how large T must be for them to be accurate

to (i) 1 per cent, (ii) 0.01 per cent.

6.20. The algebraic system

Once the coefficient matrix and the vector from the right-hand side of the integral equations have been calculated the algebraic system must be solved. Since the matrix is not sparse in general, direct methods are appropriate with the proviso that the matrix is not too large. As the size of the matrix grows the computation of the solution of the algebraic system will equal and surpass the effort devoted to the preparation of the matrix because the former expands as the cube of the order while the latter augments as the square. It may then be profitable to shunt to iterative methods.

The Gauss–Seidel and SOR iterative methods (§1.13) demand for convergence that the spectral radius be less than unity (Theorem 1.13). It has also been seen in §2.5 that the possession of Young's Property A is a desirable asset. Unfortunately, it is difficult, if not impossible, to verify that the integral equations of scattering lead to systems with the desired perquisites. Even Gerschgorin's theorem (1.11*c*) is inadequate for the purpose.

In view of these deficiencies Burton (1976) has suggested that large matrices should be dealt with by the conjugate gradient method of optimization (§4.6). The particular version recommended is due to Hestenes (1956).

Let the system of linear equations be

$$M\mathbf{u} = \mathbf{v}$$

with the usual inner product $(\mathbf{u},\mathbf{v}) = \Sigma_{i=1}^{N} u_i v_i^*$ when the vectors are of order N. An initial approximation $\mathbf{u}_0$ is first assumed. Then set $\mathbf{q}_{-1} = \mathbf{0}$, $c_{-1} = 1$ and $\mathbf{r}_0 = \mathbf{v} - M\mathbf{u}_0$. The ith iteration now consists of the operations

$$c_i = \|\mathbf{r}_i\|^2, \quad b_{i-1} = c_i/c_{i-1}, \quad \mathbf{q}_i = \mathbf{r}_i + b_{i-1}\mathbf{q}_{i-1},$$

$$\mathbf{p}_i = M^{\mathrm{A}}\mathbf{q}_i, \quad d_i = \|\mathbf{p}_i\|^2, \quad a_i = c_i/d_i,$$

$$\mathbf{u}_{i+1} = \mathbf{u}_i + a_i\mathbf{p}_i, \qquad \mathbf{r}_{i+1} = \mathbf{r}_i - a_i M\mathbf{p}_i \ .$$

Despite every iteration depending on M and its adjoint the process is feasible if the matrix is stored by columns and transfers from the backing store are carried out in fairly large blocks for one may get away with one complete matrix transfer per iteration.

The efficiency of the method stems from the tendency of the eigenvalues of M to cluster round a single point when it is drawn from an integral equation of the second kind. Suppose the operator involved is $I-T$ where T is compact. Then the eigenvalues of the operator equation will be the values of λ such that

$$(I-T)w = \lambda w \ .$$

The only limit point of the eigenvalues of T is zero so the eigenvalues of $I-T$ have the unique limit point 1. Since M reproduces approximately the properties of $I-T$ we expect that many of its eigenvalues will be near unity.

Exercises

In the following exercises use first 5 and then 10 knots per wavelength in the spline approximation as a check on whether phase variations are adequately taken care of.

44. Calculate the scattering cross-section of a sphere of radius a as a function of wavelength by means of the EFIE and the MFIE paying particular attention to what happens when ka is near 2.75 and 4.50. Do you get the same results from (6.123)?

45. Repeat Exercises 26–31 using (*a*) EFIE, (*b*) MFIE, and (*c*) (6.123). How would you rate these methods in comparison with the wire-grid model?

46. Instead of making most of the collocation points coincide with the spline knots, choose them to be all different from the spline knots. Use scattering from a sphere as a test bed. It will be found that the algebraic system becomes ill conditioned, possibly even singular.

47. The method described finds the coefficients in the spline expansion of the unknown. Reformulate it so that the values of the unknown at the spline knots are the quantities to be found.

48. For scattering by a sphere examine how the number of iterations in

the conjugate gradient method depends upon the distribution of knots and the alternative representations of Exercise 47. Do the eigenvalues of the algebraic system display a tendency to cluster?

49. Instead of collocating set up the algebraic system by Galerkin's method (§5.1) using some simple weight functions and repeat the calculations of Exercise 45. Assess the merits and defects of the various numerical methods tried.

6.21. The null-field method

An alternative to the surface integral equation and its variants is to focus attention on an interior integral equation and especially the equivalent set of equations (6.122). This was first developed systematically by Waterman (1971) under the name of *extended boundary condition* but the description *null-field method* (Bates 1968) seems preferable.

In the method j is expanded in terms of a convenient set of basis functions on the surface. By limiting the number of terms in the expansion to the same number of equations (6.122) as are imposed a linear system is obtained for the unknown coefficients. Increasing the number of coefficients and equations supplies an audit of the accuracy.

In this connection it should be remarked that in solving an integral equation by expansion in known functions and truncation the accuracy can be improved and an error bound derived by modification of the kernel (Jones 1973*a*) (see §5.6(*c*)).

There is, of course, no reason why ψ should be expanded in terms of the spherical wavefunctions ψ_{mn} and Ψ_{mn} only. Any other handy set such as circular cylinder or elliptic functions could be employed. This fact has been exploited (Bates and Wong 1974, Wall and Bates 1975) and some successful results achieved. None the less, the theoretical backing for the null-field method must be regarded as less satisfactory until *a priori* bounds are discovered for the error when the geometry of the surface bears little resemblance to the natural geometry inherent in the expansion of ψ.

Exercise

50. Try the null-field method on a selection of the problems 26–31 and

compare the results with those of previous methods.

6.22. The impedance boundary condition

Frequently, the surface of the antenna will not be perfectly conducting but the field will not penetrate deeply into the body, particularly if it is coated. Then it may be enough to account for the properties by imposing an impedance boundary condition on the surface.

Suppose that the field $\mathbf{E}_0$, $\mathbf{H}_0$ falls on the body from the exterior and that the boundary condition on S is

$$\mathbf{n} \wedge \mathbf{E} = Z_1 Z\mathbf{n} \wedge (\mathbf{n}\wedge\mathbf{H}) \tag{6.127}$$

so that Z_1 is the impedance relative to the characteristic impedance $Z = (\mu/\varepsilon)^{\frac{1}{2}}$. Then the representation to replace (6.85) and (6.86) is

$$\mathbf{E}(P) = \mathbf{E}_0(P) + \int_S [-Z_1 Z\{\mathbf{n}_q \wedge \mathbf{j}(q)\} \wedge \operatorname{grad}_q \psi(P,q) -$$

$$- \frac{\rho(q)}{\varepsilon} \operatorname{grad}_q \psi(P,q) + i\omega\mu\mathbf{j}(q)\psi(P,q)]\,dS_q, \tag{6.128}$$

$$\mathbf{H}(P) = \mathbf{H}_0(P) - \int_S [\mathbf{j}(q) \wedge \operatorname{grad}_q \psi(P,q) +$$

$$+ \frac{Z \operatorname{div}_0\{Z_1 \mathbf{n}_q \wedge \mathbf{j}(q)\}\operatorname{grad}_q \psi(P,q)}{i\omega\mu} +$$

$$+ i\omega\varepsilon Z Z_1 \mathbf{n}_q \wedge \mathbf{j}(q)\psi(P,q)]\,dS_q \tag{6.129}$$

where advantage has been taken of (6.127) to substitute for the tangential electric field in terms of $\mathbf{j}$.

Now let $P \to p$ and apply (6.127). There results

$$\mathbf{n}_p \wedge \left(\mathbf{n}_p \wedge \int_S [\mathrm{i}\omega\mu\mathbf{j}(q)\psi(p,q) - Z_1 Z\{\mathbf{n}_q \wedge \mathbf{j}(q)\} \wedge \operatorname{grad}_q \psi(p,q) - \right.$$

$$\left. - \frac{\rho(q)}{\varepsilon} \operatorname{grad}_q \psi(p,q)] \mathrm{d}S_q\right) -$$

$$- Z_1 Z\mathbf{n}_p \wedge \int_S [\mathbf{j}(q) \wedge \operatorname{grad}_q \psi(p,q) + \frac{Z \operatorname{div}_0\{Z_1 \mathbf{n}_q \wedge \mathbf{j}(q)\} \operatorname{grad}_q \psi(p,q)}{\mathrm{i}\omega\mu} +$$

$$+ \mathrm{i}\omega\varepsilon Z Z_1 \mathbf{n}_q \wedge \mathbf{j}(q)\psi(p,q)] \mathrm{d}S_q = - \mathbf{n}_p \wedge (\mathbf{n}_p \wedge \mathbf{E}_0) - Z_1 Z\mathbf{n}_p \wedge \mathbf{H}_0 . \tag{6.130}$$

This is the analogue of the EFIE and reduces to it when $Z_1 = 0$. There is also a companion to the MFIE obtained by asking that $\mathbf{n} \wedge \mathbf{H}_- = \mathbf{0}$; it is

$$-\tfrac{1}{2}\mathbf{j}(p) + \mathbf{n}_p \wedge \int_S [\mathbf{j}(q) \wedge \operatorname{grad}_q \psi(p,q) + \frac{Z \operatorname{div}_0\{Z_1 \mathbf{n}_q \wedge \mathbf{j}(q)\} \operatorname{grad}_q \psi(p,q)}{\mathrm{i}\omega\mu} +$$

$$+ \mathrm{i}\omega\varepsilon Z Z_1 \mathbf{n}_q \wedge \mathbf{j}(q)\psi(p,q)] \mathrm{d}S_q = \mathbf{n}_p \wedge \mathbf{H}_0(p). \tag{6.131}$$

A parallel to (6.123) is formed by combining (6.130) and (6.131) so as to achieve

$$\mathbf{n}_p \wedge \left(\mathbf{n}_p \wedge \int_S \left[\mathrm{i}\omega\mu\mathbf{j}(q)\psi(p,q) - Z_1 Z\{\mathbf{n}_q \wedge \mathbf{j}(q)\} \wedge \operatorname{grad}_q \psi(p,q) - \right.\right.$$

$$\left.\left. - \frac{\rho(q)}{\varepsilon} \operatorname{grad}_q \psi(p,q) \right] \mathrm{d}S_q\right) -$$

$$- \tfrac{1}{2}\beta\mathbf{j}(p) + (\beta - Z_1 Z)\mathbf{n}_p \wedge \int_S \left[\mathbf{j}(q) \wedge \operatorname{grad}_q \psi(p,q) + \mathrm{i}\omega\varepsilon Z Z_1 \mathbf{n}_q \wedge \mathbf{j}(q)\psi(p,q)\right.$$

$$\left. + \frac{Z \operatorname{div}_0\{Z_1 \mathbf{n}_q \wedge \mathbf{j}(q)\} \operatorname{grad}_q \psi(p,q)}{\mathrm{i}\omega\mu}\right] \mathrm{d}S_q = \text{known}. \tag{6.132}$$

These integral equations are more complicated than those for perfect conductivity but otherwise there are no essentially fresh features. As far as uniqueness is concerned only (6.132) will be examined. The vanishing of the left-hand side implies that

$$\mathbf{n}_p \wedge \{\mathbf{n}_p \wedge \mathbf{E}_-(p)\} + (Z_1 - \beta) Z \mathbf{n}_p \wedge \mathbf{H}_-(p) = \mathbf{0} .$$

Comparing this with (6.124) we see that $\mathbf{n} \wedge \mathbf{E}_- = \mathbf{n} \wedge \mathbf{H}_- = \mathbf{0}$ so long as $\beta - Z_1$ is neither zero nor pure imaginary. Since $\mathbf{n} \wedge (\mathbf{n} \wedge \mathbf{E}) + Z_1 Z \mathbf{n} \wedge \mathbf{H}$ is continuous across S it follows that

$$\mathbf{n}_p \wedge \{\mathbf{n}_p \wedge \mathbf{E}_+(p)\} + Z_1 Z \mathbf{n}_p \wedge \mathbf{H}_+(p) = \mathbf{0} . \tag{6.133}$$

Therefore, assuming an exterior uniqueness theorem for an impedance boundary condition, $\mathbf{n}_p \wedge \mathbf{H}_+(p) = \mathbf{0}$ and $\mathbf{j}$ is identically zero. Thus, (6.132) possesses a unique solution provided that β is chosen so that $\beta - Z_1$ is not zero or pure imaginary.

The exterior uniqueness theorem just assumed must now be verified. Let $\mathbf{E}, \mathbf{H}$ be an electromagnetic field which has no sources outside S and satisfies the radiation conditions (6.73). Suppose it also complies with the boundary condition (6.133). Let Ω_R be a large sphere of radius R. Then

$$\int_{\Omega_R} (\mathbf{E} \wedge \mathbf{H}^* + \mathbf{E}^* \wedge \mathbf{H}) \cdot \mathbf{n} \, d\Omega = \int_S (\mathbf{E} \wedge \mathbf{H}^* + \mathbf{E}^* \wedge \mathbf{H}) \cdot \mathbf{n} \, dS . \tag{6.134}$$

The left-hand side of (6.134) can be written as

$$-\frac{1}{2Z} \int_{\Omega_R} \{ |\mathbf{E} - Z\mathbf{H} \wedge \hat{\mathbf{R}}|^2 + |Z\mathbf{H} - \hat{\mathbf{R}} \wedge \mathbf{E}|^2 - 2|\mathbf{E}|^2 - 2Z^2 |\mathbf{H}|^2 + Z^2 |\mathbf{H} \cdot \hat{\mathbf{R}}|^2 + |\mathbf{E} \cdot \hat{\mathbf{R}}|^2 \} d\Omega .$$

On account of the radiation conditions (6.73) the first two terms of the integrand give a zero contribution as $R \to \infty$. Further, a scalar product with $\hat{\mathbf{R}}$ of the second pair of (6.73) indicates that the last two terms can be ignored as $R \to \infty$. Hence, as $R \to \infty$,

$$\int_{\Omega_R} (|\mathbf{E}|^2+Z^2|\mathbf{H}|^2)\mathrm{d}\Omega = Z\int_S (\mathbf{E}\wedge\mathbf{H}^*+\mathbf{E}^*\wedge\mathbf{H}) \cdot \mathbf{n}\mathrm{d}S$$

$$= -\int_S \frac{|\mathbf{n}\wedge\mathbf{E}|^2(Z_1+Z_1^*)\mathrm{d}S}{|Z_1|^2} .$$

The two sides have opposite sign if $\mathcal{R}Z_1 \geqslant 0$ and so can be consistent only if both are zero, but that is impossible unless $\mathbf{E} \equiv \mathbf{0}$ and $\mathbf{H} \equiv \mathbf{0}$. Consequently, *the exterior uniqueness theorem is valid if* $\mathcal{R}Z_1 \geqslant 0$.

An important observation is that neither the integral equations nor the exterior uniqueness theorem depend on Z_1 being a constant. Hence (6.132) will provide the unique solution for variable impedance subject to $\mathcal{R}Z_1 \geqslant 0$ at all points of S and β being picked so that $\beta - Z_1$ is not zero nor pure imaginary, e.g. $\beta = -5$ would be a suitable choice.

Exercises

51. Prove that it is not necessary for β to be constant so long as the conditions on $\beta - Z_1$ are still met.

52. Reformulate the theory of this section so as to be applicable to two-dimensional fields around cylindrical obstacles parallel to the z axis. Allow the longitudinal impedance (for E_z) to be different from the transverse impedance.

53. Is the theory of Exercise 52 adaptable to fields whose z dependence is $\exp(-ikz\cos\theta)$?

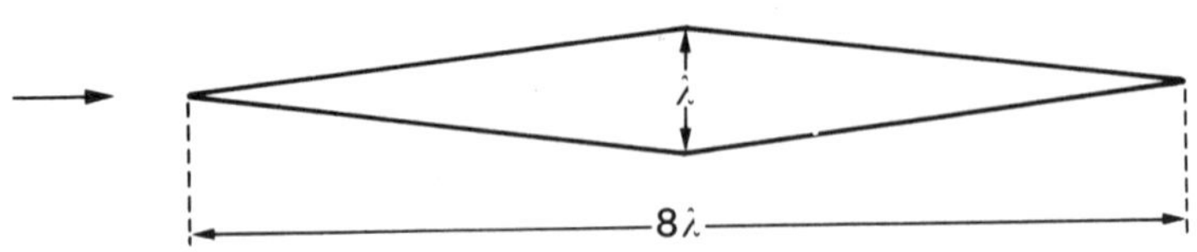

FIG. 6.20. Double wedge.

54. A plane wave with its electric vector perpendicular to the z axis is incident on the symmetric double wedge of Fig. 6.20. The surface impedance is constant and purely reactive, i.e. $\mathcal{R}Z_1 = 0$. Calculate the radar cross-

section and the analogous quantity for the forward scattering in the direction of incidence as $|Z_1|$ increases from 0 to 1. Is there any evidence for the existence of a surface wave on the wedge when $Z_1 \approx \pm 0.36i$? How do you account for the behaviour of the radar cross-section between $|Z_1| = 0.5$ and $|Z_1| = 0.6$?

55. The wedge on the right of Fig. 6.20 is replaced by a portion of circular cylinder which fits smoothly on to the left-hand wedge. Repeat the investigation of Exercise 54.

56. Undertake a study of the scattering by a sphere with a surface impedance.

DIELECTRIC ANTENNAS

Antennas can be constructed from dielectric materials which cannot be accounted for adequately by means of an impedance boundary condition. In that case there is no escaping from the introduction of interior fields. The complexity of the resulting analysis can be considerable. As a guide to what happens we commence with the propagation along circular cylinders.

6.23. The infinite dielectric circular rod

Let the dielectric rod be of radius a with material constants ε,μ and have no conductivity. The surrounding medium will also be assumed to be non-conducting with material constants ε_0,μ_0. Choose cylindrical polar coordinates r,ϕ,z with the z axis along the axis of the rod. Assume that the dependence on z of all fields is $\exp(-i\alpha z)$ when α is a real constant. In order that the field be finite at $r = 0$ Bessel functions of the first kind must be employed inside the cylinder. Outside the cylinder the Hankel function $H^{(2)}$ is appropriate in order to preserve the proper behaviour at infinity. The expansions compatible with these injunctions and Maxwell's equations are

$$E_z = \sum_{m=-\infty}^{\infty} a_m J_m(\kappa r)\exp(im\phi - i\alpha z) \qquad (r \leqslant a)$$

$$= \sum_{m=-\infty}^{\infty} c_m H_m^{(2)}(\kappa_0 r)\exp(im\phi - i\alpha z) \qquad (r \geqslant a),$$

$$H_z = \sum_{m=-\infty}^{\infty} b_m J_m(\kappa r)\exp(im\phi - i\alpha z) \qquad (r \leqslant a)$$

$$= \sum_{m=-\infty}^{\infty} d_m H_m^{(2)}(\kappa_0 r)\exp(im\phi - i\alpha z) \qquad (r \geqslant a)$$

where $\kappa^2 = \omega^2\mu\varepsilon - \alpha^2$, $\kappa_0^2 = \omega^2\mu_0\varepsilon_0 - \alpha^2$. The transverse components $\mathbf{E}_t, \mathbf{H}_t$ of the field can be deduced from (§2.2)

$$\kappa^2\mathbf{E}_t = \mathbf{k} \wedge i\omega\mu \, \text{grad}_t \, H_z - i\alpha \, \text{grad}_t \, E_z ,$$

$$\kappa^2\mathbf{H}_t = -\frac{ik^2\mathbf{k} \wedge \text{grad}_t \, E_z}{\omega\mu} - i\alpha \, \text{grad}_t \, H_z$$

where $\mathbf{k}$ is a unit vector along the z axis and grad_t is the gradient in the (x,y) plane.

The continuity of the tangential components of $\mathbf{E}$ and $\mathbf{H}$ at $r = a$ requires the satisfaction of the four equations

$$\frac{i\omega\mu}{\kappa}J_m'(\kappa a)b_m + \frac{\alpha m}{\kappa^2 a}J_m(\kappa a)a_m = \frac{i\omega\mu_0}{\kappa_0}H_m^{(2)\prime}(\kappa_0 a)d_m + \frac{\alpha m}{\kappa_0^2 a}H_m^{(2)}(\kappa_0 a)c_m ,$$

$$a_m J_m(\kappa a) = c_m H_m^{(2)}(\kappa_0 a) ,$$

$$\frac{i\omega\varepsilon}{\kappa}J_m'(\kappa a)a_m - \frac{\alpha m}{\kappa^2 a}J_m(\kappa a)b_m = \frac{i\omega\varepsilon_0}{\kappa_0}H_m^{(2)\prime}(\kappa_0 a)c_m - \frac{\alpha m}{\kappa_0^2 a}H_m^{(2)}(\kappa_0 a)d_m ,$$

$$b_m J_m(\kappa a) = d_m H_m^{(2)}(\kappa_0 a) .$$

These are homogeneous equations for the coefficients a_m, b_m, c_m, and d_m. They will have a non-trivial solution only if the determinant of the coefficients is zero, i.e.

$$\left\{\frac{\mu_0}{\kappa_0}\frac{H_m^{(2)\prime}(\kappa_0 a)}{H_m^{(2)}(\kappa_0 a)} - \frac{\mu}{\kappa}\frac{J_m'(\kappa a)}{J_m(\kappa a)}\right\}\left\{\frac{\varepsilon_0}{\kappa_0}\frac{H_m^{(2)\prime}(\kappa_0 a)}{H_m^{(2)}(\kappa_0 a)} - \frac{\varepsilon}{\kappa}\frac{J_m'(\kappa a)}{J_m(\kappa a)}\right\}$$

$$= \frac{\alpha^2 m^2}{\omega^2 a^2}\left(\frac{1}{\kappa_0^2} - \frac{1}{\kappa^2}\right)^2 . \tag{6.135}$$

When the rod is stimulated symmetrically $m = 0$ and (6.135) decouples into two equations. Since $J_0' = -J_1$ and $H_0^{(2)\prime} = -H_1^{(2)}$ they may be written as

$$\frac{\kappa_0}{\varepsilon_0}\frac{H_0^{(2)}(\kappa_0 a)}{H_1^{(2)}(\kappa_0 a)} = \frac{\kappa}{\varepsilon}\frac{J_0(\kappa a)}{J_1(\kappa a)} \tag{6.136}$$

for the TM modes and

$$\frac{\kappa_0}{\mu_0}\frac{H_0^{(2)}(\kappa_0 a)}{H_1^{(2)}(\kappa_0 a)} = \frac{\kappa}{\mu}\frac{J_0(\kappa a)}{J_1(\kappa a)} \tag{6.137}$$

for the TE modes.

Both (6.136) and (6.137) have synonymous structure so that it will be sufficient to discuss one of them in drawing conclusions about the qualitative behaviour (see also Arnbak 1969) of the values of α which satisfy them.

The first query to be answered is whether an undamped wave can propagate at the speed of light in the medium round the rod. If so, $\alpha = \omega(\mu_0\varepsilon_0)^{\frac{1}{2}}$ and $\kappa_0 = 0$. The left-hand side of (6.136) then dissolves and so $J_0(\kappa a)$ must be satisfied since $\kappa = 0$ is not permitted. Hence, if j_{0p} is the pth zero of J_0, there are symmetric TM modes with $\alpha^2 = \alpha_{0p}^2 = \omega^2\mu\varepsilon - j_{0p}^2/a^2$. Their frequencies are given by $\omega^2(\mu\varepsilon-\mu_0\varepsilon_0) = j_{0p}^2/a^2$ and thus *there is a cut-off frequency* $j_{01}/2\pi a(\mu\varepsilon-\mu_0\varepsilon_0)^{\frac{1}{2}}$ below which no TM modes of this type will propagate. For α near α_{0p} the wave outside the dielectric decays very slowly with r and so *only a small part of the energy flow takes place within the rod.*

An important practical regime is when $\varepsilon \gg \varepsilon_0$. The right-hand side of (6.136) will then be diminutive unless J_1 is near zero whereas the left-hand side will not be small unless $\kappa_0 a$ is as well. Therefore, when $\kappa_0 a$ is not small, i.e. the mode is well away from cut-off, the match necessitated by (6.136) can be attained only if $J_1(\kappa a) = 0$. Hence, in this case,

$$\alpha^2 = \omega^2\mu\varepsilon - j_{1p}^2/a^2 \tag{6.138}$$

where j_{1p} is the pth non-zero root of J_1. The zero root must be omitted because of the factor κ in the numerator of the right-hand side of (6.136). The largeness of $\varepsilon/\varepsilon_0$ assures that the formula (6.138) will not make $\kappa_0 a$ small unless $\omega^2\mu\varepsilon$ is not far from j_{1p}^2/a^2. When $\omega^2\mu\varepsilon$ is well above j_{1p}^2/a^2, κ_0 will be negative imaginary and of large modulus. The field outside the cylinder will wane rapidly as the distance from the rod mounts. *Most of the energy flow will now be within the dielectric.* This is borne out by the perception that (6.138) applies to a waveguide whose walls insist on a vanishing tangential magnetic field.

A graph of α against $\omega a(\mu_0\varepsilon_0)^{\frac{1}{2}}$ when $\varepsilon \gg \varepsilon_0$ consists essentially of the straight line $\alpha = \omega a(\mu_0\varepsilon_0)^{\frac{1}{2}}$ for the lower values of ωa and the curve $\alpha = (\omega^2\mu\varepsilon - j_{1p}^2/a^2)^{\frac{1}{2}}$ for the higher values of ωa. The transition between the two is effected by a smooth curve in the neighbourhood of the point of intersection where $\omega^2(\mu\varepsilon - \mu_0\varepsilon_0) = j_{1p}^2/a^2$.

Some notion of the shape of the transition curve can be realized by considering the solutions of (6.136) when $|\kappa_0 a|$ is small. In such a situation κ may be approximated by the constant κ_1 where $\kappa_1^2 = \omega^2(\mu\varepsilon - \mu_0\varepsilon_0)$. Eqn. (6.136) now reduces to

$$(\kappa_0 a)^2 \ln\{\tfrac{1}{2}\kappa_0 a \exp(\gamma + \tfrac{1}{2}\pi i)\} = -2/\eta \tag{6.139}$$

where $\eta = 2\varepsilon J_1(\kappa_1 a)/\varepsilon_0\kappa_1 a J_0(\kappa_1 a)$. The constant η is a measure of the *effective permittivity* since it collapses to $\varepsilon/\varepsilon_0$ when $|\kappa_1 a| \ll 1$. For negative imaginary values of κ_0 put $\kappa_0 = -i\kappa'$ where κ' is positive. Then

$$(\kappa' a)^2 \ln \tfrac{1}{2}\kappa' a e^{\gamma} = 2/\eta. \tag{6.140}$$

The smallness of $\kappa' a$ compels η to be negative for a solution of (6.140) to exist. Therefore $j_{0p} < \kappa_1 a < j_{1p}$. Note, however, that $\kappa_1 a$ cannot be close to j_{1p} because the left-hand side of (6.140) can never be large according to the assumptions on which it is based.

The modes in which $m \neq 0$ cannot be couched so that either E_z or H_z is zero. The asymmetric modes are therefore *hybrid* and the full equation (6.135) has to be solved. The cut-off characteristics are determined by making $\kappa_0 a$ small. The position is different according as m is or is not unity. If $m > 1$

$$\frac{H_m^{(2)\prime}(\kappa_0 a)}{H_m^{(2)}(\kappa_0 a)} \approx -\frac{m}{\kappa_0 a}\left\{1 - \frac{(\kappa_0 a)^2}{2m(m-1)}\right\} .$$

The dominant singularity in (6.135) is ostensibly $1/(\kappa_0 a)^4$ but the coefficient of this term vanishes automatically. So the terms in $1/(\kappa_0 a)^2$ have to be brought in and they lead to

$$(\mu_0\varepsilon+\mu\varepsilon_0)\kappa a\,\frac{J_{m-1}(\kappa a)}{J_m(\kappa a)} = m(\varepsilon-\varepsilon_0)(\mu_0-\mu) + \frac{\mu_0\varepsilon_0\kappa^2 a^2}{m-1}$$

as the equation for ω when $J_m'(z) = J_{m-1}(z) - mJ_m(z)/z$ is employed. The solutions of this transcendental equation for κa supplies the cut-off frequencies for the hybrid modes in which $m > 1$; in general $\kappa = 0$ will not be a solution.

When $m = 1$ we have

$$\frac{H_1^{(2)\prime}(\kappa_0 a)}{H_1^{(2)}(\kappa_0 a)} \approx -\frac{1}{\kappa_0 a}(1+\kappa_0^2 a^2 \ln \kappa_0 a) .$$

The highest-order singularity $1/(\kappa_0 a)^4$ is again removed by cancellation. The next-largest terms are of order $(\ln \kappa_0 a)/(\kappa_0 a)^2$ and these cannot be eliminated unless $\kappa J_1(\kappa a)$ is small. To agree with this, one possibility is $\omega^2(\mu\varepsilon-\mu_0\varepsilon_0) \approx 0$ *and there is then no cut-off frequency for the hybrid wave with $m = 1$.*

In general, (6.135) has to be solved numerically though a graphical approach is a helpful adjunct. For example, κa and $i\kappa_0 a$ are taken as Cartesian coordinates and the Hankel function is replaced via $H_m^{(2)}(z) = (2i/\pi)\exp(\frac{1}{2}m\pi i)K_m(iz)$. Then the curve whose equation is (6.135) is drawn. The curve $(\kappa a)^2 + (i\kappa_0 a)^2 = \omega^2 a^2(\mu\varepsilon-\mu_0\varepsilon_0)$ is then traced. The required values of α may be deduced from the intersections of the two curves.

There is no difficulty in confirming that for all these modes the Poynting vector is normal to the cylindrical surface when κ_0 is purely imaginary. Consequently there is no net energy flow out of the cylinder. The modes can therefore be called *surface waves*, i.e. they are waves which propagate along the interface between two different media without any transfer of energy across the dividing surface (other than that necessary to make good resistive losses).

The possibility of complex α should not be ignored. That such values are feasible may be recognized by returning to the approximation (6.139) and allowing κ_0 to be arbitrarily complex, but keeping η as a real constant. Let $\kappa_0 a = de^{i\delta}$ where $d \ll 1$. For $0 \leqslant \delta < \frac{1}{2}\pi$

$$d^2 e^{2i\delta}(\tfrac{1}{2}\pi i+\delta i+\ln \tfrac{1}{2}d) = -2/\eta$$

if the term involving γ is dropped. The real and imaginary parts of this equation may be rearranged as

$$\tfrac{1}{2}d = \exp\{-(\tfrac{1}{2}\pi+\delta)\cot 2\delta\}, \tag{6.141}$$

$$d^2 = \frac{2}{\eta}\,\frac{\sin 2\delta}{\frac{1}{2}\pi + \delta}. \tag{6.142}$$

If d is plotted as a function of δ, first from (6.141) and then from (6.142), the intersections of the curves will provide any solutions in the ranges under consideration. Evidently, for consistency of (6.142) $\eta > 0$ so that $j_{1p} < \kappa_1 a < j_{0,p+1}$ and any solutions occur in a different regime from those of (6.140). The curve of (6.141) increases monotonically with δ, starting from the value 0 at $\delta = 0$. In contrast, that of (6.142) increases steadily from 0 at $\delta = 0$, passes through a maximum for some $\delta < \frac{1}{2}\pi$, and then falls steadily to zero at $\delta = \frac{1}{2}\pi$. Hence there is precisely one solution and it lies in the interval $0 < \delta < \frac{1}{4}\pi$; as $\eta \to \infty$ it converges on $\delta = 0$ while simultaneously $d \to 0$.

In this wave κ_0 has a positive imaginary part and so the field grows as $r \to \infty$. Such waves have, on occasion, been deemed *leaky waves*. As $\kappa_1 a \to j_{0,p+1}$, $\eta \to \infty$ and $\delta \to 0$. This suggests that the leaky wave which originates for $\kappa_1 a$ just below $j_{0,p+1}$ converts to the $(p+1)$th surface wave as $\kappa_1 a$ traverses the surface-mode cut-off frequency.

Furthermore, $\cos 2\delta$ is positive at the solution of (6.141) and (6.142) so that $\mathcal{R}\alpha^2 < \omega^2\mu\varepsilon$. For δ and d near zero, $\mathcal{I}\alpha$ will be of the order of d^2 and we deduce that $\mathcal{R}\alpha < \omega(\mu\varepsilon)^{\frac{1}{2}}$. Consequently, the leaky wave travels with a phase speed faster than light along the z axis. There is, in addition, a radiation loss associated with this fast wave.

Exercise

57. An *optical fibre* consists of a circularly cylindrical inner dielectric core of refractive index N_1 surrounded by an annular cladding of smaller refractive index N_2. Surrounding the cladding is a dielectric with a refractive index N_3 not far from N_2. Set up the equation governing the modes of propagation.

If the outer medium is constructed of black glass so as to be highly lossy while the core and cladding are only slightly lossy are there any convenient approximations which can be made?

At a wavelength of 0.9 μm with $N_1 = 1.61$ extensive numerical investigations for (*a*) core radius 20 μm, cladding thickness 5 μm, $N_2 = N_3 = 0.99N_1$, (*b*) core radius 40 μm, cladding thickness 5 μm, $N_2 = N_3 = 0.96N_1$ have been carried out by Roberts (1972, 1973, 1975).

6.24. Modal excitation

The initiation of modes on an infinite dielectric rod by a given source will be the topic investigated within this section. To fix ideas, the source will be taken as the magnetic frill of §6.3 (Figs. 6.3 and 6.4). The incident field is then expressed as in eqns (6.17)–(6.22) with κ replaced by κ_0 to conform to the notation of the preceding section. The scattered field may be assumed to be represented by similar integrals but with the integrands being expansions of the type at the beginning of §6.23. In fact, for the excitation considered here, it is sufficient to limit the series to $m = 0$ and put $b_0 = 0$ and $d_0 = 0$. Apply the boundary condition of the continuity of the tangential components of the field at $r = a$. Then it will be discovered that the *total* field (i.e. the sum of the incident and scattered fields) is given in $r > b$ by (cf. Becker and Meister 1973)

$$H_\phi = -\tfrac{1}{4}\omega\varepsilon_0 b \int_{-\infty}^{\infty} \left[\frac{H_1^{(2)}(\kappa_0 b)\{\varepsilon_0\kappa J_1(\kappa_0 a)J_0(\kappa a) - \kappa_0\varepsilon J_0(\kappa_0 a)J_1(\kappa a)\}}{\varepsilon\kappa_0 J_1(\kappa a)H_0^{(2)}(\kappa_0 a) - \varepsilon_0\kappa H_1^{(2)}(\kappa_0 a)J_0(\kappa a)} + J_1(\kappa_0 b)\right] H_1^{(2)}(\kappa_0 r)\exp(-\mathrm{i}\alpha z)\,\mathrm{d}\alpha \tag{6.143}$$

while on $r = a$

$$H_\phi = \tfrac{1}{2}\omega\varepsilon\varepsilon_0 \frac{\mathrm{i}b}{\pi a}\int_{-\infty}^{\infty} \frac{H_1^{(2)}(\kappa_0 b)J_1(\kappa a)\exp(-\mathrm{i}\alpha z)}{\varepsilon\kappa_0 J_1(\kappa a)H_0^{(2)}(\kappa_0 a) - \varepsilon_0\kappa H_1^{(2)}(\kappa_0 a)J_0(\kappa a)}\,\mathrm{d}\alpha. \tag{6.144}$$

At first glance it may look as though (6.143) and (6.144) have branch points at the zeros of κ and κ_0. However, study of the numerator and denominator reveals that the behaviour is such that the zeros of κ do not correspond to branch points so that the only singularities are due to κ_0 (and any poles which the denominator may produce). To ensure that no poles lie on the path of integration (Fig. 6.5) slight dissipation is assumed to be present so that poles for $\mathcal{R}\alpha > 0$ are displaced downwards and those for $\mathcal{R}\alpha < 0$ upwards.

One might venture to calculate the field on the rod approximately by deforming the contour of integration into that of Fig. 6.6. The drawback to such a procedure is that in the deformation to the position shown by broken lines poles corresponding to leaky waves might be passed over. To avoid their consideration it is convenient to redraw the branch line from $\alpha = k$ $(k^2 = \omega^2\mu_0\varepsilon_0)$ as shown in Fig. 6.21 and deform the contour of integration into C_1. Of course, such a deformation will introduce contributions from poles answerable for surface waves. For large $|z|$ the important part of C_1 is near $\alpha = k$ and so, to a first approximation, κ may be treated as the constant κ_1. Thus (6.144) becomes approximately

$$H_\phi = \frac{1}{4\pi}\,\omega\varepsilon_0 b\eta\mathrm{i}\int_{C_1}\frac{H_1^{(2)}(\kappa_0 b)\exp(-\mathrm{i}\alpha z)}{\frac{1}{2}a\kappa_0\eta H_0^{(2)}(\kappa_0 a) - H_1^{(2)}(\kappa_0 a)}\,\mathrm{d}\alpha + \omega\varepsilon\varepsilon_0\frac{b}{a}\sum_p\left[\frac{H_1^{(2)}(\kappa_0 b)J_1(\kappa a)}{\mathrm{d}L/\mathrm{d}\alpha}\exp(-\mathrm{i}\alpha z)\right]_{\alpha=\alpha_p} \tag{6.145}$$

where

$$L = \varepsilon\kappa_0 J_1(\kappa a)H_0^{(2)}(\kappa_0 a) - \varepsilon_0\kappa H_1^{(2)}(\kappa_0 a)J_0(\kappa a) \tag{6.146}$$

and α_p is a typical zero of L which becomes real when the dissipation is removed. For the surface waves κ_0 is negative imaginary and the Hankel function may be befittingly replaced by the modified Bessel function K as in the previous section.

When $b = a$, the integral in (6.145) can be simplified by introducing the variable of integration $t = \kappa_0/k$ and combining the integrals on the two sides of the branch cut. The result

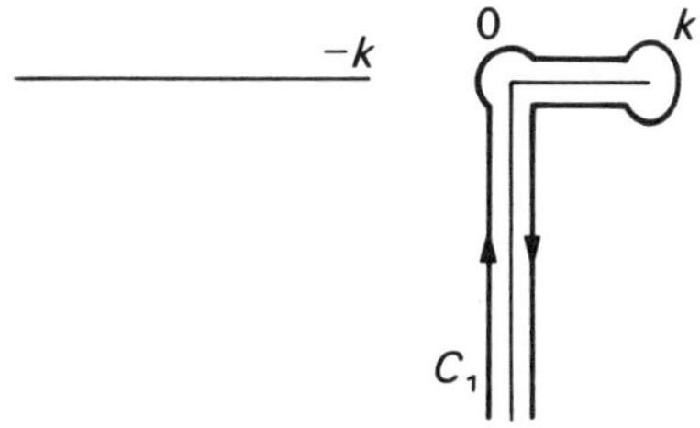

FIG. 6.21. Contour to avoid leaky waves.

is

$$H_\phi = -\frac{1}{2\pi^2}\omega\varepsilon_0\eta^2 ka^2\int_0^\infty \frac{t\exp\{-ikz(1-t^2)^{\frac{1}{2}}\}}{(1-t^2)^{\frac{1}{2}}M}\,dt + $$

$$+\ \omega\varepsilon\varepsilon_0\sum_p\left[\frac{H_1^{(2)}(\kappa_0 a)J_1(\kappa a)}{dL/d\alpha}\exp(-i\alpha z)\right]_{\alpha=\alpha_p} \tag{6.147}$$

where

$$M = \{\tfrac{1}{2}ak\eta tJ_0(kat)-J_1(kat)\}^2+\{\tfrac{1}{2}ak\eta tY_0(akt)-Y_1(kat)\}^2 \tag{6.148}$$

and $(1-t^2)^{\frac{1}{2}}$ is negative imaginary when $t > 1$.

The far field where $kr \gg 1$ may be evaluated from (6.143) by means of the method of stationary phase, the same device as was utilized for (6.33) being adopted. Any poles encountered in moving the contour to the path of stationary phase can be neglected because they will have $\mathscr{I}\kappa_0 < 0$ and so make a contribution exponentially small in comparison with the point of stationary phase. Hence

$$Z_0H_\phi \sim \tfrac{1}{2}kb\frac{\exp(-ikR)}{R}\left[\frac{H_1^{(2)}(kb\sin\theta)\{\varepsilon_0\kappa J_1(ka\sin\theta)J_0(\kappa a)-\varepsilon k\sin\theta J_0(ka\sin\theta)J_1(\kappa a)}{\varepsilon k\sin\theta J_1(\kappa a)H_0^{(2)}(ka\sin\theta)-\varepsilon_0\kappa H_1^{(2)}(ka\sin\theta)J_0(\kappa a)}\right.$$

$$\left. +\ J_1(kb\sin\theta)\right] \tag{6.149}$$

where now $\kappa = (\omega^2\mu\varepsilon-k^2\cos^2\theta)^{\frac{1}{2}}$. Any consistent choice of the radical in κ may be employed since no branch point is involved. If $b = a$, (6.149) simplifies to

$$Z_0H_\phi \sim -\frac{i}{\pi}\varepsilon k\frac{\exp(-ikR)}{R}J_1(\kappa a)\{\varepsilon k\sin\theta J_1(\kappa a)H_0^{(2)}(ka\sin\theta) -$$

$$-\ \varepsilon_0\kappa H_1^{(2)}(ka\sin\theta)J_0(\kappa a)\}^{-1}\ . \tag{6.150}$$

The radiation conductance G may, as pointed out at the end of §6.3, be calculated from the real part of (6.147) with $z = 0$. Thus

$$G = \frac{4}{\pi}\frac{(\eta ka)^2}{Z_0}\int_0^1 \frac{t\,dt}{(1-t^2)^{\frac{1}{2}}M} - 8\pi\omega\varepsilon\varepsilon_0 \mathcal{R} \sum_p \left[\frac{H_1^{(2)}(\kappa_0 a)J_1(\kappa a)}{dL/d\alpha}\right]_{\alpha=\alpha_p}. \tag{6.151}$$

Some general observations can be made on the strength of these formulae, especially when $\varepsilon\mu \gg \varepsilon_0\mu_0$. Suppose, firstly, that $k_0 a(\varepsilon\mu-\varepsilon_0\mu_0)^{\frac{1}{2}} \ll (\varepsilon_0\mu_0)^{\frac{1}{2}}$. Then no surface wave or leaky wave can be initiated and the residue terms disappear from (6.147) and (6.151). Also (6.150) becomes roughly

$$Z_0 H_\phi \sim \tfrac{1}{4}\eta(ka)^2 \sin\theta\, \frac{\exp(-ikR)}{R}$$

while (6.151) gives

$$G = \frac{2\pi\eta^2(ka)^4}{3Z_0}$$

on approximating M by $4/(\pi kat)^2$. Consequently, the pattern is that of a small loop, the main effect of the dielectric being to alter the magnitude of the field.

For $1 < ka(\varepsilon\mu/\varepsilon_0\mu_0-1)^{\frac{1}{2}} < j_{01} \approx 2.4$, ka will be quite small. The denominator of (6.150) will have a complex zero corresponding to a solution of (6.139) and the phenomenon of a leaky wave is manifest. When θ is near the real part of the complex angle, the radiation pattern will have a maximum. The rod is acting as a fairly efficient radiator, the leaky wave on the rod decaying exponentially with distance from the magnetic frill. Confirmation of the radiation is further provided by G which is still essentially equal to the integral term in (6.151).

As $ka(\varepsilon\mu/\varepsilon_0\mu_0-1)^{\frac{1}{2}}$ approaches 2.4, $|\eta| \to \infty$ and the cut-off frequency of the surface wave is near. The radiation pattern and conductance transform to those of a metal wire of the same diameter. The current now falls very slowly along the rod and the radiation of energy is inefficient. In fact, the conductance behaves as if the rod were metal for quite a wide range on either side of 2.4.

With $2.4 < ka(\varepsilon\mu/\varepsilon_0\mu_0-1)^{\frac{1}{2}} < j_{11} \approx 3.8$, there is a surface wave which can carry power along the rod without loss of amplitude. Again the radiation is relatively poor and becomes a minimum at the upper limit where the launching efficiency of the surface wave is at its highest.

Further augmentation of ka repeats these qualitative features as the successive leaky and surface waves come into play.

Exercises

58. Find the analogues of (6.143) and (6.144) when the source is not cylindrically symmetric so that hybrid modes may be generated.

59. For $z \gg 1$, what is the analogue of (6.25) deduced from (6.144)?

6.25. The finite rod

When the rod is not infinite in length the device of §6.5 suggests that a terminating impedance should be introduced at the end $z = l$ related to the upcoming wave by $-2\pi aH_\phi = Y_e V_1$. However, Y_e cannot be taken as independent of the position of the source on account of the different decays of the various waves that can be excited on the dielectric. Therefore, in general, Y_e must be a function of l and the position of the source. Once this dependence has been settled the various formulae connecting admittances in §6.5 can be carried over. For example, in place of (6.44) we have

$$Y = Y_\infty - \frac{2\{Y(l)\}^2}{Y_e(l) + Y(2l)Y_e(l)Y_e(2l)} \tag{6.152}$$

where Y_∞ is the admittance of the infinite antenna and

$$Y(z) = \frac{2(\eta ka)^2}{\pi Z_0} \int_0^\infty \frac{t \exp\{-ikz(1-t^2)^{\frac{1}{2}}\}}{(1-t^2)^{\frac{1}{2}}M} \, dt$$

from (6.147).

Exercise

60. If there is a well-established surface wave Y_e can be regarded as independent of l. Show that (6.152) is then similar to (6.44) with α for k.

6.26. Integral equations for general shapes

When the dielectric is not a circular cylinder there is little hope of finding analytical solutions unless the obstacle has a very simple shape like a sphere or special conditions (e.g. low frequencies or small refractive index) are applicable (Jones 1964). General shapes, therefore, can be tackled only by numerical techniques. In this section, the appropriate integral equations will be set up on the assumptions that the dielectric is finite in size and homogeneous.

It will be presumed that the dielectric occupies the region S of Fig. 6.15 and that the constants μ and ε supply measures of its permeability and permittivity respectively. In S_+ the constants are taken as μ_0 and ε_0 respectively. To accommodate fundamental solutions in both regions it will be convenient to make a slight change of notation and put $k^2 = \omega^2\mu\varepsilon$, $k_0^2 = \omega^2\mu_0\varepsilon_0$ with

$$\psi_0(\mathbf{x},\boldsymbol{\xi}) = \frac{\exp(-ik_0|\mathbf{x}-\boldsymbol{\xi}|)}{4\pi|\mathbf{x}-\boldsymbol{\xi}|} .$$

An incident wave $\mathbf{E}_0$, $\mathbf{H}_0$ strikes the dielectric obstacle from S_+. As a result, the body generates in S_+ a scattered field $\mathbf{E}_s$, $\mathbf{H}_s$ which has to satisfy the radiation conditions at infinity. In addition, a field $\mathbf{E}_t$, $\mathbf{H}_t$ is transmitted into S_-. The tangential components of the total field are to be continuous across S, i.e.

$$\mathbf{n} \wedge (\mathbf{E}_0+\mathbf{E}_s)_+ = \mathbf{n} \wedge (\mathbf{E}_t)_-, \quad \mathbf{n} \wedge (\mathbf{H}_0+\mathbf{H}_s)_+ = \mathbf{n} \wedge (\mathbf{H}_t)_-. \tag{6.153}$$

Inside the dielectric Maxwell's equations can be written as

$$\operatorname{curl} \mathbf{E} + i\omega\mu_0\mathbf{H} = i\omega(\mu_0-\mu)\mathbf{H},$$

$$\operatorname{curl} \mathbf{H} - i\omega\varepsilon_0\mathbf{E} = i\omega(\varepsilon-\varepsilon_0)\mathbf{E}.$$

In this context, the influence of the dielectric is represented as attributable to volume distributions of electric and magnetic current placed in the same medium as the exterior. Consequently, the total field **E**, **H** can be expressed as

$$\mathbf{E}(P) = \mathbf{E}_0(P) + (\text{grad div} + k_0^2)\int_{S_-} (\varepsilon/\varepsilon_0-1)\mathbf{E}(Q)\psi_0(P,Q)\mathrm{d}\mathbf{x}_Q - $$

$$- i\omega \operatorname{curl}\int_{S_-} (\mu-\mu_0)\mathbf{H}(Q)\psi_0(P,Q)\mathrm{d}\mathbf{x}_Q, \tag{6.154}$$

$$\mathbf{H}(P) = \mathbf{H}_0(P) + (\text{grad div} + k_0^2)\int_{S_-} (\mu/\mu_0-1)\mathbf{H}(Q)\psi_0(P,Q)\mathrm{d}\mathbf{x}_Q + $$

$$+ i\omega \operatorname{curl}\int_{S_-} (\varepsilon-\varepsilon_0)\mathbf{E}(Q)\psi_0(P,Q)\mathrm{d}\mathbf{x}_Q. \tag{6.155}$$

This representation has continuous tangential components and therefore automatically satisfies the boundary conditions (6.153). The application of (6.154) and (6.155) with P in S_- delivers integral equations to determine **E** and **H** inside the dielectric. Substitution of these values in the integrals of (6.154) and (6.155) when P is in S_+ then provides the field outside the dielectric and the problem is solved.

One tremendous advantage of these volume integral equations is that there is no necessity for ε and μ to be constant. They are therefore available for arbitrary inhomogeneous dielectrics. Indeed, replacing **E**, **H** by $\mathbf{E}_0$, $\mathbf{H}_0$ in the integrals is the standard way of obtaining Rayleigh scattering as a first approximation to an iteration procedure which may be expected to converge if the combination of frequency and material deviation is suitably low.

Unfortunately, (6.154) and (6.155) suffer from a grave defect from a practical point of view. In effect, six scalar three-dimensional integral equations have to be solved simultaneously. The effort demanded of the computer is therefore at least two orders of magnitude greater than that for a metallic scatterer. While there is little prospect of any alternative if ε and μ are variable, it is worth attempting to manufacture surface integral equations for constant ε and μ, thereby saving at least an order of magnitude in computational requirements. From this point onwards, therefore, ε and μ will be regarded as unvarying in S_-.

From (6.75) and (6.76)

$$\mathrm{E}_s(P) = \int_S [\{\mathbf{n}_q \wedge \mathrm{E}_s(q)\} \wedge \mathrm{grad}_q \psi_0(P,q) + \{\mathbf{n}_q . \mathrm{E}_s(q)\}\mathrm{grad}_q \psi_0(P,q) - i\omega\mu_0\{\mathbf{n}_q \wedge \mathrm{H}_s(q)\}\psi_0(P,q)]\mathrm{d}S_q \,, \tag{6.156}$$

$$\mathrm{H}_s(P) = \int_S [\{\mathbf{n}_q \wedge \mathrm{H}_s(q)\} \wedge \mathrm{grad}_q \psi_0(P,q) + \{\mathbf{n}_q . \mathrm{H}_s(q)\}\mathrm{grad}_q \psi_0(P,q) + i\omega\varepsilon_0\{\mathbf{n}_q \wedge \mathrm{E}_s(q)\}\psi_0(P,q)]\mathrm{d}S_q \tag{6.157}$$

where $P \in S_+$ and both right-hand sides vanish identically for P in S_-. Similarly

$$\mathrm{E}_t(P_i) = -\int_S [\{\mathbf{n}_q \wedge \mathrm{E}_t(q)\} \wedge \mathrm{grad}_q \psi(P_i,q) + \{\mathbf{n}_q . \mathrm{E}_t(q)\}\mathrm{grad}_q \psi(P_i,q) - i\omega\mu\{\mathbf{n}_q \wedge \mathrm{H}_t(q)\}\psi(P_i,q)]\mathrm{d}S_q \,, \tag{6.158}$$

$$\mathbf{H}_t(P_i) = -\int_S [\{\mathbf{n}_q \wedge \mathbf{H}_t(q)\} \wedge \operatorname{grad}_q \psi(P_i,q) + \{\mathbf{n}_q . \mathbf{H}_t(q)\} \operatorname{grad}_q \psi(P_i,q)$$
$$+ i\omega\varepsilon\{\mathbf{n}_q \wedge \mathbf{E}_t(q)\}\psi(P_i,q)]\, dS_q \tag{6.159}$$

for P_i in S_- and the pair of right-hand sides is identically zero when P_i is in S_+. Let

$$\mathbf{n} \wedge \mathbf{H}_t = -\mathbf{j}_t, \quad \mathbf{n} \wedge \mathbf{E}_t = \mathbf{j}_t' ,$$

$$i\omega\rho_t = -\operatorname{div}_0 \mathbf{j}_t, \quad i\omega\rho_t' = -\operatorname{div}_0 \mathbf{j}_t .$$

Then, from (6.81)–(6.84), $\mathbf{n} . \mathbf{E}_t = -\rho_t/\varepsilon$ and $\mathbf{n} . \mathbf{H}_t = -\rho_t'/\mu$. Also (6.153) implies that

$$\mathbf{n} \wedge \mathbf{E}_s = \mathbf{j}_t' - \mathbf{n} \wedge \mathbf{E}_0, \quad \mathbf{n} \wedge \mathbf{H}_s = -\mathbf{j}_t - \mathbf{n} \wedge \mathbf{H}_0 ,$$

$$\mathbf{n} . \mathbf{E}_s = -\rho_t/\varepsilon_0 - \mathbf{n} . \mathbf{E}_0, \quad \mathbf{n} . \mathbf{H}_s = -\mathbf{n} . \mathbf{H}_0 - \rho_t'/\mu_0$$

from (6.79) and (6.80).

Now, if $\mathbf{E}_s$ and $\mathbf{H}_s$ are replaced by $\mathbf{E}_0$ and $\mathbf{H}_0$ in the right-hand sides of (6.156) and (6.157) the integrals give $-\mathbf{E}_0$ and $-\mathbf{H}_0$ for P in S_- because the sources of the incident field are outside S. Combining these integrals with those of (6.156) and (6.157) for a point inside S_- we obtain

$$\mathbf{n}_p \wedge \lim_{P_i \to p} \int \{\mathbf{j}_t' \wedge \operatorname{grad}_q \psi_0(P_i,q) - \frac{\rho_t}{\varepsilon_0} \operatorname{grad}_q \psi_0(P_i,q) +$$
$$+ i\omega\mu_0 \mathbf{j}_t \psi_0(P_i,q)\}\, dS_q = -\mathbf{n}_p \wedge \mathbf{E}_0 , \tag{6.160}$$

$$\mathbf{n}_p \wedge \lim_{P_i \to p} \int \{-\mathbf{j}_t \wedge \operatorname{grad}_q \psi_0(P_i,q) - \frac{\rho_t'}{\mu_0} \operatorname{grad}_q \psi_0(P_i,q) +$$
$$+ i\omega\varepsilon_0 \mathbf{j}_t' \psi_0(P_i,q)\}\, dS_q = -\mathbf{n}_p \wedge \mathbf{H}_0 . \tag{6.161}$$

The next objective is to rid the integrands of ρ_t and ρ_t'.

With P in S_+

$$\mathbf{n}_p \wedge \{\lim_{P_i \to p} \int_S \rho_t \operatorname{grad}_q \psi_0(P_i,q)\mathrm{d}S_q - \lim_{P \to p} \int_S \rho_t \operatorname{grad}_q \psi(P,q)\mathrm{d}S_q\}$$

$$= \mathbf{n}_p \wedge \int_S \rho_t \operatorname{grad}_q\{\psi_0(p,q) - \psi(p,q)\}\mathrm{d}S_q$$

because (6.90) and (6.91) hold for both ψ and ψ_0. But

$$\int_S \rho_t \operatorname{grad}_q\{\psi_0(p,q)-\psi(p,q)\}\mathrm{d}S_q = -\operatorname{grad}_p \int_S \rho_t\{\psi_0(p,q)-\psi(p,q)\}\mathrm{d}S_q$$

$$= \operatorname{grad}_p \int_S \{\psi_0(p,q)-\psi(p,q)\}\operatorname{div} \mathbf{j}_t \frac{\mathrm{d}S_q}{\mathrm{i}\omega}$$

$$= -\operatorname{grad}_p \int_S \mathbf{j}_t \cdot \operatorname{grad}_q\{\psi_0(p,q)-\psi(p,q)\}\frac{\mathrm{d}S_q}{\mathrm{i}\omega}$$

from (6.78).

The right-hand of (6.158) is zero in S_+. Therefore, if we take the multiple $\varepsilon\mathbf{n}\wedge$ of it and add it to ε_0 times (6.160) we derive via (6.90) and (6.91)

$$-\tfrac{1}{2}(\varepsilon+\varepsilon_0)\mathbf{j}_t' + \mathbf{n}_p \wedge \fint_S \Big[\mathbf{j}_t'\wedge\operatorname{grad}_q\{\varepsilon_0\psi_0(p,q)-\varepsilon\psi(p,q)\} +$$

$$+ \mathrm{i}\mathbf{j}_t\{k_0^2\psi_0(p,q) - k^2\psi(p,q)\}/\omega -$$

$$-(\mathbf{j}_t\cdot\operatorname{grad}_q)\,\frac{\operatorname{grad}_q\{\psi_0(p,q)-\psi(p,q)\}}{\mathrm{i}\omega}\Big]\mathrm{d}S_q$$

$$= -\varepsilon_0\mathbf{n}_p \wedge \mathbf{E}_0\ . \tag{6.162}$$

Similar operations with (6.161) and (6.159) lead to

$$\tfrac{1}{2}(\mu+\mu_0)\mathbf{j}_t + \mathbf{n}_p \wedge \int_S \Big[\mathbf{j}_t\wedge\mathrm{grad}_q\{\mu\psi(p,q)-\mu_0\psi_0(p,q)\} +$$

$$+ \mathrm{i}\mathbf{j}_t'\{k_0^2\psi_0(p,q) - k^2\psi(p,q)\}/\omega -$$

$$- (\mathbf{j}_t'.\mathrm{grad}_q)\,\frac{\mathrm{grad}_q\{\psi_0(p,q)-\psi(p,q)\}}{\mathrm{i}\omega}\Big]\mathrm{d}S_q$$

$$= -\mu_0\mathbf{n}_p \wedge \mathbf{H}_0 \; . \qquad (6.163)$$

The integral equations (6.162) and (6.163) are the duo sought. They constitute four scalar simultaneous linear inegral equations and are appreciably easier to tackle than (6.154) and (6.155) but substantially harder than if S were a perfect conductor.

It can be shown that the operators in (6.162) and (6.163) are compact, but the proof will not be given here. The matter of uniqueness will be settled in the next section.

6.27. Uniqueness for dielectrics

If the right-hand sides of (6.162) and (6.163) are placed equal to zero the integral equations may possess a solution $\mathbf{j}_0$, $\mathbf{j}_0'$. If so, (6.162) and (6.163) are not uniquely soluble. Therefore, to guarantee uniqueness we want to show that $\mathbf{j}_0$ and $\mathbf{j}_0'$ must be identically zero.

Define ρ_0 and ρ_0' by $\mathrm{i}\omega\rho_0 = -\,\mathrm{div}_0\,\mathbf{j}_0$, $\mathrm{i}\omega\rho_0' = -\,\mathrm{div}_0\,\mathbf{j}_0'$. Then construct the following fields:

$$\mathbf{E}_1 = \int_S \{-\mathrm{i}\omega\mu\mathbf{j}_0\psi(P,q) - \mathbf{j}_0' \wedge \mathrm{grad}_q\psi(P,q) +$$

$$+ \frac{\rho_0}{\varepsilon}\,\mathrm{grad}_q\psi(P,q)\}\mathrm{d}S_q\,, \qquad (6.164)$$

$$\mathbf{H}_1 = \int_S \{-i\omega\varepsilon \mathbf{j}_0'\psi(P,q) + \mathbf{j}_0 \wedge \mathrm{grad}_q\psi(P,q) +$$

$$+ \frac{\rho_0'}{\mu} \mathrm{grad}_q\psi(P,q)\}\mathrm{d}S_q \ , \tag{6.165}$$

$$\mathbf{E}_2 = \int_S \{-i\omega\mu_0\mathbf{j}'\psi_0(P,q) \wedge \mathbf{j}_0' \wedge \mathrm{grad}_q\psi_0(P,q) +$$

$$+ \frac{\rho_0}{\varepsilon_0} \mathrm{grad}_q\psi_0(P,q)\}\mathrm{d}S_q \ , \tag{6.166}$$

$$\mathbf{H}_2 = \int_S \{-i\omega\varepsilon_0\mathbf{j}_0'\psi_0(P,q) + \mathbf{j}_0 \wedge \mathrm{grad}_q\psi_0(P,q) +$$

$$+ \frac{\rho_0'}{\mu_0} \mathrm{grad}_q\psi_0(P,q)\}\mathrm{d}S_q \ . \tag{6.167}$$

The integral equations satisfied by $\mathbf{j}_0$ and $\mathbf{j}_0'$ have been built in such a way that

$$\varepsilon_0\mathbf{n} \wedge (\mathbf{E}_2)_- = \varepsilon\mathbf{n} \wedge (\mathbf{E}_1)_+, \qquad \mu_0\mathbf{n} \wedge (\mathbf{H}_2)_- = \mu\mathbf{n} \wedge (\mathbf{H}_1)_- . \tag{6.168}$$

Now the equations satisfied by (6.164) and (6.165) in S_+ can be expressed as

$$\mathrm{curl}\ \varepsilon\mathbf{E}_1 + i\omega\varepsilon(\mu\mathbf{H}_1) = \mathbf{0}, \qquad \mathrm{curl}\ \mu\mathbf{H}_1 - i\omega\mu(\varepsilon\mathbf{E}_1) = \mathbf{0}$$

whereas, in S_-, (6.166) and (6.167) give

$$\mathrm{curl}\ \varepsilon_0\mathbf{E}_2 + i\omega\varepsilon_0(\mu_0\mathbf{H}_2) = \mathbf{0}, \qquad \mathrm{curl}\ \mu_0\mathbf{H}_2 - i\omega\mu_0(\varepsilon_0\mathbf{E}_2) = \mathbf{0}\ .$$

Moreover, the radiation conditions at infinity may be set forth as

$$R\{\mu\mathbf{H}_1 + \left(\frac{\mu}{\varepsilon}\right)^{\frac{1}{2}}\varepsilon\mathbf{E}_1 \wedge \hat{\mathbf{R}}\} \to 0 \ ,$$

$$R\{-\varepsilon\mathbf{E}_1 - \hat{\mathbf{R}} \wedge \mu\mathbf{H}_1 \left(\frac{\mu}{\varepsilon}\right)^{-\frac{1}{2}}\} \to 0.$$

Therefore the field $\mu\mathbf{H}_1$, $-\varepsilon\mathbf{E}_1$ in S_+ and $\mu_0\mathbf{H}_2$, $-\varepsilon_0\mathbf{E}_2$ in S_- is an electromagnetic field which satisfies the radiation conditions and, by (6.168), has continuous tangential components.

Let us assume that such a field must be identically zero. Then $(\mathbf{E}_1)_+ = \mathbf{0}$ and $(\mathbf{H}_1)_+ = \mathbf{0}$. Hence, from the jump in the representations across S,

$$(\mathbf{n}\wedge\mathbf{E}_1)_- = \mathbf{j}_0', \qquad (\mathbf{n}\wedge\mathbf{H}_1)_- = -\mathbf{j}_0 \ .$$

Similarly

$$(\mathbf{n}\wedge\mathbf{E}_2)_+ = -\mathbf{j}_0', \qquad (\mathbf{n}\wedge\mathbf{H}_2)_+ = \mathbf{j}_0 .$$

Consider the field $\mathbf{E}_2$, $\mathbf{H}_2$ in S_+ and $-\mathbf{E}_1$, $-\mathbf{H}_1$ in S_-. By what has just been said its tangential components are continuous through S and it obeys the radiation conditions at infinity. The assumption at the beginning of the paragraph makes the field identically zero. It follows that $\mathbf{j}_0 \equiv \mathbf{0}$, $\mathbf{j}_0' = 0$ and the uniqueness of (6.162) and (6.163) is established.

The numerical solution of (6.153) and (6.154) exhibits no novel features as compared with earlier integral equations other than intricacy, so nothing further will be said in this respect.

The assumption in the last paragraph but one can be justified as follows. If $\mathbf{E}$, $\mathbf{H}$ is a field with the given properties

$$\int_{\Omega_R} (\mathbf{E}\wedge\mathbf{H}^*+\mathbf{E}^*\wedge\mathbf{H}) \ . \ \mathbf{n}\,d\Omega = 0 \ .$$

Rewriting the integral as in §6.22 we deduce that $\int_{\Omega_R} |\mathbf{E}|^2 d\Omega \to 0$ as $R \to \infty$ and this is impossible unless $\mathbf{E}$ and $\mathbf{H}$ are identically zero.

Exercises

61. Show that the uniqueness demonstrated in the last paragraph holds for complex permeability and permittivity provided that $\mathcal{R}(i\omega\varepsilon) \geqslant 0$, $\mathcal{R}(i\omega\mu) \geqslant 0$ and $0 \geqslant \text{ph}\ \omega^2\mu_0\varepsilon_0 > -\pi$. Deduce that (6.162) and (6.163) have a unique solution for complex material constants.

62. Prove that the unique solution of (6.162) and (6.163) does furnish a solution of the dielectric scattering problem.

63. Reformulate the theory of this section for two-dimensional scattering.

6.28. Hybrid methods

The reduction of the computational labour in solving dielectric integral equations is a theme of prime importance. The allure of economy has tempted investigators to try marrying some of the concepts from integral equations with those of finite differences (Mei 1974, Chang and Mei 1974, Morgan and Mei 1974; for dielectric obstacles in waveguides see Mur, Quak and van Dijk 1976). Such methods might be termed *hybrid*, though in fact the main aim is to dispense with the integral equations to the greatest imaginable extent. The price to be paid in compensation is by no means trivial even if irregular variations in the media are encompassed.

Choose some convenient surface which encloses all inhomogeneities. For definiteness, a sphere will be chosen, but this choice is not obligatory. Outside the sphere the scattered field may be represented as a series of radiating spherical harmonics. Suppose that N of these are considered to give an adequate approximation; the scattered field will be known as soon as the N coefficients are determined. Adding the incident wave gives the total field outside. On the interior surface of the sphere expand the tangential electric field in terms of N elements of a set $\{\mathbf{E}_n\}$ of specified tangential vectors. The $\mathbf{E}_n$ may be picked freely so long as they are complete when $N \to \infty$.

In an integral equation method the tangential magnetic vector would be expanded likewise and the coefficients made to satisfy the integral equation. In the hybrid method the field resulting from the specified $\mathbf{E}_n$ inside the sphere is calculated by finite differences, variational methods, finite elements, or any other tolerable scheme. The $\mathbf{H}_n$ corresponding to $\mathbf{E}_n$ is now known.

The N interior and N exterior unknown coefficients are now determined by formulating $2N$ equations which impose the continuity of the tangential components of the total field across the spherical surface. This might be done by collocation at suitable points on the sphere or by multiplying the expansions by members of a pertinent set of weighting functions and integrating over the sphere. The problem is thereby solved.

It is plain that the virtue of the method lies in its sparse matrices and the simplicity of its program. Moreover, it has the ability to cope with a wide variety of inhomogeneities. The snag is in the determination of the $\mathbf{H}_n$ from the $\mathbf{E}_n$. In principle, this interior problem can be tackled by methods described in earlier chapters, but we have seen that it is quite difficult to produce reliable answers. The finite-difference method is inherently less stable than the integral equation. There is an additional aspect of the trade-off which must not be forgotten. It might be that the spherical surface was placed in such a position that the interior had an electrical mode of resonance at or near the frequency under consideration. Even if the difference equations could be solved at all serious errors would be expected.

Exercise

64. The following is a selection of problems for trying out methods: (*a*) the sphere with invariable interior; (*b*) the sphere with the material changing radially; (*c*) a sphere composed of segments of constant materials; (*d*) an annular circular cylinder; (*e*) an elliptic cylinder; (*f*) a square cylinder; (*g*) a biconical antenna; (*h*) a lens.

The incident wave might be plane or come from a point dipole.

7

TRANSIENT PHENOMENA

Most electromagnetic transmitters operate long enough at a single frequency for the analysis by time-harmonic waves to be appropriate. However, it is possible to produce short pulses with a broad frequency spectrum so that predictions of events for general time variations is desirable. The effects of the radiation from lightning discharges and of high-powered optical pulses then become amenable to investigation. This chapter is, therefore, devoted to the problem of transients.

7.1. Finite differences

The general problem is to solve equations of the form

$$\text{curl } \mathbf{E} + \mu \frac{\partial \mathbf{H}}{\partial t} = -\mathbf{J}', \qquad \text{curl } \mathbf{H} - \varepsilon \frac{\partial \mathbf{E}}{\partial t} = \mathbf{J} \tag{7.1}$$

where $\mathbf{J}$ and $\mathbf{J}'$ are known electric and magnetic currents. The quantities μ and ε may depend on the space variables but do not vary with the time t.

Causality insists that there shall be no disturbance until the source is switched on. Thereafter the field propagates behind wavefronts which travel with speeds and in directions characteristic of the medium. No energy flow can be detected by an observer before such a wavefront has passed over him. It is this property that suggests that it is worth re-opening our minds to the possibility of finite-difference methods which were discarded for harmonic scattering problems in §6.12.

The reason for disposing of finite differences in the harmonic case was the difficulty of satisfying the radiation conditions at infinity with a mesh of finite size. This perplexity does not arise for electromagnetic pulses. There is now a clear-cut wavefront, ahead of which there is no disturbance. Therefore the mesh does not have to go off to infinity; it merely has to extend as far as the most distant wavefront which has emanated from the obstacle. The progress of each wavefront in time and space can be traced from its initiation and no artificial boundary to account for the behaviour at infinity is needed. Thus finite-difference methods for initial-value problems are feasible for pulse propagation.

Some success has been reported (J. Morris 1974, private communication) in the application of finite differences to acoustic scattering but little effort seems to have been made in the electromagnetic domain. It would appear to be opportune to pursue this idea further.

7.2. Integral equations in the time domain

In principle, solutions in the time domain can be obtained from those of the harmonic variation in Chapter 6 by means of a Fourier transform. The process is assisted by the fast Fourier transform (§2.14). Nevertheless, the transformation is not a trivial exercise since the harmonic problem must be solved for a large range of frequencies and it is evident from Chapter 6 that the computation of results for a single frequency can be quite formidable. An examination of what is involved in working directly in the time domain is therefore vindicated.

Let S be a closed surface such that the normal components of $\mathbf{J}$ and $\mathbf{J}'$ are always zero on it. For sources of finite extent this can be arranged simply by making S large enough. Then, if ε and μ are constant within S, the general solution of (7.1) inside S is (Jones 1964)

$$\mathbf{E}(P,t) = -\frac{\partial}{\partial t}\mu\int \mathbf{J}(Q,T)\Psi(P,Q)\,\mathrm{d}\mathbf{x}_Q - \mathrm{grad}\,\frac{1}{\varepsilon}\int \rho(Q,T)\Psi(P,Q)\,\mathrm{d}\mathbf{x}_Q$$

$$- \mathrm{curl}\int \mathbf{J}'(Q,T)\Psi(P,Q)\,\mathrm{d}\mathbf{x}_Q - \int_S [\{\mathbf{n}_q\wedge\mathbf{E}(q,T)\}\wedge\mathrm{grad}_q\Psi(P,q)$$

$$- \mathrm{grad}_q\Psi(P,q)\wedge\frac{\{\mathbf{n}_q\wedge\partial\mathbf{E}(q,T)/\partial t\}}{v\Psi} + \{\mathbf{n}_q\cdot\mathbf{E}(q,T)\}\mathrm{grad}_q\Psi(P,q)$$

$$+ \{\mathbf{n}_q \cdot \frac{\partial}{\partial t}\mathbf{E}(q,T)\}\frac{\{\text{grad}_q \Psi(P,q)\}}{v\Psi} - \mu\Psi(P,q)\mathbf{n}_q \wedge \frac{\partial}{\partial t}\mathbf{H}(q,T)]\,dS_q, \tag{7.2}$$

$$\mathbf{H}(P,t) = -\frac{\partial}{\partial t}\varepsilon\int \mathbf{J}'(Q,T)\Psi(P,Q)\,d\mathbf{x}_Q - \text{grad}\,\frac{1}{\mu}\int \rho'(Q,T)\Psi(P,Q)\,d\mathbf{x}_Q$$

$$+ \text{curl}\int \mathbf{J}(Q,T)\Psi(P,Q)\,d\mathbf{x}_Q - \int_S [\{\mathbf{n}_q \wedge \mathbf{H}(q,T)\}\wedge \text{grad}_q \Psi(P,q)$$

$$- \text{grad}_q \Psi(P,q)\wedge\{\mathbf{n}_q \wedge \frac{\partial}{\partial t}\mathbf{H}(q,T)\}/v\Psi + \{\mathbf{n}_q \cdot \mathbf{H}(q,T)\}\text{grad}_q \Psi(P,q)$$

$$+ \{\mathbf{n}_q \cdot \frac{\partial}{\partial t}\mathbf{H}(q,T)\}\frac{\{\text{grad}_q \Psi(P,q)\}}{v\Psi} + \varepsilon\Psi(P,q)\mathbf{n}_q \wedge \frac{\partial}{\partial t}\mathbf{E}(q,T)]\,dS_q \tag{7.3}$$

where $v^2 = 1/\mu\varepsilon$, $\text{div}\,\mathbf{J} + \partial\rho/\partial t = 0$, $\text{div}\,\mathbf{J}' + \partial\rho'/\partial t = 0$,

$$\Psi(P,Q) = \frac{1}{4\pi|\mathbf{x}_P - \mathbf{x}_Q|}$$

and $T = t - |\mathbf{x}_P - \mathbf{x}_Q|/v$ in the volume integrals, while $\mathbf{x}_Q$ is replaced by $\mathbf{x}_q$ in the surface integrals. Eqns (7.2) and (7.3) express the field in terms of the *retarded time* T, exemplifying the fact that disturbances move at speed v and do not occur before any sources are switched on.

In special circumstances it may occur that the charges and currents are confined to the interior of the surface Σ, which itself is entirely within S, but the normal components of $\mathbf{J}$ and $\mathbf{J}'$ do not vanish on Σ. In that case, the terms

$$\frac{1}{\varepsilon}\int_\Sigma \mathbf{n}_q \cdot \mathbf{J}_1(q,T)\,\text{grad}_q \Psi(P,q)\,dS_q$$

and

$$\frac{1}{\mu}\int_{\Sigma} \mathbf{n}_q \cdot \mathbf{J}_1'(q,T)\operatorname{grad}_q \Psi(P,q)\,dS_q$$

where

$$\frac{\partial \mathbf{J}_1(q,t)}{\partial t} = \mathbf{J}(q,t), \qquad \frac{\partial \mathbf{J}_1'(q,t)}{\partial t} = \mathbf{J}'(q,t)$$

must be added to (7.2) and (7.3) respectively. Equivalently, the negative volume integrals involving ρ and ρ' can be replaced by

$$\operatorname{grad}\operatorname{div}\frac{1}{\varepsilon}\int \mathbf{J}_1(Q,T)\Psi(P,Q)\,d\mathbf{x}_Q$$

and

$$\operatorname{grad}\operatorname{div}\frac{1}{\mu}\int \mathbf{J}_1'(Q,T)\Psi(P,Q)\,d\mathbf{x}_Q \ .$$

The derivation of the integral equation for the thin wire follows similar lines to those of §6.2. Again, the current is assumed to be concentrated on the axis of the wire and the magnetic current is dropped. The charge density is also condensed to the axis. In view of the general continuity relation

$$\operatorname{div}\mathbf{J}(\mathbf{x},t) + \frac{\partial\rho}{\partial t} = 0$$

the condensed charge may be expressed in terms of the current $I(\boldsymbol{\xi},t)$ at the point $\boldsymbol{\xi}$ where the unit tangent is $\boldsymbol{\tau}(\boldsymbol{\xi})$ by writing it as

$$-\int_{-\infty}^{t}\{\boldsymbol{\tau}(\boldsymbol{\xi}).\operatorname{grad}_{\xi}\}I(\boldsymbol{\xi},u)\,du \ .$$

The surface S in (7.2) may be removed to infinity. Its contribution is then zero because no fields produced by the wire reach there in finite time. Hence, if the incident wave is $\mathbf{E}_0$, the total electric intensity is given by

$$\mathbf{E}(P,t) = \mathbf{E}_0(P,t) - \frac{\partial}{\partial t}\mu\int_0^l \mathbf{I}(\boldsymbol{\xi},T)\Psi(P,\boldsymbol{\xi})\mathrm{d}\zeta$$

$$+ \operatorname{grad}\frac{1}{\varepsilon}\int_0^l\int_{-\infty}^T\{\boldsymbol{\tau}(\boldsymbol{\xi}).\operatorname{grad}_\xi \mathbf{I}(\boldsymbol{\xi},u)\mathrm{d}u\Psi(P,\boldsymbol{\xi})\mathrm{d}\zeta \tag{7.4}$$

where ζ is the arc length and l is the length of the antenna. It has been assumed that the current vanishes at the ends of the wire.

The boundary condition that the axial component of $\mathbf{E}$ vanishes on the surface of the wire may now be applied. If the wire is straight and along the z axis the total field is

$$\mathbf{E}(P,t) = \mathbf{E}_0(P,t) - \frac{\partial}{\partial t}\mu\,\hat{\mathbf{z}}\int_0^l I(\zeta,T)\Psi(P,\zeta)\mathrm{d}\zeta$$

$$+ \operatorname{grad}\frac{1}{\varepsilon}\int_0^l\int_{-\infty}^T\frac{\partial}{\partial\zeta}I(\zeta,u)\mathrm{d}u\Psi(P,\zeta)\mathrm{d}\zeta \tag{7.5}$$

and the resulting integral equation is

$$\frac{E_{0z}(z,t)}{\mu} = \frac{\partial}{\partial t}\int_0^l I(\zeta,T)\Psi'(z,\zeta)\mathrm{d}\zeta$$

$$- \frac{\partial}{\partial z}v^2\int_0^l\int_{-\infty}^T\frac{\partial}{\partial\zeta}I(\zeta,u)\mathrm{d}u\Psi'(z,\zeta)\mathrm{d}\zeta \tag{7.6}$$

where

$$\Psi'(z,\zeta) = \frac{1}{4\pi\{(z-\zeta)^2+a^2\}^{\frac{1}{2}}}$$

and

$$T = \frac{t - \{(z-\zeta)^2+a^2\}^{\frac{1}{2}}}{v}.$$

If the derivatives outside the integrals in (7.6) are taken inside we obtain

$$\frac{E_{0z}(z,t)}{\mu} = \int_0^l \Psi'(z,\zeta)\left[\frac{\partial}{\partial t} I(\zeta,T) + \right.$$

$$\left. + \frac{z-\zeta}{R_1} v\left\{\frac{\partial}{\partial\zeta} I(\zeta,T) + \frac{v}{R_1}\int_{-\infty}^{T}\frac{\partial}{\partial\zeta} I(\zeta,u)\mathrm{d}u\right\}\right]\mathrm{d}\zeta \qquad (7.7)$$

where $R_1 = \{(z-\zeta)^2+a^2\}^{\frac{1}{2}}$. An integration by parts gives the alternative form

$$\frac{E_{0z}(z,t)}{\mu} = \int_0^l \Psi'(z,\zeta)\frac{\partial}{\partial t} I(\zeta,T)\mathrm{d}\zeta + v^2\int_{-\infty}^{T_l}\frac{\partial}{\partial l} I(l,u)\mathrm{d}u\Psi'(z,l) -$$

$$- v^2\int_{-\infty}^{T_0}\left[\frac{\partial}{\partial\zeta} I(\zeta,u)\right]_{\xi=0}\mathrm{d}u\Psi'(z,0) -$$

$$- v^2\int_0^l\int_{-\infty}^{T}\frac{\partial^2}{\partial\zeta^2} I(\zeta,u)\mathrm{d}u\Psi'(z,\zeta)\mathrm{d}\zeta \qquad (7.8)$$

where T_0 and T_l are the values of T at $\zeta = 0$ and $\zeta = l$ respectively.

If the current $I(\zeta,t) = I_0(t-\zeta/v)$ flows in a straight wire the integrals in (7.5) can be evaluated explicitly with the results

$$E_x(P,t) - E_{0x}(P,t) = -\frac{x}{4\pi\varepsilon v}\left[\frac{I_0\{t-(R_2+l)/v\}}{u_2 R_2} - \frac{I_0(t-R_1/v)}{u_1 R_1}\right], \qquad (7.9)$$

$$E_z(P,t) - E_{0z}(P,t) = \frac{\mu v}{4\pi}\left[\frac{I_0\{t-(R_2+l)/v\}}{R_2} - \frac{I_0(t-R_1/v)}{R_1}\right] \qquad (7.10)$$

where the various parameters are defined in §6.2 (Fig. 6.2). However, it must be remembered that (7.5) does not correctly represent the field when the current fails to be zero at the ends as it does here. According to the paragraph after eqns (7.2) and (7.3) it is then necessary to add to the right-hand side of (7.5)

$$-\frac{1}{\varepsilon}\operatorname{grad}\{I_1(l,T_l)\Psi(P,l)-I_1(0,T_0)\Psi(P,0)\}$$

where $\partial I_1(x,t)/\partial t = I(x,t)$. The effect of this addition is to increase the right-hand side of (7.9) by

$$\frac{x}{4\pi\varepsilon}\left[\frac{1}{vR_2^2}\,I_0\left\{t-\frac{(R_2+l)}{v}\right\}+\right.$$

$$\left.+\frac{1}{R_2^3}\,I_1\left\{t-\frac{(R_2+l)}{v}\right\}-\frac{1}{vR_1^2}\,I_0\left(t-\frac{R_1}{v}\right)-\frac{1}{R_1^3}\,I_1\left(t-\frac{R_1}{v}\right)\right] \quad (7.11)$$

and to supplement the right-hand side of (7.10) by

$$-\frac{1}{4\pi\varepsilon}\frac{\partial}{\partial z}\left[\frac{1}{R_2}\,I_1\left\{t-\frac{(R_2+l)}{v}\right\}-\frac{1}{R_1}\,I_1\left(t-\frac{R_1}{v}\right)\right]. \quad (7.12)$$

When the wire is curved the appropriate integral equation is

$$\frac{\boldsymbol{\tau}(\mathbf{x}).\mathbf{E}(\mathbf{x})}{\mu}=\int_0^l I(\zeta,T)\boldsymbol{\tau}(\mathbf{x}).\boldsymbol{\tau}(\boldsymbol{\xi})\Psi_1(\mathbf{x},\boldsymbol{\xi})\,d\zeta\;-$$

$$-\,v^2\int_0^l \boldsymbol{\tau}(\mathbf{x}).\operatorname{grad}\left[\int_{-\infty}^T\{\boldsymbol{\tau}(\boldsymbol{\xi}).\operatorname{grad}_\xi\}\{\boldsymbol{\tau}(\boldsymbol{\xi})I(\zeta,u)\}du\,\Psi_1(\mathbf{x},\boldsymbol{\xi})\right]d\zeta \quad (7.13)$$

where $\Psi_1(\mathbf{x},\boldsymbol{\xi}) = 1/4\pi|\mathbf{x}-\boldsymbol{\xi}+\mathbf{a}|$ and $\mathbf{a}$ has the same connotation as in §6.7. Once more it has been presumed that the current disappears at the ends of the wire. If this is not so, an appropriate adjustment must be made to (7.13) as in the case of the straight wire.

As soon as the axial current is known the far field can be determined from

$$\mathbf{E}(\mathbf{x},t) \sim \mathbf{E}_0(\mathbf{x},t) - \frac{\mu}{4\pi R}\frac{\partial}{\partial t}\int_0^l \left[\boldsymbol{\tau}(\boldsymbol{\xi}) - \left\{\boldsymbol{\tau}(\boldsymbol{\xi}).\frac{\mathbf{x}}{R}\right\}\frac{\mathbf{x}}{R}\right] I(\zeta,T)\,d\zeta \qquad (7.14)$$

where now

$$T = t - \frac{\{R - (\mathbf{x}.\boldsymbol{\xi})/R\}}{v}\;.$$

The formula (7.14) is valid whether the current vanishes at the end-points or not.

Exercise

1. Find the analogues of (7.9)—(7.12) when $I(\zeta,t) = I_0(t+\zeta/v)$.

7.3. Numerical methods for thin wires in the time domain

The numerical solution of the integral equations of the preceding section might be thought to present no radical problems which have not been encountered for the case of harmonic excitation. From one point of view they might be regarded as the same apart from containing an extra parameter — the time — so that the arguments about the desirable properties of the basis functions would still apply, though with redoubled force because of the necessity to allow for the additional parameter. (The presence of the infinite integral is not a genuine difficulty because the actual integration is only over the finite time that has elapsed from the stimulation of the current.) One might think of using basis functions with argument $t - \zeta/v$ so as to avoid numerical integration by taking advantage of (7.9)—(7.12). However, this procedure is likely to be efficacious only in particularly favourable situations and must be rejected for curved antennas.

Instead one must place space and time variables on the same footing, but the interpolation with respect to time must be done rather carefully. The integral equation will be imposed at the times t_1, t_2, ... in increasing order of magnitude. There is no loss of generality in assuming that the antenna is not excited before $t = 0$ so that it is permissible to have $t_1 \geqslant 0$. Suppose that the axis of the wire is represented by cubic B-splines as in §6.8. The points on the axis corresponding to the knots will be separated by certain distances; let $R_{\min}$ be the minimum distance which occurs between any pair. Then the time intervals must be arranged so that $t_k - t_{k-1} \leqslant R_{\min}/v$. The time retardation in the integral equation will then ensure that, when matching at a particular knot, the current distribution in all axial segments but those immediately adjacent to the knot will be that of an earlier time interval. Normally, the subdivision of the filament will be selected so that the geometric representation is good and the subsections are sufficiently small to resolve the spatial variation of the pulse on the wire. The inequality $t_k - t_{k-1} \leqslant R_{\min}/v$ then prescribes an upper limit to the length of time interval which can be tolerated. In order to achieve the necessary temporal resolution one may, of course, be forced to choose even smaller time intervals. Usually, the points t_k are equispaced in time though the computational efficiency would be improved if the intervals were lengthened as the response settled down.

Once the interpolation points have been decided on, the integral equation is enforced so that, in simplified notation,

$$\int I(\sigma', t_k - f(\sigma,\sigma'))K(\sigma,\sigma')\mathrm{d}\sigma' = g(\sigma, t_k)$$

where σ is the spline variable and f accounts for the retardation. The integral is approximated by a quadrature rule so that for a typical subsection

$$\int_m^{m+1} I(\sigma', t_k - f(\sigma,\sigma'))K(\sigma,\sigma')\mathrm{d}\sigma' \approx$$

$$\approx \sum_{r=1}^{n} w_r I(m+\sigma_r, t_k - f(\sigma, m+\sigma_r))K(\sigma, m+\sigma_r) \; .$$

If all the time intervals are of length t_0 and $t_k = kt_0$ the current is now replaced in the interval $((p-\frac{1}{2})t_0, (p+\frac{1}{2})t_0)$ by

$$I(\sigma',t) = \sum_{i=2} \sum_{j=p-2}^{p} I_{ij}N_i(\sigma')N_j\left(\frac{t}{t_0} - \frac{1}{2}\right). \tag{7.15}$$

Point matching at the values of σ now leads to an algebraic system to determine the I_{ij}.

For any space segment not adjoining the matching knot, f will exceed unity and the I_{ij} will come from times t_{k-1} or earlier. The only I_{ij} involving t_k will be those in a segment next door to a matching knot and here f can be set zero to a first approximation. If the I_{ij} at t_{k-1} and earlier are known we are thus led to a relatively simple algebraic system for them at t_k in terms of previous values. Hence, by marching in time from $t = 0$ we can determine succeeding values of the coefficients — the interpolation in time (7.15) has been made backward to warrant this possibility. It also ensures that the matrix inverse, once calculated for one time, need not be recomputed because it is essentially geometric in character.

The effect of a gap when the antenna is fed at a point of its length causes no further complication than in the harmonic case. A typical response for a centre-fed antenna with a feed time variation $\exp\{-(1-t/t')^2\}$, with t' being about 10^{-10} s, is shown in Fig. 7.1. The impedance was calculated by a Fourier transform of the time response. Fig. 7.2 depicts the behaviour of the far field in the plane perpendicular to the antenna for both the time and frequency domains (again obtained by Fourier transforms).

Exercises

2. Calculate the driving-point current of a centre-fed straight-wire antenna fed by a unit voltage step when $l/2a = 74.2$ and compare your results with those of Sayre and Harrington (1968).

3. A unit step of electric field is incident on the antenna of Exercise 2. Calculate the induced current and far field in the plane perpendicular to the wire.

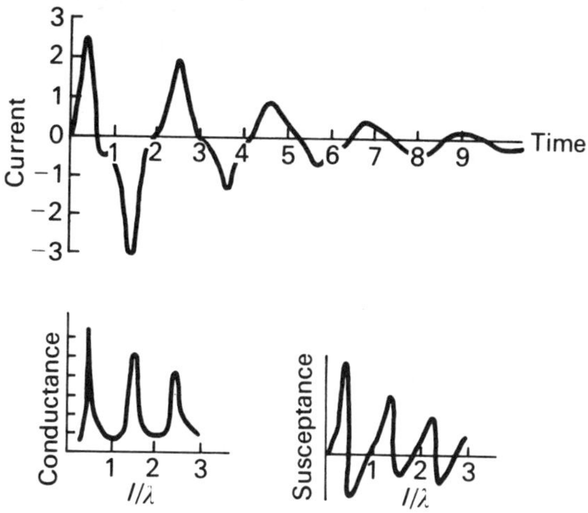

FIG. 7.1. Centre-fed antenna with Gaussian source.

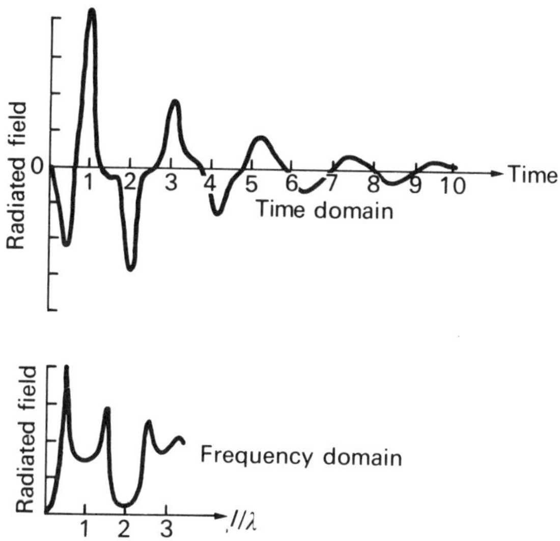

FIG. 7.2. Far field of antenna in Fig. 7.1.

4. Use the excitation of Fig. 7.1 to find how the input admittance of a V-antenna varies with frequency.

5. A circular ring is struck by a Gaussian pulse (§7.5). Find the radiated field in the time domain and show that the radar cross-section is a maximum when the wavelength is practically equal to the perimeter.

6. Choose a log-periodic antenna and find its frequency response by making a Gaussian pulse fall on it in the time domain.

7.4. Perfectly conducting bodies

For perfectly conducting bodies a return to (7.2) and (7.3) is pertinent, but the volume integrals can now be removed. To shorten the writing, only the receiving antenna subject to the incident illumination $\mathbf{E}_0$, $\mathbf{H}_0$ will be discussed but the modifications to handle the transmission regime are straightforward. On the boundary S, define

$$\mathbf{n}_q \wedge \mathbf{H}(q,t) = -\mathbf{j}(q,t), \quad -\varepsilon\mathbf{n}_q \cdot \mathbf{E}(q,t) = \rho(q,t). \tag{7.16}$$

The surface conservation of charge differs from (6.82) and now takes the form

$$\operatorname{div}_0 \mathbf{j}(q,t) + \frac{\partial\rho(q,t)}{\partial t} = 0\ . \tag{7.17}$$

The total field can now be expressed as

$$\mathbf{E}(P,t) = \mathbf{E}_0(P,t) + \int_S \Big[\mu \frac{\partial}{\partial t}\mathbf{j}(q,T)\Psi(P,q) - \Big\{\rho(q,T) + \frac{\partial}{\partial t}\frac{\rho(q,T)}{v\Psi}\Big\}\frac{\operatorname{grad}_q\Psi(P,q)}{\varepsilon}\Big]\,\mathrm{d}S_q\ , \tag{7.18}$$

$$\mathbf{H}(P,t) = \mathbf{H}_0(P,t) - \int_S \{\mathbf{j}(q,T) + \frac{\partial}{\partial t}\mathbf{j}(q,T)/v\Psi\} \wedge \operatorname{grad}_q\Psi(P,q)\,\mathrm{d}S_q \tag{7.19}$$

when P is in S_+ (Fig. 6.15). The change of sign in (7.18) and (7.19) as compared with (7.2) and (7.3) is caused by the point of observation now being outside S.

The requirement that the tangential complement of the electric intensity be zero gives, when applied to (7.18), the EFIE

$$\mathbf{n}_p \wedge \int_S \left[\mu \frac{\partial}{\partial t} \mathbf{j}(q,T)\Psi(p,q) - \left\{\rho(q,T) + \frac{\partial}{\partial t}\frac{\rho(q,T)}{v\Psi}\right\}\frac{1}{\varepsilon}\operatorname{grad}_q\Psi(p,q)\right]\mathrm{d}S_q$$

$$= -\mathbf{n}_p \wedge \mathbf{E}_0(p,t) \ . \tag{7.20}$$

It should be remarked that the difference in sign between (7.20) and (6.97) is caused by the boundary condition in Chapter 6 being (6.74) $\mathbf{n} \wedge \mathbf{E} = \mathbf{n} \wedge \mathrm{E}_0$ which is a reversal of the sign of $\mathbf{E}_0$ as compared with here.

Annexed to the EFIE is (7.17). It would enable us to replace $\partial\rho(q,T)/\partial t$ by $-\operatorname{div}_0 \mathbf{j}(q,T)$ but one ρ would still remain in (7.20). This awkwardness can be overcome by taking a time derivative of (7.20) after which ρ can be completely eliminated by means of (7.17).

The MFIE may be constructed in the same way as in §6.14. Note that the second integrand of (7.19) is not discontinuous in crossing S because the Ψ in the denominator renders it less singular than the first. We obtain

$$-\frac{1}{2}\mathbf{j}(p,t) + \mathbf{n}_p \wedge \int_S \{\mathbf{j}(q,T) + \frac{\partial}{\partial t}\,\mathbf{j}(q,T)/v\Psi\} \wedge \operatorname{grad}_q\Psi(p,q)\mathrm{d}S_q$$

$$= \mathbf{n}_p \wedge \mathbf{H}_0(p,t) \ . \tag{7.21}$$

Other integral equations are rarely considered because it is felt that the time-marching procedure already described for the thin wire prevents the occurrence of the uniqueness problems which are so obnoxious in the frequency domain.

The MFIE has two big advantages over the EFIE. Firstly, it is an integral equation of the second kind. Secondly, it does not have to be complemented by the continuity equation (7.17). Admittedly, this is evaded by the EFIE in time-derivative form but at the cost of second-order derivatives which will tend to engender less accuracy in any interpolation process than for the MFIE. The derivatives might, naturally, be taken outside the integral, thereby reducing the singularity of the integrand, and commuted to finite differences, but whether any improvement in accuracy will follow is dubious.

The MFIE further brings out with great clarity the step-by-step operation in time. Since the integral is a principal value the contribution of the self-patch (where there is virtually no retardation in time) may be neglected in the first place (cf. §6.19). Then (7.21) tells us that **j** is $-2\mathbf{n} \wedge \mathbf{H}_0$ corrected by an integral which depends only upon the past of **j**. Thus the MFIE can be regarded as an updating formula for **j** starting from known initial values. The mechanism is generated as follows. At the instant the incident wave strikes the body, put $\mathbf{j} = \mathbf{0}$ in the integral so $\mathbf{j} = -2\mathbf{n} \wedge \mathrm{H}_0$ at the points of impact. At the next time step, replace **j** by **0** or $-2\mathbf{n} \wedge \mathbf{H}_0$ depending on whether the point of integration was not or was one of the points of impact previously. Then (7.21) supplies **j** at the new time step. The procedure can obviously be repeated and **j** is obtained at all times.

Should the wavefront of the incident field be such that a shadow region should be expected on the obstacle, there can be no disturbance in it until the creeping waves (§8.21) arrive. Up to this event the field in such a region should remain small. This observation provides one internal check at any rate on how well the updating formula is operating.

Exercise

7. Derive integral equations analogous to the EFIE and MFIE for two-dimensional scattering.

7.5. Numerical matters

Nothing much needs to be added to the discussion of interpolation in §7.3 other than remarking the further complication of an extra space variable. Notwithstanding, a rather more detailed examination of the choice of patch size and step length must be undertaken.

From a theoretical point of view, it would be ideal if the incident wave were a Dirac δ function because the response to any incident field could then be synthesized by a simple convolution. However, this is impracticable in numerical work; instead it is more common to work with a *Gaussian pulse* of the type

$$\mathbf{E}_0, \mathbf{H}_0 \propto \frac{b}{\pi^{\frac{1}{2}}} \exp\left\{- b^2\left(t - \frac{x}{v}\right)^2\right\} . \tag{7.22}$$

The right-hand side of (7.22) approaches the δ function as $b \to \infty$ and is tiny enough at sufficiently negative times to be regarded as negligible.

The Fourier transform with respect to time of $(b/\pi^{\frac{1}{2}})\exp(-b^2t^2)$ is effectively $\exp(-\omega^2/4b^2)$. Roughly speaking, therefore, the solution to the pulse problem for (7.22) will supply solutions in the frequency domain for all ω less than $2b$. Consequently, the larger b is, the more harmonic frequencies will be covered. Unfortunately, the effective width of the pulse in time is $2/b$ and less than five time intervals over this width will scarcely manage to attain the temporal resolution of the pulse. This means that the time interval should not exceed $2/5b$. The inequality to allow for retardation will be met if the side of a patch is about $2v/5b$. Any size much greater will, in any case, not resolve the spatial width of the pulse which is $2v/b$. It follows from these considerations that the wavelengths in the frequency domain that will be adequately treated will not be less than four or five times the side of a surface patch.

As an illustration suppose that b is $10^{10}\,\mathrm{s}^{-1}$, which is a typical sort of value. The spatial width of the pulse is then about 0.1 m and its upper cut-off frequency about 10^9 Hz. The above considerations suggest a time step of around 10^{-11} s and a patch side of 1 cm approximately. Thus, for a body 1 m across, roughly 10^4 patches would be required. Unless the body were highly elongated or wire shaped, a wider pulse, with consequent improvement in step length and patch size, would probably be adequate. Although the guide lines are likely to be conservative, they indicate that there is a trade off between the choice of b and the demands on the computer.

There is another aspect. If the width of the scatterer is l_s, the range of wave numbers that can be handled should be within $\frac{1}{2}kl_s \leqslant 2\pi l_s/l_p$ where l_p is the spatial width of the incoming pulse. The right-hand side of the inequality is 2π for a pulse the width of the scatterer. A reduction in pulse width with the intention of extending the range of frequency response and of obtaining more accurate information on the scattering carries with it the concomitant of a proportionate increase in the number of time steps while the number of surface elements augments by a square law. Narrowing the pulse soon exposes the fact that high-frequency information has to be bartered for rapidly rising computational cost.

The Gaussian pulse is never actually zero away from infinity and this presents another slight difficulty. So long as b is reasonably large its rise is sufficiently quick to take all induced currents and charges as zero until the incident field has reached 10^{-5} of its maximum value. No criterion can be given for when the computation should cease since this depends on the response of the obstacle, which is frequently oscillatory.

Without an error analysis the only way of checking whether a program is performing adequately is by comparison with analytical results for one of the few shapes for which these are available or with experiment. The standard test-bed is the sphere and some results (Bennett and Weeks 1968) for the reaction to a Gaussian input are shown in Fig. 7.3. These are based on the MFIE. The corresponding backscatter in the frequency domain is compared in Fig. 7.4 with calculations from the exact solution in a Mie series. It is evident that the agreement is good for wave numbers which do not exceed $5/a$. On the other hand, if the Mie series is converted to the time domain and the comparison made there Fig. 7.5 is obtained. Again the agreement is highly satisfactory.

Nevertheless, a word of caution is in order. The sphere with its high degree of symmetry is not a good pointer to what will happen to asymmetric bodies, especially if they have edges. Getting the answers right for a sphere is a necessary

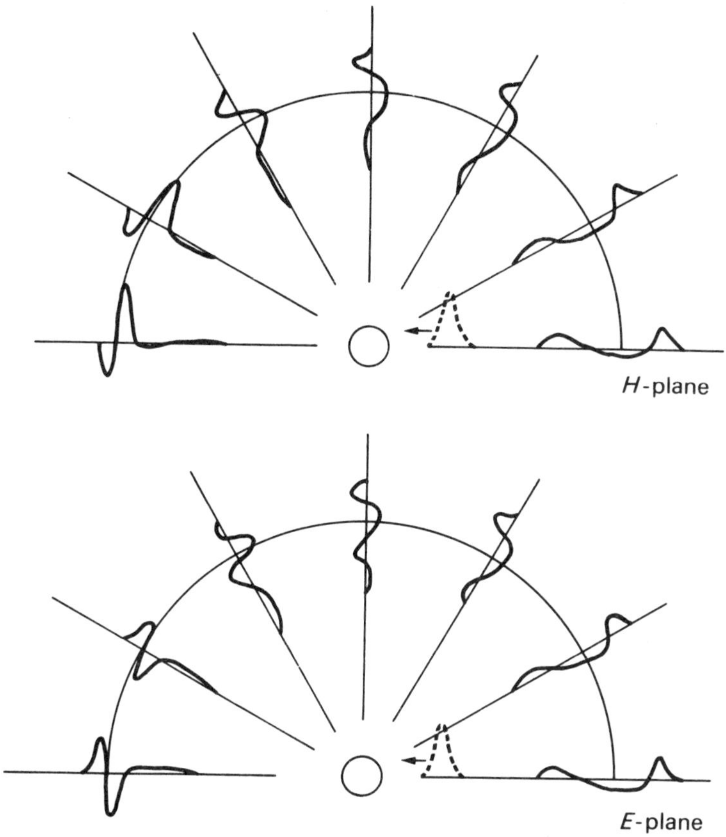

FIG. 7.3. The impulse response of a sphere.

but not sufficient condition for the validation of a numerical procedure.

Exercises

8. Formulate time-dependent integral equations for two-dimensional scatterers.

9. Obtain the integral equation in the time domain analogous to those of §6.26 for dielectric obstacles.

10. Compute the scattering from a sphere and compare your calculations with Figs. 7.3–7.5.

11. A cone of 15° half-angle is terminated smoothly by a portion of spherical surface. Find the impulse response under an incident Gaussian pulse whose spatial width is the same as the length of the scatterer when

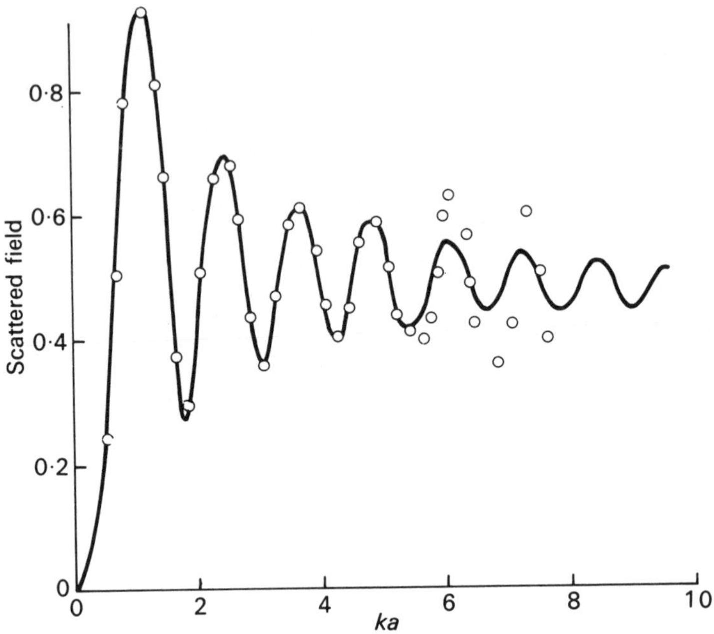

FIG. 7.4. Backscatter in the frequency domain of a sphere: curve, Mie series; points, time-domain calculation.

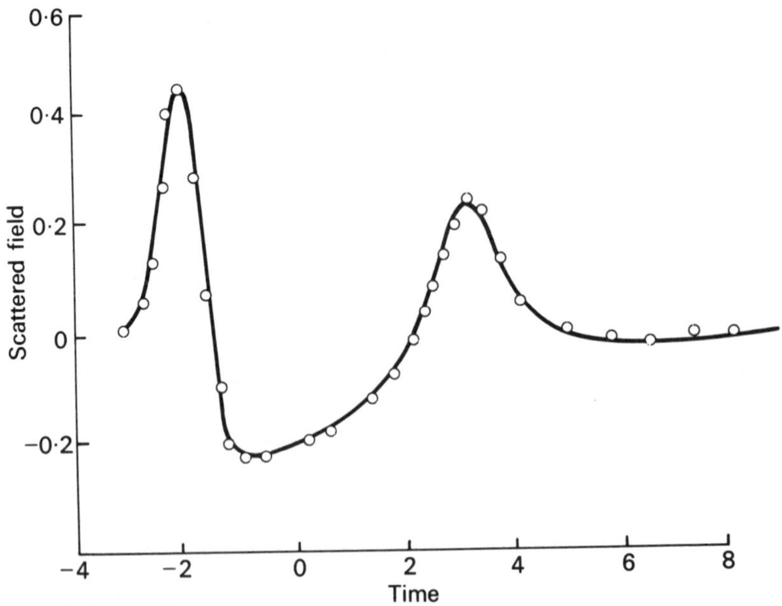

FIG. 7.5. Backscattering impulse response of a sphere: curve, Mie series converted to time domain; points, direct time-domain calculation.

the direction of the incoming radiation is (*a*) along the axis of the cone towards the cone, (*b*) along the axis of the cone towards the sphere, and (*c*) perpendicular to the axis of the cone. Use the MFIE and plot your results in the plane containing the axis of the cone and the direction of incidence in the form of Fig. 7.3.

Compare the current induced on the obstacle with the position of the incident pulse at various times as it moves over the scatterer.

12. Repeat Exercise 11 for the case when the obstacle is a length of circular cylinder capped by a hemisphere at each end.

13. In Exercises 11 and 12 make a Fourier transform from the time domain and compare your results with those derived directly in the frequency domain.

14. Carry out the process of Exercise 12 but in the opposite direction, i.e. make a Fourier transform of the harmonic results and compare in the time domain.

15. Replace the structures in Exercises 11 and 12 by wire-grid models and see if these models give appreciably different results from those already obtained.

7.6. The harmonic approach versus the impulse response

The Fourier transform enables one to transfer from the time to the frequency domain and vice versa, but the choice of the domain in which to undertake the computations calls for delicate judgement. Naturally, if the input is a pulse and the immediate transient is wanted the time domain is the correct selection, but other cases are much less clear-cut.

In harmonic waves a matrix inverse is determined at a given frequency. The inverse is independent of the incident field so that, once it has been found, a change of incident excitation can be dealt with in a simple manner. However, as soon as the frequency is changed, the matrix inverse has to be recomputed.

In contrast, the pulse solution combined with a fast Fourier transform provides results over a wide frequency band but only for a single direction of excitation. Whenever the position of the source is altered the entire computation has to be repeated. Running counter to this disadvantage is the tendency for the pressure on computer storage to be lighter by a factor of 10 or more in the time domain than in the frequency domain, though the storage grows with the time interval being studied.

The amount of computer time involved is a further consideration but the evidence is inconclusive. In some numerical experiments (Miller 1972) where the complete response was obtained the frequency domain required about five times as much time as the pulse approach for wires. Set against this is the behaviour for surfaces where the harmonic solution was faster by a factor of five than the time domain. Another observation was that the time domain was always worse than the frequency domain in monostatic calculations for both wires and surfaces. For the bistatic response the time domain was more efficient than the harmonic approach for wires, but the reverse was true for surfaces.

In these confusing circumstances only the following broad guidance can be offered.

(i) For early transients and short times use the time domain.

(ii) If many incident fields at few frequencies are to be investigated employ the harmonic approach.

(iii) If few incident fields at many frequencies are needed work via incident pulses.

7.7. The Laplace transform

A classic method in analysis for tackling transients is the Laplace transform. Normally, the inversion back to the time domain is not dissimilar to the Fourier transform already mentioned in the preceding sections. However, the conjecture that the time response of a body might be represented as a series of exponentials led to the idea that a body might be characterized by the singularities in the *complex* frequency domain of the inverse Laplace transform (Baum 1971). Since these singularities are often simple poles and depend solely on the geometry of the obstacle they need be found only once. Their storage requirements are minimal and the computation of even broadband frequency response is relatively straightforward. As soon as the relation between the singularities and the radar pulse signature is known, either can be characterized by the other.

Supporting theory for the idea has been forthcoming (Marin and Latham 1972, Tesche 1973, Marin 1973, 1974) and it is the aim of this section to explain the foundations.

The Laplace transform **e** of **E** is defined by

$$\mathbf{e}(s) = \int_{-\infty}^{\infty} \mathbf{E}(t) e^{-st} dt . \tag{7.23}$$

So long as **E** is non-zero only for t greater than some finite constant and is bounded by an exponential as t becomes infinite, the Laplace transform exists as a regular function of the complex variable s provided that $\mathcal{R}s$ is greater than some constant σ_0. Taking σ_0 to be positive does not incur any loss of generality. Then, for $\mathcal{R}s > \sigma_0$, Maxwell's equations in a time-independent isotropic medium go over to

$$\text{curl } \mathbf{e} + s\mu\mathbf{h} = \mathbf{0}, \quad \text{curl } \mathbf{h} - s\varepsilon\mathbf{e} = \mathbf{0} \tag{7.24}$$

when there are no sources.

If a time-dependent solution is sought which is outgoing at infinity and is such that $\mathbf{n} \wedge \mathbf{E} = \mathbf{n} \wedge \mathbf{E}_0$ on the surface S, then (7.24) must be solved in such a way that $\mathbf{n} \wedge \mathbf{e} = \mathbf{n} \wedge \mathbf{e}_0$ on S. The condition at infinity is that **e** and **h** must decay at least exponentially there since $\sigma_0 > 0$. The problem is then very similar to that of §6.14 and the MFIE can be formulated as

$$-\tfrac{1}{2}\mathbf{j}(p) + \mathbf{n}_p \wedge \int_S \mathbf{j}(q) \wedge \text{grad}_q \phi(p,q) \mathrm{d}S_q =$$

$$\mathbf{n}_p \wedge \int_S \{(\mathbf{n}_q \cdot \mathbf{h}_0) \text{grad}_q \phi(p,q) + s\varepsilon \mathbf{n}_q \wedge \mathbf{e}_0 \phi(p,q)\} \mathrm{d}S_q \tag{7.25}$$

where

$$\phi(\mathbf{x},\mathbf{y}) = \frac{\exp\{-s(\mu\varepsilon)^{\frac{1}{2}}|\mathbf{x}-\mathbf{y}|\}}{4\pi|\mathbf{x}-\mathbf{y}|} . \tag{7.26}$$

The quantity $\mathbf{n} \cdot \mathbf{h}_0$ can, of course, be expressed in terms of $\mathbf{n} \wedge \mathbf{e}_0$. The integral equation (7.25) holds for $\mathcal{R}s > \sigma_0$ and it will be seen presently that this restriction guarantees a unique solution.

Despite this asset we now wish to drop the restriction and consider the solution of (7.25) for any complex value of s. The proof of §6.16 that the integral operator is compact may be carried over and may also be adapted to show that the kernel associated with the square of the operator is square integrable. This observation enables an explicit solution of the integral equation to be written down. While this formula is not of great value in actual computation it is of tremendous theoretical importance and has some consequences which are of practical significance.

To simplify the writing matrix notation will be adopted and the integral equation cast into the form

$$\mathbf{j}(x) - \int C(x,y)\mathbf{j}(y)\mathrm{d}y = \mathbf{g}(x) \tag{7.27}$$

where $\mathbf{g}$ is a specified vector and C is a known matrix. Explicit forms for C and $\mathbf{g}$ can be deduced from (7.25) but these are not necessary to the current discussion. In order to comply with the known properties of (7.25) it is assumed that the integral operator in (7.27) is compact and that $\int C(x,t)C(t,y)\mathrm{d}t$ is square integrable.

The first iteration of (7.27) is

$$\mathbf{j}(x) - \int H(x,y)\mathbf{j}(y)\mathrm{d}y = \mathbf{g}(x) + \int C(x,y)\mathbf{g}(y)\mathrm{d}y \tag{7.28}$$

with $H(x,y) = \int C(x,t)C(t,y)\mathrm{d}t$.

An integral equation of the type (7.28) can always be converted to a single scalar equation (Goursat 1942). For our purposes it will be sufficient to demonstrate the technique in two-dimensional space, denoting components by subscripts. Suppose (7.28) holds on the set S; then obviously a constant vector x_0 can be chosen such that the set S_0 formed by $x + x_0$, $x \in S$, is disjoint from S. Define the functions j' and K on the union of S and S_0 by

$$j'(x) = j_1(x) \quad (x \in S)$$

$$= j_2(x - x_0) \quad (x \in S_0),$$

$$K(x,y) = H_{11}(x,y) \quad (x \in S,\ y \in S)$$

$$= H_{12}(x, y - x_0) \quad (x \in S,\ y \in S_0)$$

$$= H_{21}(x - x_0, y) \quad (x \in S_0, y \in S)$$

$$= H_{22}(x - x_0,\ y - y_0) \quad (x, y \in S_0)\ .$$

Similarly, if the right-hand side of (7.28) be signified by **G**, specify G' by

$$G'(x) = G_1(x) \quad (x \in S)$$

$$= G_2(x - x_0) \quad (x \in S_0)\ .$$

Then (7.28) transforms to a scalar integral equation on $S \cup S_0$, namely

$$j'(x) - \int K(x,y) j'(y) \mathrm{d}y = G'(x)$$

the integration now being over $S \cup S_0$. This scalar equation is exactly equivalent to (7.28) because S and S_0 are disjoint. Since K is square integrable, standard Fredholm theory (Cochran 1972) allows one to say that, when the solution of (7.28) is unique,

$$j'(x) = G'(x) + \int \Gamma(x,y) G'(y) \mathrm{d}y \tag{7.29}$$

where the *resolvent* Γ is given by

$$\Gamma(x,y) = D(x,y)/D \ .$$

The number D is defined by $D = \Sigma_{n=0}^{\infty} d_n$ where $d_0 = 1$ and

$$d_n = \frac{(-)^n}{n!}\int\cdots\int \begin{vmatrix} 0 & K(t_1,t_2) & \cdots & K(t_1,t_n) \\ K(t_2,t_1) & 0 & \cdots & K(t_2,t_n) \\ \cdots & \cdots & \cdots & \cdots \\ K(t_n,t_1) & K(t_n,t_2) & \cdots & 0 \end{vmatrix} dt_1\ldots dt_n$$

for $n \geqslant 1$. The function $D(x,y)$ has the structure

$$D(x,y) = \sum_{n=0}^{\infty} D_n(x,y)$$

where $D_0(x,y) = K(x,y)$ while, for $n \geqslant 1$,

$$D_n(x,y) = d_n K(x,y) + \int D_{n-1}(x,t)K(t,y)\mathrm{d}t$$

$$= d_n K(x,y) + \int K(x,t)D_{n-1}(t,y)\mathrm{d}t \ .$$

In addition, the resolvent satisfies

$$\Gamma(x,y) = K(x,y) + \int K(x,t)\Gamma(t,y)\mathrm{d}t \qquad (7.30)$$

$$= K(x,y) + \int \Gamma(x,t)K(t,y)\mathrm{d}t \ . \qquad (7.31)$$

Any solution of (7.27) satisfies (7.28). Therefore, when the solution of (7.28) is unique, (7.27) has only a single solution and it is given by (7.29). This does not mean that, if the solution of (7.27) is known to be unique but no other information about (7.28) is available, the desired solution is (7.29), for (7.28) may not have a unique solution. Because of the assumption about (7.27) any solution of the homogeneous form of (7.28) must satisfy

$$\mathbf{j}(x) + \int C(x,y)\mathbf{j}(y)\mathrm{d}y = \mathbf{0}. \tag{7.32}$$

It is only when there is no solution of (7.32) other than the trivial one and the solution of (7.27) is unique that it can be asserted that (7.28) has uniqueness with (7.29) being the sought solution. Verification that in the scattering problem the uniqueness of (7.27) enforces $\mathbf{j} \equiv \mathbf{0}$ as the sole solution of (7.32) will come later.

The kernel K which originates from (7.25) contains the parameter s, and the behaviour of the resolvent as s varies in the complex plane is of interest. Results from the customary theory of integral equations cannot be taken over straight away because the parametric dependence in the conventional theory is through multiplication by a complex constant whereas C, H, and K involve s via an exponential factor which also includes the variable of integration.

H is a matrix whose elements are regular functions of s in the complex s plane and so K has the same regularity. Therefore, d_n and $D_n(x,y)$ depend upon s in a regular manner. Hence, D and $D(x,y)$, being composed of uniformly convergent series of regular functions, are regular. The implication is that D and $D(x,y)$ are entire functions in the s plane so that the only possible singularities of Γ are poles at zeros of D.

It will first be shown that the solution of (7.28), or equivalently (7.27) in the scattering problem, is not unique at $s = s_n$ if and only if $D = 0$ at $s = s_n$. In the course of the proof it will be helpful to indicate the dependence of a quantity on s explicitly; for example, $d_n(s)$ will be written in place of d_n on such occasions. Now, it is readily checked from the defining formulae that

$$(n+1)d_{n+1} = \int \{d_n K(x,x) - D_n(x,x)\}\mathrm{d}x$$

with which the special formula $d_1 = 0$ is in harmony. Hence

$$\int \{\Gamma(x,x) - K(x,x)\}\,\mathrm{d}x = -\sum_{n=0}^{\infty} \frac{(n+1)d_{n+1}}{D}. \tag{7.33}$$

Consider now the series $\Sigma_{r=0}^{\infty} d_r(s_n)(s/s_n)^r$. It is certainly convergent for $|s/s_n| \leqslant 1$ and approaches D as $s \to s_n$ from $|s/s_n| < 1$. Therefore, it behaves like $(s-s_n)^m$ $(m \geqslant 1)$ where m is the order of the zero at s_n. The derivative with respect to s commands like properties and so has the form $m(s-s_n)^{m-1}$ near s_n. Consequently, the right-hand side of (7.33) has a simple pole at $s = s_n$. However, K has no singularity in the s plane. Therefore Γ cannot be regular at $s = s_n$ and so (7.29) is unable to provide a unique solution.

Conversely, suppose that the homogeneous integral equation has a non-trivial solution at $s = s_n$ but Γ is regular there. By the second part of Theorem 6.15 there is a non-trivial solution j_1 of

$$j_1(x) = \int K^*(y,x)j_1(y)\,\mathrm{d}y$$

at $s = s_n$. Let s be near s_n and consider the integral equation

$$j(x) - \int K(x,y,s)j(y) = j_1(x) \tag{7.34}$$

the dependence of the kernel on s being explicitly noted. Eqn (7.34) has a unique solution which can be expressed in the form (7.29); if this form be substituted in (7.34) we have the identity

$$j_1(x) + \int \Gamma(x,y,s)j_1(y)\mathrm{d}y - \int K(x,y,s)\{j_1(y) +$$

$$+ \int \Gamma(y,t,s)j_1(t)\mathrm{d}t\}\mathrm{d}y = j_1(x) .$$

Because of the assumed regularity of Γ, the limit can be taken as $s \to s_n$ and then $j_1(x) + \int\Gamma(x,y,s_n)j_1(y)\mathrm{d}y$ is a solution of (7.34) with $s = s_n$. However, that is contrary to Theorem 6.15 which declares that the right-hand side of (7.34) must be orthogonal to j_1 when $s = s_n$ for a solution to exist. Hence Γ cannot be regular at $s = s_n$ and the converse has been proved.

Thus the zeros of D or, equivalently, the poles of Γ identify completely the values of s for which (7.27) is non-unique. The zeros of D will be isolated points unless D is identically zero for all s. If D were identically zero for all s there would not be a unique solution for any s in contradiction to what has been proved already in §6.16 for the static kernel. The conclusion to be drawn is that D possesses only isolated zeros.

The general theory of §6.15 tells us that at a zero of D there is only a finite number of linearly independent solutions of the homogeneous equations, i.e. for $s = s_n$ the independent eigenelements are finite in number. There is a connection between the eigenelements and the order of the pole in Γ. Let m be the order of the pole so that when s is near s_n

$$\Gamma(x,y) = \frac{u_m(x,y)}{(s-s_n)^m} + \frac{u_{m-1}(x,y)}{(s-s_n)^{m-1}} + \ldots$$

where u_m, u_{m-1}, ... are independent of s. Substitution in (7.30) gives

$$\frac{u_m(x,y)}{(s-s_n)^m} + \frac{u_{m-1}(x,y)}{(s-s_n)^{m-1}} + \ldots = K(x,y,s) + \int K(x,t,s)\left\{\frac{u_m(t,y)}{(s-s_n)^m} + \ldots\right\}\mathrm{d}t$$

where the dependence of K on s has been explicitly indicated.

Equating the dominant terms on either side we obtain

$$u_m(x,y) = \int K(x,t,s_n)u_m(t,y)\mathrm{d}t .$$

Thus, for fixed y, u_m must be an eigenelement of K at $s = s_n$. The next-order terms give

$$u_{m-1}(x,y) = \int \{K(x,t,s_n)u_{m-1}(t,y) + + u_m(t,y)\partial K(x,t,s_n)/\partial s_n\}\mathrm{d}t \tag{7.34a}$$

if $m > 1$. If $m = 1$, $K(x,y,s_n)$ must be added to the right-hand side.

According to Theorem 6.15, u_{m-1} can exist for $m > 1$ only if the term involving u_m in (7.34a) (regarded as a function of x) is orthogonal to all the solutions of the adjoint equation. In other words, if v is any solution of

$$v(x) = \int K^*(t,x,s_n)v(t)\mathrm{d}t \tag{7.35}$$

it is necessary for $m > 1$ that

$$\int v^*(x) \int u_m(t,y) \frac{\partial K}{\partial s_n} (x,t,s_n)\mathrm{d}t\mathrm{d}x = 0 . \tag{7.36}$$

Conversely, if (7.36) holds and $m = 1$, then u_0 can exist only if

$$\int v^*(x)K(x,y,s_n)\mathrm{d}x = 0$$

which implies from (7.35) that $v \equiv 0$. However, this contradicts what has been proved above, namely that a pole of Γ entails non-trivial solutions of the homogeneous integral equation and hence of the adjoint. It follows that $m > 1$. Thus (7.36) is a necessary and sufficient condition for $m > 1$.

If all the poles of the resolvent are simple the kernel is said to be *semi-simple*. Clearly, (7.36) must fail for a semi-simple kernel. Since u_m itself is an eigenelement of K, (7.36) can be cast into a slightly simpler form and we can state the following:

A kernel is semi-simple if and only if there are eigenelements u, v of the integral equation and its adjoint respectively such that

$$\int v^*(x) \int u(t) \frac{\partial}{\partial s_n} K(x,t,s_n)\mathrm{d}t\mathrm{d}x \neq 0 \tag{7.37}$$

for every s_n for which the kernel possesses eigenelements.

Since the resolvent of the adjoint is the adjoint of the resolvent of the original equation, we could interchange the roles of the adjoint and original in determining whether a kernel is semi-simple. However, the criterion takes exactly the same form as (7.37).

We may note in passing that when the dependence of K on s is linear, i.e. $K(x,y,s) = sK_0(x,y)$, as in the classical eigenvalue problem though not in the scattering integral equation, $\partial K/\partial s = K_0$. Then (7.37) reduces to

$$\int v^*(x)u(x)\mathrm{d}x \neq 0. \tag{7.38}$$

Hence the existence of eigenelements satisfying (7.38) guarantees that a kernel is semi-simple in the classical eigenvalue problem (Lalesco 1912). If the kernel is not semi-simple, the application of the operator $I - K_0$ to (7.34*a*) reveals that

$$(I-K_0)^2 u_{m-1} = 0.$$

A solution of such an equation is often known as a *generalized eigenelement*. If (7.38) is valid there can be no generalized eigenelements. To put it another way, a semi-simple kernel does not possess generalized eigenelements in the classical eigenvalue problem.

Further properties of u_m may be derived by working with the adjoint. It is then evident that

$$u_m(x,y) = \sum_r a_r u_r(x) v_r^*(y)$$

where u_r and v_r are eigenelements of the original and adjoint equations respectively. Hence, for s near s_n when the kernel is semi-simple

$$j'(x) = G'(x) + (s-s_n)^{-1} \sum_r a_r u_r(x) \int G'(y) v_r^*(y) \mathrm{d}y \; . \qquad (7.39)$$

However,

$$\int G'(x) v_r^*(x) \mathrm{d}x = \int \{j'(x) - \int K(x,y,s) j'(y) \mathrm{d}y\} v_r^*(x) \mathrm{d}x$$

$$= \int \left[\int \{K(x,y,s_n) - K(x,y,s)\} j'(y) \mathrm{d}y \right] v_r^*(x) \mathrm{d}x$$

$$\approx - (s-s_n) \int \int \frac{\partial}{\partial s_n} K(x,y,s_n) j'(y) \mathrm{d}y \, v_r^*(x) \mathrm{d}x \; . \qquad (7.40)$$

The $u_r(x)$ can be arranged to be orthonormal. When this has been done the v_r can be selected so that they are orthogonal to $\int (\partial/\partial s_n) K(x,y,s_n) u_p(y) \mathrm{d}y$ unless $p = r$ when a non-zero result ensues because the kernel is semi-simple. Then, if (7.40) be substituted in (7.39) and the coefficients of u_r compared, we discover that consistency demands that

$$a_r = -1 / \int\int \frac{\partial}{\partial s_n} K(x,y,s_n) u_r(y) \mathrm{d}y v_r^*(x) \mathrm{d}x \, . \qquad (7.41)$$

A similar, but more complicated, route may be followed to determine the coefficient when the kernel is not semi-simple.

7.8. The location of the poles

Having described the general theory we want to show now that (7.27) and (7.32) go together as far as uniqueness is concerned in the scattering context, so that *the only singularities of the resolvent are poles*. In doing so some information about the location of the poles will appear.

The homogeneous form of (7.25) states that $\mathbf{n} \wedge \mathbf{h}_- = \mathbf{0}$ on S. By taking the complex conjugate we discover that if s_n is a pole so is s_n^*. It will therefore be sufficient to limit consideration to $\mathcal{I}s \geqslant 0$. Now

$$\mathcal{R} \int_S (\mathbf{n}\wedge\mathbf{h}.\mathbf{e}^*)_- \mathrm{d}S = (s+s^*) \int_{S_-} (\varepsilon|\mathbf{e}|^2 + \mu|\mathbf{h}|^2) \mathrm{d}\mathbf{x} \, .$$

Therefore, the left-hand side can be zero only if either $\mathcal{R}s = 0$ or the electromagnetic field is identically zero in S_-. In the latter case $\mathbf{n} \wedge \mathbf{e}_- = \mathbf{0}$ on S and hence $\mathbf{n} \wedge \mathbf{e}_+ = \mathbf{0}$ since the tangential components of the representation of $\mathbf{e}$ in terms of $\mathbf{j}$ are continuous across S. If $\mathcal{R}s \geqslant 0$ the field will be radiating at infinity ($\mathcal{R}s = 0$) or decaying exponentially ($\mathcal{R}s > 0$). In either case the standard exterior uniqueness theorem obliges the electromagnetic field to be identically zero in the exterior S_+. Therefore $\mathbf{n} \wedge \mathbf{h}_+ = \mathbf{0}$ and hence $\mathbf{j} = \mathbf{0}$.

Hence it has been demonstrated that (7.25) is non-unique only when either the field produced by $\mathbf{j}$ in S_- is non-zero and $\mathcal{R}s = 0$ or the field in S_- is identically zero and $\mathcal{R}s < 0$. The former corresponds to interior modes of magnetic resonance whereas the latter are exterior modes of electric resonance since $\mathbf{n} \wedge \mathbf{e}_+ = \mathbf{0}$. On account of $\mathcal{R}s < 0$ the exterior modes grow at infinity rather than diminish. Both types of mode exist for the sphere and so must be allowed for in scattering by general obstacles.

Consider now the electromagnetic field in which

$$\left.\begin{aligned} \mathbf{e}(P) &= \int_S \mathbf{j}' \wedge \operatorname{grad}_q \phi(P,q)\,dS_q \,, \\ \mathbf{h}(P) &= \int_S \{\varepsilon s \mathbf{j}' \phi(P,q) + \frac{1}{s\mu} \operatorname{div}_0 \mathbf{j}' \operatorname{grad}_q \phi(P,q)\} dS_q \,. \end{aligned}\right\} \quad (7.42)$$

If $\mathbf{j}'$ is the tangential component of the electric field in an interior magnetic mode $\mathbf{e}$ vanishes identically in S_+ by the discussion after (6.106). Then $\mathbf{n} \wedge \mathbf{e}_+ = 0$ and

$$\tfrac{1}{2}\mathbf{j}' + \mathbf{n}_p \wedge \int_S \mathbf{j}' \wedge \operatorname{grad}_q \phi(p,q)\,dS_q = \mathbf{0} \,. \quad (7.43)$$

If (7.42) is used as the representation of the exterior electric mode then (7.43) also results on account of the boundary condition. Thus with each solution of the homogeneous form of (7.25) can be associated a solution of (7.43), the $\mathbf{j}'$ being the tangential component of the interior electric field of the magnetic mode ($\mathcal{R}s=0$) and of (7.42) ($\mathcal{R}s<0$). We wish to show that the reverse is also true so that (7.25) and (7.43) (and thereby (7.27) and (7.32)) stand or fall together in the matter of uniqueness.

The dimension of the space of solutions of (7.43) is the same as that of its adjoint which, as has been seen in §6.14, has elements $\mathbf{n} \wedge \mathbf{j}^*$. Therefore (7.25) and (7.43) have the same dimension. Hence the solutions of the homogeneous form of (7.25) are not one-to-one with those of (7.43) only if there is one such that the electric field produced by it satisfies $\mathbf{n} \wedge \mathbf{e}_- = \mathbf{0}$ when $\mathcal{R}s = 0$. However, $\mathbf{n} \wedge \mathbf{e}_+ = \mathbf{0}$ with the consequence $\mathbf{n} \wedge \mathbf{h}_+ = \mathbf{0}$ from which follows $\mathbf{j} = \mathbf{0}$. Hence the relationship must be one-to-one and the case $\mathcal{R}s = 0$ has been dealt with.

When $\mathcal{R}s < 0$ let the field be defined by (7.42) with $\mathbf{j}'$ satisfying (7.43). Then, from (6.86)

$$\mathbf{h}(P_i) = -\int_S \left\{(\mathbf{n}\wedge\mathbf{h})\wedge\mathrm{grad}_q\phi(P_i,q)+s\varepsilon\mathbf{n}\wedge\mathbf{e}\_\phi(P_i,q) + \right.$$

$$\left. + \frac{\mathrm{div}_0(\mathbf{n}\wedge\mathbf{e}\_)\,\mathrm{grad}_q\phi(P_i,q)}{s\mu}\right\}\mathrm{d}S_q \qquad (7.44)$$

for P_i in $S\_$. However, by virtue of (7.43), $-\mathbf{n}\wedge\mathbf{e}\_ = \mathbf{j}'$. Comparison of (7.42) and (7.44) indicates that when $s = s_n$

$$\int_S (\mathbf{n}\wedge\mathbf{h})\wedge\mathrm{grad}_q\phi(P_i,q)\mathrm{d}S_q = \mathbf{0} \qquad (P_i \in S\_) \;. \qquad (7.45)$$

Allowing P_i to tend to a point of S we deduce that $\mathbf{n}\wedge\mathbf{h}$ satisfies (7.25). Remark that if $\mathbf{n}\wedge\mathbf{h} = \mathbf{0}$ then $\mathbf{n}\wedge\mathbf{e}\_ = 0$ so that $\mathbf{j}' = \mathbf{0}$. Once again, the one-to-one correspondence has been established.

The kernel of (7.25) is semi-simple according to (7.37) if, for each s_n, there is a solution of the homogeneous version of (7.25) and a solution of (7.43) such that

$$\int_S \mathbf{j}'(p)\cdot\int_S \mathbf{j}(q)\wedge\mathrm{grad}_q\{-s_n(\mu\varepsilon)^{\frac{1}{2}}|\mathbf{x}_p-\mathbf{x}_q|\}\mathrm{d}S_q\mathrm{d}S_p \neq 0 \;. \quad (7.46)$$

Suppose that $\mathcal{R}s_n = 0$. Then, from the foregoing analysis, one $\mathbf{j}'$ can be associated with the electric intensity of the interior magnetic mode generated by $\mathbf{j}$. Therefore, (7.46) will be met for $\mathcal{R}s_n = 0$, if it can be shown that

$$\int_S \mathbf{n}\cdot\frac{\partial\mathbf{h}\_}{\partial s}\wedge\mathbf{e}\_\mathrm{d}S \neq 0 \;. \qquad (7.47)$$

This will be proved by assuming the contrary, i.e. it will be imagined that the left-hand side of (7.47) is zero, and demonstrating that the field must vanish identically. Since $\mathbf{n}\wedge\mathbf{h}\_ = \mathbf{0}$, the assumption gives

$$0 = \int_S \mathbf{n} \cdot \left\{\frac{\partial \mathbf{h}_-}{\partial s} \wedge \mathbf{e}_- - \mathbf{h}_- \wedge \frac{\partial \mathbf{e}_-}{\partial s}\right\} dS$$

$$= \int_{S_-} \left\{\mathbf{e}.\left(s\varepsilon \frac{\partial \mathbf{e}}{\partial s} + \varepsilon \mathbf{e}\right) + \frac{\partial \mathbf{h}}{\partial s} . s\mu \mathbf{h} - \frac{\partial \mathbf{e}}{\partial s} . s\varepsilon \mathbf{e} - \mathbf{h} . \left(s\mu \frac{\partial \mathbf{h}}{\partial s} + \mu \mathbf{h}\right)\right\} d\mathbf{x}$$

$$= \int_{S_-} (\varepsilon \mathbf{e} . \mathbf{e} - \mu \mathbf{h} . \mathbf{h}) d\mathbf{x} \tag{7.48}$$

from the divergence theorem and (7.24), together with the derivative with respect to s. The theory of §3.5 informs us that in an interior mode of oscillation $\mathbf{e}$ may be taken as purely real and $\mathbf{h}$ as purely imaginary. Hence (7.48) implies that $\mathbf{e}$ and $\mathbf{h}$ are identically zero. Thus the failure of (7.47) entails the disappearance of the electromagnetic field. Accordingly, (7.47) and thereby (7.46) must hold for a magnetic mode. In other words, all the poles of the resolvent which have $\mathscr{R}s_n = 0$ are simple.

The exterior electric mode is relevant when $\mathscr{R}s_n < 0$ and then the analogue of (7.47) is

$$\int_S \mathrm{n} \cdot \mathbf{h}_+ \wedge \frac{\partial \mathbf{e}_+}{\partial s} \, dS \neq 0 . \tag{7.49}$$

No proof has so far been forthcoming as to whether or not (7.49) is true for general obstacles. It is certainly valid for the sphere and one may conjecture that it has universal veracity.

From now on it will be assumed that the scattering kernel is semi-simple. Then advantage may be taken of (7.41). Let $\mathbf{j}_r$ be an orthonormal set of eigenelements of (7.25) when $s = s_n$ and select associated solutions $\mathbf{j}'_m$ of (7.43) (remember $\mathbf{n} \wedge \mathbf{j}'^*_m$ satisfies the adjoint of (7.25)) such that

$$(\mu\varepsilon)^{\frac{1}{2}} \int_S \mathbf{j}'_m(p) \cdot \int_S \mathbf{j}_r(q) \wedge \mathrm{grad}_q \exp\{-s_n(\mu\varepsilon)^{\frac{1}{2}}|\mathbf{x}_p - \mathbf{x}_q|\} dS_q dS_p = 0$$

if $m \neq r$ but is non-zero if $m = r$. More briefly this may be expressed in inner product notation as

$$\langle A\mathbf{j}_r, \mathbf{j}_m'^*\rangle = 0 \quad (r \neq m)$$

$$\neq 0 \quad (r = m)$$

with an obvious significance for the operator A. If the right-hand side of (7.25) be denoted by $\mathbf{j}_i$ the solution of (7.25) may be expressed as

$$\mathbf{j}(p) = -2\mathbf{j}_i(p) - \int_S \underline{\Gamma}(p,q) \cdot \mathbf{j}_i(q)\mathrm{d}S_q \tag{7.50}$$

where the resolvent $\underline{\Gamma}$ is a dyadic. The dyadic has only simple poles in the s plane and near a pole $s = s_n$

$$\underline{\Gamma}(p,q) = (s-s_n)^{-1} \sum_r \frac{\mathbf{j}_r(p)\{\mathbf{n}_q \wedge \mathbf{j}_r'(q)\}}{\langle A\mathbf{j}_r, \mathbf{j}_r'^*\rangle} \tag{7.51}$$

on account of (7.39) and (7.41). Since $\underline{\Gamma}$ is meromorphic a representation valid in the whole s plane could be constructed for $\underline{\Gamma}$ by Mittag–Leffler's theorem (Goursat 1942) but that is unnecessary for our purposes.

Note that the effect of (7.51) is to give a term in $\mathbf{j}_r$ via the coupling between the incident field and the interior tangential electric field accompanying the natural modes of oscillation. If $\mathbf{e}$ is the field of (7.42) and $\mathbf{j}_i = \mathbf{n} \wedge \mathbf{h}_i$, the coefficient is proportional to $\int_S \mathbf{n} \wedge \mathbf{e}_- \cdot \mathbf{h}_i \mathrm{d}S$. Now, if $\mathbf{h}_i$ has no sources inside S this is the same as $\int_S \mathbf{n} \wedge \mathbf{e}_i \cdot \mathbf{h}\mathrm{d}S$ at $s = s_n$. This last integral vanishes if s_n is pure imaginary because then $\mathbf{n} \wedge \mathbf{h} = \mathbf{0}$. Thus the residue of $\underline{\Gamma}$ disappears at any pole corresponding to an interior magnetic resonance when the incident field comes from outside S. In other words there is no coupling between the exterior incident field and the magnetic field inside at frequencies of interior resonance.

Before leaving this theoretical investigation we derive

some properties which are valuable when making deformations in the complex s plane. For a surface vector **g** define **f** by

$$\mathbf{f}(p) = -\frac{1}{2}\mathbf{g}(p) + \mathbf{n}_p \wedge \int_S \mathbf{g}(q) \wedge \mathrm{grad}_q \phi(p,q)\,dS_q$$

$$= -\frac{1}{2}\mathbf{g}(p) + T\mathbf{g}$$

in the notation of §6.16. According to §6.16, T can be split as $T_1 + T_2$ where T_2 comes from a domain of S of diameter 2δ surrounding p and, for any fixed s, δ can be chosen small enough for $\|T_2\mathbf{g}\|^2 \leqslant \varepsilon_1\|\mathbf{g}\|^2$ for any $\varepsilon_1 > 0$. The proof of this statement is unaffected if $\mathcal{R}s$ is replaced by any larger value. Therefore, as $\mathcal{R}s \to \infty$, we first choose δ so that $\|T_2\mathbf{g}\|^2 \leqslant \varepsilon_1^2\|\mathbf{g}\|^2$ and then make $\mathcal{R}s$ so large that $\|T_1\mathbf{g}\|^2 \leqslant \varepsilon_1^2\|\mathbf{g}\|^2$. Evidently, as $\mathcal{R}s \to \infty$,

$$-\frac{1}{2}I + T \to -\frac{1}{2}I\,. \tag{7.52}$$

Further

$$\|(I-2T)^{-1}\mathbf{g}-\mathbf{g}\| \leqslant \sum_{r=1}^{\infty}(2\|T\mathbf{g}\|)^r \leqslant \sum_{r=1}^{\infty}(4\varepsilon_1\|\mathbf{g}\|)^r \leqslant 4\varepsilon_1\|\mathbf{g}\|(1-4\varepsilon_1\|\mathbf{g}\|)^{-1}$$

so long as $4\varepsilon_1\|\mathbf{g}\| < 1$. Allowing $\varepsilon_1 \to 0$ we see that

$$(-\tfrac{1}{2}I+T)^{-1} \to -\tfrac{1}{2}I \tag{7.53}$$

as $\mathcal{R}s \to \infty$.

The behaviour as $|s| \to \infty$ in general needs more elaborate analysis. Let **g** be continuous on S and let the maximum value of $|\mathbf{g}|$ attained on S be denoted by $\max|\mathbf{g}|$. Then, for fixed $\delta(<1)$, there is a finite K such that

$$|T_1\mathbf{g}| \leqslant K \exp\{|s|d(\mu\varepsilon)^{\frac{1}{2}}\}\max|\mathbf{g}|$$

where d is the maximum separation between any pair of points of S. On the other hand, (6.114) implies that

$$|T_2\mathbf{g}| \leqslant K\delta \exp\{|s|d(\mu\varepsilon)^{\frac{1}{2}}\}\max|\mathbf{g}|.$$

Thus, as $|s| \to \infty$,

$$|\mathbf{f}| \leqslant K \exp\{|s|d(\mu\varepsilon)^{\frac{1}{2}}\}\max|\mathbf{g}| \ . \tag{7.54}$$

Now considering only the variations of T with s, we observe that, for given p, $\mathbf{f}$ is an entire function of s which, by virtue of (7.54), is of exponential type and of order 1. Hence, by the properties of the minimum modulus (Titchmarsh 1934), there are, for given p, arbitrarily large circles of radius $|s|$ on which

$$|\mathbf{f}(p)| > \exp\{-(d+\eta)|s|(\mu\varepsilon)^{\frac{1}{2}}\}\max|\mathbf{g}|$$

for any $\eta > 0$. Writing $\mathbf{g} = (-\frac{1}{2}+T)^{-1}\mathbf{f}$ this may be expressed as

$$|(-\tfrac{1}{2}+T)^{-1}\mathbf{f}| < \exp\{(d+\eta)|s|(\mu\varepsilon)^{\frac{1}{2}}\}|\mathbf{f}(p)| \tag{7.55}$$

on some circles of arbitrarily large radius. Broadly speaking, (7.55) may be interpreted as stating that $\underline{\Gamma}$ is meromorphic and of order 1, its growth at infinity being dictated by the largest distance between any brace of points of S. Meromorphic functions of order 1 can be factorized as the ratio of entire functions of order 1 but that fact will not be needed here.

7.9. The impulse response

Suppose that the antenna is acting as a perfectly conducting receptor under the influence of the illumination $\mathbf{e}_0$, $\mathbf{h}_0$ from outside. Then the total field outside may be expressed as

$$\mathbf{h}(P) = \mathbf{h}_0(P) - \int_S \mathbf{j}(q) \wedge \operatorname{grad}_q\phi(P,q)\mathrm{d}S_q \tag{7.56}$$

leading to the integral equation

$$-\tfrac{1}{2}\mathbf{j}(p) + \mathbf{n}_p \wedge ⨍_S \mathbf{j}(q) \wedge \operatorname{grad}_q\phi(p,q)\mathrm{d}S_q = \mathbf{n}_p \wedge \mathbf{h}_0(p) \ . \tag{7.57}$$

The same equation could be obtained from (7.25) by the device of §6.14, relabelling $\mathbf{j} + \mathbf{n} \wedge \mathbf{h}_0$ as $\mathbf{j}$ and then changing the sign of $\mathbf{h}_0$.

From the theory of the preceding section the solution of (7.57) is, from (7.50),

$$\mathbf{j}(p) = -2\mathbf{n}_p \wedge \mathbf{h}_0(p) - \int_S \underline{\boldsymbol{\Gamma}}(p,q) \cdot \mathbf{n}_q \wedge \mathbf{h}_0(q)\mathrm{d}S_q \tag{7.58}$$

which enables the determination of the total field from (7.56). The integral in (7.56), indeed, supplies an expansion in terms of the natural modes of the body but only those representing the exterior electric modes arise in view of the remarks of the preceding section after (7.51).

Typical conduct emerges when the incident wave is plane and impulsive, its time dependence being decreed by a δ function. Naturally, the performance for other incident fields can then be deduced by convolution. To focus thoughts let $\mathbf{H}_0 = \mathbf{I}_0\delta\{t-(\mu\varepsilon)^{\frac{1}{2}}x\}$ where $\mathbf{I}_0$ is a constant vector. Forcing the plane wave to travel along the x axis causes no loss of generality since the orientation of the axes is at our disposal. The assumed form of H_0 make $\mathbf{h}_0 = \mathbf{I}_0\exp\{-s(\mu\varepsilon)^{\frac{1}{2}}x\}$. With $\mathbf{n}_p \wedge \mathbf{I}_0 = -\mathbf{j}_0(p)$, (7.58) gives

$$\mathbf{j}(p) = 2\mathbf{j}_0(p)\exp\{-s(\mu\varepsilon)^{\frac{1}{2}}x\} + \int_S \underline{\boldsymbol{\Gamma}}(p,q) \cdot \mathbf{j}_0(q)\exp\{-s(\mu\varepsilon)^{\frac{1}{2}}x_q\}\mathrm{d}S_q \tag{7.59}$$

where x_q is the value of x at the point of integration.

Returning to the time domain we obtain

$$\mathbf{J}(p,t) = -2\mathbf{n}_p \wedge \mathrm{H}_0(p) +$$

$$+ \frac{1}{2\pi\mathrm{i}}\int_{c-\mathrm{i}\infty}^{c+\mathrm{i}\infty} \exp(st) \int_S \underline{\boldsymbol{\Gamma}}(p,q) \cdot \mathbf{j}_0(q)\exp\{-s(\mu\varepsilon)^{\frac{1}{2}}x_q\}\mathrm{d}S_q\mathrm{d}s \tag{7.60}$$

as the current induced on the obstacle by an impulse. The constant c is positive so that the contour of integration in the s plane lies to the right of all singularities.

Inferences of some importance can be drawn from (7.60). Properties (7.52) and (7.53) indicate that, as $\mathcal{R}s \to \infty$, a formula for $\underline{\boldsymbol{\Gamma}}$ can be derived by iteration of the original kernel. This steers one to

$$\int_S \underline{\boldsymbol{\Gamma}}(p,q)\cdot\mathbf{j}(q)\mathrm{d}S_q = 4\mathbf{n}_p \wedge \int_S \mathbf{j}(q)\wedge \operatorname{grad}_q \phi(p,q)\mathrm{d}S_q + \ldots \tag{7.61}$$

as $\mathcal{R}s \to \infty$. It is then evident, by deforming the contour to the right, that the first term of (7.61) gives zero contribution in (7.60) if $t/(\mu\varepsilon)^{\frac{1}{2}} < x_q + |x_p - x_q|$ and it is readily verified that the higher iterates in (7.61) make no subscription under the same condition. If the body is confined to $x_1 \leqslant x \leqslant x_2$, the inequality is certainly satisfied if $t < x_1(\mu\varepsilon)^{\frac{1}{2}}$. Thus $\mathbf{J}(p,t) = \mathbf{0}$ for all p if $t < x_1(\mu\varepsilon)^{\frac{1}{2}}$, i.e. no current is induced on the obstacle until the pulse strikes it. In other words the dictates of causality are obeyed. Furthermore, if $t < x_p(\mu\varepsilon)^{\frac{1}{2}}$, the inequality holds for all $x_q < t/(\mu\varepsilon)^{\frac{1}{2}}$ and also for $x_q \geqslant t/(\mu\varepsilon)^{\frac{1}{2}}$. Consequently, there is no current at a point of the surface before the incident pulse reaches it.

Another interpretation is that, if $x_1 < t/(\mu\varepsilon)^{\frac{1}{2}} < x_2$, the integration in (7.60) can be limited to that part of S on which $x_q + |x_p - x_q| < t/(\mu\varepsilon)^{\frac{1}{2}}$. Once $t > x_2(\mu\varepsilon)^{\frac{1}{2}}$ the whole of S will be involved and the first term on the right-hand side of (7.60) may be removed.

If $t/(\mu\varepsilon)^{\frac{1}{2}} > x_2 + d$, the limitation on the growth of $\underline{\boldsymbol{\Gamma}}$ demonstrated in the last section empowers deformation of the contour to the left in (7.60). Only the poles of $\underline{\boldsymbol{\Gamma}}$ need to be taken into account and

$$\mathbf{J}(p,t) = \sum_n \exp(s_n t)\sum_r \frac{\mathbf{j}_{nr}(p)}{\langle A_n \mathbf{j}_{nr}, \mathbf{j}'^{*}_{nr}\rangle}\int_S \mathbf{n}_q \wedge \mathbf{j}'_{nr}(q)\cdot \mathbf{j}_0(q)\exp\{-s_n(\mu\varepsilon)^{\frac{1}{2}}x_q\}\mathrm{d}S_q \tag{7.62}$$

when $t/(\mu\varepsilon)^{\frac{1}{2}} > x_2 + d$. The subscript n has been added to the quantities in (7.51) in order to identify a particular pole. The summation over r may depend on n since the number of eigenelements of (7.25) may vary with the pole under consideration.

Nominally, all the poles of $\underline{\Gamma}$ may be allowed for in (7.62). However, it has already been pointed out, after (7.52), that the interior magnetic resonances can be ignored when the sources of the incident field are outside S. Therefore, the summation in (7.62) can be limited to those values of n for poles corresponding to the exterior electric modes and so all terms are exponentially damped in time.

At first sight (7.62) does not appear to be real, but it must be remembered that if s_n is a pole so is s_n^* with associated eigenelement $\mathbf{j}_{nr}^*$. Thus each term of (7.62) has a matching complex conjugate so that $\mathbf{J}$ is real, as it needs to be.

The weight to be attached to the expansion (7.62) is considerable. It demonstrates that after a certain interval of time the induced current at any point has a particularly simple time dependence, being a series of decaying exponentials whose coefficients do not vary with time. Moreover, the exponentials, though not the coefficients, are independent of the position of the point on the surface.

The delay which has to be accepted before (7.62) is valid is quite short in practice. For example, with a sphere of radius ¼ m in air the delay is a little more than 3 ns. For an ellipsoid whose largest and smallest semi-axes are a and c respectively, the delay does not exceed $12a$ ns roughly and may be as little as $6(a+c)$ ns, depending upon the direction of incidence of the incoming pulse.

Despite these advantages the formulae (7.60) and (7.62) will be valuable practically only if there are ways of calculating the positions of the poles and their associated residues. This problem will be examined in the next section.

7.10. Practical determination of the positions of the poles

To locate the poles it is necessary to solve the integral equation

$$-\tfrac{1}{2}\mathbf{j}(p) + \mathbf{n}_p \wedge \int_S \mathbf{j}(q) \wedge \operatorname{grad}_q \phi(p,q)\mathrm{d}S_q = \mathbf{0}\ . \qquad (7.63)$$

If this can be undertaken analytically then there is little more to be said. However, the analysis is likely to be intractable for all except the very simplest shapes and so recourse to other methods is inevitable.

The direct approach to (7.63) is by numerical approximation, replacing (7.63) by a finite algebraic system based on any appropriate method from Chapter 6. A solution will exist only if the determinant of the coefficients vanishes, leading to an equation in s which will be satisfied by the s_n or rather by approximations of the s_n. Once the s_n are known approximation to the eigenelements of (7.63) can be constructed from the solutions of the algebraic system.

The procedure may be clearer if (7.63) is replaced by its scalar equivalent as in §7.6, so consider

$$j'(x) - \int K(x,y)j'(y)\mathrm{d}y = 0\ .$$

The algebraic system corresponding to this is

$$\sum_{n=1}^{N} Z_{mn}(s)j_n' = 0 \qquad (m = 1, \ldots, N) \qquad (7.64)$$

where the dependence of the coefficients on s, due to K involving s, has been explicitly displayed. The equations have a nontrivial solution only if

$$\det[Z_{mn}(s)] = 0\ . \qquad (7.65)$$

The N roots of this equation are regarded as approximations to the s_n. Obviously, not all of the s_n can be covered but the dominant ones should be there if N is large enough and the accuracy should improve as N increases. In any case, the s_n

which have a much more negative real part than the others are not generally of much significance. Practical methods for solving (7.65) are those of Newton (§1.8) and of Muller (§1.8).

Having found a solution of (7.65) the corresponding j_n' is determined from (7.64) and regarded as an approximation to j'. Alternatively, since we expect Z^{-1} (Z being the matrix with elements Z_{mn}) to have a pole at $s = s_n$ we can try to discover the residue directly by evaluating

$$\eta\{Z(s_n+\eta)\}^{-1}$$

for small η until the result does not alter appreciably with η.

Results of reasonable accuracy have been obtained in this way (Tesche 1973). Notwithstanding, the effort to be expended is not trifling. Not only have the algebraic equations (7.65) and (7.64) to be solved but also the integrals in (7.60) and/or (7.62) have to be evaluated, so the labour is at least of the same order of magnitude as a solution in the frequency domain and probably substantially more. Moreover, the method is not as efficient in the early time stages as the updating process described in the initial sections of this chapter and, besides, fails to make the most of (7.62). Another strategy will, therefore, be set forth now.

7.11. Prony's method

The fact that the surface current after a period of transient behaviour, which lasts at most for the time to transmit twice the greatest diameter of the obstacle, settles down to the exponential evanescence of (7.62) suggests the following tactics. First, calculate the current at a point by working in the time domain with an updating integral equation until sufficient time has elapsed for a reasonable stretch where (7.62) is valid to be available. By sampling in this stretch, arrive at the values of the coefficients for all subsequent times. The input pulse is taken as a Gaussian approximation to the δ function as in §7.5.

The new problem is to discover the coefficients in the expansion in exponentials but this can be tackled efficaciously by Prony's method (Whittaker and Robinson 1952). Suppose that we are going to adopt the representation

$$\mathbf{J}(t) = \sum_{n=1}^{N} \mathbf{A}_n \exp(s_n t) \tag{7.66}$$

where the $\mathbf{A}_n$ and s_n are to be found from observations on $\mathbf{J}$ at $t = 0, \tau, \ldots, (M+N-1)\tau$. A simple change of time origin permits the sampling to start at any given time step. Denote $\mathbf{J}(r\tau)$ by $\mathbf{J}_r$. Then (7.66) implies that $\mathbf{J}(t)$ satisfies the linear difference equation

$$\sum_{n=0}^{N} a_n \mathbf{J}_{r+n} = \mathbf{0} \tag{7.67}$$

where the roots of the algebraic equation

$$a_N z^N + a_{N-1} z^{N-1} + \ldots + a_0 = 0 \tag{7.68}$$

are $\exp(s_1\tau)$, $\exp(s_2\tau)$, ..., $\exp(s_N\tau)$. Now apply (7.67) at $r = 0, 1, \ldots, M-1$. The quantities $\mathbf{J}_{r+n}$ are known from the sample and so there are $2M$ scalar linear equations to determine the a_n. Full generality is preserved by choosing $a_N = 1$. If $2M = N$ straightforward inversion of a real matrix has to be carried out whereas, if $2M > N$, the method of least squares is fitting. With the a_n known, (7.68) can be solved by any technique for finding the roots of a polynomial (§1.8) and thus $s_n\tau$ is unveiled. Substitute these $s_n\tau$ in (7.66) and then, from the data, we have $M + N$ linear vector equations for the $\mathbf{A}_n$ and these may be solved by the method of least squares.

As a check on the calculations we note that the s_n and $\mathbf{A}_n$ should occur in complex conjugate pairs.

To locate N poles, Prony's method requires sampling at $\frac{3}{2}N$ equally spaced data points for a vector current or $2N$ sampling points if a single current component is under consideration.

Prony's method replaces the infinite series of (7.62) by the finite sum (7.66) and so cannot be expected to reproduce the first N poles in precisely the positions they would occupy in the physical problem. Some will have to take care of the neglected part of the infinite series. Nevertheless, if N is large enough, most of the poles should not be far from their proper places. Usually, if N is 20 or more, only two or three exhibit discrepant behaviour (Van Blaricum and Mittra 1975).

A Fourier transform of (7.66), assumed to be valid for $t > 0$, gives

$$\mathbf{j} = \sum_{n=1}^{N} \frac{\mathbf{A}_n}{\mathrm{i}\omega - s_n}. \tag{7.69}$$

Since (7.66) certainly holds as $t \to \infty$ we anticipate that (7.69) will be reliable at low frequencies though it cannot be trusted at high frequencies where the transient behaviour of $\mathbf{J}$ at the onset of the incident pulse plays an essential role.

The great boon of Prony's method is that it is only necessary to calculate the $\mathbf{A}_n$ and s_n once and then store them. A relatively simple computation then provides the pulse signature and the behaviour in the frequency domain provided that the wavelength is not too short. To put it another way, the basic characteristics of a scatterer are sustained by storing a small quantity of data.

Exercises

16. A scalar current $I(t)$ is observed to have the following values at the times shown:

t	10	20	30	40	50	60	70	80	90
$I(t)$	6460	6090	5642	5049	4417	3623	2401	983	142

If $I(t)$ is approximated by

$$A_1 \exp(s_1 t) + A_2 \exp(s_2 t) + A_3 \exp(s_3 t)$$

show that $\exp(s_1)$, $\exp(s_2)$, and $\exp(s_3)$ are the roots of

$$z^3 - 3.029z^2 + 3.788z - 1.687 = 0.$$

17. A scalar current $I(t)$ provides the data

t	0	8	16	24	32	40
$I(t)$	248	345	421	481	529	569

If Prony's method with three terms is used show that an approximation to $I(t)$ is $629.4(1.001)^t - 381.4(0.9670)^t + 0.0005(1.236)^t$.

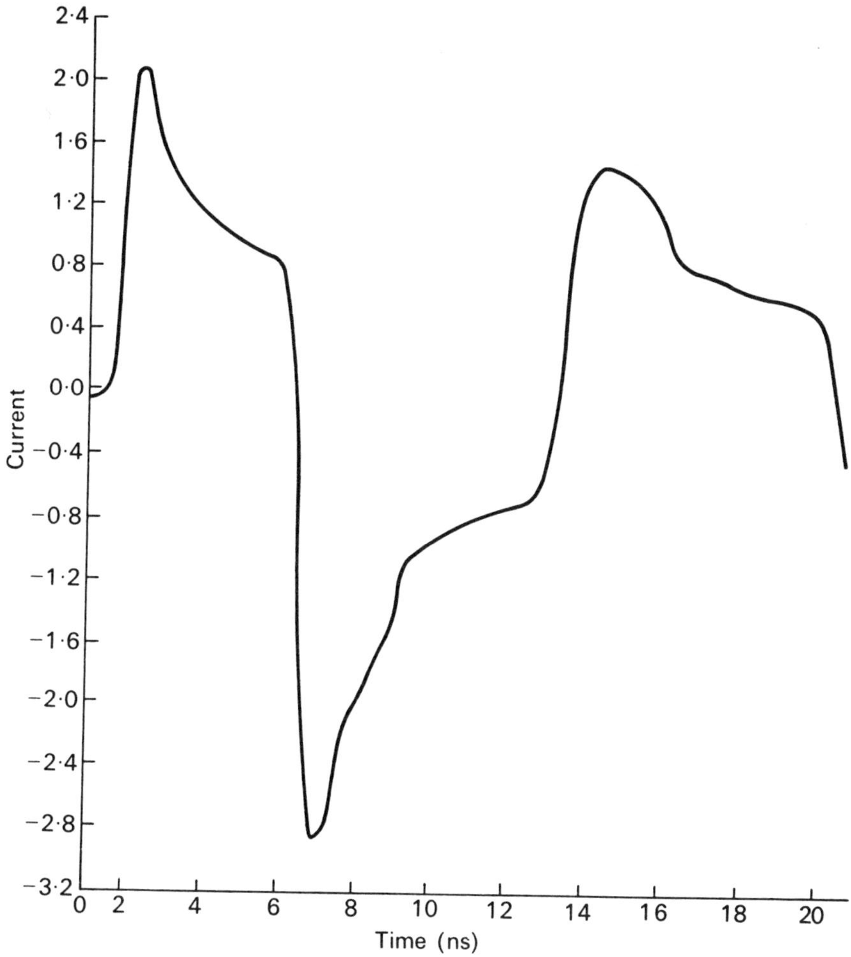

FIG. 7.6. Data to be sampled.

18. A scalar current is found to have the waveform displayed in Fig. 7.6. Take samples at intervals of $\frac{1}{5}$ ns and use Prony's method to locate 40 poles. Show that most of the poles are not very distant from the imaginary axis. Calculate the waveform of the exponential series over 0–100 ns and decide whether it is sufficiently accurate in the first 20 ns for one to have faith in the extrapolation.

19. Calculate the impulse response of a perfectly conducting sphere by

the methods of the last two sections and compare the computer requirements of the two critically.

20. A centre-fed dipole of length 1 m and radius ½ cm is excited by the Gaussian pulse $\exp\{-25 \times 10^{18} (t - 5.556 \times 10^{-10})^2\}$. Determine the induced current not far from the feed as in §7.3 during the first 30 ns. Extrapolate from there by Prony's method. Find the input impedance in the frequency domain from (7.69) and show that it gives acceptable values up to about 2 GHz.

21. A prolate spheroid with semi-axes a, b is illuminated along its axis of symmetry. Trace the movement of poles as a/b varies.

22. A finite circular cylinder of length l and radius a is subject to a plane impulse from outside. The positions of the poles depend upon l, a and the angle of incidence. Plot their behaviour for various typical situations. By taking l very small obtain the response of a circular disc.

23. Assuming that the theory applies to two coaxial circular cylinders obtain the impulse response of a single vertical cylinder above a horizontal perfectly conducting ground plane.

8

GEOMETRIC THEORY OF DIFFRACTION

The advent of the computer has meant that the small number of analytical solutions to electromagnetic scattering problems can be supplemented by numerical investigation. Antecedent chapters have indicated ways in which answers can be fabricated numerically and reveal that, despite the power of the computer, the accuracy falls off as the frequency increases. Another route must therefore be pioneered to handle the scattering from bodies which are large compared with the wavelength. The method will be one of approximate analytical technique, often conjoined to numerical methods as well.

The classical high-frequency approach is that of *geometrical optics*, but it fails to account adequately for the behaviour in shadows and the influence of edges. To cope with these problems an extended version, known as the *geometric theory of diffraction* (GTD), has been devised (Keller 1957, 1962). It is the purpose of this chapter to describe the main features of the geometric theory of diffraction.

8.1. The high-frequency approximation

It will be assumed that the fields vary harmonically in time according to the factor $\exp(i\omega t)$ where ω is real. Then Maxwell's equations take the form

$$\text{curl } \mathbf{E} + i\omega\mu\mathbf{H} = \mathbf{0}, \quad \text{curl } \mathbf{H} - i\omega\varepsilon\mathbf{E} = \mathbf{0} \tag{8.1}$$

where the real μ and ε account for the permeability and permittivity of the medium. The *refractive index* N of the medium is defined by $N = (\mu\varepsilon/\mu_0\varepsilon_0)^{\frac{1}{2}}$ where μ_0 and ε_0 refer to free space. The real quantity $\omega(\mu_0\varepsilon_0)^{\frac{1}{2}}$ will be denoted by k_0.

If the medium is not isotropic μ and ε must be replaced in (8.1) by tensors. The resulting analysis then becomes much more complicated in detail without affecting the basic principles of the method. For simplicity, therefore, only the case of scalar μ and ε will be treated.

When the wavelength is so small that significant medium changes occur only over distances large compared with it, a reasonable assumption is that in local regions the field behaves as if it were in a homogeneous medium. Thus, locally, the wave may be expected to look like a plane wave. This suggests that $\mathbf{E}$ might have the form $\mathbf{E}_0 \exp(-ik_0 L)$. However, it is more likely that this offers a first approximation. Therefore, we introduce the more recondite assumption or *Ansatz* that

$$\left.\begin{aligned} \mathbf{E} &= \exp(-ik_0L)\left(\mathbf{E}_0 + \frac{\mathbf{E}_1}{k_0} + \frac{\mathbf{E}_2}{k_0^2} + \ldots\right), \\ \mathbf{H} &= \exp(-ik_0L)\left(\mathbf{H}_0 + \frac{\mathbf{H}_1}{k_0} + \frac{\mathbf{H}_2}{k_0^2} + \ldots\right) \end{aligned}\right\} \qquad (8.2)$$

where L, $\mathbf{E}_0$, $\mathbf{E}_1$, ..., $\mathbf{H}_0$, $\mathbf{H}_1$, ... may depend upon position but are independent of k_0. There is no *a priori* knowledge that the expansion (8.2) is valid whether as a convergent or as an asymptotic series. Even in special circumstances it can be extremely difficult to prove the legitimacy of the expansion. Be that as it may, experience is overwhelmingly in favour of its introduction. It tenders the possibility of progress when exact or numerical solutions are out of the question. Even when the expansion is confuted there may be others of a similar nature which will succeed as will be seen.

If (8.2) is substituted in (8.1) and it is agreed to take derivatives term by term then

$$\sum_{r=0} \frac{\text{curl}\,\mathbf{E}_r}{k_0^r} - ik_0 \,\text{grad}\, L \wedge \sum_{r=0} \frac{\mathbf{E}_r}{k_0^r} + i\omega\mu \sum_{r=0} \frac{\mathbf{H}_r}{k_0^r} = \mathbf{0}\,, \qquad (8.3)$$

$$\sum_{r=0} \text{curl}\,\frac{\mathbf{H}_r}{k_0^r} - ik_0 \,\text{grad}\, L \wedge \sum_{r=0} \frac{\mathbf{H}_r}{k_0^r} - i\omega\varepsilon \sum_{r=0} \frac{\mathbf{E}_r}{k_0^r} = \mathbf{0}\,. \qquad (8.4)$$

If L, $\mathbf{E}_r$, $\mathbf{H}_r$, and their derivatives are finite and vary significantly only over several wavelengths, the coefficients of separate powers of k_0 may be equated to zero in (8.3) and (8.4).

There results

$$\left.\begin{aligned}(\mu_0\varepsilon_0)^{\frac{1}{2}}\text{grad } L \wedge \mathbf{E}_0 &= \mu\mathbf{H}_0,\\ (\mu_0\varepsilon_0)^{\frac{1}{2}}\text{grad } L \wedge \mathbf{H}_0 &= -\varepsilon\mathbf{E}_0,\end{aligned}\right\} \tag{8.5}$$

$$\text{grad } L \wedge \mathbf{E}_m - \frac{\mu\mathbf{H}_m}{(\mu_0\varepsilon_0)^{\frac{1}{2}}} = -\text{ i curl } \mathbf{E}_{m-1}\ , \tag{8.6}$$

$$\text{grad } L \wedge \mathbf{H}_m + \frac{\varepsilon\mathbf{E}_m}{(\mu_0\varepsilon_0)^{\frac{1}{2}}} = -\text{ i curl } \mathbf{H}_{m-1} \tag{8.7}$$

for $m = 1, 2, \ldots$.

From (8.5) it is evident that, so long as μ and ε are non-zero,

$$\mathbf{E}_0 \cdot \text{grad } L = 0, \quad \mathbf{H}_0 \cdot \text{grad } L = 0, \quad \mathbf{H}_0 \cdot \mathbf{E}_0 = 0. \tag{8.8}$$

Furthermore, if $\mathbf{H}_0$ is eliminated from (8.5),

$$(\text{grad}^2\ L - N^2)\mathbf{E}_0 = \mathbf{0}$$

when account is taken of (8.8). If $\mathbf{E}_0$ is not to be identically zero it is necessary that

$$\text{grad}^2\ L = N^2. \tag{8.9}$$

Eqn (8.9) is a partial differential equation for L, the function which defines the surfaces of constant phase, i.e. the *wavefronts*. The function L is known as the *eikonal* and (8.9) is called the *eikonal equation*. From (8.8) it can be seen that $\mathbf{E}_0$ and $\mathbf{H}_0$ are transverse to grad L, i.e. are *transverse to the direction of propagation of the wavefront*. Since $\mathbf{E}_0$ is also perpendicular to $\mathbf{H}_0$ the Poynting vector is parallel to grad L and so the direction of energy flow is normal to the wavefront.

Consequently, the field due to $\mathbf{E}_0$, $\mathbf{H}_0$ has locally all the properties of a plane wave and may be expected to be a reasonable first approximation at high frequencies. It will, however, require close scrutiny at points where N becomes small or changes discontinuously and at points where $\mathbf{E}_0$ or $\mathbf{H}_0$ is predicted to be large.

The curves which have at each point the direction of the energy flow in the field are known as *rays*. It has just been shown that the rays are normal to the wavefronts.

In general media there are two characteristic velocities. One is the wave velocity which is the rate of displacement of the wavefront in the direction normal to itself. The other is the energy or ray velocity. For the isotropic medium just considered the two velocities can be identified with one another. For anisotropic media the wave and ray velocities may not coincide; the distinction between the two is then vital and must not be forgotten.

From (8.6)

$$\operatorname{curl}\left(\frac{1}{\mu}\operatorname{grad} L \wedge \mathbf{E}_m\right) - \frac{\operatorname{curl}\mathbf{H}_m}{(\mu_0\varepsilon_0)^{\frac{1}{2}}} = -\,i\operatorname{curl}\left(\frac{1}{\mu}\operatorname{curl}\mathbf{E}_{m-1}\right)$$

while, from (8.7),

$$\begin{aligned}\operatorname{div}(\varepsilon\mathbf{E}_m) &= -(\mu_0\varepsilon_0)^{\frac{1}{2}}\operatorname{div}(\operatorname{grad} L \wedge \mathbf{H}_m) = (\mu_0\varepsilon_0)^{\frac{1}{2}}\operatorname{grad} L \,.\, \operatorname{curl}\mathbf{H}_m \\ &= i\varepsilon \operatorname{grad} L \,.\, \mathbf{E}_{m+1}\end{aligned}$$

when (8.7) with m replaced by $m + 1$ is noted. Hence

$$\begin{aligned}\mu\operatorname{curl}\left(\frac{1}{\mu}\operatorname{curl}\mathbf{E}_{m-1}\right) - \operatorname{grad}\left(\frac{1}{\varepsilon}\operatorname{div}\varepsilon\mathbf{E}_{m-1}\right) &= i\operatorname{grad} L\operatorname{div}\mathbf{E}_m - \\ &\quad - i\,\frac{\mu\operatorname{curl}\mathbf{H}_m}{(\mu_0\varepsilon_0)^{\frac{1}{2}}} - i(\operatorname{grad} L) \wedge \operatorname{curl}\mathbf{E}_m - \\ &\quad - i\mu\mathbf{E}_m\operatorname{div}\left(\frac{1}{\mu}\operatorname{grad} L\right) - 2i(\operatorname{grad} L \,.\, \operatorname{grad})\mathbf{E}_m - \\ &\quad - \left(\frac{i}{\mu}\right)(\mathbf{E}_m \,.\, \operatorname{grad}\mu)\operatorname{grad} L. \qquad (8.10)\end{aligned}$$

Multiply (8.6) by grad $L \wedge$ and add $\mu/(\mu_0\varepsilon_0)^{\frac{1}{2}}$ times (8.7) to obtain

$$\frac{1}{\varepsilon}(\text{div } \varepsilon\mathbf{E}_{m-1})\text{grad } L = \frac{\mu \text{ curl } \mathbf{H}_{m-1}}{(\mu_0\varepsilon_0)^{\frac{1}{2}}} + \text{grad } L \wedge \text{curl } \mathbf{E}_{m-1} \quad (8.11)$$

by virtue of (8.9). Replace $m - 1$ by m in (8.11) and then substitute in (8.10). There results

$$2i(\text{grad } L \,.\, \text{grad})\mathbf{E}_m + i\mu\mathbf{E}_m\text{div}\left\{\frac{1}{\mu} \text{ grad } L\right\} + \frac{i}{\mu\varepsilon}\{\mathbf{E}_m \,.\, \text{grad}(\mu\varepsilon)\}\text{grad } L$$

$$= \text{grad}\left(\frac{1}{\varepsilon} \text{ div } \varepsilon\mathbf{E}_{m-1}\right) - \mu \text{ curl } \frac{1}{\mu} \text{ curl } \mathbf{E}_{m-1} \,. \quad (8.12)$$

Similarly, or by observing that the equations for $\mathbf{H}_m$ are the same as those for $\mathbf{E}_m$ with μ and ε replaced by $-\varepsilon$ and $-\mu$ respectively, it may be shown that

$$2i(\text{grad } L.\, \text{grad})\mathbf{H}_m + i\varepsilon\mathbf{H}_m\text{div}\left(\frac{1}{\varepsilon} \text{ grad } L\right) + \frac{i}{\mu\varepsilon}\{\mathbf{H}_m \,.\, \text{grad}(\mu\varepsilon)\}\text{grad } L$$

$$= \text{grad}\left(\frac{1}{\mu} \text{ div } \mu\mathbf{H}_{m-1}\right) - \varepsilon \text{ curl}\left(\frac{1}{\varepsilon} \text{ curl } \mathbf{H}_{m-1}\right). \quad (8.13)$$

The only derivatives of $\mathbf{E}_m$ and $\mathbf{H}_m$ which occur in (8.12) and (8.13) are normal to the wavefront. Therefore (8.12) and (8.13) are ordinary differential equations for $\mathbf{E}_m$ and $\mathbf{H}_m$ along a ray. They are known as *transport equations*.

When $\mathbf{E}_{m-1} = \mathbf{0}$ and $\mathbf{H}_{m-1} = \mathbf{0}$ the right-hand sides of (8.12) and (8.13) vanish. The scalar product of (8.12) with $\mathbf{E}_m$ then gives

$$\text{grad } L \,.\, \text{grad } \mathbf{E}_m^2 + \mu\mathbf{E}_m^2 \text{ div}(\tfrac{1}{\mu} \text{ grad } L) = 0$$

since now $\mathbf{E}_m \,.\, \text{grad } L = 0$ from (8.7). The equation may be written as

$$\operatorname{div}\left(\frac{1}{\mu}\,\mathbf{E}_m^2 \operatorname{grad} L\right) = 0 . \tag{8.14}$$

Similarly

$$\operatorname{div}\left(\frac{1}{\varepsilon}\,\mathbf{H}_m^2 \operatorname{grad} L\right) = 0 . \tag{8.15}$$

In particular, it is always true that

$$\operatorname{div}\left(\frac{1}{\mu}\,\mathbf{E}_0^2 \operatorname{grad} L\right) = 0, \qquad \operatorname{div}\left\{\frac{1}{\varepsilon}\,\mathbf{H}_0^2 \operatorname{grad} L\right\} = 0 . \tag{8.16}$$

Exercises

1. In a homogeneous anisotropic medium $\mathbf{D} = \underline{\varepsilon} \cdot \mathbf{E}$, $\mathbf{B} = \underline{\mu} \cdot \mathbf{H}$ where $\underline{\varepsilon}$, $\underline{\mu}$ are real symmetric positive definite tensors. Assuming an approximation of the form $\mathbf{E} = \mathbf{E}_0 \exp(-ik_0 L)$, etc. show that $\mathbf{D}_0$ is perpendicular to $\mathbf{H}_0$ and grad L while $\mathbf{B}_0$ is perpendicular to $\mathbf{E}_0$ and grad L.

If $\mathbf{S} = \mathbf{E}_0 \wedge \mathbf{H}_0$ define the energy velocity as $2\mathbf{S}/(\mathbf{E}_0\cdot\mathbf{D}_0+\mathbf{H}_0\cdot\mathbf{B}_0)$. Show that the phase speed v of the wavefront is equal to the projection of the energy velocity on the wave normal.

If μ is a scalar and $\underline{\varepsilon}$ has elements ε_1, ε_2, ε_3 with respect to its principal axes show that v satisfies *Fresnel's equation*

$$\frac{n_1^2}{v^2 - v_1^2} + \frac{n_2^2}{v^2 - v_2^2} + \frac{n_3^2}{v^2 - v_3^2} = 0$$

where $v_i^2 = 1/\mu\varepsilon_i$ and n_1, n_2, n_3 are the components of grad L along the principal axes. If $\mathbf{D}_0'$, $\mathbf{D}_0''$ correspond to two different solutions of Fresnel's equation show that $\mathbf{D}_0' \cdot \mathbf{D}_0'' = 0$.

2. In Exercise 1 assume that $\mathbf{S}$ is parallel to neither $\underline{\varepsilon} \cdot \operatorname{grad} L$ nor $\underline{\mu} \cdot \operatorname{grad} L$. Prove that $\operatorname{div}(\mathbf{E}_0\wedge\mathbf{H}_0) = 0$.

8.2. Geometrical optics

In an isotropic medium the approximation $\mathbf{E} = \mathbf{E}_0 \exp(-ik_0 L)$ is often known as *geometrical optics*, though the name is sometimes applied to the full expansion (8.2). Since the rays are normal to the wavefront the tangent to a ray must be parallel to grad L. Hence a unit vector $\mathbf{s}$ along the tangent is

$$\mathbf{s} = \frac{1}{N}\,\text{grad}\,L \tag{8.17}$$

when cognisance is taken of the eikonal equation (8.9). Let κ be the curvature at the same point and $\boldsymbol{\nu}$ a unit vector in the direction of the radius of curvature. Then $\kappa\boldsymbol{\nu} = \mathrm{d}\mathbf{s}/\mathrm{d}s$ where s is the arc length measured along the ray. Now

$$\frac{\mathrm{d}\mathbf{s}}{\mathrm{d}s} = (\mathbf{s}.\text{grad})\mathbf{s} = -\,\mathbf{s}\wedge\text{curl}\,\mathbf{s}$$

and so

$$\kappa = -\,\boldsymbol{\nu}\,.\,\mathbf{s}\wedge\text{curl}\,\mathbf{s}\;.$$

On using (8.17) and replacing grad$(1/N)$ by $-(1/N)$grad ln N we obtain

$$\kappa = \boldsymbol{\nu}\,.\,\text{grad}\,\ln N \tag{8.18}$$

since $\boldsymbol{\nu}$ is perpendicular to grad L.

Eqn (8.18) carries the information that the rate of change of N in the direction of the centre of curvature is positive. This means that a *ray bends towards the region of higher refractive index* N or, equivalently, towards a region with a lower speed of light.

In a homogeneous medium N is independent of position and the right-hand side of (8.18) vanishes. Thus the curvature κ is zero and the rays are straight lines.

The behaviour of the magnitude of $\mathbf{E}_0$ can be found by considering what happens to a tube of rays. Let the tube of rays intersect the wavefronts L_1 and L_2 in the surface elements $\mathrm{d}S_1$ and $\mathrm{d}S_2$ respectively (Fig. 8.1). No power will flow across the sides of the tube because the energy moves along a ray; the flow through any normal cross-section of the tube must therefore be constant. Hence

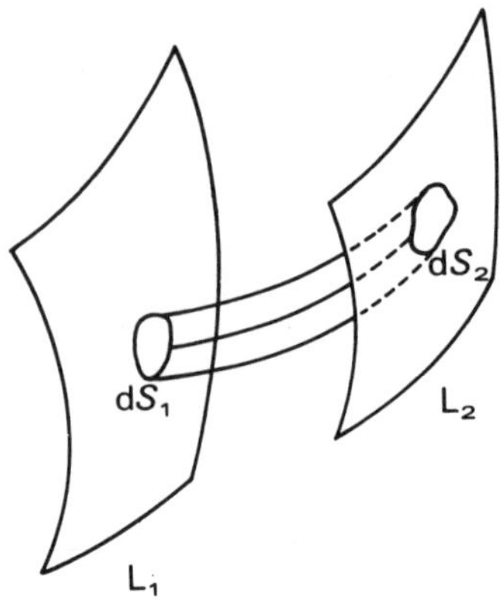

FIG. 8.1. Propagation along a ray tube.

$$|\mathbf{E}_0 \wedge \mathbf{H}_0|_1 \mathrm{d}S_1 = |\mathbf{E}_0 \wedge \mathbf{H}_0|_2 \mathrm{d}S_2$$

or, from (8.5),

$$\left(\frac{\varepsilon_1}{\mu_1}\right)^{\frac{1}{2}} |\mathbf{E}_0|_1^2 \mathrm{d}S_1 = \left(\frac{\varepsilon_2}{\mu_2}\right)^{\frac{1}{2}} |\mathbf{E}_0|_2^2 \mathrm{d}S_2 \ . \qquad (8.19)$$

This is known as *the intensity law of geometrical optics*. It may also be obtained by integrating (8.16) over a ray tube and remembering the eikonal equation. Similarly, $(\mu/\varepsilon)^{\frac{1}{2}}\mathbf{H}_0$ satisfies the intensity law of geometrical optics.

The intensity law permits the determination of the field amplitude at any point on the ray when it is known as a single point. The form (8.19) is often valuable though other representations are sometimes helpful (§8.3). If μ is independent of position (8.19) can be expressed in terms of the refractive index N as

$$N_1 |\mathbf{E}_0|_1^2 \mathrm{d}S_1 = N_2 |\mathbf{E}_0|_2^2 \mathrm{d}S_2 \ . \qquad (8.20)$$

When the medium is homogeneous an illuminating formula can be deduced from (8.19). Let the ray through the point A of the wavefront L_1 be the z axis. Take A as origin and choose the (x,z) and (y,z) planes to be the planes containing the principal

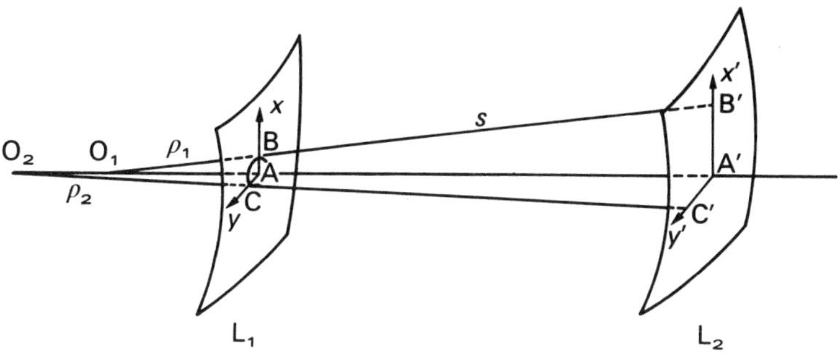

FIG. 8.2. Propagation of energy in a homogeneous isotropic medium.

radii of curvature ρ_1 and ρ_2 of the wavefront at A (Fig. 8.2). A ray through B, a point on the x axis adjacent to A, will intersect the z axis at O_1 and $O_1B = \rho_1$. Similarly, a ray through the adjacent point C on the y axis will intersect the z axis at O_2 where $O_2C = \rho_2$. Take a radius of curvature as positive when the centre of curvature is on the negative z axis as in Fig. 8.2; otherwise it is negative. Continue the rays through A, B and C until they meet the wavefront L_2 at A′, B′, and C′ respectively. Often the ray AA′ which passes through the middle of the small element of L_1 is known as an *axial ray*, whereas rays such as BB′ through the periphery of the element are called *paraxial rays*. The paraxial rays form the boundary of a tube.

The rays are normal to L_2 and so the normals at A′ and B′ intersect at O_1, i.e. the (x',z') plane contains a principal radius of curvature. This radius of curvature can be no other than $\rho_1 + s$ where s is the constant length of ray cut off by the wavefront L_1 and L_2. Similarly, the (y',z') plane contains the other principal radius of curvature $\rho_2 + s$.

Let dS_1 be the element of area surrounding A and let the paraxial rays produce the element of area dS_2 about A′. Corresponding points in the two areas are related by

$$x' = \left|\frac{\rho_1+s}{\rho_1}\right| x, \quad y' = \left|\frac{\rho_2+s}{\rho_2}\right| y .$$

Hence

$$\mathrm{d}S_2 = \left|\frac{(\rho_1+s)(\rho_2+s)}{\rho_1\rho_2}\right|\mathrm{d}S_1 . \tag{8.21}$$

Since $N_1 = N_2$, (8.20) gives

$$|\mathbf{E}_0|_2 = \left|\frac{\rho_1\rho_2}{(\rho_1+s)(\rho_2+s)}\right|^{\frac{1}{2}} |\mathbf{E}_0|_1 \tag{8.22}$$

in a homogeneous medium. When $\rho_1 = \rho_2$ the wavefront is spherical and we have the customary inverse square law of decrease of intensity with distance.

It will be observed that (8.22) will give trouble when ρ_1 or ρ_2 is negative, i.e. geometrical optics becomes suspect when a principal centre of curvature of a wavefront is in the region into which energy is propagating. Then focusing effects can be anticipated (*astigmatism* will occur if $\rho_1 \neq \rho_2$) and some modification of the theory will be necessary.

More generally, we can deduce from (8.21) the relation between $\mathrm{d}S_1$ and $\mathrm{d}S_2$ when the medium is inhomogeneous, for, when the two surfaces are close together, we may expect (8.21) to continue to hold. In other words, on a small displacement the change in an element of area $\mathrm{d}S_1$ is $\{1/\rho_1(s)+1/\rho_2(s)\}\mathrm{d}S_1\mathrm{d}s$ where ρ_1 and ρ_2 are the principal radii of curvature at s. Hence, in general,

$$\mathrm{d}S_2 = \mathrm{d}S_1 \exp \int_{s_1}^{s_2}\left\{\frac{1}{\rho_1(s)} + \frac{1}{\rho_2(s)}\right\}\mathrm{d}s \tag{8.23}$$

which expresses the expansion of the element of area as an integral of the mean curvature of the wavefront. Eqn (8.23) can be inserted in (8.19) to provide the change in intensity from s_1 to s_2, as

$$|\mathbf{E}_0|_2^2 = \left(\frac{\mu_2\varepsilon_1}{\mu_1\varepsilon_2}\right)^{\frac{1}{2}} |\mathbf{E}_0|_1^2\exp\left[-\int_{s_1}^{s_2}\left\{\frac{1}{\rho_1(s)} + \frac{1}{\rho_2(s)}\right\}\mathrm{d}s\right] . \tag{8.24}$$

8.3. The ray and transport equations

A standard method of generating solutions of a partial differential equation such as the eikonal equation (8.9) is to introduce the curves which satisfy the ordinary differential equations

$$\frac{dx}{d\sigma} = \frac{\partial L}{\partial x}, \quad \frac{dy}{d\sigma} = \frac{\partial L}{\partial y}, \quad \frac{dz}{d\sigma} = \frac{\partial L}{\partial z}. \tag{8.25}$$

The parameter σ varies along a curve of solution and is connected to the arc length s by

$$\left(\frac{ds}{d\sigma}\right)^2 = N^2 \tag{8.26}$$

on account of the eikonal equation.

It is obvious from (8.25) that the curves are normal to the wavefronts and therefore we can identify them with rays. Yet the most convenient form is not furnished by (8.25) because the unknown L is involved. To eliminate L, observe that

$$\frac{d^2x}{d\sigma^2} = \frac{d}{d\sigma}\left(\frac{\partial L}{\partial x}\right) = \frac{\partial^2 L}{\partial x^2}\frac{dx}{d\sigma} + \frac{\partial^2 L}{\partial x \partial y}\frac{dy}{d\sigma} + \frac{\partial^2 L}{\partial x \partial z}\frac{dz}{d\sigma} = \frac{1}{2}\frac{\partial}{\partial x}(\text{grad}^2 L)$$

from (8.25). We infer from the eikonal equation (8.9) that

$$\frac{d^2x}{d\sigma^2} = N\frac{\partial N}{\partial x}, \quad \frac{d^2y}{d\sigma^2} = N\frac{\partial N}{\partial y}, \quad \frac{d^2z}{d\sigma^2} = N\frac{\partial N}{\partial z}. \tag{8.27}$$

These are differential equations for the rays which can be tracked directly and, in particular, by numerical methods. They can be solved when suitable initial data are supplied for some value of σ (see also §8.24).

The differential equations for the rays can be expressed in terms of the arc length via (8.26). They become

$$\frac{d}{ds}\left(N\,\frac{dx}{ds}\right)=\frac{\partial N}{\partial x},\qquad \frac{d}{ds}\left(N\,\frac{dy}{ds}\right)=\frac{\partial N}{\partial y},\qquad \frac{d}{ds}\left(N\,\frac{dz}{ds}\right)=\frac{\partial N}{\partial z} \tag{8.28}$$

subject to

$$\left(\frac{dx}{ds}\right)^2+\left(\frac{dy}{ds}\right)^2+\left(\frac{dz}{ds}\right)^2=1\ . \tag{8.29}$$

Once the rays have been found the eikonal L can be determined by remarking that its change between two points on a ray is $\int N\mathrm{d}s$. The magnitude of $\mathbf{E}_0$ can be found from the intensity law and so the geometrical optics field is known completely as soon as the direction of $\mathbf{E}_0$ is discovered.

To unravel the change in polarization along a ray a return is made to the transport equation (8.12) which now reduces to

$$2(\operatorname{grad} L.\operatorname{grad})\mathbf{E}_0+\mu\mathbf{E}_0\operatorname{div}(\tfrac{1}{\mu}\operatorname{grad} L)+\frac{1}{\mu\varepsilon}\{\mathbf{E}_0.\operatorname{grad}(\mu\varepsilon)\}\operatorname{grad} L=\mathbf{0}. \tag{8.30}$$

Introduce the vector $\mathbf{P}=\mathbf{E}_0/|\mathbf{E}_0|$; then

$$\frac{d\mathbf{P}}{d\sigma}=\frac{1}{|\mathbf{E}_0|}\frac{d\mathbf{E}_0}{d\sigma}-\frac{\mathbf{E}_0}{|\mathbf{E}_0|^3}\,\mathbf{E}_0\cdot\frac{d\mathbf{E}_0}{d\sigma}\ .$$

But, from (8.30), (8.25), and (8.8)

$$\mathbf{E}_0\cdot\frac{d\mathbf{E}_0}{d\sigma}=-\ \tfrac{1}{2}\mu|\mathbf{E}_0|^2\operatorname{div}\left(\tfrac{1}{\mu}\operatorname{grad} L\right)\ .$$

Hence (8.30) gives

$$\frac{d\mathbf{P}}{d\sigma}+\frac{1}{N}\,\mathbf{P}\,.\operatorname{grad} N\operatorname{grad} L=0. \tag{8.31}$$

It may be verified readily that $\mathbf{H}_0/|\mathbf{H}_0|$ satisfies (8.31) also.

One consequence of (8.31) is that $\mathbf{P} \cdot d\mathbf{P}/d\sigma = 0$ so that $|\mathbf{P}|^2$ is independent of σ, which is consistent with $\mathbf{P}$ being a unit vector.

Let $\mathbf{b}$ be a unit vector in the direction of the binormal to the ray. Then $\mathbf{b} = \mathbf{s} \wedge \boldsymbol{\nu}$ and the Serret–Frenet formulae are

$$\kappa\boldsymbol{\nu} = \frac{d\mathbf{s}}{ds}, \quad \frac{d\boldsymbol{\nu}}{ds} = -\kappa\mathbf{s} + \tau\mathbf{b}, \quad \frac{d\mathbf{b}}{ds} = -\tau\boldsymbol{\nu} \tag{8.32}$$

where τ is the torsion of the ray. Then, from (8.28),

$$\text{grad } N = \frac{d(N\mathbf{s})}{ds} = N\kappa\boldsymbol{\nu} + \mathbf{s}\frac{dN}{ds} .$$

Since $\mathbf{P} \cdot \mathbf{s} = 0$ it follows from (8.26) that (8.31) can be written as

$$\frac{d\mathbf{P}}{ds} + \kappa\mathbf{P} \cdot \boldsymbol{\nu}\mathbf{s} = \mathbf{0} .$$

Because $\mathbf{P}$ is perpendicular to $\mathbf{s}$, it is a linear combination of $\boldsymbol{\nu}$ and $\mathbf{b}$, say $P = \alpha\boldsymbol{\nu} + \beta\mathbf{b}$. Further $\alpha^2 + \beta^2 = 1$ because $\mathbf{P}$ is a unit vector. Hence

$$\alpha\frac{d\boldsymbol{\nu}}{ds} + \boldsymbol{\nu}\frac{d\alpha}{ds} + \beta\frac{d\mathbf{b}}{ds} + \mathbf{b}\frac{d\beta}{ds} + \kappa\alpha\mathbf{s} = \mathbf{0}$$

or

$$\left(\frac{d\alpha}{ds} - \tau\beta\right)\boldsymbol{\nu} + \left(\frac{d\beta}{ds} + \alpha\tau\right)\mathbf{b} = \mathbf{0}$$

on account of (8.32). Since $\mathbf{b}$ and $\boldsymbol{\nu}$ are perpendicular, their coefficients must vanish. So, if $\alpha = \cos\theta$ and $\beta = \sin\theta$,

$$\frac{d}{ds}\exp(i\theta) + i\tau\exp(i\theta) = 0$$

whence

$$\exp(i\theta) = \exp(i\theta_1)\exp\left(-i\int_{s_1}^{s}\tau ds\right)$$

where θ_1 is the value of θ at s_1. Hence

$$\mathbf{E}_0 = |\mathbf{E}_0|\{\boldsymbol{\nu}\cos(\theta_1 - \int_{s_1}^{s}\tau ds) + \mathbf{b}\sin(\theta_1 - \int_{s_1}^{s}\tau ds)\}\ . \qquad (8.33)$$

From (8.5)

$$\mathbf{H}_0 = |\mathbf{H}_0|\{\mathbf{b}\cos(\theta_1 - \int_{s_1}^{s}\tau ds) - \boldsymbol{\nu}\sin(\theta_1 - \int_{s_1}^{s}\tau ds)\}\ . \qquad (8.34)$$

Eqns (8.33) and (8.34) show how the polarization varies as one moves along a ray. The vectors $\mathbf{E}_0$, $\mathbf{H}_0$ start in the plane of $\boldsymbol{\nu}$ and $\mathbf{b}$ with $\mathbf{E}_0$ making an angle θ_1 with $\boldsymbol{\nu}$ and then rotate through the angle $-\int\tau ds$ with respect to $\boldsymbol{\nu}$ as $\mathbf{E}_0$, $\mathbf{H}_0$, $\boldsymbol{\nu}$, and $\mathbf{b}$ travel down the ray (Fig. 8.3).

If the ray is a plane curve the torsion τ disappears. The vector $\mathbf{E}_0$ then stays in the plane of $\boldsymbol{\nu}$ and $\mathbf{b}$ keeping at a constant angle to $\boldsymbol{\nu}$. In particular, if the medium is homogeneous the vector $\mathbf{E}_0$ always lies in a fixed plane containing the ray.

Once $\mathbf{E}_0$ is known we may contemplate the possibility of determining higher-order approximations from the transport equation (8.12). Write (8.12) as

$$2(\text{grad}\, L.\text{grad})\mathbf{E}_m + \mu\mathbf{E}_m\,\text{div}\left(\frac{1}{\mu}\,\text{grad}\, L\right) +$$

$$+ \frac{1}{\mu\varepsilon}\{\mathbf{E}_m.\text{grad}(\mu\varepsilon)\}\text{grad}\, L = \mathbf{C}_m$$

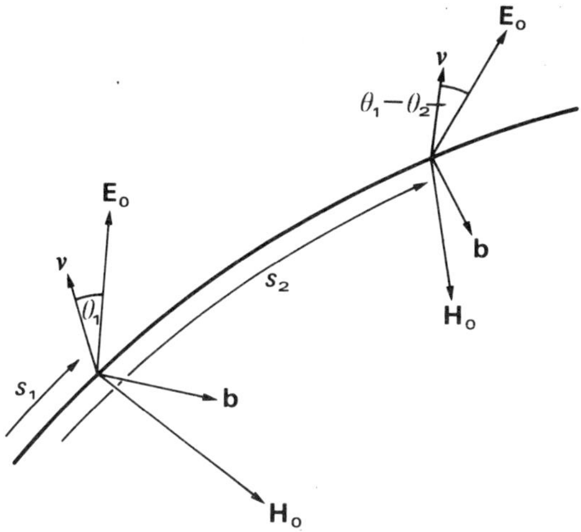

FIG. 8.3. Rotation of polarization along a ray.

where

$$\mathbf{C}_m = i\mu \operatorname{curl}\left(\frac{1}{\mu} \operatorname{curl} \mathbf{E}_{m-1}\right) - i \operatorname{grad}\left(\frac{1}{\varepsilon} \operatorname{div} \varepsilon \mathbf{E}_{m-1}\right)$$

is supposed to be known. Let $\mathbf{P}_m = \mathbf{E}_m/|\mathbf{E}_0|$. Then the following equation is obtained by the same means as (8.31) was derived:

$$\frac{d\mathbf{P}_m}{ds} + \frac{(\mathbf{P}_m \cdot \operatorname{grad} N)\mathbf{s}}{N} = \frac{\mathbf{C}_m}{2N|\mathbf{E}_0|} \quad . \tag{8.35}$$

A scalar product with $\mathbf{s}$ yields

$$\frac{d(N\mathbf{s}.\mathbf{P}_m)}{ds} = \frac{1}{2}\frac{\mathbf{C}_m.\mathbf{s}}{|\mathbf{E}_0|} \; .$$

Hence

$$(N\mathbf{s}.\mathbf{P}_m)_2 = \int_{s_1}^{s_2} \frac{1}{2}\frac{\mathbf{C}_m.\mathbf{s}\,ds}{|\mathbf{E}_0|} + (N\mathbf{s}.\mathbf{P}_m)_1 \quad (m \geqslant 1) \tag{8.36}$$

which supplies the change in the component of $\mathbf{P}_m$ normal to the wavefront in going along a ray from s_1 to s_2 since $\mathbf{C}_m$ and $\mathbf{E}_0$ are supposed to be known. If (8.36) is supplemented by an equation for the components of $\mathbf{P}_m$ tangential to the wavefront then $\mathbf{P}_m$ has been found.

Take a scalar product of (8.35) with $\mathbf{P}$. Then

$$\mathbf{P} \cdot \frac{d\mathbf{P}_m}{ds} = \frac{\mathbf{P} \cdot \mathbf{C}_m}{2N|\mathbf{E}_0|} .$$

On the other hand, a scalar product of (8.31) with $\mathbf{P}_m$ leads to

$$\mathbf{P}_m \cdot \frac{d\mathbf{P}}{ds} + \kappa\mathbf{P} \cdot \boldsymbol{\nu}\mathbf{P}_m \cdot \mathbf{s} = 0 .$$

By addition

$$\frac{d(\mathbf{P}.\mathbf{P}_m)}{ds} = \frac{\mathbf{P} \cdot \mathbf{C}_m}{2N|\mathbf{E}_0|} - \kappa\mathbf{P} \cdot \boldsymbol{\nu}\mathbf{P}_m \cdot \mathbf{s}$$

and so

$$(\mathbf{P}_m.\mathbf{P})_2 = \int_{s_1}^{s_2} \left\{\frac{\mathbf{P}.\mathbf{C}_m}{2N|\mathbf{E}_0|} - \kappa\mathbf{P}.\boldsymbol{\nu}\mathbf{P}_m.\mathbf{s}\right\} ds + (\mathbf{P}_m.\mathbf{P})_1 . \tag{8.37}$$

The integrand is known because $\mathbf{P}$ has already been found while $\mathbf{P}_m \cdot \mathbf{s}$ is given by (8.36). Accordingly, the tangential component of $\mathbf{P}_m$ parallel to $\mathbf{E}_0$ has been determined. In (8.37), $\mathbf{P}$ can be replaced by $\mathbf{H}_0/|\mathbf{H}_0|$ without altering its validity and so all components of $\mathbf{P}_m$ are now available.

A similar investigation of (8.13) provides the components of $\mathbf{H}_m$.

It should be remarked that the determination of $\mathbf{C}_m$ may not be trivial in practice. The procedure described supplies the field at points of rays, i.e. in ray coordinates. The vector operators in $\mathbf{C}_m$ have therefore to be expressed in terms of these coordinates or the ray coordinates have to be inverted to Cartesian form. In either case the formulae may be complicated although there is no difficulty of principle.

There is another way of writing the differential equations for a ray which can be revealing. In vector notation (8.25) is abbreviated to

$$\frac{\mathrm{d}\mathbf{x}}{\mathrm{d}\sigma} = \operatorname{grad} L. \tag{8.38}$$

Also

$$\frac{\mathrm{d}L}{\mathrm{d}\sigma} = N^2 . \tag{8.39}$$

$$\frac{\mathrm{d}(\operatorname{grad} L)}{\mathrm{d}\sigma} = N \operatorname{grad} N \tag{8.40}$$

as in the derivation of (8.27). Eqns (8.38)–(8.40) constitute an autonomous system of first-order differential equations for $\mathbf{x}$, grad L, and L. For some purposes the fact that they are first order may make them preferable to the second-order system of (8.27).

Let $\mathbf{k}_0 = k_0 \operatorname{grad} L$. Then (8.38) and (8.40) can be expressed as

$$\frac{\mathrm{d}\mathbf{x}}{\mathrm{d}\sigma} = \frac{\mathbf{k}_0}{k_0}, \qquad \frac{\mathrm{d}\mathbf{k}_0}{\mathrm{d}\sigma} = k_0 N \operatorname{grad} N.$$

Introduce the speed $c = 1/(\mu\varepsilon)^{\frac{1}{2}}$ and put $\mathrm{d}s/\mathrm{d}t = c$, the symbol t indicating that it is of the dimensions of time. Then

$$\frac{\mathrm{d}\mathbf{x}}{\mathrm{d}t} = \frac{c\mathbf{k}_0}{k_0 N}, \qquad \frac{\mathrm{d}\mathbf{k}_0}{\mathrm{d}t} = -\,k_0 N \operatorname{grad} c . \tag{8.41}$$

Before leaving the subject we derive an alternative, but equivalent, form for (8.24). Suppose that $\mathbf{x}$, L, and grad L are given on some surface S. On the surface they will be functions of two parameters, say σ_2 and σ_3. Off S values are determined by the integration of (8.38)–(8.40). They will be functions of σ, σ_2, and σ_3 with σ_2 and σ_3 constant on any particular ray. Thus $\mathbf{x}$ can be regarded as a function of σ, σ_2, and σ_3. To assist temporarily with the notation replace σ by σ_1 and let $\boldsymbol{\sigma}$ be the vector with components σ_1, σ_2, and σ_3.

Form the Jacobian

$$J = \frac{\partial(\mathbf{x})}{\partial(\boldsymbol{\sigma})} = \frac{\partial(x_1, x_2, x_3)}{\partial(\sigma_1, \sigma_2, \sigma_3)} \quad .$$

Then

$$\frac{\partial J}{\partial \sigma_1} = \sum_{j=1}^{3} \sum_{i=1}^{3} \frac{\partial}{\partial \sigma_1}\left(\frac{\partial x_i}{\partial \sigma_j}\right) \operatorname{cof} \frac{\partial x_i}{\partial \sigma_j}$$

where cof a_{ij} stands for the co-factor of a_{ij} in J. Hence

$$\frac{\partial J}{\partial \sigma_1} = \sum_{j=1}^{3} \sum_{i=1}^{3} \frac{\partial}{\partial \sigma_j}\left(\frac{\partial L}{\partial x_i}\right) \operatorname{cof} \frac{\partial x_i}{\partial \sigma_j} \quad .$$

Now $\operatorname{cof}(\partial x_i/\partial \sigma_j) = J\partial \sigma_j/\partial x_i$ and so

$$\frac{\partial J}{\partial \sigma_1} = J \sum_{j=1}^{3} \sum_{i=1}^{3} \frac{\partial}{\partial \sigma_j}\left(\frac{\partial L}{\partial x_i}\right)\frac{\partial \sigma_j}{\partial x_i} = J \sum_{i=1}^{3} \frac{\partial^2 L}{\partial x_i^2} = J\nabla^2 L \quad . \tag{8.42}$$

Reverting to our original notation and substituting for $\nabla^2 L$ in (8.16), we obtain

$$\frac{\mathrm{d}}{\mathrm{d}\sigma}\left(\frac{J}{\mu}\, \mathbf{E}_0^2\right) = 0 \tag{8.43}$$

i.e. $J\mathbf{E}_0^2/\mu$ is independent of σ on a ray. Changes of J account for the variation of the cross-section of a tube of rays during motion along the tube. Instead of calculating the integral of mean curvature as in (8.24) we have now to evaluate the Jacobian determinant of the relations between ray and Cartesian coordinates (see also §8.24 and R.M. Jones 1968).

8.4. The stratified medium

A medium in which μ and ε are constant on any one of a family of parallel planes is said to be *stratified*. The speed of propagation and refractive index will then vary only in a direction perpendicular to these planes.

Let the family of planes be parallel to $z = 0$. There is no loss of generality in taking the z axis vertically upwards. Then N is a function of z only. Only rays in the (x,z) plane need to be attended to because those in other planes through the z axis can be obtained by rotation about the z axis.

The ray equations (8.28) reduce to

$$\frac{\mathrm{d}}{\mathrm{d}s}\left(N\,\frac{\mathrm{d}x}{\mathrm{d}s}\right) = 0, \qquad \frac{\mathrm{d}}{\mathrm{d}s}\left(N\,\frac{\mathrm{d}z}{\mathrm{d}s}\right) = \frac{\partial N}{\partial z}\,.$$

Hence

$$\frac{\mathrm{d}x}{\mathrm{d}s} = \frac{C}{N} \tag{8.44}$$

where C is a constant. From (8.29)

$$\frac{\mathrm{d}z}{\mathrm{d}s} = \pm\left(1 - \frac{C^2}{N^2}\right)^{\frac{1}{2}}\,. \tag{8.45}$$

Assume that the ray starts from the origin and that $N(0)$ is positive. Measure s positively along the ray from the origin. Then C will be positive on rays which are propagating to the right. The upper sign in (8.45) will assign a ray travelling upwards, whereas the lower sign furnishes a ray going downwards. To fix ideas, let us concentrate on the upward ray. Then, after division of (8.45) by (8.44),

$$\frac{dz}{dx} = \left(\frac{N^2}{C^2} - 1\right)^{\frac{1}{2}} . \tag{8.46}$$

Consequently, the choice of C decides the slope of the ray as it leaves the origin. Clearly, the ray will be real only if $N^2 > C^2$; it will be assumed that this condition holds.

Integration of (8.46) discloses the equation of a ray through the origin as

$$x = \int_0^z \left\{\frac{N^2(w)}{C^2} - 1\right\}^{-\frac{1}{2}} dw . \tag{8.47}$$

Explicit evaluation of the integral is rarely possible and numerical integration will have to be resorted to. Fortunately, some properties can be inferred without knowledge of the explicit value of the integral.

If the tangent to the ray makes an angle θ with the z axis, $dx/ds = \sin\theta$. Hence (8.44) implies that

$$N(z)\sin\theta = C = N(0)\sin\theta_0 \tag{8.48}$$

where θ_0 is the value of θ at the origin. In a region where N increases with z, θ decreases and the ray bends towards the vertical. In contrast, if N decreases with z, i.e. the phase speed increases with height, the ray bends away from the vertical. There then arises the possibility that the ray is bent so far round that it starts to come down. In essence, the ray is being reflected and returned to its original level. Such reflection can take place only if at some height h the ray is horizontal, i.e. $\theta = \frac{1}{2}\pi$. Thus the phenomenon of reflection requires the existence of $N(h)$ such that

$$N(h) = N(0)\sin\theta_0 . \tag{8.49}$$

No real angle of launching is allowed unless $N(h) < N(0)$, i.e. the refractive index at altitude h is less than that at launching.

Suppose that, in fact, the refractive index decreases to a minimum at $z = m$ and then increases. Let θ_m be the angle such

$$\sin \theta_m = \frac{N(m)}{N(0)} .$$

If $\theta_0 < \theta_m$, the condition (8.49) can never be met for any value of h. A ray launched at such an angle will always travel vertically upwards and steadily depart from the level $z = 0$ (Fig. 8.4). If $\theta_0 > \theta_m$ the ray will be reflected and turn back at the height h given by (8.49). The horizontal distance x_h to the point of turning is, from (8.47) and (8.48), supplied by

$$x_h = N(0)\sin \theta_0 \int_0^h \{N^2(w) - N^2(0)\sin^2\theta_0\}^{-\frac{1}{2}} \mathrm{d}w .$$

In the critical case $\theta_0 = \theta_m$ we have $h = m$. Now, for w near m, $N^2(w) - N^2(m) = O\{(w-m)^2\}$ because of the minimum of $N(w)$ at $w = m$. The integrand in x_h thus becomes non-integrable and $x_h \to \infty$ as $\theta_0 \to \theta_m$. Therefore, the limiting ray $\theta_0 = \theta_m$ approaches the altitude $z = m$ asymptotically.

The equation for the reflected ray is not delivered by

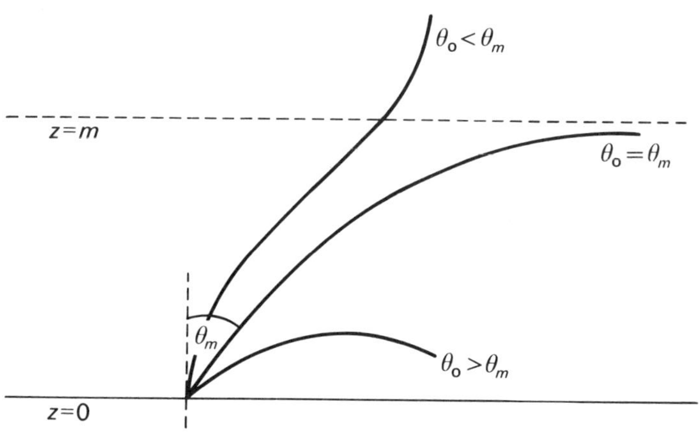

FIG. 8.4. Rays when N has a minimum at $z = m$.

(8.47) when $x > x_h$ since in this range the lower sign must be adopted in (8.45) to make the ray migrate downwards. The appropriate formula is

$$x = x_h - N(h) \int_h^z \{N^2(w) - N^2(h)\}^{-\frac{1}{2}} \mathrm{d}w \; . \qquad (8.50)$$

when $x \geqslant x_h$.

The eikonal or phase variation on a ray is estimated by

$$L = \int N \mathrm{d}s \; ,$$

the integration being along a ray. For an upgoing ray

$$L = \int_0^z \frac{N^2(w)}{\{N^2(w) - C^2\}^{\frac{1}{2}}} \mathrm{d}w = Cx + \int_0^z \{N^2(w) - C^2\}^{\frac{1}{2}} \mathrm{d}w \qquad (8.51)$$

when $L = 0$ at the origin. For $\theta_0 > \theta_m$, (8.51) is still correct while the ray is journeying upwards but, after reflection, must be replaced by

$$L = N(h)x + \int_0^h \{N^2(w) - N^2(h)\}^{\frac{1}{2}} \mathrm{d}w - \int_h^z \{N^2(w) - N^2(h)\}^{\frac{1}{2}} \mathrm{d}w \qquad (8.52)$$

where now x is given by (8.50).

The rays under consideration are plane curves and so, according to the theory of the preceding section, the polarization of the geometrical optics field is invariant. It therefore remains to evaluate the amplitude of the field. Formulae (8.19), (8.24), and (8.43) are available. For the sake of illustration we shall employ (8.43) but we must take care to acknowledge its three-dimensional character. Let r, ϕ, z be cylindrical polar coordinates. The cylindrical symmetry of the propagation guarantees that the equation of a ray in ϕ = constant can be obtained from (8.47) and (8.50) by substituting r for x. Therefore, in three dimensions, the form of the equation of a ray is

$$f(r, z, \theta_0) = 0 \ . \tag{8.53}$$

Characterize a ray by θ_0 and ϕ; then, in the definition of J, $\sigma_2 = \theta_0$ and $\sigma_3 = \theta$. Hence

$$J = r\left(\frac{\partial r}{\partial \theta_0}\frac{\partial z}{\partial \sigma} - \frac{\partial r}{\partial \sigma}\frac{\partial z}{\partial \theta_0}\right) .$$

From (8.53),

$$\frac{\partial f}{\partial r}\frac{\partial r}{\partial \sigma} + \frac{\partial f}{\partial z}\frac{\partial z}{\partial \sigma} = 0$$

so that

$$\frac{\partial r}{\partial \sigma} = -\lambda\frac{\partial f}{\partial z}, \qquad \frac{\partial z}{\partial \sigma} = \lambda\frac{\partial f}{\partial r} .$$

Hence $J = -r\lambda\, \partial f/\partial \theta_0$. On account of (8.26)

$$\lambda^2\left\{\left(\frac{\partial f}{\partial r}\right)^2 + \left(\frac{\partial f}{\partial z}\right)^2\right\} = N^2 .$$

Then $|\mathbf{E}_0|$ is determined from

$$\frac{r(\varepsilon/\mu)^{\frac{1}{2}}|\mathbf{E}_0|^2|\partial f/\partial \theta_0|}{\{(\partial f/\partial r)^2 + (\partial f/\partial z)^2\}^{\frac{1}{2}}} = \text{constant} \tag{8.54}$$

on a ray.

The geometrical optics field is therefore completely known. There is, however, a difficulty in that $\partial f/\partial \theta_0$ is zero on a reflected ray at the point of turning. Hence, geometrical optics ceases to hold near the point of turning. Indeed, there is also a change of phase of $\frac{1}{2}\pi$ introduced by passing through the point of turning as will be seen later (§8.14) when we come back to the consideration of how to cope with the failure of geometrical optics.

Exercises

3. A horizontally stratified medium has $N^2 = 1 + z$. Show that an up-going ray through $z = z_0$ is

$$x = 2C(1+z-C^2)^{\frac{1}{2}} - 2C(1+z_0-C^2)^{\frac{1}{2}}$$

and find the equation of a downgoing ray. If $1 + z_0 > C^2 > 1$ show that the downgoing ray starts to rise eventually and then has equation

$$x = 2C\{(1+z_0-C^2)^{\frac{1}{2}} + (1+z-C^2)^{\frac{1}{2}}\} .$$

Prove also that the point of turning lies on the curve

$$x = 2(1+z)^{\frac{1}{2}}(z_0-z)^{\frac{1}{2}} .$$

4. In Exercise 3 show that on the downgoing ray

$$L = Cx + \frac{2}{3}(1+z_0-C^2)^{3/2} \mp \frac{2}{3}(1+z-C^2)^{3/2}$$

indicating where the different signs are to be employed.

5. In a horizontally stratified medium $N^2 = N_1^2(1+a^2z^2)$. A downward ray leaves $x = 0$, $z = z_0$ at an angle $\cos^{-1}\{C/N_1(1+a^2z_0^2)^{\frac{1}{2}}\}$ to the x axis. If $C > N_1$ show that the ray turns at $z = h = (C^2-N_1^2)^{\frac{1}{2}}/aN_1$ and that its equation thereafter is

$$x = \frac{C}{N_1 a} \ln \frac{\{z+(z^2-h^2)^{\frac{1}{2}}\}\{z_0+(z_0^2-h^2)^{\frac{1}{2}}\}}{h^2} .$$

Prove that the change of phase along the ray is

$$\frac{1}{2} N_1 a\{z_0(z_0^2-h^2)^{\frac{1}{2}} - z(z^2-h^2)^{\frac{1}{2}}\} \pm \frac{1}{2} \frac{x(N_1^2+C^2)}{C}$$

the upper or lower sign being used according as the rising or falling part of the ray is under consideration.

6. By commencing with the vector form of (8.28), prove that the differential equations for rays in two-dimensional cylindrical polar coordinates r,ϕ are

$$N(\ddot{r}-r\dot{\phi}^2) + \dot{N}\dot{r} = \frac{\partial N}{\partial r},$$

$$N(r\ddot{\phi}+2\dot{r}\dot{\phi}) + \dot{N}r\dot{\phi} = \frac{\partial N}{r\partial \phi}$$

the dots signifying derivatives with respect to s.

In a *radially stratified medium* $\partial N/\partial\phi = 0$. Prove that in such a medium the rays have equations of the form

$$\phi = \int^{r} \frac{C\mathrm{d}r}{r(r^2N^2-C^2)^{\frac{1}{2}}} .$$

8.5. Fermat's principle

Draw any curve joining the points P_0 and P. Then the line integral $\int_{P_0}^{P} N\mathrm{d}s$ is known as the *optical path length* between P_0 and P along the curve. In general, the optical path length will depend not only upon the end-points P_0 and P but also on the actual path traced between them. If the curve is a ray, the optical path length is $L(P) - L(P_0)$ on account of (8.39). An immediate deduction is that the optical path length is the same for every ray between two wavefronts.

Comparison of (8.28) with the equations of motion of a particle in the theory of mechanics informs us that the optical path length is stationary on a ray. This way of characterizing a ray is embodied in

Fermat's principle: The rays between P_0 and P are those curves along which the optical path length is stationary with respect to infinitesimal variations in the path.

There may, of course, be several rays joining P_0 and P because of the various routes which can be open.

The inferences to be drawn from this guiding principle will now be examined. From the outset it must be recognized that the principle does not require the optical path length to be a minimum; only stationarity with respect to local variations of path is asked for. If this point is forgotten significant rays may be omitted.

When the medium is homogeneous the optical path length is a constant multiple of the geometrical path length. A straight line donates a minimum for both; therefore Fermat's principle asserts that in a homogeneous medium the rays are straight lines, in harmony with previous conclusions.

Consider the problem of an interface separating two different homogeneous media with P_0 and P on the same side of the interface (Fig. 8.5). Of the paths joining P_0 and P, the straight line P_0P is one path where the arc length is stationary. In fact, it is an absolute minimum of all paths between P_0 and P. It corresponds to the direct ray from P_0 to P and carries the primary radiation as if there were no interface present.

The next eventuality to investigate is whether there is a point O on the interface for which $P_0O + OP$ is stationary for small displacements of O on the interface. The connecting paths P_0O and OP must be straight lines because the medium is

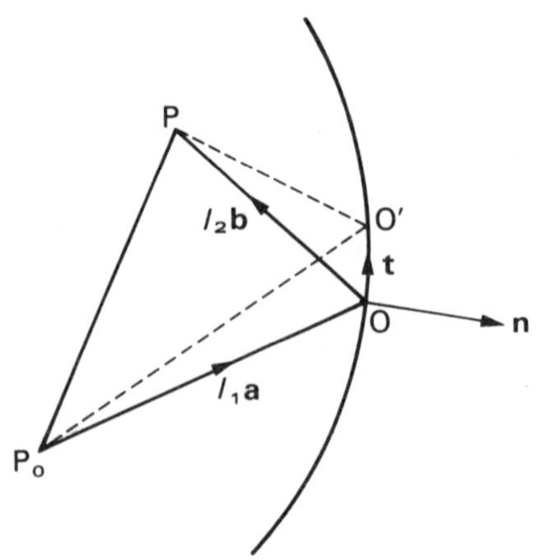

FIG. 8.5. Fermat's principle applied to reflection.

homogeneous. Let O′ be a point near to O with its position specified by $\mathbf{t}\delta s$ where $\mathbf{t}$ is a unit vector tangential to the interface at O. Draw $\mathbf{n}$ the unit vector along the normal to the interface at O and let $\mathbf{a}$, $\mathbf{b}$ be unit vectors along P_0O and OP respectively. Then the vectors P_0O and OP may be expressed as $l_1\mathbf{a}$ and $l_2\mathbf{b}$ respectively. Let the lengths of P_0O' and O′P be $l_1 + \delta l_1$ and $l_2 + \delta l_2$ respectively. The difference between the optical path length along $P_0O'P$ and that along P_0OP is $N(\delta l_1+\delta l_2)$. According to Fermat's principle this must vanish to the first order.

Now, if $\mathbf{a} + \delta\mathbf{a}$ is a unit vector along P_0O',

$$(l_1+\delta l_1)(\mathbf{a}+\delta \mathbf{a}) = l_1\mathbf{a} + \mathbf{t}\delta s$$

or

$$l_1\delta\mathbf{a} + \mathbf{a}\delta l_1 = \mathbf{t}\delta s$$

correct to the first order. But $\mathbf{a} \cdot \delta\mathbf{a} = 0$ because $\mathbf{a} + \delta\mathbf{a}$ is a unit vector. Hence

$$\delta l_1 = \mathbf{t} \cdot \mathbf{a}\delta s \, .$$

Similarly, $\delta l_2 = -\mathbf{t} \cdot \mathbf{b}\delta s$. As a consequence, Fermat's principle enforces

$$(\mathbf{a}-\mathbf{b}) \cdot \mathbf{t} = 0$$

for all $\mathbf{t}$ in the tangent plane at O. Therefore, the plane containing $\mathbf{a}$ and $\mathbf{b}$ is perpendicular to the tangent plane, i.e. the incident ray, reflected ray, and surface normal are all in the same plane. Also the incident and reflected rays make equal angles with the interface normal. This is in accordance with the customary laws of reflection. Note that no assertion is made that there is only a single point of reflection; there could be several relevant points depending on the relative positions of P_0 and P.

Snell's laws of refraction may also be verified by placing P_0 and P on opposite sides of the interface but the details will be omitted.

These results might have been anticipated since the

approximation of geometrical optics is one of a locally plane wave, and they suggest that even when the media are inhomogeneous Snell's laws are valid for reflection and refraction at a discontinuity. Nevertheless they must be regarded as an unexpected bonus.

The reason for that statement is that geometrical optics cannot be expected to be valid near the interface because the discontinuity in N violates the hypothesis on which the theory is based — namely, that the material characteristics shall be slowly varying. Actually, there is no real paradox because it is not genuinely affirmed that the rays are reflected and refracted at an interface according to Snell's laws. What is contended is that, away from the interface, the reflected and refracted waves can be thought of as composed of rays and that, if the rays were continued back to the interface (despite the ray picture being no longer a proper description there), they would comply with Snell's laws (see also §8.28).

Exercise

7. By placing P_0 and P on opposite sides of the interface in Fig. 8.5 verify that the incident ray, refracted ray, and surface normal lie in a plane. Show also that, if $\cos(\mathbf{a},\mathbf{t})$ is the cosine of the angle between $\mathbf{a}$ and $\mathbf{t}$.

$$N_1 \cos(\mathbf{a},\mathbf{t}) = N_2 \cos(\mathbf{b},\mathbf{t})$$

where N_1 and N_2 are the refractive indices of the two media.

Fermat's principle tells us the trajectory followed by a ray on reflection or refraction but says nothing about what happens to the amplitude of the field. The resolution of this difficulty depends upon the exact solution of reflection by an interface. However, it will be enough to discuss a plane wave and extract the reflection and refraction coefficients from that theory because of the local nature of the geometrical optics approximation. In other words, supplementary information is abstracted from a *canonical problem*.

Let the incident, reflected, and refracted rays be distinguished by the superscripts i, r, and t respectively. Let the incident and reflected rays make an angle θ with the normal to the interface (Fig. 8.6) whereas the refracted ray is inclined at an angle θ_1, where $N \sin\theta = N_1 \sin\theta_1$, N and N_1 being the refractive indices on each side of the interface. It is already known that the three rays through O lie in a plane, called the *plane of incidence*. The electric field may therefore be split into a component orthogonal to the plane of incidence (such a wave will be designated *electrically polarized*) and a component in the plane of incidence (giving rise to a *magnetically polarized* wave). For an electrically polarized wave the reflection coefficient R^{E} and transmission coefficient T^{E} are given by

$$R^{\mathrm{E}} = \frac{\mu_1 N\cos\theta - \mu(N_1^2 - N^2\sin^2\theta)^{\frac{1}{2}}}{\mu_1 N\cos\theta + \mu(N_1^2 - N^2\sin^2\theta)^{\frac{1}{2}}}, \tag{8.55}$$

$$T^{\mathrm{E}} = \frac{2\mu_1 N\cos\theta}{\mu_1 N\cos\theta + \mu(N_1^2 - N^2\sin^2\theta)^{\frac{1}{2}}} \tag{8.56}$$

while in the case of the magnetically polarized wave

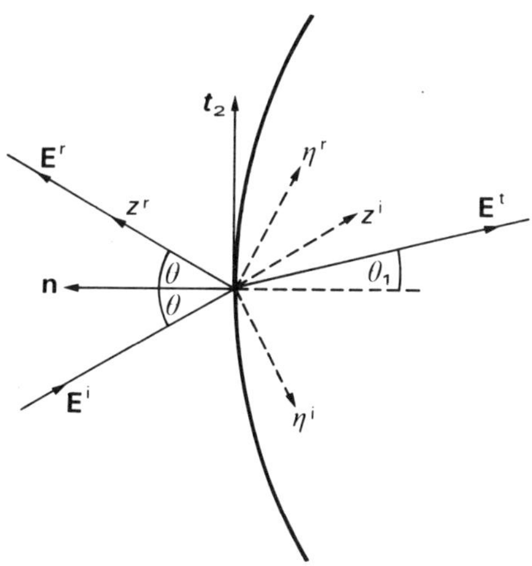

FIG. 8.6. Reflection and refraction at an interface.

$$R^{\mathrm{M}} = \frac{\mu N_1^2 \cos\theta - \mu_1 N(N_1^2 - N^2 \sin^2\theta)^{\frac{1}{2}}}{\mu N_1^2 \cos\theta + \mu_1 N(N_1^2 - N^2 \sin^2\theta)^{\frac{1}{2}}}, \tag{8.57}$$

$$T^{\mathrm{M}} = \frac{(\mu_1\varepsilon/\mu\varepsilon_1)^{\frac{1}{2}}\, 2\mu N_1^2 \cos\theta}{\mu N_1^2 \cos\theta + \mu_1 N(N_1^2 - N^2 \sin^2\theta)^{\frac{1}{2}}}. \tag{8.58}$$

In (8.57) the components are in the directions of η^{i} and η^{r} in Fig. 8.6 while the extra factor $(\mu_1\varepsilon/\mu\varepsilon_1)^{\frac{1}{2}}$ in (8.58) takes account of the fact that the electric intensity rather than the magnetic intensity is involved. If the boundary should be perfectly conducting these simplify to $R^{\mathrm{E}} = -1$, $R^{\mathrm{M}} = 1$.

In homogeneous media the field on a reflected ray may now be taken as

$$\mathbf{E}^{\mathrm{r}} = \left|\frac{\rho_1^{\mathrm{r}}\rho_2^{\mathrm{r}}}{(\rho_1^{\mathrm{r}}+s^{\mathrm{r}})(\rho_2^{\mathrm{r}}+s^{\mathrm{r}})}\right|^{\frac{1}{2}} \exp(-\mathrm{i}k_0 L^{\mathrm{r}}) R\mathbf{E}^{\mathrm{i}}(0) \tag{8.59}$$

from (8.22). The superscript r indicates quantities appropriate to the reflected wavefront and R is a reflection matrix. When the field is dissected into its electrically and magnetically polarized parts

$$R = \begin{pmatrix} R^{\mathrm{E}} & 0 \\ 0 & R^{\mathrm{M}} \end{pmatrix} \tag{8.60}$$

with R^{E} and R^{M} given by (8.55) and (8.57). There is a similar representation for the transmitted wave.

If the media are not homogeneous we can still apply

$$\mathbf{E}^{\mathrm{r}}(0) = R\mathbf{E}^{\mathrm{i}}(0), \qquad \mathbf{E}^{\mathrm{t}}(0) = T\mathbf{E}^{\mathrm{i}}(0)$$

but now N, N_1, μ, and μ_1 must have their values at O in R and T. Away from the boundary, the field has to be determined by the laws already established for tubes and the possibility of a rotation of polarization cannot be ignored.

In (8.59) the curvature of the reflected wavefront must be known before calculations can be carried out and we now investigate the affiliation with the curvatures of the incident wavefront and interface. Let the z axis in Fig. 8.2 go from A towards A′ and let ξ, η be any Cartesian coordinates in the (x, y) plane with origin A. Then, on the axial and paraxial rays,

$$L(\boldsymbol{\xi}, z) = L(\mathbf{0}, z) + \frac{1}{2} N\boldsymbol{\xi}^{\mathrm{T}} C(z)\boldsymbol{\xi} \tag{8.61}$$

correct to the second order. Here $\boldsymbol{\xi}$ is a vector in the (x, y) plane and C is a symmetric matrix governing the behaviour of the curvature. In fact,

$$\det C(0) = \frac{1}{\rho_1 \rho_2}, \qquad \operatorname{trace} C(0) = \frac{1}{\rho_1} + \frac{1}{\rho_2} \tag{8.62}$$

Since $\mathrm{grad}^2 L$ is constant a derivative with respect to z, followed by putting $\boldsymbol{\xi} = \mathbf{0}$, $z = 0$, gives $(\partial^2 L/\partial z^2)_0 = 0$. Hence (8.61) reduces to

$$L(\boldsymbol{\xi}, z) = L(\mathbf{0}, 0) + Nz + \frac{1}{2} N\boldsymbol{\xi}^{\mathrm{T}} C(0)\boldsymbol{\xi} \tag{8.63}$$

so long as z is correct to the second order.

Now choose the origin to be the point O of the frontier in Fig. 8.6. Let t_1, t_2 be coordinates perpendicular to the unit normal $\mathbf{n}$ with t_1 perpendicular to the plane of incidence. If $\boldsymbol{\xi}^{\mathrm{i}}$, z^{i} and $\boldsymbol{\xi}^{\mathrm{r}}$, z^{r} are the coordinates relevant to the incident and reflected waves select ξ^{i} and ξ^{r} as perpendicular to the plane of incidence. Let $\mathbf{t}$ be the tangential position of a point on the boundary near O; the displacement of the point in the direction of $\mathbf{n}$ will be $-\frac{1}{2}\mathbf{t}^{\mathrm{T}} C_{\mathrm{b}} \mathbf{t}$ where C_{b} is the curvature matrix of the interface at O. Hence, for results correct to the second order, we may take

$$z^{\mathrm{i}} = t_2 \sin\theta + \frac{1}{2}\mathbf{t}^{\mathrm{T}} C_{\mathrm{b}} \mathbf{t} \cos\theta, \qquad \xi^{\mathrm{i}} = t_1, \qquad \eta^{\mathrm{i}} = -t_2 \cos\theta .$$

Hence, from (8.63),

$$L^{\mathrm{i}} = L^{\mathrm{i}}(\mathbf{0},0) + Nt_2 \sin\theta + \tfrac{1}{2} N\mathbf{t}^{\mathrm{T}}C_{\mathrm{b}}\mathbf{t}\cos\theta + \tfrac{1}{2} N\boldsymbol{\xi}^{\mathrm{iT}}C^{\mathrm{i}}\boldsymbol{\xi}^{\mathrm{i}} .$$

For the reflected ray

$$z^{\mathrm{r}} = t_2 \sin\theta - \tfrac{1}{2}\mathbf{t}^{\mathrm{T}}C_{\mathrm{b}}\mathbf{t}\cos\theta, \qquad \xi^{\mathrm{r}} = t_1, \qquad \eta^{\mathrm{r}} = t_2\cos\theta ,$$

$$L^{\mathrm{r}} = L^{\mathrm{r}}(\mathbf{0},0) + Nt_2 \sin\theta - \tfrac{1}{2} N\mathbf{t}^{\mathrm{T}}C_{\mathrm{b}}\mathbf{t}\cos\theta + \tfrac{1}{2} N\boldsymbol{\xi}^{\mathrm{rT}}C^{\mathrm{r}}\boldsymbol{\xi}^{\mathrm{r}} .$$

The phases of the incident and reflected waves must agree on the interface, i.e. $L^{\mathrm{i}} = L^{\mathrm{r}}$. That demands firstly that $L^{\mathrm{i}}(\mathbf{0},0) = L^{\mathrm{r}}(\mathbf{0},0)$ which fixes the constant in the eikonal. Secondly,

$$\mathbf{t}^{\mathrm{T}}C_{\mathrm{b}}\mathbf{t}\cos\theta + \tfrac{1}{2}\boldsymbol{\xi}^{\mathrm{iT}}C^{\mathrm{i}}\boldsymbol{\xi}^{\mathrm{i}} = \tfrac{1}{2}\boldsymbol{\xi}^{\mathrm{rT}}C^{\mathrm{r}}\boldsymbol{\xi}^{\mathrm{r}} .$$

This equation holds for arbitrary small $\mathbf{t}$ and so

$$C^{\mathrm{r}} = \begin{pmatrix} 2C_{\mathrm{b}_{11}}\cos\theta + C^{\mathrm{i}}_{11} & 2C_{\mathrm{b}_{12}} - C^{\mathrm{i}}_{12} \\ 2C_{\mathrm{b}_{12}} - C^{\mathrm{i}}_{12} & 2C_{\mathrm{b}_{22}}\sec\theta + C^{\mathrm{i}}_{22} \end{pmatrix} \tag{8.64}$$

where C_{ij} is a typical element of the matrix C.

Eqn (8.64) specifies the curvature matrix of the reflected wavefront at O and the principal radii of curvature may be deduced from (8.62). There is now enough information to infer the curvature elsewhere.

As regards the transmitted wave a parallel analysis throws up $L^{\mathrm{t}}(\mathbf{0},0) = L^{\mathrm{i}}(\mathbf{0},0)$ and

$$N_1C^{\mathrm{t}} = \begin{pmatrix} C_{\mathrm{b}11}(N\cos\theta - N_1\cos\theta_1) + NC^{\mathrm{i}}_{11} & C_{\mathrm{b}12}(N\cos\theta\sec\theta_1 - N_1) + NC^{\mathrm{i}}_{12}\cos\theta\sec\theta_1 \\ C_{\mathrm{b}12}(N\cos\theta\sec\theta_1 - N_1) + NC^{\mathrm{i}}_{12}\cos\theta\sec\theta_1 & C_{\mathrm{b}22}\sec\theta_1(N\cos\theta\sec\theta_1 - N_1) + NC^{\mathrm{i}}_{22}\cos^2\theta\sec^2\theta_1 \end{pmatrix} . \tag{8.65}$$

It will be remarked that (8.64) gives trouble when $\theta = \frac{1}{2}\pi$, i.e. when the incident ray is tangential to the interface. Also (8.65) is unsatisfactory when the transmitted ray is tangential to the boundary. To deal with such exceptional points more canonical problems have to be solved. We shall return to these matters later (§8.21). In the meantime, we observe that, apart from exceptional points, geometrical optics converts the problem of propagation to one of solving a system of ordinary differential equations such as (8.38)–(8.40). It is therefore appropriate to spend some time on the practical solution of ordinary differential equations.

NUMERICAL SOLUTION OF ORDINARY DIFFERENTIAL EQUATIONS

A vast literature exists concerning solving ordinary differential equations, and so it will be necessary to restrict what follows to the salient points. It is hoped that in the process the reader will garner sufficient knowledge of the terminology and properties to be able to pick out the main features of the literature. The reader is quite likely to end up searching papers for methods on differential equations because, at the moment, the state of the art is such that there is a strong probability that one will find that the routines on the local machines are not performing what one wants done reasonably efficiently and inexpensively.

Should the reader discover that he needs to write his own program he should bear in mind that, to make a good quality efficient program available to others without difficulty, it must be easy to implement at other computer installations. Adequate documentation is essential and aspects which should be included are as follows.

(i) Comment cards in the program identifying arguments and providing general instructions for the user in case the program becomes separated from other documentation.

(ii) A document with examples showing how to use the program and illustrating user-oriented features.

(iii) Substantial examples of the performance of the program over a wide range of problems.

(iv) Description of any actions which are known to result in misuse or failure of the program.

(v) Explanation of how to convert the code to other machines with a clear enumeration of any machine-dependent ingredients.

(vi) Comment cards in the program explaining what is being done.

(vii) An account of the theoretical support for the program as well as a rationale for any heuristic parts.

Full attention to (i)–(v) will enable the casual user, who wants to regard the operation of the program as if it were in a black box, to solve his problem satisfactorily and without undue effort. Indeed, the designer of any program, whether for differential equations or not, should never lose sight of the above criteria for documentation.

8.6. Multistep methods

A typical first-order differential equation is $y' = f(x,y)$ where the prime indicates a derivative with respect to x. Usually, a solution is required on some interval (a,b) such that $y(a) = \eta$, i.e. an *initial-value problem* has to be solved. By employing vector notation, we may place a system of first-order differential equations in the same form, namely

$$\mathbf{y}' = \mathbf{f}(x,\mathbf{y}), \quad \mathbf{y}(a) = \boldsymbol{\eta}.$$

The initial-value problem for a higher-order differential equation can always be transformed into one for a first-order system, so confining our discussion to the differential equation in the first sentence will not be a serious limitation in bringing out the basic principles that will be encountered.

All numerical methods solve a differential equation by steps, i.e. values are found at $x_n = a + nh$ where h is the *steplength*. An efficient computer program should select the steplength automatically at any stage subject to overrule by the user. The target will be to make the steps as large as possible while consistent with any error bounds imposed by the user.

Let y_n be the value found by the computer at x_n. In general it will differ from the value $y(x_n)$ attained by the exact solution there. If f_n is written for $f(x_n, y_n)$ an equation

$$\sum_{j=0}^{k} \alpha_j y_{n+j} = h \sum_{j=0}^{k} \beta_j f_{n+j} \tag{8.66}$$

in which $\alpha_k = 1$ is said to produce a *linear multistep method of stepnumber k* or a *linear k-step method*. If $\beta_k = 0$ the method is *explicit* because the current value y_{n+k} can be computed immediately from earlier values. If $\beta_k \neq 0$ the method is *implicit* and finding y_{n+k} obliges one to solve a non-linear equation of the form

$$y_{n+k} = h\beta_k f(x_{n+k}, y_{n+k}) + g$$

g being known. Usually, the iterative techniques of §1.8 (§4.5 for a system) will be involved.

By Taylor's theorem

$$y(x_n+h) = y(x_n) + hy'(x_n) + \frac{1}{2} h^2 y''(x_n) + \ldots .$$

Truncating this expansion after two terms and substituting for y' from the differential equation, we obtain

$$y(x_n+h) = y(x_n) + hf(x_n, y(x_n)) . \tag{8.67}$$

If $y(x_n)$ is now replaced by its approximation y_n, there ensues

$$y_{n+1} = y_n + hf_n . \tag{8.68}$$

The formula (8.68) provides an explicit linear one-step method, known as *Euler's rule*. Since the inaccuracy in (8.67) is $O(h^2)$, the error in Euler's rule may be said to be of $O(h^2)$.

The inclusion of the third term in the Taylor expansion together with the expansion of $y(x_n-h)$ yields the *mid-point rule*

$$y_{n+2} - y_n = 2hf_{n+1} .$$

with an error of $O(h^3)$. Another method with error of $O(h^3)$ is the *trapezoidal rule*

$$y_{n+1} - y_n = \frac{1}{2} h(f_n + f_{n+1}) .$$

A different way of generating multistep methods is to start from the exact equation

$$y(x) - y(x_n) = \int_{x_n}^{x} y'(t)\mathrm{d}t = \int_{x_n}^{x} f(t,y(t))\mathrm{d}t . \tag{8.69}$$

The integral is evaluated by approximating f by a polynomial. For example, if $P(x)$ is the quadratic polynomial which takes the values f_{n+1}, f_n, and f_{n-1} at x_{n+2}, x_n, and x_{n-1} respectively, we write

$$y_{n+1} - y_n = \int_{x_n}^{x_{n+1}} P(t)\mathrm{d}t = \frac{h}{12}(5f_{n+1} + 8f_n - f_{n-1}) \tag{8.70}$$

which is called a *two-step Adams–Moulton method* and is implicit. The Newton–Cotes quadrature formulae (§5.12), which are based on polynomial approximation, all lead to multistep methods, for instance, *Simpson's rule*

$$y_{n+2} - y_n = \frac{1}{3} h(f_{n+2} + 4f_{n+1} + f_n) \tag{8.71}$$

which is an implicit two-step method.

There are thus many multistep methods (for a selection see Lambert 1973) and their proliferation accounts for some of the difficulty in assessing their relative merits. One essential requirement is that as $h \to 0$ the numerical solution should tend to the exact solution. Since we wish to compare them at a fixed x we must also ask that, as $h \to 0$, $n \to \infty$ in such a way that nh does not alter. If $nh = c$ in this procedure, we may say that a linear multistep method is *convergent* if, for all initial values, $y_n \to y(a+c)$ for any c satisfying $0 \leqslant c \leqslant b-a$ and for all solutions y_n complying with the starting conditions $y_m = \eta_m(h)$ for which

$$\lim_{h\to 0} \eta_m(h) = \eta \qquad (m = 0, 1, \ldots, k-1) .$$

With a linear multistep method is associated the linear difference operator $\mathcal{L}$ defined by

$$\mathcal{L}[z(x);h] = \sum_{j=0}^{k} \{\alpha_j z(x+jh) - h\beta_j z'(x+jh)\} \tag{8.72}$$

where z is any function which is continuously differentiable on $[a,b]$. If z and z' are expanded as Taylor series about x,

$$\mathcal{L}[z(x);h] = C_0 z(x) + C_1 hz'(x) + \ldots \tag{8.73}$$

where

$$C_0 = \alpha_0 + \alpha_1 + \ldots + \alpha_k ,$$

$$C_1 = \alpha_1 + 2\alpha_2 + \ldots + k\alpha_k - (\beta_0+\beta_1+\ldots+\beta_k) , \tag{8.74}$$

$$C_q = \frac{(\alpha_1+2^q\alpha_2+\ldots+k^q\alpha_k)}{q!} - \frac{(\beta_1+2^{q-1}\beta_2+\ldots+k^{q-1}\beta_k)}{(q-1)!}$$

for $q = 2, 3, \ldots$. A multistep method is said to be of *order* p if $C_0 = C_1 = \ldots = C_p = 0$ but $C_{p+1} \neq 0$; C_{p+1} is then known as the *error constant*. If y is the theoretical solution of the differential equation $\mathcal{L}[y(x);h]$ is called the *local truncation error*.

Exercise

8. Find the order and error constant of (*a*) Euler's rule, (*b*) *Quade's method*

$$y_{n+4} - \frac{8}{19}(y_{n+3}-y_{n+1}) - y_n = \frac{6h}{19}(f_{n+4}+4f_{n+3}+4f_{n+1}+f_n) .$$

Suppose now that a multistep method is convergent, i.e. as $h \to 0$, $n \to \infty$ with $nh = c$, $y_n \to y(a+c)$. Then $y_{n+j} \to y(a+c)$ for $j = 0, 1, \ldots, k$. Hence $y(a+c) = y_{n+j} + \theta_{jn}(h)$ where $\theta_{jn} \to 0$ for $j = 0, 1, \ldots, k$. Then

$$y(a+c) \sum_{j=0}^{k} \alpha_j = h \sum_{j=0}^{k} \beta_j f_{n+j} + \sum_{j=0}^{k} \alpha_j \theta_{jn}(h) .$$

In the limit as $h \to 0$, both terms on the right-hand side vanish. Since $y(a+c)$ is not in general zero $\Sigma_{j=0}^{k}\alpha_j = 0$. Thus a necessary condition for convergence is, from (8.74), $C_0 = 0$, i.e. the method is of order zero at least.

In addition, $(y_{n+j}-y_n)/jh \to y'(a+c)$ for $j = 1, \ldots, k$ necessitates

$$y_{n+j} - y_n = jhy'(a+c) + jh\phi_{jn}(h)$$

where $\phi_{jn} \to 0$. Hence

$$\sum_{j=0}^{k} \alpha_j(y_{n+j}-y_n) = hy'(a+c) \sum_{j=0}^{k} j\alpha_j + h \sum_{j=0}^{k} j\alpha_j\phi_{jn}(h) .$$

Since $\Sigma_{j=0}^{k}\alpha_j = 0$, division by h gives

$$\sum_{j=0}^{k} \beta_j f_{n+j} = y'(a+c) \sum_{j=0}^{k} j\alpha_j + \sum_{j=0}^{k} j\alpha_j\phi_{jn}(h)$$

or

$$f(a+c,y(a+c)) \sum_{j=0}^{k} \beta_j = y'(a+c) \sum_{j=0}^{k} j\alpha_j$$

after the limiting process. The differential equation can be satisfied if and only if $\Sigma_{j=0}^{k} j\alpha_j = \Sigma_{j=0}^{k} \beta_j$, i.e. $C_1 = 0$.

A multistep method of order $p \geqslant 1$ is said to be *consistent*. Thus, the above demonstration shows that y_n *converges to a solution of the differential equation only if the multistep method is consistent*. Although consistency is necessary to convergence it is not sufficient.

The stability of the numerical scheme is connected with the polynomials

$$\rho(\zeta) = \sum_{j=0}^{k} \alpha_j \zeta^j, \qquad \sigma(\zeta) = \sum_{j=0}^{k} \beta_j \zeta^j \tag{8.75}$$

which are called the *first* and *second characteristic polynomials* respectively. A multistep method can be consistent if and only if $\rho(1) = 0$, $\rho'(1) = \sigma(1)$. Consequently, in a consistent method the polynomial $\rho(\zeta)$ always has a zero at $\zeta = 1$. This zero is called the *principal zero* and the remaining zeros $\zeta_2, \ldots, \zeta_k$ are designated *spurious zeros*. If f is identically zero, the multistep method becomes the linear difference equation $\Sigma_{j=0}^{k} \alpha_j y_{n+j} = 0$ with a solution

$$y_n = d_1 + d_2 \zeta_2^n + \ldots + d_k \zeta_k^n$$

when the zeros of ρ are distinct. Clearly, a crisis will develop if $|\zeta_s| > 1$ for any s but zeros with $|\zeta_s| \leqslant 1$ entail no trouble. Multiple zeros give rise to secular terms which do not affect the situation unless $|\zeta_s| = 1$, when there is blow up. Thus, parasitic solutions are avoided by insisting that $|\zeta_s| \leqslant 1$ with any ζ_s such that $|\zeta_s| = 1$ being simple. For this reason a multistep method is said to be *zero-stable* if no zero of $\rho(\zeta)$ has modulus greater than unity and if every zero of modulus 1 is simple.

In a consistent zero-stable method $\sigma(1) \neq 0$. Otherwise $\rho'(1) = 0$ which implies a double zero at least at $\zeta = 1$.

It can be proved that a multistep method is convergent if and only if it is consistent and zero-stable. Indeed, consistency keeps the local truncation error down while zero-stability controls the propagation of error as the calculation proceeds. There is no point in considering multistep methods which are not both consistent and zero-stable. In this connection it should be noted that no zero-stable method can have order exceeding $k + 1$ (k odd) or $k + 2$ (k even).

Adams methods are those in which $\rho(\zeta) = \zeta^k - \zeta^{k-1}$. The group is subdivided into the explicit or *Adams–Bashforth* methods and the implicit or *Adams–Moulton* methods. Methods in which $\rho(\zeta) = \zeta^k - \zeta^{k-2}$ are known as *Nyström* when explicit and *Milne–Simpson* when implicit. All the spurious roots in Adams methods are at the origin and so the family is zero-stable. The Nyström and Milne–Simpson groups are also zero-stable because they have one spurious root at -1 and the rest at the origin.

Exercises

9. The initial value problem $y' = 4xy^{\frac{1}{2}}$, $y(0) = 1$ is to be solved on $0 \leqslant x \leqslant 2$ by

$$y_{n+2} - y_{n+1} = \frac{1}{3}\, h(3f_{n+1} - 2f_n) \;.$$

Show that the method is zero-stable but not consistent. By taking $y_1 = (1+h^2)^2$ calculate the numerical solution for $h = 0.1$ and show that it differs substantially from the exact solution at $x = 2$ ($y(2) = 25$). Repeat the calculation with $h = 0.05$, $h = 0.025$ to demonstrate that the situation worsens as the step length decreases.

10. Prove that the method

$$y_{n+2} - (1+a)y_{n+1} + ay_n = \frac{1}{2}\, h\{(3-a)f_{n+1} - (1+a)f_n\}$$

is consistent. Show that it is zero-stable for $a = 0$ but not for $a = -5$. For these two values of a solve numerically the initial value problem of Exercise 9 for the same values of h and $y_1 = (1+h^2)^2$ to exemplify the effects of instability. Are the effects mitigated by reducing the step length?

11. Is Quade's method (Exercise 8) zero-stable?

12. Show that the method

$$y_{n+2} + (a-1)y_{n+1} - ay_n = \frac{1}{4}\, h\{(a+3)f_{n+2} + (3a+1)f_n\}$$

has order 2 if $a \neq -1$ and order 3 if $a = -1$. Is the method convergent when $a = -1$?

13. Prove that, for any value of a for which

$$y_{n+2} + ay_{n+1} - ay_n - y_{n-1} = \frac{1}{2}(3+a)h(f_{n+1}-f_n)$$

is zero-stable, the order cannot exceed 2.

A k-step method in which $k > 1$ requires more starting data than provided by the initial value. Therefore, if k is kept fixed, means must be discovered whereby additional data are created. If f is differentiable a sufficient number of times, derivatives of the differential equation permit the determination of y_1 by a Taylor expansion about $x = a$ with an error of order $O(h^{p+1})$ if the method is of order p. Repetition of the process at $x = a + h,\ a + 2h,\ \ldots$ will then supply the requisite number of starting values. The execution is cumbersome at best and fails completely if f does not possess enough derivatives. Most practical programs, therefore, vary k. At the first step $k = 1$ and then k is increased as more data become available. There is the accompanying advantage that the degree of interpolation can be adjusted as well as the step length although the optimal choice of k and step length is a complicated matter.

To assist with this choice we need some idea of the errors being committed by the algorithm. It can be proved that for all Adams methods the local truncation error satisfies

$$|\mathcal{L}[y(x_n);h]| \leqslant h^{p+1}K \max_{x\in[a,b]} |y^{(p+1)}(x)| \tag{8.76}$$

where the method is of order p and

$$K = \frac{\left|\int_0^k \sum_{j=0}^{k} \{\alpha_j(j-t)_+^p - p\beta_j(j-t)_+^{p-1}\}dt\right|}{p!} = |C_{p+1}| .$$

A bound similar to (8.76) prevails for other methods but the definition of K is different.

Inequality (8.76) instructs us how the error is behaving locally but leaves open the accumulation of error as the computation progresses. Round-off error may have to be accommodated; suppose on each application of the method it does not exceed $K_1 h^{q+1}$. Let the maximum error in the k starting values be δ and assume that, for $a \leqslant x \leqslant b$, $|f(x,y)-f(x,Y)| \leqslant L|y-Y|$. Then, for an Adams–Bashforth or Nyström method, the *global error* satisfies

$$|y_n - y(x_n)| \leqslant \{\delta + (x_n - a)(h^p K \max|y^{(p+1)}| + h^q K_1)\}\exp\{L(x_n - a)\sum_{j=0}^{k-1}|\beta_j|\}. \tag{8.77}$$

The bound for any zero-stable method has a similar structure.

The snag with both bounds (8.76) and (8.77) is being compelled to estimate $y^{(p+1)}$. Frequently, no analytical bound is available and one must be constructed from the computed solution, probably by numerical differentiation — a notoriously inaccurate undertaking. Even then bounds in (8.76) and (8.77) can be very conservative. One can, however, assert with confidence that if the bound for the local error is $O(h^{p+1})$ that for the global error will be not better than $O(h^p)$.

Two ways suggest themselves for circumventing the impediment in error estimation: (i) adopt a criterion which forces the global error to die out as the numerical reckoning proceeds; (ii) invent a method for which the evaluation of an error bound is feasible. The possibility (i) will be considered firstly, (ii) being left on one side until later.

The theoretical solution of the initial value problem satisfies

$$\sum_{j=0}^{k} \alpha_j y(x_{n+j}) = h \sum_{j=0}^{k} \beta_j f(x_{n+j}, y(x_{n+j})) + \mathcal{L}[y(x_n);h]$$

the last term being the local truncation error. If R_{n+k} is the round-off error introduced at the nth application of the method, the numerical solution provides $\tilde{y}_n$ where

$$\sum_{j=0}^{k} \alpha_j \tilde{y}_{n+j} = h \sum_{j=0}^{k} \beta_j f(x_{n+j}, \tilde{y}_{n+j}) + R_{n+k} .$$

Assume that $\partial f/\partial y$ has the constant value λ and that $\mathcal{L}[y(x_n);h] - R_{n+k} = a$ where a is a constant. Then, from subtraction of the above equations, the global error $e_n = y(x_n) - \tilde{y}_n$ satisfies

$$\sum_{j=0}^{k} (\alpha_j - h\lambda\beta_j) e_{n+j} = a .$$

The general solution of this linear difference equation is, when the multistep method is consistent and zero-stable,

$$e_n = \sum_{s=1}^{k} c_s r_s^n - \frac{a}{h\lambda \sum_{j=0}^{k} \beta_j} . \tag{8.78}$$

where c_s is an arbitrary constant and the r_s are the roots of

$$\rho(r) - h\lambda\sigma(r) = 0. \tag{8.79}$$

It has been supposed in (8.78) that the roots of this polynomial equation are simple.

Put $h_1 = h\lambda$. As $h_1 \to 0$, (8.79) approaches $\rho(r) = 0$ and $r_s \to \zeta_s$; in particular, $r_1 \to 1$. No other r_s tends to 1 because of zero-stability. Now, when the method is of order p, $\mathcal{L}[z(x);h] = O(h^{p+1})$ so that, choosing $z(x) = \mathrm{e}^{\lambda x}$, we obtain

$$\rho(\exp h_1) - h_1 \sigma(\exp h_1) = O(h_1^{p+1})$$

or

$$(\exp h_1 - r_1)(\exp h_1 - r_2) \ldots (\exp h_1 - r_k) = O(h_1^{p+1}) \ .$$

Since $\exp h_1 - r_s$ does not tend to zero as $h_1 \to 0$ for $s \neq 1$ we conclude that

$$r_1 = \exp h_1 + O(h_1^{p+1}) \tag{8.80}$$

as $h_1 \to 0$. An immediate deduction is that $r_1 > 1$ for sufficiently small h_1.

The derivation of (8.80) is based on $\partial f/\partial y = \lambda$ and so holds strictly only for the differential equation $y' = \lambda y + g(x)$. In general, a solution will contain a multiple of $e^{\lambda x}$. The error equation (8.78) embraces a term r_1^n which, on account of (8.80), is $\exp\{\lambda(x_n - a)\} + O(h_1^{p+1})$. Therefore, if e_n is dominated by r_1^n the error grows no more rapidly than the true solution, i.e. the relative error remains roughly invariable. Consequently, a multistep method is said to be *relatively stable* for any h_1 for which $|r_s| < |r_1|$, $s \neq 1$.

The derivation of the notion of relative stability has been founded on the assumption that $\partial f/\partial y = \lambda$ which will be untrue in general. The restriction can be dispensed with if λ is picked as a bound or typical value of $\partial f/\partial y$. The choice may be relevant only to some particular subinterval. A change in subinterval may cause an alteration in the h_1 for which relative stability is available.

Given this flexibility in the selection of λ, the step length can be controlled by requiring h_1 to be such that relative stability occurs.

Exercises

14. For the method of Exercise 10 with $a \geqslant -1$ show that in (8.76) $K = \frac{1}{12}(a+5)$. Show that bound given by (8.76) for the initial value problem of Exercise 9 is $20h^3$ when $a = 0$.

15. In applying Euler's rule to $y' = \lambda y$ $(\lambda < 0)$, $y(0) = 1$ show that, when $h < -1/\lambda$, the global error is non-negative and does not exceed $\frac{1}{2}x_n h \max y''$ when there is no round-off error. Thus (8.77) overestimates by the factor $\exp(|\lambda| x_n)$.

16. If $-1 \leqslant a < 1$ find for what h_1 the method of Exercise 10 is relatively stable.

17. If $-1 \leqslant a < 1$ show that

$$y_{n+2} - (1+a)y_{n+1} + ay_n = \frac{h}{12}\{(5+a)f_{n+2}+8(1-a)f_{n+1}-(1+5a)f_n\}$$

is relatively stable for $h_1 > 3(a+1)/2(a-1)$.

The other device to combat awkwardness in error estimation has already been alluded to and consists of combining explicit and implicit methods in a particular fashion. First, calculate y_{n+k} by an explicit method. Relabel it as $y_{n+k}^{[0]}$ and regard it as a prediction which we wish to improve. The amelioration is achieved by iterating an implicit method, say

$$y_{n+k}^{[s+1]} + \sum_{j=0}^{k-1} \alpha_j y_{n+j} = h\beta_k f(x_{n+k}, y_{n+k}^{[s]}) + h \sum_{j=0}^{k-1} \beta_j f_{n+j}$$

with $y_{n+k}^{[0]}$ as starting value. After m cycles $y_{n+k}^{[m]}$ is reached and this is deemed to be a corrected value which is to be taken as the appropriate value of y_{n+k}. The whole manoeuvre is then copied at the next step and so on.

Thus, at each step there is a prediction followed by a correction. The explicit method is the *predictor* while the implicit performs as *corrector* and the amalgamation is known as a *predictor—corrector* method.

The number of iterations m may be fixed in advance, which has the merit of limiting the demands on the computer but the disadvantage of making the error and stability characteristics depend upon both the predictor and corrector, or m may be left free and the iteration stopped when $|y_{n+k}^{[m]}-y_{n+k}^{[m-1]}|$ is less than some pre-assigned tolerance. This can always be arranged if the iteration converges, which will be guaranteed if $h < 1/L|\beta_k|$. Since now $y_{n+k}^{[m]}$ is independent of $y_{n+k}^{[0]}$ the stability and error traits will be those of the corrector alone. The pressure on the computer may, however, be heavy because there may be many iterations at each step.

A priori there is no reason why the predictor and corrector should not have different step numbers and different orders.

Notwithstanding, substantial benefit accrues from allotting them the same step number k and the same order p, for then it can be shown that, subject to certain conditions, the local truncation error is

$$\frac{C_{p+1}}{C^*_{p+1} - C_{p+1}}(y^{[m]}_{n+k} - y^{[0]}_{n+k}) + O(h^{p+2}) \tag{8.81}$$

where C^*_{p+1} and C_{p+1} are the error constants of the predictor and corrector respectively. The conditions for (8.81) are met if an Adams–Bashforth method is used as predictor, followed by an Adams–Moulton method as corrector. This is the combination employed by many practical programs. Clearly (8.81) evades the problem of estimating $y^{(p+1)}$ and predictor–corrector methods offer a businesslike policy of coping with errors.

The strategy for the control of step length with a predictor–corrector has the following stages. Estimate $\partial f/\partial y$ at each step and then form $h_1 = h\partial f/\partial y$. Then make h as large as will not transcend (*a*) $h < 1/L|\beta_k|$ where β_k refers to the corrector, (*b*) h_1 corresponding to relative stability, and (*c*) the estimate (8.81), ignoring the order term, being within prescribed limits. If m is free, the relative stability in (*b*) is that of the corrector but, if m is fixed, it is that of the combined predictor–corrector for which formulae can be found in Lambert (1973).

8.7. Runge–Kutta methods

All Runge–Kutta methods are of the form

$$y_{n+1} - y_n = hg(x_n, y_n, h)$$

for some suitably chosen g. These are all explicit and are the only ones that will be discussed here, though implicit methods have been investigated.

A Runge–Kutta method is said to be of *order* p if p is the largest integer for which

$$y(x+h) - y(x) - hg(x,y(x),h) = O(h^{p+1})$$

and *consistent* if $g(x,y,0) \equiv f(x,y)$. If g is continuous in all its variables and Lipschitz in y it can be proved that a *Runge–Kutta method is convergent if and only if it is consistent.*

If

$$g(x,y,h) = \sum_{m=1}^{M} w_m k_m ,$$

$$k_1 = f(x,y) ,$$

$$k_r = f(x+ha_r, y+h \sum_{s=1}^{m-1} b_{rs} k_s),$$

$$a_r = \sum_{s=1}^{r-1} b_{rs}$$

for $r = 2, 3, \ldots, M$ the Runge–Kutta method is said to have *M stages*. Note that $\Sigma_{m=1}^{M} w_m = 1$ is necessary for consistency and therefore for convergence.

There are numerous Runge–Kutta methods, matching the fecundity of parameters available. One two-stage method of the second order is obtained by putting $w_1 = \frac{1}{2}$, $w_2 = \frac{1}{2}$, $a_2 = 1$ with the result

$$y_{n+1} - y_n = \tfrac{1}{2}h\{f(x_n,y_n)+f(x_n+h,y_n+hf(x_n,y_n))\} . \tag{8.82}$$

Another two-stage method stems from $w_1 = 0$, $w_2 = 1$, $a_2 = \frac{1}{2}$. *Heun's third-order three-stage* method is

$$y_{n+1} - y_n = \tfrac{1}{4}h(k_1+3k_3)$$

with

$$k_1 = f(x_n,y_n), \quad k_2 = f(x_n+\tfrac{1}{3}h, y_n+\tfrac{1}{3}hk_1), \quad k_3 = f(x_n+\tfrac{2}{3}h, y_n+\tfrac{2}{3}hk_2).$$

Perhaps the most popular Runge–Kutta method is the four-stage fourth-order

$$y_{n+1} - y_n = \tfrac{1}{6}h(k_1+2k_2+2k_3+k_4) , \qquad (8.83)$$

$$k_1 = f(x_n,y_n), \qquad k_2 = f(x_n+\tfrac{1}{2}h, y_n+\tfrac{1}{2}hk_1) ,$$

$$k_3 = f(x_n+\tfrac{1}{2}h, y_n+\tfrac{1}{2}hk_2), \qquad k_4 = f(x_n+h, y_n+hk_3) .$$

It is perhaps worth pointing out that when there are five or more stages the order is always less than the number of stages.

The local truncation error is

$$y(x_{n+1}) - y(x_n) - hg(x_n, y(x_n)h) .$$

For a method of order p it is $O(h^{p+1})$. If

$$|f(x,y)| < Q, \qquad \left|\frac{\partial^{i+j} f(x,y)}{\partial x^i \partial y^j}\right| < \frac{P^{i+j}}{Q^{j-1}} \qquad (i + j \leqslant p)$$

the modulus of the local truncation error does not exceed

$$\left(\frac{1}{3} + \left|\frac{2}{3} - a_2\right|\right) h^3 P^2 Q$$

for a method of order 2. In a method of order 3, the bound is a multiple of h^4P^3Q while for (8.83) it is $73h^5P^4Q/720$.

If the local truncation error is bounded by Kh^{p+1} the global error satisfies

$$|y(x_n)-y_n| \leqslant \frac{h^pK}{L}[\exp\{L(x_n-a)\}-1]$$

in the absence of round-off error.

These error bounds are usually difficult to apply in practice and direct control of step length by them is out of the question. Instead, results are usually based on two distinct runs. For example, if the local truncation error with step length h is

$$y(x_{n+1}) - y_{n+1} = \psi(x_n, y(x_n))h^{p+1} + O(h^{p+2})$$

the corresponding error when the method is applied at x_{n-1}

with step length $2h$ is

$$y(x_{n+1}) - Y_{n+1} = \psi(x_{n-1}, y(x_{n-1}))(2h)^{p+1} + O(h^{p+2})$$

$$= \psi(x_n, y(x_n))(2h)^{p+1} + O(h^{p+2}) \ .$$

Hence

$$y_{n+1} - Y_{n+1} = (2^{p+1}-1)\psi(x_n, y(x_n))h^{p+1} + O(h^{p+2})$$

which suggests $(y_{n+1}-Y_{n+1})/(2^{p+1}-1)$ as an estimate of the error. Thus error estimation for Runge–Kutta methods involves substantial computational effort and has a tendency to be expensive, much more so than that for predictor–corrector methods where the additional computation is negligible.

As regards stability, try the general Runge–Kutta method on the test equation $y' = \lambda y$. There results

$$y_{n+1} - y_n = P_M(h_1)y_n$$

where $h_1 = h\lambda$ and P_M is a polynomial of degree M. Consequently

$$y_n = c_1 r_1^n$$

where $r_1 = 1 + P_M(h_1)$. Therefore, we say that the method is *absolutely stable* if $|r_1| < 1$. When M is the same as the order p the formula for r_1 simplifies to

$$r_1 = 1 + h_1 + \frac{h_1^2}{2!} + \ldots + \frac{h_1^p}{p!} \ .$$

From this it can be deduced that a first-order one-stage method is absolutely stable if $-2 < h_1 < 0$, a second-order two-stage method is absolutely stable for $-2 < h_1 < 0$, a third-order three-stage method is absolutely stable for $-2.51 < h_1 < 0$, and a fourth-order four-stage method is absolutely stable for $-2.78 < h_1 < 0$. In particular, the popular variety (8.83) is absolutely stable for $-2.78 < h_1 < 0$. It is unwise to use a

Runge–Kutta method outside the region of absolute stability.

Efforts have been made to choose the parameters in Runge–Kutta methods so as to improve the stability characteristics when $M \geqslant 5$. Choices have also been made to reduce the local truncation bounds since the traditional values were selected to suit mechanical desk calculators rather than electronic machines. However, the theoretical errors were not greatly diminished although the actual computed errors were sensitive to alterations in the parameters.

Exercises

18. Show that (8.83) applied to $y' = \lambda y$ gives $y_{n+1}/y_n = e^{\lambda h} + O((\lambda h)^5)$.

19. Prove that (8.82) is the same as a predictor–corrector in which the predictor is Euler's rule and the corrector one application of the trapezoidal rule.

20. The method (8.83) is applied to the initial-value problem $y' = \exp(10(x-y))$, $y(0) = 0.1$ on $[0,1]$. Show that $P = 10$ and that an estimate of Q is 0.37. Deduce that the step length should be less than 10^{-2} for the local truncation error to be less than 10^{-8}.

If $h = 0.1$, find the global error and demonstrate that it is very conservative by comparing the numerical and analytical solutions.

21. Use the estimate for $(y_{n+1}-Y_{n+1})/(2^{p+1}-1)$ to adjudge the error in employing (8.83) with $h = 0.1$ for $y' = 1 + y^2$, $y(0) = 1$ and compare your estimate with the actual error.

22. The initial-value problem

$$y' = -5xy^2 + \frac{5}{x} - \frac{1}{x^2}, \qquad y(1) = 1$$

is to be solved on $[1,10]$ by means of (8.83). Assuming that a reasonable estimate for $\partial f/\partial y$ is -10, find which of $h = 0.2$, 0.3, 0.4 correspond to absolute stability. Compare the numerical solutions for these three step lengths.

8.8. Comparison of Runge–Kutta and predictor–corrector methods

The first observation is that Runge–Kutta methods, being one step and explicit, do not require the extra starting values necessary for the predictor–corrector. Accordingly, Runge–Kutta methods are sometimes deployed to find the essential data for a predictor–corrector and then discarded. The expense in error estimation is thereby minimized.

It is customarily understood that Runge–Kutta methods are more stable and accurate than multistep methods, but the true situation is by no means clearcut. While it is correct that the regions of stability are substantially larger for Runge–Kutta methods than for Adams methods it must not be overlooked that this is balanced by an increased number of function evaluations. If a predictor–corrector method is compared with a Runge–Kutta method of the same order and the step length of each is adjusted so that they have the same number of evaluations over a given interval the predictor–corrector often turns out to be the more accurate. Again, if a predictor–corrector program has a variable step length option it will automatically bring itself into line with stability requirements. Indeed, the relative stability characteristics and ease of controlling the step length (through monitoring the local truncation error) of predictor–corrector methods suggest that they are on the whole preferable to Runge–Kutta methods, though the Runge–Kutta is much more easily programmed for a fixed step length.

8.9. Extrapolation

Polynomial extrapolation is sometimes adopted to refine the accuracy of previous methods. Suppose that one of the described methods has yielded the approximation $y(x;h)$ with the step length h. A *basic step length* H is selected (typically $H \gg h$) and then polynomial extrapolation is employed to furnish approximations at $x_0 + jH$, $j = 0, 1, \ldots$, where x_0 is the starting point. Let N_0 be a positive integer. Apply the numerical method N_0 times with step length H/N_0 to generate the approximation $y(x_0+H;h_0)$ where $h_0 = H/N_0$. Replace N_0 by the positive integer N_1 ($N_1>N_0$) and repeat the exercise, thereby obtaining $y(x_0+H;h_1)$ with $h_1 = H/N_1$. Suppose that for the given numerical method

$$y(x;h) = y(x) + A_1h + A_2h^2 + \ldots \qquad (8.84)$$

as $h \to 0$, the coefficients $A_1, A_2, \ldots$ being independent of h.

Then

$$\frac{h_1 y(x_0+H;h_0) - h_0 y(x_0+H;h_1)}{h_1 - h_0} = y(x) + O(h^2) \tag{8.85}$$

so that the left-hand side of (8.85) ought to be nearer the theoretical solution than either of the two approximations from which it is built. If the left-hand side is denoted by $y_1(x+H;H)$ the whole process is repeated for $y' = f(x,y)$, $y(x_0+H) = y_1(x+H;H)$ to determine a value at $x_0 + 2H$.

Actually, (8.85) is the simplest of the possibilities. Further intermediate steps $h_2, h_3, \ldots, h_r$ could be introduced with $h_0 > h_1 > h_2 > \ldots > h_r$ and then a linear combination formed in which the error was $O(h^{r+1})$. In this way, high accuracy might be attained when r is sufficiently large.

To implement the method of extrapolation it must be ensured that the numerical method satisfies (8.84) or preferably

$$y(x;h) = y(x) + A_2 h^2 + A_4 h^4 + \ldots$$

because of the enhanced accuracy that flows from the same numerical effort. Methods which have this property have been investigated and can be classified in terms of the $\rho(\zeta)$ and $\sigma(\zeta)$ of §8.6, but details will be omitted here. (The interested reader should consult Lambert (1973).)

In order to have numerical stability in the extrapolation procedure the step sizes $h_1, h_2, \ldots$ must decrease fairly rapidly. While it is true that the accuracy improves swiftly with increasing r so does the workload, and values of r greater than 5 are rarely practicable. In some ways there is a kinship with a Runge—Kutta of variable step. The size of H must be relatively large for the extrapolation to be efficient with the consequence that the production of results is expensive.

8.10. Systems of differential equations

Most of the preceding theory carries through for systems if y is replaced by $\mathbf{y}$ and statements such as $y_n \to y(a+c)$ are understood to mean $\|\mathbf{y}_n - \mathbf{y}(a+c)\| \to 0$ in some appropriate norm. One difference is that $\partial f/\partial y$ must be replaced by the Jacobian matrix $\partial\mathbf{f}/\partial\mathbf{y}$ (§4.1) so that $L = \|\partial\mathbf{f}/\partial\mathbf{y}\|$ and then the constant λ used in the scalar stability theory is replaced by an eigenvalue of the Jacobian. Thus conditions on h_1 become conditions on $h\lambda_i$ for every eigenvalue λ_i of the Jacobian.

If the Jacobian is a constant matrix, all of whose eigenvalues have a negative real part, the solution consists of a *transient*, which decays exponentially and a *steady state*. The solution will take a long time to settle down if the real part of one of the eigenvalues is small. On the other hand, if $|\mathcal{R}\lambda_i| \gg 1$ small step lengths will be necessary to ensure stability. If both possibilities occur simultaneously so that

$$\max_i |\mathcal{R}\lambda_i| \gg \min_i |\mathcal{R}\lambda_i|$$

we are forced when finding the steady state to integrate over a long range with tiny step lengths. In an unsatisfactory situation like this the system is said to be *stiff*.

What to do with stiff systems is an area of vigorous research so that, if they are met, it is advisable to consult the literature to learn the latest recommended tactics.

8.11. The 'best' numerical method

The simple answer to a general query 'which is the best method?' is that there is none because so much depends on the user of the program and why he wishes to make calculations. Obviously, a user will repudiate a program if it has inadequate documentation or insufficient robustness for his purposes or if it does not have the capability to provide the sort of output he desires. Again, high accuracy may be vital but the cost of function evaluations of no concern or, conversely, the expense of evaluation may be a prime consideration but medium accuracy may be acceptable. If storage is at a premium, that may affect the choice.

Perhaps the most important point to realize is that the same algorithm prepared by two different programmers can perform in quite different ways. Even a modest difference in program can be reflected in a significant change in output. Every effort should therefore be made to test a program on a wide variety of problems and incorporate the conclusions in the associated documentation. The task of the user seeking the program best suited to his needs will then be eased. For a recent comparison of some routines see Shampine, Watts, and Davenport (1976).

CANONICAL PROBLEMS

8.12. Geometrical optics revisited

The high-frequency approximation of geometrical optics as enunciated in the first sections of this chapter involves finding the rays and then determining the amplitude by the intensity law supplemented by the rotation of polarization (if necessary). Only the solution of ordinary differential equations arises and we can now regard this as always possible in principle either by analytical or numerical means, so geometrical optics offers a practical scheme of calculating the field at short wavelengths. However, if the cross-section of a tube of rays shrinks to zero, the intensity law predicts an infinite amplitude which is obviously unsatisfactory. Therefore, complementary analysis is required to cope with such a contingency. There may be other circumstances in which a predicament materializes and so some generalities are in order.

The general technique is to introduce a *canonical problem* which is susceptible to analysis and which reproduces the main features of the ray picture. Then predictions from the analysis are interpreted so as to explain the modifications to be incorporated in the ray theory. Since geometrical optics is a local approximation it will usually be sufficient to set the canonical problem in a homogeneous medium and often the consideration can be restricted to what happens to a plane wave. This approach has already been typified in §8.5 where the reflection of a tube of rays at an interface has been treated by means of the canonical problem in which a plane wave is reflected by a plane interface.

The general method of extending the validity of geometrical

optics by the judicious involvement of canonical problems could well be called the *geometrical theory of diffraction*, although it was originally developed for scattering by edges and curved boundaries. Various canonical problems in the geometrical theory of diffraction will now be discussed.

8.13. Focusing

The particular problem mentioned in the preceding section occurs when two adjacent rays intersect so that the tube collapses momentarily. In a homogeneous medium two adjacent rays from a wavefront intersect at a principal centre of curvature. The loci of the two principal centres of curvature of a given wavefront are two surfaces known as *caustics* or *focal surfaces*. Every normal to a wavefront is tangential to both caustics.

In the special case of a spherical wavefront all the rays of any pencil intersect in a common point; such a pencil is said to be *homocentric*. A point source of light produces a homocentric pencil but the pencil does not remain homocentric in general after reflection or refraction. Instead the light becomes concentrated on the caustics and astigmatism occurs.

A suitable canonical problem is a wavefront with a known field which, if translated into rays, would have a caustic in the region of interest. It is convenient to formulate this in terms of an irradiated perfectly conducting obstacle in a medium in which the refractive index N has the constant value of unity.

Let the incident field be $\mathbf{E}_0$. Then the total electric intensity can be expressed as

$$\mathbf{E}(P) = \mathbf{E}_0 + (\text{grad div} + k_0^2)\int_S \frac{\mathbf{n}_q \wedge \mathbf{H}\psi(P,q)\,\mathrm{d}S_q}{\mathrm{i}\omega\varepsilon_0}$$

where $\psi(P,q) = \exp(-\mathrm{i}k_0|\mathbf{x}_P-\mathbf{x}_q|)/4\pi|\mathbf{x}_P-\mathbf{x}_q|$. Although $\mathbf{H}$ is not known exactly on S we can take advantage of the high frequency. The canonical problem for reflection suggests that we shall not go far wrong in assuming, to a first approximation, that the current distribution over the illuminated part of S is the same

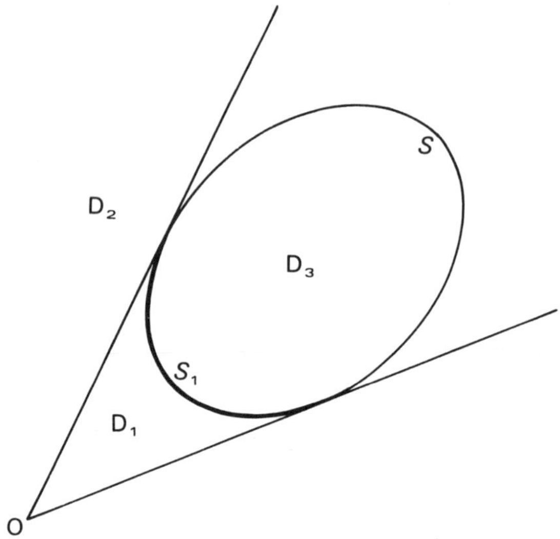

FIG. 8.7. The illuminated and shadow regions of an obstacle.

as if at every point the incident wave were reflected as if it impinged on the infinite (perfectly conducting) tangent plane. On the shadow part the current distribution is assumed to be zero (Macdonald 1913). Thus, on the illuminated part S_1 (Fig. 8.7) $\mathbf{n} \wedge \mathrm{H} = 2\mathbf{n} \wedge \mathrm{H}_0$ and elsewhere $\mathbf{n} \wedge \mathrm{H} = \mathbf{0}$.

The approximation should be valid when the radii of curvature of S and of the incident wavefront are large compared with the wavelength. The condition for the incident wave can always be met if its source is not too close to S.

Let the source of the primary field be an electric dipole of moment $\mathbf{p}_0$ at the point O with coordinates $\mathbf{x}_0$. Then

$$\mathbf{E}_0 = (\text{grad div} + k_0^2)\mathbf{p}_0 \frac{\exp(-ik_0R)}{4\pi\varepsilon_0 R},$$

$$\mathbf{H}_0 = \frac{\omega(\mathbf{x}-\mathbf{x}_0) \wedge \mathbf{p}_0}{4\pi R^2}\left(k - \frac{i}{R}\right)\exp(-ik_0R)$$

where $R = |\mathbf{x}-\mathbf{x}_0|$. Hence

$$\mathbf{E} = \mathbf{E}_0 - \frac{1}{2\pi\varepsilon_0}(\text{grad div} + k_0^2)\int_{S_1}\left(\frac{1}{R_q} + ik\right)\mathbf{n}_q \wedge \{(\mathbf{x}_q - \mathbf{x}_0) \wedge \mathbf{p}_0\}\frac{\exp(-ik_0 R_q)}{R_q^2}\psi \mathrm{d}S_q$$

where $R_q = |\mathbf{x}_q - \mathbf{x}_0|$. In view of the short wavelength the integral can be evaluated asymptotically by the method of stationary phase and the main contribution will come from those points of S_1 where $R_q + |\mathbf{x} - \mathbf{x}_q|$ is stationary. The close connection with Fermat's principle is immediately exposed.

Split the domain surrounding O into three portions D_1, D_2, and D_3. D_2 constitutes the exterior of the tangent cone from O to S (Fig. 8.7) whereas D_1 is the interior of the tangent cone from O to S_1. The rest of the interior of the tangent cone is designated D_3. If S has the shape shown in Fig. 8.7 and the point of observation is in $D_1 + D_2$ the stationary point Q is the point of contact of a prolate spheroid with O and P as foci and touching S. If the point of observation is in D_3 it will be denoted by P′ and the stationary point Q is the point of intersection of OP′ and S_1.

It is now convenient to choose Q as origin, with the z axis along the outward normal to S_1 and the η axis in the plane of incidence (Fig. 8.8). Let θ be the angle of incidence so that O is the point $(0, -R_0 \sin\theta, R_0 \cos\theta)$ with respect to the axes at Q. Assume, firstly, that the point of observation is

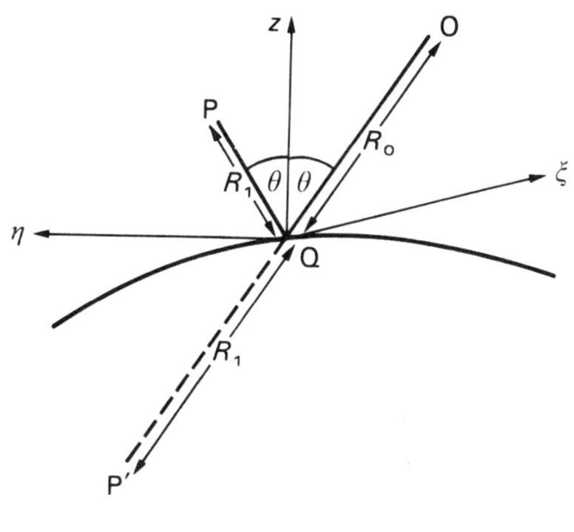

FIG. 8.8. Axes with the stationary point as origin.

in D_3 so that it will be at $(0, R_1 \sin\theta, -R_1 \cos\theta)$ on the broken line in Fig. 8.8. When the variable of integration (ξ,η) is near Q

$$R_q + |\mathbf{x}-\mathbf{x}_0| \approx R_0 + R_1 + \frac{1}{2}\left(\frac{1}{R_0}+\frac{1}{R_1}\right)(\xi^2+\eta^2\cos^2\theta).$$

Hence

$$\int \exp\{-ik_0(R_q+|\mathbf{x}-\mathbf{x}_0|)\}dS_q$$

$$\sim \exp\{-ik_0(R_0+R_1)\}\int_{-\infty}^{\infty}\int_{-\infty}^{\infty}\exp\{-\tfrac{1}{2}ik_0\left(\frac{1}{R_0}+\frac{1}{R_1}\right)(\xi^2+\eta^2\cos^2\theta)\}d\xi d\eta$$

$$\sim \frac{2\pi R_0R_1\exp\{-ik_0(R_0+R_1)\}}{ik_0(R_0+R_1)\cos\theta}.$$

In this case $R_0 + R_1 = R$ and $\mathbf{n}_q \cdot \mathbf{R}_q = -R_0 \cos\theta$ so that

$$\mathbf{E} \sim \mathbf{E}_0 - (\text{grad div} + k_0^2)\left\{\mathbf{p}_0 - \frac{(\mathbf{n}_Q\cdot\mathbf{p}_0)\mathbf{R}_0}{R_0}\right\}\frac{\exp(-ik_0R)}{4\pi\varepsilon_0R}$$

on neglecting terms of smaller order. When k_0R_1 is large the second term under the vector operator can be omitted and so, to this order of approximation, **E** vanishes in the shadow in conformity with geometrical optics.

For P in $D_1 + D_2$ the analysis is more complicated. Let the equation of S_1 near Q be

$$z \approx -\tfrac{1}{2}\boldsymbol{\xi}^T C_b\boldsymbol{\xi} - \tfrac{1}{6}(D\xi^3+3E\xi^2\eta+3F\xi\eta^2+G\eta^3)$$

where C_b is the curvature matrix as in §8.5. For points near Q

$$R_q + |\mathbf{x}-\mathbf{x}_0| \approx R_0 + R_1 + \tfrac{1}{2}\boldsymbol{\xi}^T C\boldsymbol{\xi}$$

when only quadratic terms are retained, C being the matrix defined by

$$C = 2C_b \cos\theta + \left(\frac{1}{R_0} + \frac{1}{R_1}\right)\begin{pmatrix}1 & 0\\ 0 & \cos^2\theta\end{pmatrix}.$$

Hence

$$\int \exp\{-ik_0(R_q + |\mathbf{x}-\mathbf{x}_0|)\}dS_q$$

$$\sim \exp\{-ik_0(R_0+R_1)\}\int_{-\infty}^{\infty}\int_{-\infty}^{\infty}\exp(-\tfrac{1}{2}ik_0\boldsymbol{\xi}^{\mathrm{T}}C\boldsymbol{\xi})d\xi d\eta$$

$$\sim \frac{2\pi\exp\{-ik_0(R_0+R_1)-i\delta\}}{k_0|\det C|^{\frac{1}{2}}}$$

where $\delta = 0, \frac{1}{2}\pi$, or $-\frac{1}{2}\pi$ according as the eigenvalues of C have opposite signs, are both positive, or are both negative. It is a simple matter to check that

$$\det C = \left(\kappa_1^{\mathrm{r}} + \frac{1}{R_1}\right)\left(\kappa_2^{\mathrm{r}} + \frac{1}{R_1}\right)\cos^2\theta$$

where κ_1^{r}, κ_2^{r} are the principal curvatures of the reflected wavefront, obtained from the curvature matrix C^{r} of (8.64) because the incident wave is spherical so that C^{i} has only the diagonal elements $1/R_0$. If we adopt the rule that

$$\left(\kappa^{\mathrm{r}} + \frac{1}{R_1}\right)^{\frac{1}{2}} = \left|\kappa^{\mathrm{r}} + \frac{1}{R_1}\right|^{\frac{1}{2}}\exp(-\tfrac{1}{2}\pi i)$$

when $R_1\kappa^{\mathrm{r}} + 1 < 0$ the phase change δ is automatically taken care of. Then, if k_0R_0 and k_0R_1 are both large, we obtain

$$\mathbf{E} \sim \mathbf{E}_0 + \left(\kappa_1^{\mathrm{r}} + \frac{1}{R_1}\right)^{-\frac{1}{2}}\left(\kappa_2^{\mathrm{r}} + \frac{1}{R_1}\right)^{-\frac{1}{2}} R_0\{2(\mathbf{E}_0.\mathbf{n})\mathbf{n}-\mathbf{E}_0\}_Q\,\frac{\exp(-ik_0R_1)}{R_1} \quad (8.86)$$

where $\{\ \}_Q$ signifies the value at Q.

If the disposition of the surface is as in Fig. 8.8, $C_{b11} > 0$ and $C_{b11}C_{b22} > C_{b12}^2$. Then both κ_1^r and κ_2^r are positive. Formula (8.86) holds for all R_1 and no focusing occurs. Indeed this is true whenever κ_1^r and κ_2^r are both positive. On the other hand, if either of the curvatures is negative (which can happen if S is anticlastic or synclastic but concave towards O so that $C_{b11} < 0$), there is the possibility of focusing. Now $\kappa^r + 1/R_1$ is certainly positive for small R_1 so that as R_1 increases and $\kappa^r + 1/R_1$ passes through a zero the phase in (8.86) augments by $\frac{1}{2}\pi$ according to the rule for the square root delineated above. But the amplitude in (8.86) is the same as that of geometrical optics so long as neither square root is near zero so that the following interpretation presents itself.

The formulae of geometrical optics may be used when focusing takes place provided that when a caustic of an astigmatic beam is crossed in the direction of propagation the phase is advanced by $\frac{1}{2}\pi$.

After passing two caustics the phase advance will be π and this is true even if the two caustics coincide at a point (in which case the beam has a *focus* and $\kappa_1^r = \kappa_2^r$).

The above derivation breaks down if θ is near $\frac{1}{2}\pi$. Then Q is near the *penumbral curve* (i.e. the curve where the incident rays are tangent to the obstacle) and the approximate current distribution is not satisfactory. A new canonical problem must then be proposed (§8.21).

Both (8.86) and geometrical optics fail when $R_1\kappa_1^r$ or $R_1\kappa_2^r$ is near -1. Then the cubic terms in z, hitherto neglected, become important. By keeping the cubic terms an asymptotic evaluation supplies a field which is *uniformly valid*, i.e. it makes an even transition from geometrical optics through the field at the caustic and back to geometrical optics (plus phase advance) again. The main feature is that *the field at a caustic is larger than that at other points by a factor of the order of* $k_0^{1/6}$. It is the attempt to emulate this behaviour that causes the amplitude of geometrical optics to go infinite. If the two caustics coincide at a focus the factor of increase is $k_0^{1/3}$. All these estimates depend upon the non-vanishing of certain quantities and when this is not so higher-order calculations are inevitable but the formulae are extremely unwieldy.

The formula (8.86) has assumed that there is a single stationary point Q to be considered. For an arbitrary obstacle there may be several stationary points and then the contributions from the reflection at each must be added in order to obtain the total field at P. It is also possible that a ray may suffer multiple reflection, i.e. strike the body again before reaching P. Then repeated applications of the formula will sometimes resolve the problem although, in some circumstances, a new canonical problem may be involved.

If the obstacle is not perfectly conducting (8.86) may still be employed so long as the appropriate Fresnel reflection coefficient is inserted.

Exercises

23. A circular cylinder of radius a is placed in free space with its axis along the z axis. An electrically polarized plane wave in which $E_z = \exp(-ik_0x)$ is incident along the x axis. The boundary condition on the cylinder is $\mathbf{n} \wedge \mathbf{H} + (\varepsilon_0/\mu_0)^{\frac{1}{2}}Z\mathbf{E}_t = \mathbf{0}$ where $\mathbf{E}_t$ is the tangential component of $\mathbf{E}$. Show that, according to geometrical optics, the electric intensity of the reflected wave is

$$\frac{\cos\beta + Z}{\cos\beta - Z}\left(\frac{d_1r_1}{2r_1 + d_1}\right)^{\frac{1}{2}} \frac{\exp\{-ik_0(x_1+r_1)\}}{r_1^{\frac{1}{2}}}$$

where r_1 is the distance to the point of observation from the point of reflection (x_1,y_1), β is the angle of incidence at (x_1,y_1), and $d_1 = a\cos\beta$.

24. A parabolic cylinder has equation $y^2 = -2a^2x + a^4$. It is irradiated in $y^2 > -2a^2x + a^4$ by an electrically polarized plane wave in which $E_z = \exp\{-ik_0(x\cos\phi_0 + y\sin\phi_0)\}$ $(0 < \phi_0 \leqslant \pi)$. If the boundary condition is the same as in Exercise 23 show that the formula of Exercise 23 for the reflected wave still holds provided that $\exp(-ik_0x_1)$ is replaced by $\exp -ik_0(x_1\cos\phi_0 + y_1\sin\phi_0)$ and d_1 is understood to mean the projection of the radius of curvature at (x_1,y_1) on the incident ray.

Would this also be true if the excitation were from $y^2 < -2a^2x + a^4$ and what happens if $\phi_0 = 0$?

8.14. Reflection by stratification

In §8.4 we came across rays being reflected by a minimum of refractive index in a stratified medium. The rays will intersect one another and one might anticipate that there will be a phase advance of $\frac{1}{2}\pi$ as in the preceding section. However, the canonical problem of §8.13 is not appropriate because what we are effectively dealing with in the stratified medium is a problem in which the refractive index has a zero. To ascertain this it is sufficient to consider two-dimensional waves in a stratified medium since this does not radically affect the conclusions.

Let the fields be independent of the variable y and have E_y as the sole electric component. Let μ have the constant value μ_0; the analysis is thereby substantially simplified without loss of the chief characteristics. Then

$$\frac{\partial^2 E_y}{\partial x^2} + \frac{\partial^2 E_y}{\partial z^2} + k_0^2 N^2 E_y = 0$$

where $N^2 = \varepsilon/\varepsilon_0$. Let the x dependence be $\exp(-ik_0\alpha x)$ so that

$$\frac{\partial^2 E_y}{\partial z^2} + k_0^2(N^2-\alpha^2)E_y = 0 . \tag{8.87}$$

Thus as z varies there can be values for which $N^2 - \alpha^2 \leqslant 0$ and geometrical optics is not immediately applicable. If N has a single minimum the number of zeros is at most two and so the discussion will be limited to the case of two zeros.

Change the notation and rewrite (8.87) as

$$\frac{d^2 w}{dz^2} + k_0^2 q(z) w = 0 . \tag{8.88}$$

Without specifying branches precisely for the time being substitute $\xi = ik_0\int^z q^{\frac{1}{2}}(t)dt$, $w = q^{-\frac{1}{4}}W(\xi)$ and obtain

$$\frac{d^2W}{d\xi^2} - W = -S(\xi)W \tag{8.89}$$

where

$$S(\xi) = \left\{3\left(\frac{d^2z}{d\xi^2}\right)^2 - 2\left(\frac{dz}{d\xi}\right)\left(\frac{d^3z}{d\xi^3}\right)\right\}\Big/4\left(\frac{dz}{d\xi}\right)^2$$

This differential equation may be converted into the integral equation

$$W(\xi) = Ae^{\xi} + Be^{-\xi} + \frac{1}{2}\int^{\xi}(e^{-\xi+t}-e^{\xi-t})S(t)W(t)dt$$

since we aim to have E_y and dE_y/dz continuous.

If q is always positive, choose $q^{\frac{1}{2}}$ likewise and then a possible solution of (8.89) is

$$W(\xi) = e^{-\xi} + \frac{1}{2}\int_{-i\infty}^{\xi}(e^{-\xi+t}-e^{\xi-t})S(t)W(t)dt . \tag{8.90}$$

As $z \to -\infty$, one may expect this to approach $e^{-\xi}$. Consequently, $e^{-\xi}$ can be regarded as an incident wave of unit amplitude propagating through the medium. The integral in (8.90) represents a correction which may become significant as z increases but is often negligible. No attempt will be made here or subsequently to specify conditions on q which warrant statements that an integral is small though studies have been conducted in this matter (Heading 1975, Olver 1965).

We now turn to the case when q has two zeros. Specifically we assume that $q(z) = (z-z_1)(z-z_2)f(z)$ where z_1, z_2 are real with $z_2 > z_1$ while $f(z)$ is positive for all real z and regular in a complex z domain which includes the whole of the real axis. Zeros of q off the real axis are irrelevant to the current purpose.

We first establish an exact energy relationship. For $z > z_2$ take $q^{\frac{1}{2}}$ positive and define

$$\xi = ik_0 \int_{z_2}^{z} q^{\frac{1}{2}}(t)\,dt \quad ,$$

$$W_1(\xi) = e^{-\xi} + \frac{1}{2}\int_{i\infty}^{\xi} (e^{-\xi+t} - e^{\xi-t})S(t)W_1(t)\,dt \ .$$

From our argument for (8.90) we can regard this as a unit transmitted wave. For $z < z_1$ take $q^{\frac{1}{2}}$ positive also and define

$$\xi = ik_0 \int_{z_1}^{z} q^{\frac{1}{2}}(t)\,dt \quad ,$$

$$W_2(\xi) = e^{-\xi} + \frac{1}{2}\int_{-i\infty}^{\xi} (e^{-\xi+t} - e^{\xi-t})S(t)W_2(t)\,dt \ ,$$

$$W_3(\xi) = e^{\xi} + \frac{1}{2}\int_{-i\infty}^{\xi} (e^{-\xi+t} - e^{\xi-t})S(t)W_3(t)\,dt \ .$$

W_2 can be considered as a unit incident wave whereas W_3 is a unit reflected wave.

Let w be a solution of (8.88) for all real z such that

$$w(z) = q^{-\frac{1}{4}}\{W_2(\xi) + RW_3(\xi)\} \qquad (z < z_1)$$

$$= q^{-\frac{1}{4}}TW_1(\xi) \qquad (z > z_2) \ .$$

Now, since q is real for real z, (8.88) implies

$$ww^{*\prime\prime} - w^{*}w^{\prime\prime} = 0$$

whence $\mathscr{I} ww^{*\prime}$ is constant for all real z. But, in $z > z_2$,

$$\mathscr{I} ww^{*\prime} = \mathscr{I} q^{-\frac{1}{4}} T W_1 q^{-\frac{1}{4}} T^* \frac{\mathrm{d}W_1^*}{\mathrm{d}\xi} \mathrm{i}k_0 q^{\frac{1}{2}}$$

$$\rightarrow TT^* k_0$$

as $z \rightarrow \infty$. Similarly, from $z < z_1$, we find $\mathscr{I} ww^{*\prime} \rightarrow (1-RR^*)k_0$ as $z \rightarrow -\infty$. Since these must be the same we have conservation of energy expressed by

$$RR^* + TT^* = 1 \tag{8.91}$$

It must be emphasized that (8.91) is relevant to the exact solution. If we assumed the approximation

$$w = q^{-\frac{1}{4}}(\mathrm{e}^{-\xi} + \hat{R}\mathrm{e}^{\xi}) \qquad (z < z_1)$$

$$= q^{-\frac{1}{4}}\hat{T}\mathrm{e}^{-\xi} \qquad (z > z_2)$$

and substituted in $\mathscr{I} ww^{*\prime}$ = constant we would obtain

$$\hat{R}\hat{R}^* + \hat{T}\hat{T}^* = 1 \,. \tag{8.92}$$

However, if $\hat{T}$ is small it may be swamped by the errors in $\hat{R}$ so that all that could properly be deduced from (8.92) is that $|\hat{R}|$ is approximately 1. To obtain a better result we must determine the exact R or T.

The difficulty of relating R and T originates from their being connected by what supervenes between z_1 and z_2. In this interval a solution which is negligible at z_1 may be dominant at z_2 so that straightforward matching without appreciable error is hard to achieve. The way round the snag is to split the real axis into two parts each of which contains only a solitary zero and use solutions which can pass through one zero (Jones 1964, Olver 1974).

Let z_0 be a fixed point between z_1 and z_2. Then two uniformly valid asymptotic solutions of (8.88) on $[z_0, \infty]$ are

$$w_1 = \pi^{\frac{1}{2}} k_0^{\frac{1}{6}} \left(\frac{\zeta}{q}\right)^{\frac{1}{4}} \mathrm{Bi}(-k_0^{\frac{2}{3}}\zeta) \;,$$

$$w_2 = \pi^{\frac{1}{2}} k_0^{\frac{1}{6}} \left(\frac{\zeta}{q}\right)^{\frac{1}{4}} \mathrm{Ai}(-k_0^{\frac{2}{3}}\zeta)$$

where Ai and Bi are the standard Airy functions and

$$\zeta = \pm \left|\frac{3}{2}\int_{z_2}^{z} q^{\frac{1}{2}}(t)\,\mathrm{d}t\right|^{\frac{2}{3}}$$

the upper or lower sign being taken as $z > z_2$ or $z < z_2$. Let $z \to \infty$ so that $\zeta \to \infty$. The Airy functions may be replaced by their asymptotic expansions for large negative argument with the result

$$w_1 \sim -q^{-\frac{1}{4}} \sin\left\{k_0 \int_{z_2}^{z} q^{\frac{1}{2}}(t)\,\mathrm{d}t - \frac{1}{4}\pi\right\} \;,$$

$$w_2 \sim q^{-\frac{1}{4}} \cos\left\{k_0 \int_{z_2}^{z} q^{\frac{1}{2}}(t)\,\mathrm{d}t - \frac{1}{4}\pi\right\} \;.$$

Comparison with the behaviour of W_1 as $z \to \infty$ reveals that

$$q^{-\frac{1}{4}} W_1 = \exp(-\tfrac{1}{4}\pi\mathrm{i})(w_2 + \mathrm{i}w_1) \;.$$

In $(-\infty, z_0]$ we employ

$$w_3 = \pi^{\frac{1}{2}} k_0^{\frac{1}{6}} \left(\frac{\zeta_1}{q}\right)^{\frac{1}{4}} \mathrm{Bi}(-k_0^{\frac{2}{3}}\zeta_1) \,,$$

$$w_4 = \pi^{\frac{1}{2}} k_0^{\frac{1}{6}} \left(\frac{\zeta_1}{q}\right)^{\frac{1}{4}} \mathrm{Ai}(-k_0^{\frac{2}{3}}\zeta_1)$$

where

$$\zeta_1 = \mp \left|\frac{3}{2}\int_{z_1}^{z} q^{\frac{1}{2}}(t)\,\mathrm{d}t\right|^{\frac{2}{3}}$$

the upper or lower sign being adopted according as $z > z_1$ or $z < z_1$. Allowing $z \to -\infty$, especially through a sequence of values in which the cosine in the Airy function asymptotic expansion takes alternately the values zero and unity, permits the identification

$$q^{-\frac{1}{4}} W_2 = \exp(-\tfrac{1}{4}\pi i)(w_3 + i w_4) \ ,$$

$$q^{-\frac{1}{4}} W_3 = \exp(-\tfrac{1}{4}\pi i)(w_4 + i w_3) \ .$$

It follows that $w_3 + iw_4 + R(w_4 + iw_3)$ and $T(w_2 + iw_1)$ represent the same solution to (8.88). They and their derivatives must therefore be continuous at $z = z_0$. Hence

$$R = \frac{\mathscr{W}(w_4,w_1) - \mathscr{W}(w_3,w_2) - i\{\mathscr{W}(w_4,w_2) + \mathscr{W}(w_3,w_1)\}}{\mathscr{W}(w_4,w_2) - \mathscr{W}(w_3,w_1) + i\{\mathscr{W}(w_3,w_2) + \mathscr{W}(w_4,w_1)\}} \ ,$$

$$T = \frac{-2\mathscr{W}(w_3,w_4)}{\mathscr{W}(w_4,w_2) - \mathscr{W}(w_3,w_1) + i\{\mathscr{W}(w_3,w_2) + \mathscr{W}(w_4,w_1)\}}$$

where

$$\mathscr{W}(w_i,w_j) = w_i w_j' - w_i' w_j$$

is the Wronskian evaluated at $z = z_0$. Since we know the Wronskian is a constant independent of z we could calculate it at any convenient point. However, the advantage of $z = z_0$ is that the argument of the Airy functions is large and positive so that asymptotic expansions are available. In fact

$$w_1 \sim \gamma^{-1} \exp(k_0 f), \quad w_2 \sim \tfrac{1}{2}\gamma^{-1} \exp(-k_0 f) \ ,$$

$$w_3 \sim \gamma^{-1} \exp(k_0 g), \quad w_4 \sim \tfrac{1}{2}\gamma^{-1} \exp(-k_0 g) \ ,$$

$$w_1' \sim -k_0\gamma \exp(k_0 f), \quad w_2' \sim \tfrac{1}{2}k_0\gamma \exp(-k_0 f) \ ,$$

$$w_3' \sim k_0\gamma \exp(k_0 g), \quad w_4' \sim -\tfrac{1}{2}k_0\gamma \exp(-k_0 g)$$

where

$$\gamma = |q(z_0)|^{\frac{1}{4}}, \quad f = \int_{z_0}^{z_2} |q(t)|^{\frac{1}{2}} dt, \quad g = \int_{z_1}^{z_0} |q(t)|^{\frac{1}{2}} dt .$$

Therefore $\mathscr{W}(w_4, w_1) = 0$, $\mathscr{W}(w_3, w_2) = 0$ and

$$\mathscr{W}(w_4, w_2) \sim \tfrac{1}{2}k_0 \exp(-k_0 l), \quad \mathscr{W}(w_3, w_1) \sim -2k_0 \exp(k_0 l),$$

$$\mathscr{W}(w_3, w_4) \sim -k_0$$

where

$$l = f + g = \int_{z_1}^{z_2} |q(t)|^{\frac{1}{2}} dt .$$

Consequently

$$R = \mathrm{i}, \quad T = \exp(-k_0 l) . \tag{8.93}$$

There is no point in retaining the exponentially small terms in R because only the first term in the asymptotic expansions has been kept so that the most we can assert is that $R = \mathrm{i} + O(1/k_0)$. However, the exact relation (8.91) coupled with (8.93) for T does allow the statement

$$|R|^2 = 1 - \exp(-2k_0 l) . \tag{8.94}$$

The interpretation of these formulae is that when the refractive index possesses two zeros there is a barrier which prevents the transmission of all but a minute amount of energy. The amplitude of the reflected wave is the same as that of the incident wave to within an exponentially small factor. The phase, however, is advanced by $\frac{1}{2}\pi$. Consequently, when a ray is reflected by a minimum in refractive index *the reflected field may be evaluated by geometrical optics provided that a phase advance of* $\frac{1}{2}\pi$ *is inserted*. Of course, the result will not be legitimate very near the reflecting level but only some distance below it.

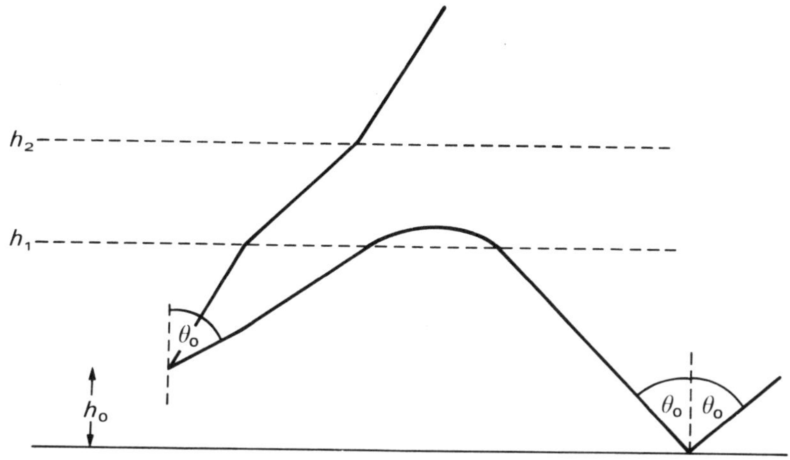

FIG. 8.9. Ray when the refractive index has a minimum above a plane earth.

One application of this theory is to propagation over a plane earth. Suppose that N is constant and equal to 1 up to a height h_1, passes through a positive minimum $N(m)$ between h_1 and h_2 and becomes constant again at altitudes greater than h_2 (Fig. 8.9). If a ray is launched at an angle θ_0 to the vertical the theory of §8.4 informs us that if $\sin\theta_0 < N(m)$ the ray will not be reflected but will continue steadily away from the earth after refraction through the layer.

If $\sin\theta_0 > N(m)$ the ray will be reflected at a height h where $N(h) = \sin\theta_0$, the abscissa of the point of reversing direction vertically being

$$x_h = \sin\theta_0 \int_{h_0}^{h} \{N^2(w) - \sin^2\theta_0\}^{-\frac{1}{2}} dw$$

where h_0 is the height of the source above the earth. (It is assumed that the source is below the layer and in the region in which $N = 1$.) As θ_0 increases a reflected ray first appears when $\sin\theta_0$ is just greater than $N(m)$ and then x_h is virtually infinite. Increasing θ_0 further decreases x_h but x_h must subsequently increase because obviously $x_h \to \infty$ as $\theta_0 \to \frac{1}{2}\pi$. Hence x_h has a minimum for some value of θ_0. This minimum can be zero only if the ray with $\theta_0 = 0$ is reflected, i.e. if $N(m) = 0$. *In general, there are at least two rays which turn round at any value of x_h greater than the minimum.*

A reflected ray will eventually strike the earth. If $N(m) > 0$, the point of impact must be beyond a certain distance because x_h cannot be below its minimum. The minimum horizontal range which must be traversed before a ray hits the earth is known as the *skip distance*. The field due to the reflected rays is often known as the *sky wave*. The sky wave is responsible for the ability to transmit signals great distances because it can penetrate to places where the direct wave from the source has been reduced to negligible proportions by horizon effects.

When the reflected ray strikes the earth it will be reflected by the earth at the same angle to the vertical. It will therefore return to the upper layer and be reflected back to the earth again. The phenomenon is therefore one of multiple reflection. If the source is a vertical transmitter which would give $\exp(-ik_0R)/R$ in free space the sky wave after m reflections at the earth and l reflections at the variable layer is

$$\frac{\exp(\frac{1}{2}l\pi i)D\{R_v(\theta_0)\}^m\exp(-ik_0L)}{d}$$

where $R_v(\theta_0)$ is the Fresnel coefficient for the reflection at the earth at an angle of incidence θ_0, L is the optical path length, d is the geometrical length of the ray, and D is a *divergence factor* representing the effect of spreading of the rays. The only influence of the layer reflection other than in D is the phase advance of $\frac{1}{2}\pi$ on each occasion.

The total field at a point is obtained by summing over all the reflected rays which pass through it and adding that which arrives directly without reflection. In the situation of Fig. 8.9 there may be rays emitted downwards which strike the earth before going to the transition layer; their contribution must be included.

If both the transmitting and receiving points are on the surface of the earth the formula for the sky wave is

$$\frac{\Sigma D\{R_v(\theta_0)\}^m\{1+R_v(\theta_0)\}^2\exp\{-ik_0L+\tfrac{1}{2}(m+1)\pi i\}}{d}$$

where m is the number of earth reflections between the transmitter and receiver. The factor $1 + R_v$ is included once because the receiver is at a point where both a downcoming and an upgoing ray occur and a second time because the transmitter is on the earth. The summation must be carried out over those values of θ_0 which supply a ray through the point of observation, i.e. which satisfy $x = 2(m+1)x_h$ where x is the distance between transmitter and receiver. Also

$$L = \int N\mathrm{d}s = 2(m+1)\int_0^h N^2(w)\{N^2(w)-\sin^2\theta_0\}^{-\frac{1}{2}}\mathrm{d}w\ .$$

With regard to D the rays in a small cone occupy an area $d^2 \sin\theta_0\,\mathrm{d}\theta_0\mathrm{d}\phi$ in free space. Here they fill an area $x\cos\theta_0\,\mathrm{d}x\mathrm{d}\phi$. Hence

$$D = d\left(\frac{\tan\theta_0|\mathrm{d}\theta_0/\mathrm{d}x|}{x}\right)^{\frac{1}{2}}$$

and the sky wave can now be computed.

In practice, the determination of rays when the positions of the transmitter and receiver are specified has to be carried out by trial and error. One way is to trace a ray from the transmitter. If it misses the receiver by being at a range x_i instead of x_R at the height h_R of the receiver, a new angle of launching based on the iteration

$$\operatorname{cosec}\theta_{i+1} = \operatorname{cosec}\theta_i - N(h_T)(x_R-x_i)/N^2(h_R)\partial x_i/\partial N(h_R)$$

is tried with h_T the altitude of the transmitter. Alternatively, and probably more effectively, rays are traced for a selected set of launching angles and then, if two consecutive ones bracket the receiver, a better approximation is derived by interpolation.

Exercises

25. Show that the change in range at constant height due to a change in launching angle satisfies

$$\frac{\partial x}{\partial \theta_0} = \frac{N_0}{N'_0} + \cot \theta_0 \int_0^x \left(1 - \frac{NN''}{N'^2}\right) dx \,.$$

Deduce the intensity of a ray.

26. Assuming that N is given only at a sequence of discrete data levels interpolate (i) $1/N$ linearly between levels, (ii) N^2 by quadratics with continuous first derivatives, and (iii) N^2 by cubic B-splines. Set up a program for tracing rays and compare the three possibilities when $N(z) = 1 + z/a$ and when $1/N(z) = 1 + z/a$. Note particularly the comparative positions of any caustics.

If N has a parabolic variation, first increasing from the surface of the earth to a maximum at height h_1, then decreasing to a height h_2, and thereafter decreasing linearly to a height h_3 where there is a perfectly reflecting boundary, draw a ray diagram.

8.15. Edges

An essential feature of the procedure in applying geometrical optics to scattering by an obstacle is the substitution of the tangent plane at a point for the local surface. Such a replacement is plausible if the principal radii of curvature at the point are large compared with the wavelength. If the wavelength is too large another canonical problem may be necessary but the more awkward situation is that in which the radii of curvature are large in some places and small in others. If one radius of curvature is large and the other small another canonical problem is involved. When one radius of curvature is actually zero as at an edge the geometrical theory of diffraction selects as its canonical problem the scattering by a semi-infinite plane. This section is concerned with elucidating the properties of this canonical problem.

Let the perfectly conducting semi-infinite plane occupy that part of $y = 0$ on which x is negative (Fig. 8.10) and lie in free space. Represent x, y by the cylindrical polar coordinates r, ϕ. Then we consider an incident electrically polarized plane wave in which the z component of the electric intensity is given by

$$E_0 = \exp\{-ik_0 r \cos(\phi-\phi_0)\}$$

where ϕ_0 is a real angle satisfying $0 < \phi_0 < \pi$. In parallel there is a magnetically polarized incident wave in which

$$H_0 = \exp\{-ik_0 r \cos(\phi-\phi_0)\} .$$

The boundary condition on the semi-infinite plane is that the total E vanishes there or, correspondingly, $\partial H/\partial y = 0$.

A detailed derivation of the solution will not be given. That the formulae displayed below do comply with all conditions of the problem may be readily confirmed by the reader. The solution is expressed in terms of the function $F(w)$ defined by

$$F(w) = \exp(iw^2) \int_w^\infty \exp(-it^2)\,dt \tag{8.95}$$

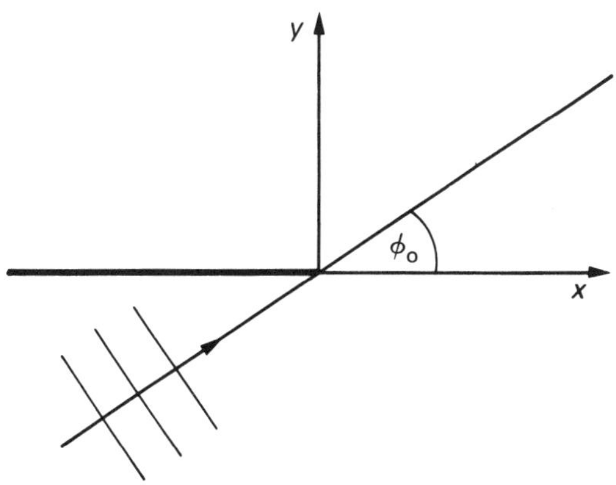

FIG. 8.10. Plane wave incident on a semi-infinite plane.

which has been tabulated (Clemmow and Mumford 1952; see also Baños and Johnston 1970) for complex w; it may also be written otherwise by Fresnel's integrals. One useful attribute is that

$$F(-w) = \pi^{\frac{1}{2}} \exp(iw^2 - \tfrac{1}{4}\pi i) - F(w) . \tag{8.96}$$

Integration by parts in (8.95) easily furnishes

$$F(w) = -\frac{i}{2w} + \frac{1}{4w^3} + O\left(\frac{1}{|w|^5}\right) \tag{8.97}$$

for $\frac{1}{2}\pi > \text{ph}w > -\pi$. Formula (8.96) caters for the asymptotic conduct in other ranges of ph$w$ (see also §8.26). By putting

$$F(w) = \exp(iw^2)\left\{\int_0^\infty \exp(-it^2)dt - \int_0^w \exp(-it^2)dt\right\}$$

we see that

$$F(w) = \tfrac{1}{2}\pi^{\frac{1}{2}} \exp(-\tfrac{1}{4}\pi i) - w + O(|w|^2) \tag{8.98}$$

for $|w| \ll 1$.

The electrically polarized solution is

$$E = \pi^{-\frac{1}{2}} \exp(-ik_0 r + \tfrac{1}{4}\pi i)[F\{(2k_0 r)^{\frac{1}{2}} \sin \tfrac{1}{2}(\phi - \phi_0)\} -$$

$$- F\{(2k_0 r)^{\frac{1}{2}} \sin \tfrac{1}{2}(\phi + \phi_0)\}] . \tag{8.99}$$

Let $r \to \infty$ and preserve only the terms which do not tend to zero. Then, from (8.96) and (8.97),

$$E = \exp\{-ik_0 r \cos(\phi - \phi_0)\} - \exp\{-ik_0 r \cos(\phi + \phi_0)\} \quad (-\pi \leqslant \phi < -\phi_0)$$

$$= \exp\{-ik_0 r\cos(\phi-\phi_0)\} \qquad (-\phi_0 < \phi < \phi_0)$$

$$= 0\ . \qquad (\phi_0 < \phi \leqslant \pi).$$

The first discloses the presence of a plane wave reflected from the face of the diffracting sheet as if it were infinite in extent; the second contains the incident wave alone while the third indicates a zero field behind the screen. These formulae bring to light geometrical optics. For there is a *shadow zone* behind the plane where there is no field (Fig. 8.11), an *illuminated region* where the incident plane wave exists, and a *reflection sector* where a reflected wave is manifest. The field of geometrical optics is discontinuous across the lines $\phi = \pm\phi_0$ which is why the more recondite expression (8.99) is wanted to secure a smooth transition.

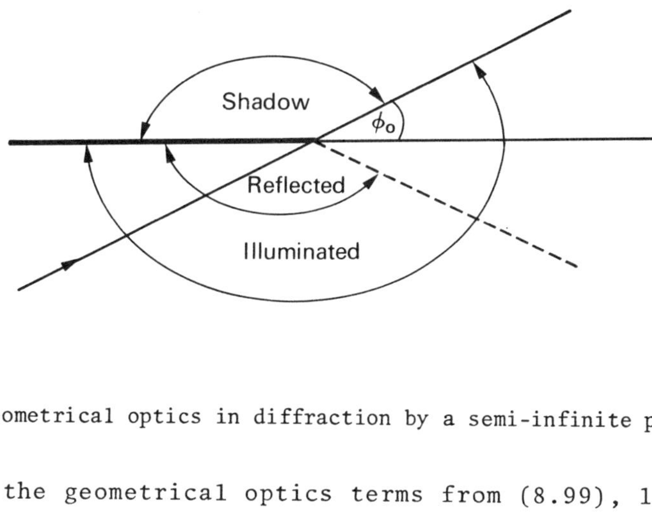

FIG. 8.11. Geometrical optics in diffraction by a semi-infinite plane.

Extract the geometrical optics terms from (8.99), leaving what is known as the *diffracted field*. When $k_0 r \gg 1$ and ϕ is not near $\pm\phi_0$ the diffracted field is given by

$$E_d \sim \frac{\cos\frac{1}{2}\phi\,\sin\frac{1}{2}\phi_0}{\cos\phi_0 - \cos\phi}\left(\frac{2}{\pi k_0 r}\right)^{\frac{1}{2}}\exp(-\tfrac{1}{4}\pi i - ik_0 r) \qquad (8.100)$$

from (8.96) and (8.97). The implication of (8.100) is that the diffraction field behaves as if it originates from a line source placed along the diffracting edge but with an amplitude factor which depends upon the angle of observation. Experimentally, this may be confirmed by measurements in the shadow zone where the electromagnetic field appears to issue from the edge.

The formula (8.100) must be dropped when ϕ is near $\pm\phi_0$ because then a small argument may occur in F even though k_0r is large. A notion of the regions of invalidity can be obtained by drawing the curves $2k_0r\sin^2\frac{1}{2}(\phi\pm\phi_0) = 10$. The curves are parabolas with foci at the origin and axes $\phi = \pm\phi_0$ (Fig. 8.12). The larger k_0r is, the narrower the parabolas are. The formula (8.100) should be avoided in the shaded regions. In these regions the full expression (8.99) must be used; there is no simpler approximation.

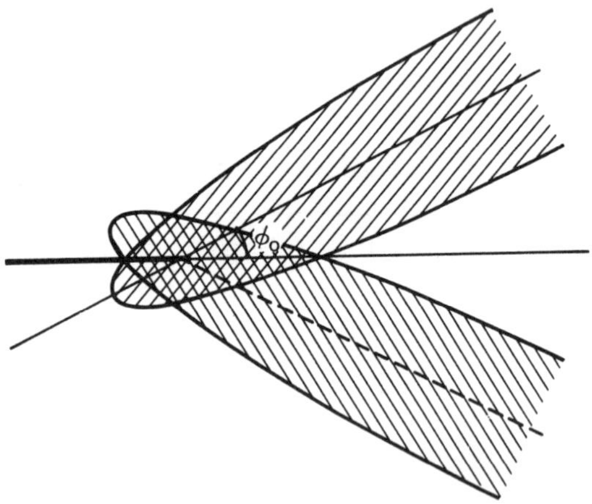

FIG. 8.12. The asymptotic formula fails in the shaded regions.

According to (8.99)

$$E = \frac{1}{2}\exp(-ik_0r) + O\{(k_0r)^{-\frac{1}{2}}\}$$

when $\phi = \phi_0$ so that the diffracted field is of the same order of magnitude as the incident wave. A similar statement is true at $\phi = -\phi_0$. This explains in another way why it is not legitimate to employ (8.100) in the shaded regions of Fig. 8.12.

In a region where the diffracted and incident waves are of

comparable magnitude there is the prospect of interference. It is borne out by calculations. A typical example is shown in Fig. 8.13 where the field three wavelengths behind the screen is plotted, i.e. on $y = 6\pi/k_0$, with an incident plane wave at normal incidence ($\phi_0=\frac{1}{2}\pi$) to the plane. The interference fringes in the illuminated region are clearly visible as well as the steady decay which transpires as one moves into the shadow.

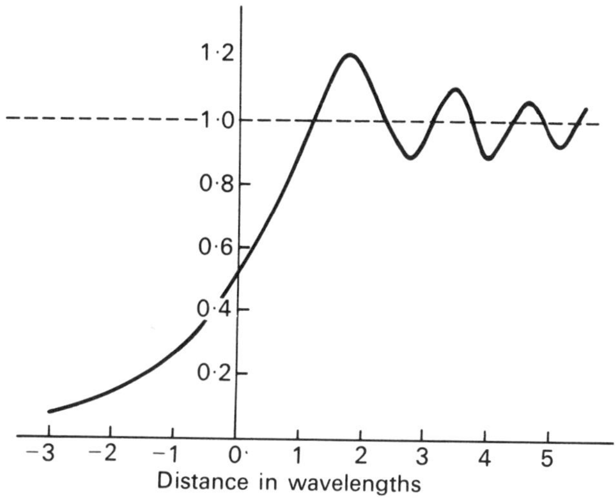

FIG. 8.13. Interference fringes for a plane wave at normal incidence.

For magnetic polarization

$$H = \pi^{-\frac{1}{2}} \exp(-ik_0r+\tfrac{1}{4}\pi i)[F\{(2k_0r)^{\frac{1}{2}} \sin \tfrac{1}{2}(\phi-\phi_0) + F\{(2k_0r)^{\frac{1}{2}} \sin \tfrac{1}{2} (\phi+\phi_0)\}] \tag{8.101}$$

which differs from (8.99) only in the sign of the second term. The general structure of the field of geometrical optics is the same as in Fig. 8.11. When $k_0r \gg 1$, the diffracted field is given by

$$H_d \sim \frac{\sin \frac{1}{2} \phi \cos \frac{1}{2} \phi_0}{\cos \phi_0 - \cos \phi} \left(\frac{2}{\pi k_0 r}\right)^{\frac{1}{2}} \exp(-ik_0r-\tfrac{1}{4}\pi i) \tag{8.102}$$

provided that ϕ is away from $\pm\phi_0$, i.e. the shaded regions of Fig. 8.12 must again be excluded. Once more, the diffracted field looks as though it stems from a line source along the diffracting edge.

So far the direction of propagation of the plane wave has been in the (x,y) plane and perpendicular to the diffracting edge. This restriction will now be lifted and the case of general oblique incidence considered. Let the electric intensity of the primary wave be

$$\mathbf{E} = (l_0\mathbf{i}+m_0\mathbf{j}+n_0\mathbf{k})\exp\{-ik_0(x \sin\theta_0 \cos\phi_0+y \sin\theta_0 \sin\phi_0+z \cos\theta_0)\}$$

where $\mathbf{i}$, $\mathbf{j}$, $\mathbf{k}$ are unit vectors parallel to the Cartesian axes and

$$(l_0 \cos\phi_0+m_0 \sin\phi_0) \sin\theta_0 + n_0 \cos\theta_0 = 0\ .$$

The direction of propagation makes an angle θ_0 with the z axis (which is along the edge) and the wave degenerates to the non-oblique case when $\theta_0 = \frac{1}{2}\pi$.

If the z dependence of all fields is taken to be $\exp(-ik_0 z \cos\theta_0)$, the governing equations assume the same guise as in two dimensions except that k_0 is replaced by $k_0 \sin\theta_0$. The solution may be deduced from (8.99) and (8.101) by such a replacement. It is found that, if $\mathbf{E}^{(3)}$ denotes the total three-dimensional field,

$$\mathbf{E}^{(3)} = n_0 \exp(-ik_0 z \cos\theta_0)\,\frac{(k_0\mathbf{k} \sin^2\theta_0-i \cos\theta_0 \operatorname{grad})E^{(3)}}{k_0 \sin^2\theta_0}$$
$$- i \exp(-ik_0 z \cos\theta_0)\,\frac{(l_0 \sin\phi_0-m_0 \cos\phi_0)\mathbf{k} \wedge \operatorname{grad} H^{(3)}}{k_0 \sin\theta_0} \tag{8.103}$$

where $E^{(3)}$ and $H^{(3)}$ are given by (8.99) and (8.101) with $k_0 \sin\theta_0$ in place of k_0. Also

$$\left(\frac{\mu_0}{\varepsilon_0}\right)^{\frac{1}{2}} \mathbf{H}^{(3)} = -\, i n_0 \exp(-ik_0 z \cos\theta_0) \frac{\mathbf{k} \wedge \operatorname{grad} E^{(3)}}{k_0 \sin^2\theta_0}$$

$$- \exp(-ik_0 z \cos\theta_0) \frac{(l_0 \sin\phi_0 - m_0 \cos\phi_0)(k_0 \mathbf{k} \sin^2\theta_0 - \mathrm{i}\cos\theta_0 \operatorname{grad}) H^{(3)}}{k_0 \sin\theta_0} \tag{8.104}$$

When $k_0 r \sin\theta_0$ is large the Fs can be estimated asymptotically provided that ϕ is not close to $\pm\phi_0$, i.e. so long as the point of observation is not near the shadow boundary or boundary of the reflected wave. Then for the diffracted field

$$E_{dr}^{(3)} \sim -\, n_0 \cot\theta_0 \frac{\cos\frac{1}{2}\phi \sin\frac{1}{2}\phi_0}{\cos\phi_0 - \cos\phi}\left(\frac{2}{\pi k_0 r \sin\theta_0}\right)^{\frac{1}{2}} \exp(-ik_0 r \sin\theta_0 - ik_0 z \cos\theta_0 - \tfrac{1}{4}\pi \mathrm{i}), \tag{8.105}$$

$$E_{d\phi}^{(3)} \sim -(l_0 \sin\phi_0 - m_0 \cos\phi_0) \frac{\sin\frac{1}{2}\phi \cos\frac{1}{2}\phi_0}{\cos\phi_0 - \cos\phi}\left(\frac{2}{\pi k_0 r \sin\theta_0}\right)^{\frac{1}{2}} \exp\{-ik_0(r \sin\theta_0 + z \cos\theta_0) - \tfrac{1}{4}\pi \mathrm{i}\}, \tag{8.106}$$

$$E_{dz}^{(3)} \sim n_0 \frac{\cos\frac{1}{2}\phi \sin\frac{1}{2}\phi_0}{\cos\phi_0 - \cos\phi}\left(\frac{2}{\pi k_0 r \sin\theta_0}\right)^{\frac{1}{2}} \exp\{-ik_0 r \sin\theta_0 - ik_0 z \cos\theta_0 - \tfrac{1}{4}\pi \mathrm{i}\} \tag{8.107}$$

whence

$$E_{dr}^{(3)} \sin\theta_0 + E_{dz}^{(3)} \cos\theta_0 \sim 0, \tag{8.108}$$

$$E_{dr}^{(3)} \cos\theta_0 - E_{dz}^{(3)} \sin\theta_0 \sim -\frac{E_{dz}^{(3)}}{\sin\theta_0}. \tag{8.109}$$

Moreover

$$\left(\frac{\mu_0}{\varepsilon_0}\right)^{\frac{1}{2}} \mathbf{H}_d^{(3)} \sim (\hat{\mathbf{r}} \sin\theta_0 + \mathbf{k} \cos\theta_0) \wedge \mathbf{E}_d^{(3)} \tag{8.110}$$

where $\hat{\mathbf{r}}$ is a unit vector in the direction of the cylindrical polar radial vector.

These formulae are valid so long as $k_0 r$ is large, i.e. points of observation almost on the edge are excluded, and $\sin\theta_0$ is not too small. Thus the incident wave must not have a direction of propagation practically coincident with the edge.

It is transparent from (8.108) and (8.110) that the electric and magnetic fields are perpendicular to each other and both are orthogonal to $\hat{\mathbf{r}}\sin\theta_0 + \mathbf{k}\cos\theta_0$. Thus, locally, the diffracted field has the appearance of a plane wave moving parallel to $\hat{\mathbf{r}}\sin\theta_0 + \mathbf{k}\cos\theta_0$ and, generally, can be vizualized as propagating along rays of this type. Hence the field looks like a progressive wave emanating from the edge. If a cone of semi-angle θ_0 and apex on the edge be drawn with its axis along the edge, the field on the cone travels along a generator, being constant on any given generator apart from variations in the distance to the apex (Fig. 8.14). The existence of such a cone has been confirmed experimentally.

If $\theta_0 = \frac{1}{2}\pi$, the cone degenerates to a plane perpendicular

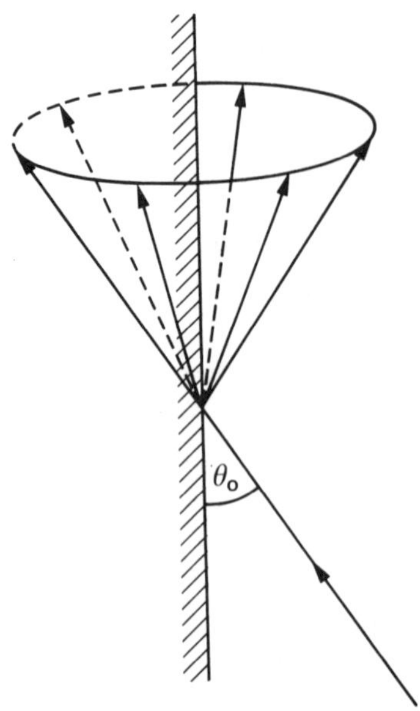

FIG. 8.14. The cone of diffracted rays produced by an edge.

to the edge and the generators are radial lines from the edge, agreeing with the two-dimensional behaviour described earlier.

It is now convenient to rewrite (8.105)–(8.110) differently. Introduce the coordinates ξ and η on a ray so that both are perpendicular to the ray and form a right-handed system with the direction of propagation. Choose ξ to lie in the plane containing the ray and edge. Denote the incident ray and a ray emitted by the edge by the superscripts i and e respectively. For example, η^{e} will lie in the opposite direction to the unit ϕ vector and $E_{\xi}^{\mathrm{e}} = E_{dz}^{(3)} \sin\theta_0 - E_{dr}^{(3)} \cos\theta_0$. Then (8.105)–(8.110) give

$$\begin{pmatrix} E_{\xi}^{\mathrm{e}} \\ E_{\eta}^{\mathrm{e}} \end{pmatrix} = \begin{pmatrix} D_E & 0 \\ 0 & D_H \end{pmatrix} \begin{pmatrix} E_{\xi 0}^{\mathrm{i}} \\ E_{\eta 0}^{\mathrm{i}} \end{pmatrix} \frac{\exp(-\mathrm{i}k_0 r \operatorname{cosec}\theta_0)}{(r \operatorname{cosec}\theta_0)^{\frac{1}{2}}} \tag{8.111}$$

where

$$D_E = \left(\frac{2}{\pi k_0}\right)^{\frac{1}{2}} \exp(-\tfrac{1}{4}\pi\mathrm{i}) \frac{\cos\frac{1}{2}\phi \sin\frac{1}{2}\phi_0}{\cos\phi_0 - \cos\phi} \operatorname{cosec}\theta_0, \tag{8.112}$$

$$D_H = \left(\frac{2}{\pi k_0}\right)^{\frac{1}{2}} \exp(-\tfrac{1}{4}\pi\mathrm{i}) \frac{\sin\frac{1}{2}\phi \cos\frac{1}{2}\phi_0}{\cos\phi_0 - \cos\phi} \operatorname{cosec}\theta_0 \tag{8.113}$$

and $E_{\xi 0}^{\mathrm{i}}$, $E_{\eta 0}^{\mathrm{i}}$ are the components of the incident field along ξ^{i} and η^{i} at the point of the edge from where the ray starts. The magnetic intensity can be obtained from (8.110). These results form the basis of the geometrical theory of diffraction.

8.16. Edge rays

To enable the geometrical theory of diffraction (which will hereinafter be abbreviated to GTD) to cope with edges we postulate that they have an effect which can be modelled by the behaviour in §8.15. In other words, an Ansatz is introduced stating that the rays behave in the same way even if the edge is curved. The incident ray from P_0 to any point O of the edge

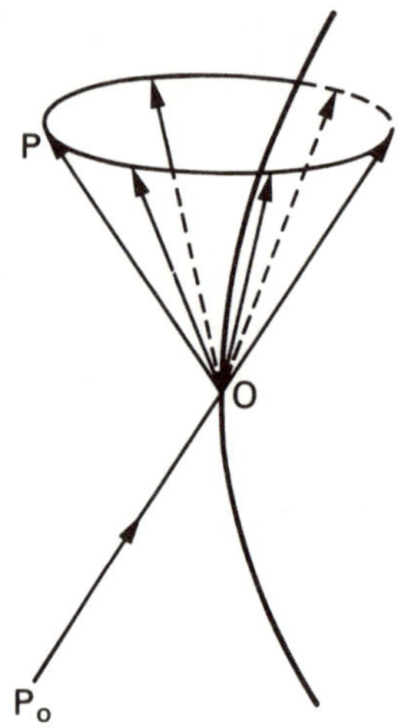

FIG. 8.15. Cone of rays at a curved edge.

is imagined to initiate a set of *edge rays* all disposed on a right circular cone with vertex O, axis the tangent to the edge at O, and with one generator the continuation of the incident ray (Fig. 8.15). Fermat's principle will then make P_0OP a contributory ray when O is chosen so that the aforementioned cone passes through P. If the incident ray happens to strike the edge at right angles the edge rays spread out all over the plane perpendicular to the edge as depicted in Fig. 8.16.

An edge ray may itself encounter an edge. It will then originate its own cone of edge rays, often called *doubly diffracted rays* to indicate that two edges have been involved. There

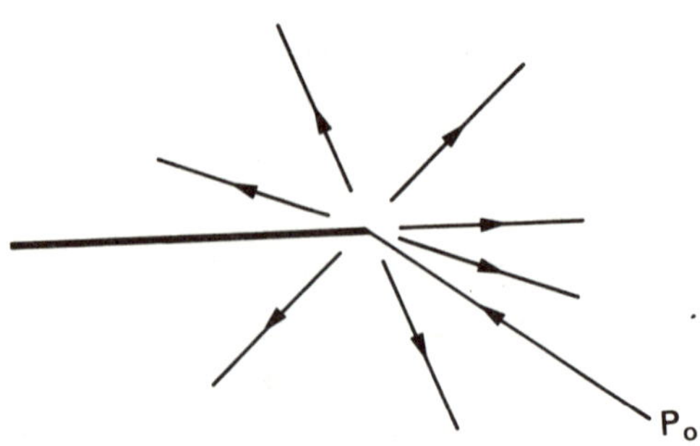

FIG. 8.16. Edge rays for perpendicular incidence.

is nothing new in principle unless the second edge is near the shadow or reflection boundaries of the first (see §8.18) but it may be extraordinarily complicated to decide which rays go through a particular point.

The knowledge of the trajectories of the rays has to be supplemented by a prescription for the amplitude of the field to make GTD complete. According to §8.2 the field on an edge ray can be computed from

$$\mathrm{E}^{e}(P) = \left\{\frac{\rho_1^e\ \rho_2^e}{(\rho_1^e+s)(\rho_2^e+s)}\right\}^{\frac{1}{2}} \mathbf{E}^{e}(Q)\ \exp\{-ik_0(L_0+s)\} \qquad (8.114)$$

when $N = 1$. Here $\mathbf{E}^{e}(Q)$ is the field at the point Q of the ray, L_0 the value of L there, s the distance from Q to P, and ρ_1^e, ρ_2^e the principal radii of curvature at Q of the wavefront produced by the edge rays.

It would be pleasant to pick Q as the point O of the edge in Fig. 8.16 so as to connect $\mathbf{E}^{e}(Q)$ with the incident ray. Unfortunately, all the rays on the cone intersect at the apex O so that the edge is a caustic of the diffracted rays. Hence, if Q coincides with O, ρ_2^e is zero and $\mathbf{E}^{e}(Q)$ would need to be infinite to provide a finite non-zero value for $\mathbf{E}^{e}(P)$ from (8.114).

Contrariwise, (8.114) declares that $(\rho_2^e)^{\frac{1}{2}}\mathbf{E}^{e}(Q)$ is finite as Q tends to O. Let its limit be $\mathbf{E}(0)$. Then

$$\mathbf{E}^{e}(P) = \left\{\frac{\rho_1^e}{(\rho_1^e+s)s}\right\}^{\frac{1}{2}} \mathbf{E}(0)\ \exp\{-ik_0(L_0+s)\}$$

where now s is measured along the edge ray from O. The radius of curvature ρ_1^e is the distance of O from the second caustic of the rays. Its evaluation is straightforward. Let θ_0 be the angle between the incident ray and the tangent OT to the edge (Fig. 8.17). Let $\boldsymbol{\nu}$ be a unit vector along the principal normal ON which lies in the osculating plane. Unit vectors along P_0O and OP are denoted by $\hat{\mathbf{s}}^{i}$ and $\hat{\mathbf{s}}^{e}$ respectively. Then, if κ_0 is the curvature of the edge at O,

$$\frac{1}{\rho_1^e} = -\kappa_0 \boldsymbol{\nu} \cdot \hat{\mathbf{s}}^e \operatorname{cosec}^2 \theta_0 - \operatorname{cosec} \theta_0 \frac{d\theta_0}{d\sigma_0} \qquad (8.115)$$

where σ_0 is the arc length of edge in the direction OT and the derivative is evaluated at O. The change in θ_0 is related to the curvature of the incident wavefront as is brought out by writing (8.115) as

$$\frac{1}{\rho_1^e} = C_{11}^i + \kappa_0 \boldsymbol{\nu} \cdot (\hat{\mathbf{s}}^i - \hat{\mathbf{s}}^e) \operatorname{cosec}^2 \theta_0 \qquad (8.116)$$

where C^i is the curvature matrix of §8.5 but based on the co-ordinates ξ^i, η^i of §8.17.

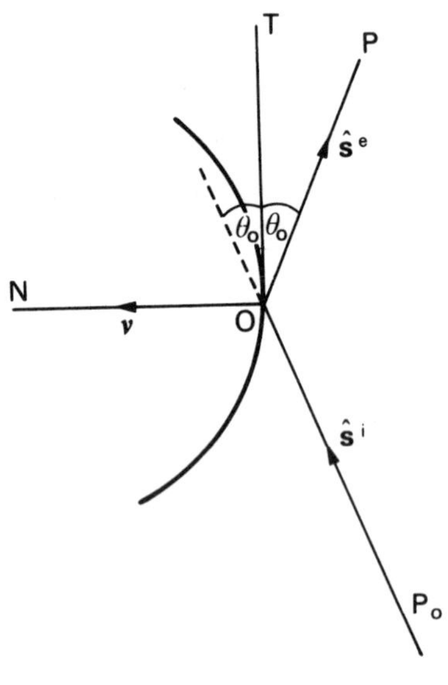

FIG. 8.17. Parameters for the calculation of ρ_1^e.

It remains to find $\mathbf{E}(0)\exp(-ik_0 L_0)$. The suggestion is that it should be joined to $\mathbf{E}^i(0)$, the incident field at O, in a similar manner to that in which they are when the edge is straight, i.e. $\mathbf{E}(0)\exp(-ik_0 L_0) = D\mathbf{E}^i(0)$ where D is the matrix with diagonal elements D_E, D_H and it is understood that the two sides are expressed in terms of their respective ξ,η ray coordinates. Now, if we do not stray too far from the edge s will be negligible compared with ρ_1^e and the diffracted field will have the form

$$s^{-\frac{1}{2}}\mathbf{E}(0)\,\exp\{-ik_0(L_0+s)\}\ .$$

Indeed, for an incident plane wave and straight edge ρ_1^e is infinite. In the notation of §8.15 $s = r\,\operatorname{cosec}\theta_0$ and (8.111) is correctly reproduced by the proposed choice for $\mathbf{E}(0)$. Hence, according to GTD, the field on an edge ray is

$$\mathbf{E}^e(P) = \left\{\frac{\rho_1^e}{(\rho_1^e+s)s}\right\}^{\frac{1}{2}}\exp(-ik_0 s)D\mathbf{E}^i(0)\ . \tag{8.117}$$

The quantities ϕ_0 and ϕ in D in (8.117) must now be taken as angles made with the continuation of NO by the projections of the incident ray P_0O and of the edge ray OP respectively on the normal plane at O.

To find higher-order terms in the expansion reversion to (8.36) and (8.37) is necessary. For example, since μ and ε are constant and $N = 1$, from (8.36)

$$\hat{\mathbf{s}}^e\cdot\mathbf{E}_m|J(s)|^{\frac{1}{2}} = -i\int_{s_1}^{s}\frac{1}{2}|J(\sigma)|^{\frac{1}{2}}\hat{\boldsymbol{\sigma}}\cdot\nabla^2\mathbf{E}_{m-1}\,d\sigma + \hat{\mathbf{s}}^e\cdot\mathbf{E}_m(s_1)|J(s_1)|^{\frac{1}{2}}$$

where J is the Jacobian of §8.3 because s and σ can be identified with each other. In fact, $J(s) = s(1+s/\rho_1^e)\sin^2\theta_0$. The iteration is commenced by replacing $\mathbf{E}_0$ by $\mathbf{E}^e$ as given by (8.117).

However, we cannot put $s_1 = 0$ because of the singularities which occur there. This difficulty is overcome by means of *Hadamard's finite part*. If $\int_\delta f(t)dt$ is expanded in powers (possibly fractional) as $\delta \to +0$, let $S(\delta)$ denote those which are negative and therefore give rise to a singularity in the limit. Then

$$\text{fin} \int_0 f(t)\mathrm{d}t = \lim_{\delta\to+0} \left\{ \int_\delta f(t)\mathrm{d}t - S(\delta) \right\} .$$

Letting s_1 tend to zero we have

$$\hat{\mathbf{s}}^{\mathrm{e}} \cdot \mathbf{E}_m |J(s)|^{\frac{1}{2}} = \hat{\mathbf{s}}^{\mathrm{e}} \cdot \mathbf{e}_m - \tfrac{1}{2}\mathrm{i}\ \text{fin} \int_0^s |J(\sigma)|^{\frac{1}{2}} \hat{\boldsymbol{\sigma}} \cdot \nabla^2 \mathbf{E}_{m-1} \mathrm{d}\sigma \tag{8.118}$$

where $\mathbf{e}_m = \text{fin}\ \mathbf{E}_m(0)|J(0)|^{\frac{1}{2}}$. Regrettably, (8.118) still cannot be implemented because there is no rule for specifying $\mathbf{e}_m$. To resolve this deficiency requires uniformly valid formulae which will be discussed in the next section.

8.17. Uniformly valid approximations

The formula (8.117) has certain failings. It is discontinuous at the shadow and reflection boundaries ($\phi = \pm\phi_0$). Indeed, an infinite field is predicted there because D is infinite there. Also the edge wave approaches infinity at the edge where $s = 0$, so that edge conditions may not be fulfilled. The higher-order terms, as exemplified by (8.118), cannot be determined. Finally, there are discontinuities in the field or its derivatives at points of free space which lie on $\phi = -\pi$ or $\phi = \pi$. The formula (8.117) is therefore *non-uniform* in that its domain of validity excludes certain regions. An expansion which removes some of these objections by being continuous across the shadow and reflection boundaries while displaying the correct edge behaviour is called *uniformly valid*.

The analysis for the uniformly valid expansion will be restricted to the case when the edge is plane. To be precise we consider a finite aperture, such as a circle, in an infinite perfectly conducting plane screen. Points will be identified by the ray coordinates s^{e}, σ_0, and ϕ where, as before, σ_0 is the arc length along the edge to the point originating the edge ray.

Assume that the incident field has an asymptotic expansion

$$\mathbf{E}^{\mathrm{i}} = \exp(-\mathrm{i}k_0 L^{\mathrm{i}}) \sum_{m=0} \frac{\mathbf{E}_m^{\mathrm{i}}}{k_0^m}, \qquad \mathbf{H}^{\mathrm{i}} = \exp(-\mathrm{i}k_0 L^{\mathrm{i}}) \sum_{m=0} \frac{\mathbf{H}^{\mathrm{i}}}{k_0^m} \tag{8.119}$$

where $\mathbf{E}_m^{\mathrm{i}}$ and $\mathbf{H}_m^{\mathrm{i}}$ satisfy the equations of §8.1. Let $L^{\mathrm{e}} = L^{\mathrm{i}}(\sigma_0) + s^{\mathrm{e}}$ so that L^{e} is the eikonal of the edge ray. Introduce the quantity ψ defined by

$$\psi^2 = L^{\mathrm{e}} - L^{\mathrm{i}}.$$

Since the distance along an incident ray to a point is never more than that in going to the point via an edge ray by Fermat's principle, $L^{\mathrm{e}} - L^{\mathrm{i}} \geqslant 0$ and so ψ is real. To fix its sign note firstly that it vanishes on the shadow boundary where $L^{\mathrm{e}} = L^{\mathrm{i}}$. Further, $\sin \frac{1}{2}(\phi-\phi_0)$ is zero on the shadow boundary; indeed, this is true whenever $\phi - \phi_0 = 2n\pi$ for integer n. Accordingly, choose ψ to have the same sign as $\sin \frac{1}{2}(\phi-\phi_0)$. Then, as ϕ varies, ψ will repeat itself at intervals of 4π but not of 2π, so ψ is double valued in the physical space.

We now seek an electromagnetic field **e**, **h** which complies with the Ansatz (Ahluwalia, Lewis, and Boersma 1968)

$$\mathbf{e} = \pi^{-\frac{1}{2}}\exp(-\mathrm{i}k_0 L^{\mathrm{e}} + \tfrac{1}{4}\pi\mathrm{i})\left\{F(k_0^{\frac{1}{2}}\psi) \sum_{m=0} \frac{\mathbf{E}_m^{\mathrm{i}}}{k_0^m} + k_0^{-\frac{1}{2}} \sum_{m=0} \frac{\mathbf{F}_m}{k_0^m}\right\}, \tag{8.120}$$

$$\mathbf{h} = \pi^{-\frac{1}{2}}\exp(-\mathrm{i}k_0 L^{\mathrm{e}} + \tfrac{1}{4}\pi\mathrm{i})\left\{F(k_0^{\frac{1}{2}}\psi) \sum_{m=0} \frac{\mathbf{H}_m^{\mathrm{i}}}{k_0^m} + k_0^{-\frac{1}{2}} \sum_{m=0} \frac{\mathbf{G}_m}{k_0^m}\right\}. \tag{8.121}$$

Since $\mathrm{d}F(w)/\mathrm{d}w = 2\mathrm{i}wF(w) - 1$, (8.120) and (8.121) will satisfy Maxwell's equations if

$$\mathrm{i}\ \mathrm{grad}\ L^{\mathrm{e}} \wedge \mathbf{F}_0 + \mathrm{grad}\ \psi \wedge \mathbf{E}_0^{\mathrm{i}} = \mathrm{i}\left(\frac{\mu_0}{\varepsilon_0}\right)^{\frac{1}{2}} \mathbf{G}_0\ , \tag{8.122}$$

$$\mathrm{i}\ \mathrm{grad}\ L^{\mathrm{e}} \wedge \mathbf{G}_0 + \mathrm{grad}\ \psi \wedge \mathbf{H}_0^{\mathrm{i}} = -\mathrm{i}\left(\frac{\varepsilon_0}{\mu_0}\right)^{\frac{1}{2}} \mathbf{F}_0\ , \tag{8.123}$$

$$\mathrm{i}\ \mathrm{grad}\ L^{\mathrm{e}} \wedge \mathbf{F}_m + \mathrm{grad}\ \psi \wedge \mathbf{E}_m^{\mathrm{i}} = \mathrm{i}\left(\frac{\mu_0}{\varepsilon_0}\right)^{\frac{1}{2}} \mathbf{G}_m + \mathrm{curl}\ \mathbf{F}_{m-1}\,, \tag{8.124}$$

$$\mathrm{i}\ \mathrm{grad}\ L^{\mathrm{e}} \wedge \mathbf{G}_m + \mathrm{grad}\ \psi \wedge \mathbf{H}_m^{\mathrm{i}} = -\mathrm{i}\left(\frac{\varepsilon_0}{\mu_0}\right)^{\frac{1}{2}} \mathbf{F}_m + \mathrm{curl}\ \mathbf{G}_{m-1} \tag{8.125}$$

for $m \geqslant 1$.

By the same method as was adopted for deriving (8.12) and (8.13), and bearing in mind that $\mathbf{E}_m^{\mathrm{i}}$, $\mathbf{H}_m^{\mathrm{i}}$ satisfy (8.5)–(8.7), we obtain

$$2(\mathrm{grad}\ L^{\mathrm{e}}\ .\ \mathrm{grad})\mathbf{F}_m + \mathbf{F}_m\nabla^2 L^{\mathrm{e}} = -\,\mathrm{i}\nabla^2\mathbf{F}_{m-1} + \\ + 2\mathrm{i}(\mathrm{grad}\ \psi\ .\ \mathrm{grad})\mathbf{E}_m^{\mathrm{i}} + \mathrm{i}\mathbf{E}_m^{\mathrm{i}}\nabla^2\psi\,, \tag{8.126}$$

$$2(\mathrm{grad}\ L^{\mathrm{e}}\ .\ \mathrm{grad})\mathbf{G}_m + \mathbf{G}_m\nabla^2 L^{\mathrm{e}} = -\,\mathrm{i}\nabla^2\mathbf{G}_{m-1} + \\ + 2\mathrm{i}(\mathrm{grad}\ \psi\ .\ \mathrm{grad})\mathbf{H}_m^{\mathrm{i}} + \mathrm{i}\mathbf{H}_m^{\mathrm{i}}\nabla^2\psi\,. \tag{8.127}$$

Write the right-hand sides of (8.126) and (8.127) as $\mathbf{f}_m$ and $\mathbf{g}_m$ respectively. Then, from (8.42),

$$\frac{2\mathrm{d}(J^{\frac{1}{2}}\mathbf{F}_m)}{\mathrm{d}s^{\mathrm{e}}} = \mathbf{f}_m J^{\frac{1}{2}}$$

whence

$$J^{\frac{1}{2}}(s)\mathbf{F}_m(s) = \frac{1}{2}\int_0^s J^{\frac{1}{2}}(\sigma)\mathbf{f}_m(\sigma)\mathrm{d}\sigma + [J^{\frac{1}{2}}\mathbf{F}_m]_0 \tag{8.128}$$

where $[\ ]_0$ means the value as $s \to 0$. Since the edge conditions force $\mathbf{F}_m$ to be no more singular than $s^{-\frac{1}{2}}$ as $s \to 0$ and J disappears like s, the last term in (8.128) is a finite quantity depending on σ_0 and ϕ. Similarly

$$J^{\frac{1}{2}}(s)\mathbf{G}_m(s) = \frac{1}{2}\int_0^s J^{\frac{1}{2}}(\sigma)\mathbf{g}_m(\sigma)\,d\sigma + [J^{\frac{1}{2}}\mathbf{G}_m]_0 . \tag{8.129}$$

The integrands in both (8.128) and (8.129) are finite so that the only singularity of $\mathbf{F}_m$ and $\mathbf{G}_m$ predicted by these equations as $s \to 0$ is of order $s^{-\frac{1}{2}}$ and stems from the last terms. If the last terms are absent $\mathbf{F}_m$ and $\mathbf{G}_m$ vanish at the edge.

Both $\mathbf{f}_m$ and $\mathbf{g}_m$ can be regarded as known, so both $\mathbf{F}_m$ and $\mathbf{G}_m$ are determined as soon as the last terms are specified. This has to be done by the edge conditions, as will be seen shortly.

Let $\mathbf{i}_2$ be a unit vector in the direction of increasing ϕ and let e_ϕ be the component of $\mathbf{e}$ along $\mathbf{i}_2$. Then, if the dependence on ϕ is explicitly indicated,

$$\mathbf{e} = e_\phi(\phi)\mathbf{i}_2 + \tilde{\mathbf{e}}(\phi), \quad \mathbf{h} = h_\phi(\phi)\mathbf{i}_2 + \tilde{\mathbf{h}}(\phi) . \tag{8.130}$$

Now if ϕ is replaced by $2\pi - \phi$ a solution of Maxwell's equations is still available provided that the signs of e_ϕ and $\tilde{\mathbf{h}}$ are reversed. Therefore consider the electromagnetic field

$$\begin{aligned} &\{e_\phi(\phi)+e_\phi(2\pi-\phi)\}\mathbf{i}_2 + \tilde{\mathbf{e}}(\phi) - \tilde{\mathbf{e}}(2\pi-\phi) , \\ &\{h_\phi(\phi)-h_\phi(2\pi-\phi)\}\mathbf{i}_2 + \tilde{\mathbf{h}}(\phi) + \tilde{\mathbf{h}}(2\pi-\phi) \end{aligned} \tag{8.131}$$

as a candidate for solving the diffraction problem in $-\pi \leqslant \phi \leqslant \pi$.

The tangential components of the electric intensity are null on the screen $\phi = \pi$. They are also zero on $\phi = -\pi$ because of the built-in 4π-periodicity. Next, by (8.96) and (8.97),

$$F(k_0^{\frac{1}{2}}\psi) \sim \pi^{\frac{1}{2}} \exp(ik_0\psi^2 - \tfrac{1}{4}\pi i)H(-\psi) - \frac{i}{2k_0^{\frac{1}{2}}\psi} + \frac{1}{4k_0^{3/2}\psi^3}$$

when ψ is not near zero, H being the usual Heaviside unit function. Thus the incident wave is properly reproduced in $\phi \leqslant \phi_0$, while the reflected wave is there in $\phi \leqslant -\phi_0$. The non-uniform diffracted wave part of **e** is

$$\pi^{-\frac{1}{2}}\exp(-ik_0L^e+\tfrac{1}{4}\pi i)\left[k_0^{-\frac{1}{2}}\left(\mathbf{F}_0 - \frac{i\mathbf{E}_0^i}{2\psi}\right) + k_0^{-3/2}\left\{\mathbf{F}_1 + \frac{\mathbf{E}_0^i}{4\psi^3} - \frac{i\mathbf{E}_1^i}{2\psi}\right\}\right] \qquad (8.132)$$

and more terms could be calculated by carrying the expansion of F further.

To complete the non-uniform expansion needs the determination of the last terms in (8.128) and (8.129). Select local Cartesian axes $\mathbf{t}_1$, $\mathbf{t}_2$, $\mathbf{t}_3$ at a point of the edge so that $\mathbf{t}_1$ and $\mathbf{t}_3$ coincide with $-\boldsymbol{\nu}$ and σ_0 in previous notation (Fig. 8.18). The components of the electric and magnetic intensities parallel to $\mathbf{t}_3$ must be bounded as $s \to 0$ in order to comply with the edge conditions. This can be arranged by requiring $\mathbf{t}_3 \cdot [\mathbf{F}_m]_0$ and $\mathbf{t}_3 \cdot [\mathbf{G}_m]_0$ to be finite. It follows that

$$\mathbf{t}_3 \cdot [J^{\frac{1}{2}}\mathbf{F}_m]_0 = 0, \quad \mathbf{t}_3 \cdot [J^{\frac{1}{2}}\mathbf{G}_m]_0 = 0 . \qquad (8.133)$$

These edge conditions carry certain implications about the other components of $\mathbf{F}_m$ and $\mathbf{G}_m$ at $s = 0$ as consequences of (8.122)–(8.125).

A scalar product of (8.122) with $\mathbf{t}_3$ gives

$$(\sin\theta_0 \cos\phi\, \mathbf{t}_2 - \sin\theta_0 \sin\phi\, \mathbf{t}_1) \cdot \left(i\mathbf{F}_0 + \frac{\mathbf{E}_0^i}{2\psi}\right)$$

$$= \sin\theta_0(\cos\phi_0\, \mathbf{t}_2 - \sin\phi_0\, \mathbf{t}_1) \cdot \frac{\mathbf{E}_0^i}{2\psi} + i\left(\frac{\mu_0}{\varepsilon_0}\right)^{\frac{1}{2}} \mathbf{t}_3 \cdot \mathbf{G}_0$$

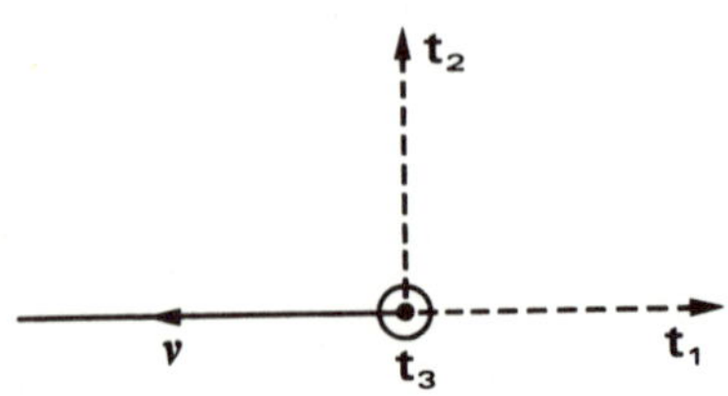

FIG. 8.18. The local Cartesian axes at a point of the edge.

since

$$\text{grad } L^{\text{e}} = \sin\theta_0 \cos\phi\, \mathbf{t}_1 + \sin\theta_0 \sin\phi\, \mathbf{t}_2 + \cos\theta_0\, \mathbf{t}_3 \; .$$

Multiplying by $J^{\frac{1}{2}}$ and applying (8.133) leads to

$$(\cos\phi\, \mathbf{t}_2 - \sin\phi\, \mathbf{t}_1)\, .\, \left\{\left(\mathbf{F}_0 - \mathrm{i}\frac{\mathbf{E}_0^{\text{i}}}{2\psi}\right)J^{\frac{1}{2}}\right\}_0$$

$$= -\mathrm{i}(\cos\phi_0\, \mathbf{t}_2 - \sin\phi_0\, \mathbf{t}_1)\, .\, \left[\frac{J^{\frac{1}{2}}\mathbf{E}_0^{\text{i}}}{2\psi}\right]_0 \tag{8.134}$$

which supplies another component of $[J^{\frac{1}{2}}\mathbf{F}_0]_0$. On the other hand, a scalar product of (8.123) with grad L^{e} provides

$$\text{grad } L^{\text{e}}\, .\, \left(\mathbf{F}_0 - \mathrm{i}\frac{\mathbf{E}_0^{\text{i}}}{2\psi}\right) = 0$$

when it is recalled that grad $L^{\text{i}} \wedge \mathbf{H}_0^{\text{i}} = -(\varepsilon_0/\mu_0)^{\frac{1}{2}}\mathbf{E}_0^{\text{i}}$. Hence

$$(\cos\phi\, \mathbf{t}_1 + \sin\phi\, \mathbf{t}_2)\, .\, \left[\left(\mathbf{F}_0 - \mathrm{i}\frac{\mathbf{E}_0^{\text{i}}}{2\psi}\right)J^{\frac{1}{2}}\right]_0 = \mathrm{i}\cot\theta_0\, \mathbf{t}_3\, .\, \left[\frac{\mathbf{E}_0^{\text{i}}J^{\frac{1}{2}}}{2\psi}\right]_0 \tag{8.135}$$

which furnishes the final part of $[J^{\frac{1}{2}}\mathbf{F}_0]_0$. For $\mathbf{G}_0$ it is only necessary to replace $\mathbf{F}_0$ and $\mathbf{E}_0^{\text{i}}$ in (8.134) and (8.135) by $\mathbf{G}_0$ and $\mathbf{H}_0^{\text{i}}$ respectively.

Since $\mathbf{F}_0 - \mathrm{i}\mathbf{E}_0^{\text{i}}/2\psi = [J^{\frac{1}{2}}(\mathbf{F}_0 - \mathrm{i}\mathbf{E}_0^{\text{i}}/2\psi)]_0/J^{\frac{1}{2}}$ we deduce that the first contribution of the non-uniform expansion (8.132) to $\mathbf{e}_\xi$ is, from (8.133) and (8.135),

$$\frac{\pi^{-\frac{1}{2}} \exp(-\mathrm{i}k_0 L^{\text{e}} - \frac{1}{4}\pi\mathrm{i})E_{0\xi}^{\text{i}}}{2^{3/2}k_0^{\frac{1}{2}} \sin\theta_0\; s^{\frac{1}{2}}(1+s/\rho_1^{\text{e}})^{\frac{1}{2}} \sin\frac{1}{2}(\phi-\phi_0)} \; .$$

When the corresponding term with $2\pi - \phi$ for ϕ is subtracted we obtain complete agreement with the appropriate part of (8.117).

Similarly (8.134) and (8.131) ensure compliance with the remainder of (8.117).

The rest of (8.132) gives the next higher-order term which defeated the non-uniform theory of the preceding section because of the lack of starting values. The uniform theory does not have this deficiency.

Take the scalar product of (8.124) with $\mathbf{t}_3$. Then

$$(\cos\phi\,\mathbf{t}_2 - \sin\phi\,\mathbf{t}_1)\cdot\left[\left(\mathbf{F}_1 - \mathrm{i}\frac{\mathbf{E}_1^{\mathrm{i}}}{2\psi}\right)J^{\frac{1}{2}}\right]_0$$

$$= -\mathrm{i}(\cos\phi_0\,\mathbf{t}_2 - \sin\phi_0\,\mathbf{t}_1)\cdot\left[\frac{J^{\frac{1}{2}}\mathbf{E}_1^{\mathrm{i}}}{2\psi}\right]_0 - \mathrm{i}\mathbf{t}_3\cdot[J^{\frac{1}{2}}\,\mathrm{curl}\,\mathbf{F}_0]_0\,\mathrm{cosec}\,\theta_0. \tag{8.136}$$

Also, from the scalar product of (8.125) with grad L^{e},

$$\mathrm{grad}\,L^{\mathrm{e}}\cdot\{\mathbf{F}_1 - \frac{\mathrm{i}\mathbf{E}_1^{\mathrm{i}}}{2\psi} + \frac{1}{2\psi}\,\mathrm{curl}\,\mathbf{H}_0^{\mathrm{i}} + \mathrm{i}\,\mathrm{curl}\,\mathbf{G}_0\} = 0$$

whence

$$(\cos\phi\,\mathbf{t}_1 + \sin\phi\,\mathbf{t}_2)\cdot[\{\mathbf{F}_1 - \frac{\mathrm{i}\mathbf{E}_1^{\mathrm{i}}}{2\psi} + \frac{1}{2\psi}\,\mathrm{curl}\,\mathbf{H}_0^{\mathrm{i}} + \mathrm{i}\,\mathrm{curl}\,\mathbf{G}_0\}J^{\frac{1}{2}}]_0$$

$$= \cot\theta_0\,\mathbf{t}_3\cdot\left[\left\{\frac{\mathrm{i}\mathbf{E}_1^{\mathrm{i}}}{2\psi} - \frac{1}{2\psi}\,\mathrm{curl}\,\mathbf{H}_0^{\mathrm{i}} - \mathrm{i}\,\mathrm{curl}\,\mathbf{G}_0\right\}J^{\frac{1}{2}}\right]_0. \tag{8.137}$$

From (8.133), (8.136) and (8.137) the finite part of $\mathbf{F}_1 + \mathbf{E}_0^{\mathrm{i}}/4\psi^3 - \mathrm{i}\mathbf{E}_1^{\mathrm{i}}/2\psi$ as $s \to 0$ can be deduced and thereby the starting values for second-order terms on account of (8.132). These beginning values will, in general, involve derivatives of lower-order terms.

Suppose, in fact, that $\phi_0 = \pi$ so that the incident ray travels towards the screen and grazes it. Then D_H of (8.113) vanishes and the evaluation of the leading term in E_η necessitates proceeding to a higher order. For simplicity in deriving the leading term it will be assumed that the incident field is horizontally polarized so that the only non-vanishing component is $E^{\mathrm{i}}_{0\eta}$. Then, from (8.133)–(8.135),

$$\mathbf{F}_0 - \frac{\mathrm{i}\mathbf{E}^{\mathrm{i}}_0}{2\psi} = \frac{\mathrm{i}\,\mathbf{i}_2 E^{\mathrm{i}}_{0\eta}(0)}{2^{3/2}J^{\frac{1}{2}}\sin\frac{1}{2}(\phi-\phi_0)} \tag{8.138}$$

with $E^{\mathrm{i}}_{0\eta}(0)$ signifying the value at $s = 0$. It follows that the next order term in the non-uniform expansion is given by

$$J^{\frac{1}{2}}\left(\mathbf{F}_1 - \frac{\mathrm{i}\mathbf{E}^{\mathrm{i}}_1}{2\psi} + \frac{\mathbf{E}^{\mathrm{i}}_0}{4\psi^3}\right) \cdot \mathbf{i}_2 = \mathrm{fin}\ J^{\frac{1}{2}}\left(\mathbf{F}_1 - \frac{\mathrm{i}\mathbf{E}^{\mathrm{i}}_1}{2\psi} + \frac{\mathbf{E}^{\mathrm{i}}_0}{4\psi^3}\right) \cdot \mathbf{i}_2 \ .$$

From (8.136), the right-hand side is

$$\frac{\mathrm{i}E^{\mathrm{i}}_{1\eta}(0)}{2^{3/2}\sin\frac{1}{2}(\phi-\phi_0)} - \mathrm{i}\mathbf{t}_3 \cdot [J^{\frac{1}{2}}\,\mathrm{curl}\,\mathbf{F}_0]_0 \operatorname{cosec}\theta_0 + \mathrm{fin}\left(\frac{J^{\frac{1}{2}}\mathbf{i}_2 \cdot \mathbf{E}^{\mathrm{i}}_0}{4\psi^3}\right)$$

Now

$$\mathbf{t}_3 \cdot \mathrm{curl}\ \mathbf{F}_0 = \mathbf{t}_3 \cdot \mathrm{curl}\left(\mathbf{F}_0 - \frac{\mathrm{i}\mathbf{E}^{\mathrm{i}}_0}{2\psi}\right) + \frac{\mathrm{i}\mathbf{t}_3}{2\psi} \cdot \mathrm{curl}\ \mathbf{E}^{\mathrm{i}}_0 - \frac{\mathrm{i}\mathbf{t}_3}{2\psi^2} \cdot \mathrm{grad}\,\psi \wedge \mathbf{E}^{\mathrm{i}}_0$$

and so an alternative expression is

$$\frac{\{\mathrm{i}E^{\mathrm{i}}_{1\eta}(0) + [\mathbf{t}_3 . \mathrm{curl}\ \mathbf{E}^{\mathrm{i}}_0]_0 \operatorname{cosec}\theta_0\}}{2^{3/2}\sin\frac{1}{2}(\phi-\phi_0)} +$$

$$+\ \mathrm{fin}\ J^{\frac{1}{2}}\mathbf{t}_3 \operatorname{cosec}\theta_0 \cdot \left\{\frac{1}{4\psi^3}\mathrm{grad}\ L^{\mathrm{i}} \wedge \mathbf{E}^{\mathrm{i}}_0 - \mathrm{i}\ \mathrm{curl}\left(\mathbf{F}_0 - \frac{\mathrm{i}\mathbf{E}^{\mathrm{i}}_0}{2\psi}\right)\right\} \ .$$

But,

$$L^{\mathrm{i}} = L^{\mathrm{i}}(0) + s\frac{\partial}{\partial s}L^{\mathrm{i}}(0) + \frac{1}{2}bs^2 + \ldots$$

where

$$b = \sum_{j=1}^{3} \sum_{k=1}^{3} \frac{\partial^2 L^{i}(0)}{\partial x_j \, \partial x_k} X_j X_k ,$$

$$X_1 = \sin\theta_0 \cos\phi, \quad X_2 = \sin\theta_0 \sin\phi, \quad X_3 = \cos\theta_0$$

and the x_j are Cartesian coordinates corresponding to the base vectors $\mathbf{t}_j$. Hence, for small s,

$$\psi = (2s)^{\frac{1}{2}} \sin\theta_0 \sin\tfrac{1}{2}(\phi-\phi_0)\left\{1 - \frac{bs}{8\sin^2\theta_0 \sin^2\frac{1}{2}(\phi-\phi_0)}\right\}.$$

Therefore

$$\text{fin}\frac{J^{\frac{1}{2}}f}{\psi^3} = \frac{\sin\theta_0}{\{2^{\frac{1}{2}}\sin\theta_0 \sin\frac{1}{2}(\phi-\phi_0)\}^3}\left[\left\{\frac{1}{2\rho_1^e} + \frac{3b}{8\sin^2\theta_0 \sin^2\frac{1}{2}(\phi-\phi_0)}\right\}f(0) + \right.$$

$$\left. + \frac{\partial}{\partial s}f(0)\right] . \tag{8.139}$$

Putting $\phi_0 = \pi$ and inserting (8.138) in the curl we obtain

$$\text{fin } J^{\frac{1}{2}}\left(\mathbf{F}_1 - \frac{i\mathbf{E}_1^i}{2\psi} + \frac{\mathbf{E}_0^i}{4\psi^3}\right) . \mathbf{i}_2 = -\left\{\frac{iE_{i\eta}^{i}(0) + [\text{curl } \mathbf{E}_0^i]_0 . \mathbf{t}_3 \text{ cosec } \theta_0}{2\cos\frac{1}{2}\phi}\right\} +$$

$$+ \frac{1}{4}\frac{\sin\theta_0}{\{2^{\frac{1}{2}}\sin\theta_0 \cos\frac{1}{2}\phi\}^3}\left[\left\{\frac{1}{2\rho_1^e} + \frac{3b}{8\sin^2\theta_0 \cos^2\frac{1}{2}\phi}\right\}E_{0\eta}^{i}(0) + \mathbf{X} \text{ . grad } E_{0\eta}^{i}(0)\right]$$

$$- \frac{\cos\theta_0}{2^{3/2}\sin^2\theta_0}\left(\frac{d\theta_0}{d\sigma_0}\cos\theta_0 + \sin\theta_0\frac{\partial}{\partial\sigma_0} + \kappa_0\cos\theta_0\cos^2\tfrac{1}{2}\phi\right)\frac{E_{0\eta}^{i}(0)}{\sin\theta_0 \cos\frac{1}{2}\phi} .$$

(8.140)

For the total field (see (8.131)) we must add on the same

quantity with ϕ replaced by $2\pi - \phi$. This removes the first and last terms on the right-hand side of (8.140). Also ρ_1^e is unaffected by the change, as can be seen from (8.115), so its contribution drops out. Moreover, because the incident rays are tangent to the screen, $\partial L^i/\partial x_2 \equiv 0$ and $\partial^2 L^i(0)/\partial x_j \partial x_2 \equiv 0$. Hence b is unaltered by the change from ϕ to $2\pi - \phi$. Accordingly, only the term involving $\mathbf{X}$ is left and

$$\mathbf{X}(\phi) \cdot \operatorname{grad} E^i_{0\eta} - \mathbf{X}(2\pi-\phi) \cdot \operatorname{grad} E^i_{0\eta} = 2 \sin\theta_0 \sin\phi \frac{\partial E^i_{0\eta}}{\partial x_2} .$$

Hence

$$E^e_{1\eta} = \left\{\frac{\rho_1^e}{(\rho_1^e+s)s}\right\}^{\frac{1}{2}} \frac{\exp(-ik_0 s)}{k_0} D'_H \frac{\partial}{\partial x_2} E^i_{0\eta}(0) \tag{8.141}$$

where

$$D'_H = -\frac{\exp(\frac{1}{4}\pi i)\sin\frac{1}{2}\phi}{2^{3/2}\pi^{\frac{1}{2}}k_0^{\frac{1}{2}} \sin^2\theta_0 \cos^2\frac{1}{2}\phi} . \tag{8.142}$$

Note that

$$D'_H = \frac{1}{i \sin\theta_0}\left[\frac{\partial D_H}{\partial \phi_0}\right]_{\phi_0=\pi} . \tag{8.143}$$

The occurrence of the derivative in (8.141) explains why it is difficult to find starting values for the higher order terms in the non-uniform expansion when the uniformly valid approximation is unknown.

In the case of grazing incidence from the screen $\phi_0 = 0$ and D_E vanishes. If the incident wave has only an E_ξ component an analysis similar to the above leads to

$$E^{\mathrm{e}}_{1\xi} = \left\{\frac{\rho^{\mathrm{e}}_1}{(\rho^{\mathrm{e}}_1+s)s}\right\}^{\frac{1}{2}} \frac{\exp(-ik_0 s)}{k_0} D'_E \frac{\partial}{\partial x_2} E^{\mathrm{i}}_{0\xi}(0) \tag{8.144}$$

where

$$D'_E = -\frac{\exp(\frac{1}{4}\pi i)\cos\frac{1}{2}\phi}{2^{3/2}\pi^{\frac{1}{2}}k_0^{\frac{1}{2}}\sin^2\theta_0\sin^2\frac{1}{2}\phi} = \frac{1}{\mathrm{i}\sin\theta_0}\left[\frac{\partial D_E}{\partial\phi_0}\right]_{\phi_0=0}. \tag{8.145}$$

It must be remembered in both (8.141) and (8.144) that ξ,η refer on the left-hand side to coordinates attached to the edge ray and on the right-hand side to those on the incident ray. Formulae with the same character as (8.141) and (8.144) will arise when the incidence is not grazing but $\mathbf{E}^{\mathrm{i}}_0$ vanishes at the edge.

8.18. Double edge diffraction

The scattering of an edge ray by an edge can be handled by the preceding theory provided that the second edge lies in a region where an expansion of the form (8.119) is valid. This will not be true if the second edge lies on the shadow or reflection boundary of the first edge. When this occurs another canonical problem must be solved.

To simplify the analysis only the case of two straight edges will be examined. This problem has been discussed rigorously (Jones 1973*b*) but here we shall be content with an approximate treatment (Boersma 1975) which is a natural extension of the method of the preceding section and whose results are in agreement with the rigorous theory.

It will be assumed that the waves are electrically polarized parallel to the edges so that only the electric vector need be considered. It will be denoted by the letter u generically. To set the scene a single half-plane will first be treated by a uniform approach. Let the incident wave be $u^{\mathrm{i}}(x,y)$, with the ray through the edge making an angle ϕ_0 with the positive x axis (Fig. 8.19). Then if $u(x,y)$ is the total field (which is zero on the conductor) we obtain from (8.131) and (8.120)

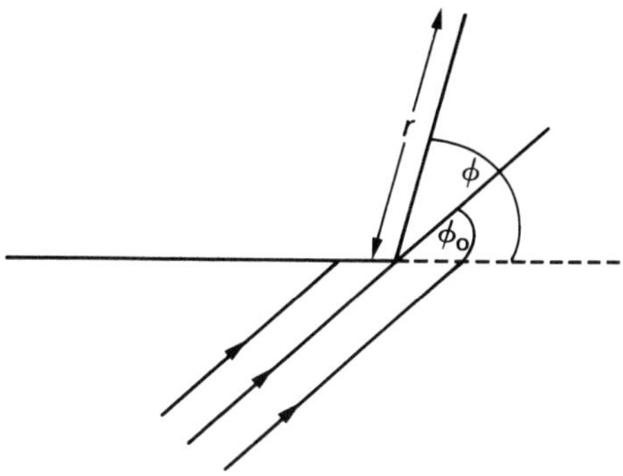

FIG. 8.19. Scattering of a wave by a perfectly conducting half-plane.

$$u(x,y) = U(r,\phi) - U(r,2\pi-\phi)$$

where

$$U(r,\phi) = \pi^{-\frac{1}{2}} \exp(-ik_0\psi^2+\tfrac{1}{4}\pi i)u^{\mathrm{i}}(x,y)F(k_0^{\frac{1}{2}}\psi) +$$

$$+ \pi^{-\frac{1}{2}} \exp\{-ik_0(L^{\mathrm{i}}+r)+\tfrac{1}{4}\pi i\}k_0^{-\frac{1}{2}} \sum_{m=0} \frac{F_m}{k_0^m} ,$$

L_0^{i} is the value of L^{i} at the edge, $L^{\mathrm{e}} = L_0^{\mathrm{i}} + r$, and $\psi^2 = L_0^{\mathrm{i}} + r - L^{\mathrm{i}}$. The edge conditions require F_m to be finite at the edge and so we deduce immediately from (8.128) (with r in place of s) that $F_m \to 0$ as $r \to 0$.

Now, as is easily verified,

$$\frac{1}{J^{\frac{1}{2}}}\frac{d}{d\sigma}\left(\frac{J^{\frac{1}{2}}E_m^{\mathrm{i}}}{\psi^p}\right) = \frac{1}{\psi^{p-1}}\{E_m^{\mathrm{i}}\nabla^2\psi+2\,\mathrm{grad}\,\psi\,.\,\mathrm{grad}\,E_m^{\mathrm{i}} + (1-p)E_m^{\mathrm{i}}\,\mathrm{grad}^2\,\psi\}-$$

$$- \frac{\mathrm{i}}{2\psi^p}\nabla^2E_{m-1}^{\mathrm{i}} . \tag{8.146}$$

Applying this result to the indefinite integral form of (8.128) we have

$$J^{\frac{1}{2}}\left\{F_m - \frac{\mathrm{i}E^{\mathrm{i}}_m}{2\psi}\right\} = \frac{1}{2}\int J^{\frac{1}{2}}\{-\mathrm{i}\nabla^2 F_{m-1} - \frac{1}{2\psi}\nabla^2 E^{\mathrm{i}}_{m-1}\}\mathrm{d}\sigma$$

$$= \frac{1}{2}\int J^{\frac{1}{2}}\left\{(-\mathrm{i})\nabla^2\left(F_{m-1} - \frac{\mathrm{i}E_{m-1}}{2\psi}\right) - \right.$$

$$\left. - \frac{(\psi\ \mathrm{grad}\ E^{\mathrm{i}}_{m-1}\cdot\mathrm{grad}\ \psi - E^{\mathrm{i}}_{m-1}\ \mathrm{grad}^2\ \psi + \frac{1}{2}\psi E^{\mathrm{i}}_{m-1}\nabla^2\psi)}{\psi^3}\right\}\mathrm{d}\sigma\ .$$

Apply (8.146) with $p = 3$ and

$$J^{\frac{1}{2}}\left\{F_m - \frac{\mathrm{i}E^{\mathrm{i}}_m}{2\psi} + \frac{E^{\mathrm{i}}_{m-1}}{4\psi^3}\right\} = \frac{1}{2}\int J^{\frac{1}{2}}\left\{(-\mathrm{i})\nabla^2\left(F_{m-1} - \frac{\mathrm{i}E^{\mathrm{i}}_{m-1}}{2\psi}\right) + \frac{\mathrm{i}}{4\psi^3}\nabla^2 E^{\mathrm{i}}_{m-2}\right\}\mathrm{d}\sigma\ .$$

Further reduction may be carried out in this way. Let

$$\hat{F}_m = F_m - \frac{1}{2}\sum_{n=0}^{m}\left(\frac{1}{2}\right)_n \frac{\mathrm{i}^{n+1}}{\psi^{2n+1}} E^{\mathrm{i}}_{m-n} \tag{8.147}$$

where

$$\left(\frac{1}{2}\right)_0 = 1,\quad \left(\frac{1}{2}\right)_n = \frac{1}{2}\left(\frac{1}{2}+1\right)\ldots\left(\frac{1}{2}+n-1\right)\quad (n = 1, 2, \ldots)\ .$$

Then, for the definite integral, we have

$$J^{\frac{1}{2}}\hat{F}_m = \mathrm{fin}\ J^{\frac{1}{2}}\hat{F}_m - \frac{1}{2}\mathrm{i}\int_0^r J^{\frac{1}{2}}\nabla^2\hat{F}_{m-1}\mathrm{d}\sigma\ ,$$

the analogue of (8.118). From (8.115), $1/\rho^{\mathrm{e}}_1 = 0$ and so $J = r$ in this case. Hence, finally

$$\hat{F}_m = r^{-\frac{1}{2}} \text{ fin } r^{\frac{1}{2}}\hat{F}_m - \tfrac{1}{2}ir^{-\frac{1}{2}} \text{ fin } \int_0^r \sigma^{\frac{1}{2}}\nabla^2\hat{F}_{m-1}d\sigma . \qquad (8.148)$$

Note that

$$\text{fin } r^{\frac{1}{2}}\hat{F}_0 = \text{fin}\left(- \frac{ir^{\frac{1}{2}}E_0^i}{2\psi}\right) = - \frac{i}{2^{3/2}} E_0^i(0,0) \text{ cosec } \tfrac{1}{2}(\phi-\phi_0). \qquad (8.149)$$

There are similar results for $\hat{F}_m$, involving a linear combination of the first m derivatives of the incident field at the edge. For example,

$$\text{fin } r^{\frac{1}{2}}\hat{F}_1 = - i2^{-3/2}E_1^i(0,0) \text{ cosec } \tfrac{1}{2}(\phi-\phi_0) +$$

$$+ \frac{3b}{32} 2^{-3/2}E_0^i(0,0) \text{ cosec}^5 \tfrac{1}{2}(\phi-\phi_0) +$$

$$+ 2^{-7/2} \frac{\partial E_0^i(0,0)}{\partial r} \text{ cosec}^3 \tfrac{1}{2}(\phi-\phi_0) \qquad (8.150)$$

where

$$b = \frac{\partial^2 L^i}{\partial x^2} \cos^2 \phi + 2 \frac{\partial^2 L^i}{\partial x \partial y} \cos \phi \sin \phi + \frac{\partial^2 L^i}{\partial y^2} \sin^2 \phi$$

evaluated at the edge. From (8.148) and (8.149)

$$\hat{F}_0 = -i2^{-3/2}r^{-\frac{1}{2}} E_0^i(0,0) \text{ cosec } \tfrac{1}{2}(\phi-\phi_0) \qquad (8.151)$$

and, since

$$\nabla^2\hat{F}_0 = -i2^{-5/2}r^{-5/2}E_0^i(0,0) \text{ cosec}^3 \tfrac{1}{2}(\phi-\phi_0) ,$$

(8.148) and (8.150) give

$$\hat{F}_1 = 2^{-3/2}r^{-\frac{1}{2}} \operatorname{cosec} \tfrac{1}{2}(\phi-\phi_0)\{- E_1^{\mathrm{i}}(0,0) + \tfrac{3b}{32}E_0^{\mathrm{i}}(0,0) \operatorname{cosec}^4 \tfrac{1}{2}(\phi-\phi_0) + \\ + \tfrac{1}{4}\frac{\partial E_0^{\mathrm{i}}(0,0)}{\partial r} \operatorname{cosec}^2 \tfrac{1}{2}(\phi-\phi_0)\} + \\ + 2^{-7/2}r^{-3/2}E_0^{\mathrm{i}}(0,0) \operatorname{cosec}^3 \tfrac{1}{2}(\phi-\phi_0). \tag{8.152}$$

In terms of $\hat{F}_m$

$$U(r,\phi) = \pi^{-\frac{1}{2}} \exp(-\mathrm{i}k_0\psi^2+\tfrac{1}{4}\pi i)u^{\mathrm{i}}(x,y)\Big\{F(k_0^{\frac{1}{2}}\psi) + \\ + \tfrac{1}{2}\mathrm{i}k_0^{-\frac{1}{2}} \sum_{n=0}^{\infty} \left(\tfrac{1}{2}\right)_n \left(\frac{\mathrm{i}}{k_0}\right)^n \psi^{-2n-1}\Big\} + \\ + \pi^{-\frac{1}{2}}\exp\{-\mathrm{i}k_0(L_0^{\mathrm{i}}+r)+\tfrac{1}{4}\pi i\} \sum_{m=0}^{\infty} \frac{\hat{F}_m}{k_0^m}. \tag{8.153}$$

This form is convenient for asymptotics because

$$F(w) \sim \pi^{\frac{1}{2}} \exp(\mathrm{i}w^2-\tfrac{1}{4}\pi i)H(-w) - \tfrac{1}{2}\mathrm{i}w^{-1} \sum_{m=0}^{\infty} \left(\tfrac{1}{2}\right)_m \left(\frac{\mathrm{i}}{w^2}\right)^m, \tag{8.154}$$

H being the Heaviside unit function, as $w \to \pm\infty$. Thus when F can be replaced by its asymptotic expansion the series in the first brace of (8.153) cancels that in (8.154).

For $0 \leqslant \phi \leqslant \pi$, this property may be used for $U(r,2\pi-\phi)$ since then the argument of F never vanishes (it is assumed that $0 < \phi_0 < \pi$). Hence, if r is not small,

$$u(x,y) \sim \pi^{-\frac{1}{2}} \exp(-ik_0\psi^2+\tfrac{1}{4}\pi i)u^{\mathrm{i}}(x,y)\left\{F(k_0^{\frac{1}{2}}\psi) + \right.$$

$$\left. + \tfrac{1}{2}ik_0^{-\frac{1}{2}}\psi^{-1} \sum_{n=0}^{\infty} \left(\tfrac{1}{2}\right)_n \left(\frac{\mathrm{i}}{k_0\psi^2}\right)^n\right\} +$$

$$+ \pi^{-\frac{1}{2}} \exp\{-ik_0(L_0^{\mathrm{i}}+r)+\tfrac{1}{4}\pi i\}k_0^{-\frac{1}{2}} \sum_{m=0}^{\infty} \frac{\{\hat{F}_m(r,\phi)-\hat{F}_m(r,2\pi-\phi)\}}{k_0^m} . \qquad (8.155)$$

The leading terms are, from (8.148) and (8.149),

$$u(x,y) \sim \pi^{-\frac{1}{2}} \exp(-ik_0\psi^2+\tfrac{1}{4}\pi i)u^{\mathrm{i}}(x,y)\{F(k_0^{\frac{1}{2}}\psi) + \tfrac{1}{2}ik_0^{-\frac{1}{2}}\psi^{-1}\} -$$

$$- \frac{\exp(-ik_0r-\frac{1}{4}\pi i)}{2(2\pi k_0 r)^{\frac{1}{2}}} u^{\mathrm{i}}(0,0)g(\phi_0,\phi) \qquad (8.156)$$

where

$$g(\phi_1,\phi_2) = \operatorname{cosec} \tfrac{1}{2}(\phi_1-\phi_2) + \operatorname{cosec} \tfrac{1}{2}(\phi_1+\phi_2) . \qquad (8.157)$$

If the incident wave is the plane wave

$$u_0 = \exp\{-ik_0(x\cos\phi_0 + y\sin\phi_0)\}$$

$\psi = (2r)^{\frac{1}{2}} \sin \frac{1}{2}(\phi-\phi_0)$ and the total field given by (8.156) is

$$u_1(x,y) \sim \pi^{-\frac{1}{2}} \exp(-ik_0r+\tfrac{1}{4}\pi i)F\{(2k_0r)^{\frac{1}{2}} \sin \tfrac{1}{2}(\phi-\phi_0)\}-$$

$$- \frac{\exp(-ik_0r-\frac{1}{4}\pi i)}{2(2\pi k_0 r)^{\frac{1}{2}}} \operatorname{cosec} \tfrac{1}{2}(\phi+\phi_0) \qquad (8.158)$$

consistent with earlier results (§8.15).

The result analogous to (8.155) when $0 \geqslant \phi \geqslant -\pi$ is

$$u(x,y) \sim u^{\mathrm{i}}(x,y) - \pi^{-\frac{1}{2}} \exp(-\mathrm{i}k_0\overline{\psi}^2+\tfrac{1}{4}\pi\mathrm{i})u^{\mathrm{i}}(x,-y)\Big\{F(k_0^{\frac{1}{2}}\overline{\psi}) + $$

$$+ \tfrac{1}{2}\mathrm{i}k_0^{-\frac{1}{2}}\overline{\psi}^{-1} \sum_{n=0}^{\infty} \left(\tfrac{1}{2}\right)_n \left(\frac{\mathrm{i}}{k_0\overline{\psi}^2}\right)^n\Big\} +$$

$$+ \pi^{-\frac{1}{2}} \exp\{-\mathrm{i}k_0(L_0^{\mathrm{i}}+r)+\tfrac{1}{4}\pi\mathrm{i}\}k_0^{-\frac{1}{2}} \sum_{m=0}^{\infty} \frac{\{\hat{F}_m(r,\phi)-\hat{F}_m(r,2\pi-\phi)\}}{k_0^m} \qquad (8.159)$$

where $\overline{\psi}$ is ψ with $2\pi - \phi$ for ϕ. Since $\overline{\psi} = \psi(2\pi-\phi) = -\psi(-\phi)$ because of the periodic property, the scattered-wave portion of (8.159) (i.e. with the incident wave deleted) is the same as would be obtained by changing ϕ to $-\phi$ in the scattered-wave part of (8.155). The leading terms of (8.159) are

$$u(x,y) \sim u^{\mathrm{i}}(x,y) - \pi^{-\frac{1}{2}} \exp(-\mathrm{i}k_0\overline{\psi}^2+\tfrac{1}{4}\pi\mathrm{i})u^{\mathrm{i}}(x,-y)\{F(k_0^{\frac{1}{2}}\overline{\psi}) + \tfrac{1}{2}\mathrm{i}k_0^{-\frac{1}{2}}\overline{\psi}^{-1}\} -$$

$$- \frac{\exp(-\mathrm{i}k_0r-\frac{1}{4}\pi\mathrm{i})}{2(2\pi k_0r)^{\frac{1}{2}}} u^{\mathrm{i}}(0,0)g(\phi_0,\phi) \qquad (8.160)$$

which reduces to

$$u_1(x,y) \sim u_0(x,y) - \pi^{-\frac{1}{2}} \exp(-\mathrm{i}k_0r+\tfrac{1}{4}\pi\mathrm{i})F\{(2k_0r)^{\frac{1}{2}} \sin \tfrac{1}{2}(\phi+\phi_0)\} -$$

$$- \frac{\exp(-\mathrm{i}k_0r-\frac{1}{4}\pi\mathrm{i})}{2(2\pi k_0r)^{\frac{1}{2}}} \operatorname{cosec} \tfrac{1}{2}(\phi_0-\phi) \qquad (8.161)$$

for an incident plane wave.

Suppose now that the plane wave falls on two half-planes as shown in Fig. 8.20. The edge of the lower half-plane is chosen as origin and the edge of the upper half-plane is at $x = h$ $y = d$. Polar coordinates (r_1,ϕ_1) and (r_2,ϕ_2) with respect to the two edges are used, the upper edge being at $(r^{(1)},\phi^{(1)})$ with respect to the lower edge. It is assumed that ϕ_0 and $\phi^{(1)}$ are not both near zero nor both near π so that grazing incidence and the second half-plane almost absent or almost the whole plane are

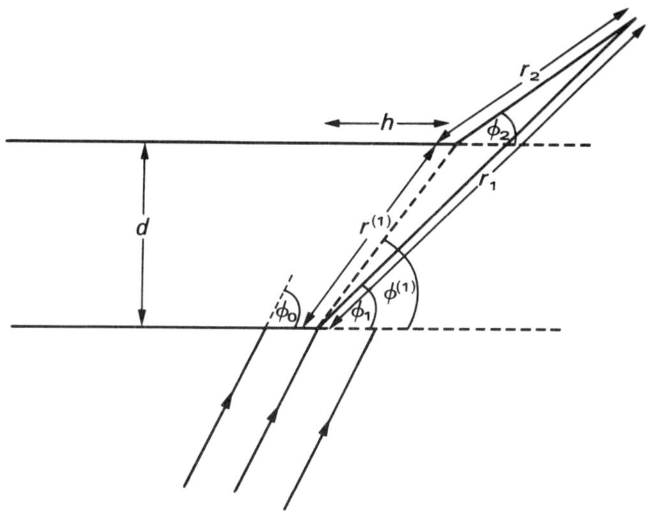

FIG. 8.20. Configuration of two half-planes.

excluded cases.

The total field produced by the lower half-plane in the plane wave is u_1 as set out in (8.158) with r_1 and ϕ_1 in place of r and ϕ respectively. Write $u_1 = u_0 + v_1$ and let the total field of the upper half-plane when v_1 is incident be $v_1 + v_2$, i.e. v_2 is the field scattered by the upper half-plane under the illumination v_1.

Suppose, firstly, that $\phi_0 < \phi^{(1)} - \delta$, $\delta > 0$ so that the upper diffracting edge lies in the shadow region $\phi_1 > \phi_0$. According to (8.158) the incident field $v_1 = u_1 - u_0$ can be approximated by substituting

$$u_1(x,y) \sim -\frac{\exp(-ik_0r_1-\frac{1}{4}\pi i)}{2(2\pi k_0r_1)^{\frac{1}{2}}}\, g(\phi_0,\phi_1) \tag{8.162}$$

when $\phi_1 > \phi_0$. Diffraction of $-u_0$ produces the total field

$$-u_0(x,y) - \exp\{-ik_0r^{(1)}\cos(\phi^{(1)}-\phi_0)\}v_1(x-h,y-d) \; .$$

The other term of v_1, which appears as a cylindrical wave issuing from the lower edge, supplies in $\phi_2 \geqslant 0$ a total field from (8.156) of

$$\pi^{-\frac{1}{2}} \exp(-ik_0\psi^2+\tfrac{1}{4}\pi i)u_1(x,y)\{F(k_0^{\frac{1}{2}}\psi) + \tfrac{1}{2}ik^{-\frac{1}{2}}\psi^{-1}\} -$$

$$- \frac{\exp(-ik_0r_2-\frac{1}{4}\pi i)}{2(2\pi k_0 r_2)^{\frac{1}{2}}} u_1(h,d)g(\phi^{(1)},\phi_2)$$

where now $\psi = (r^{(1)}+r_2-r_1)^{\frac{1}{2}}$ with ψ positive (negative) for $\phi_2 > (<) \phi^{(1)}$. We cannot replace u_1 immediately by (8.162) because its coefficient is non-zero where (8.162) is not valid. However, this can be overcome by employing (8.96) to provide a coefficient which is non-zero only where (8.162) holds. Hence, for $\phi_0 < \phi^{(1)} - \delta$,

$$v_1(x,y) + v_2(x,y) \sim \frac{\exp\{-ik_0(r^{(1)}+r_2)\}}{2\pi(2k_0r_1)^{\frac{1}{2}}} g(\phi_0,\phi_1)\Big[F\{-k_0^{\frac{1}{2}}(r^{(1)}+r_2-r_1)^{\frac{1}{2}}\}$$

$$- \frac{1}{2}ik_0^{-\frac{1}{2}}(r^{(1)}+r_2-r_1)^{-\frac{1}{2}}\Big] - \frac{i\exp\{-ik_0(r^{(1)}+r_2)\}}{8\pi k_0(r^{(1)}r_2)^{\frac{1}{2}}} g(\phi_0,\phi^{(1)})g(\phi^{(1)},\phi_2)$$

$$+ v_1(x,y) - \exp\{-ik_0r^{(1)}\cos(\phi^{(1)}-\phi_0)\}v_1(x-h,y-d) \ . \quad (8.163)$$

The formula (8.163) is valid uniformly for $0 \leqslant \phi_2 \leqslant \pi$ so long as r_2 is not small. If r_2 is small, (8.156) fails because terms in $U(r,2\pi-\phi)$ which have been dropped must be retained. The effect of the retention of these terms is to add some extra terms to (8.163).

For $\phi_0 > \phi^{(1)} + \delta$, the upper edge is illuminated by the plane wave u_0 and

$$u_1(x,y) \sim u_0(x,y) - \frac{\exp(-ik_0r_1-\frac{1}{4}\pi i)}{2(2\pi k_0r_1)^{\frac{1}{2}}} g(\phi_0,\phi_1) \ .$$

By the same process as has just been adopted

$$v_1(x,y) + v_2(x,y) \sim -\frac{\exp\{-ik_0(r^{(1)}+r_2)\}}{2\pi(2k_0r_1)^{\frac{1}{2}}}g(\phi_0,\phi_1)\Big[F\{k_0^{\frac{1}{2}}(r^{(1)}+r_2-r_1)^{\frac{1}{2}}\}$$

$$+\tfrac{1}{2}ik_0^{-\frac{1}{2}}(r^{(1)}+r_2-r_1)^{-\frac{1}{2}}\Big] - \frac{i\exp\{-ik_0(r^{(1)}+r_2)\}}{8\pi k_0(r^{(1)}r_2)^{\frac{1}{2}}}g(\phi_0,\phi^{(1)})g(\phi^{(1)},\phi_2)$$

(8.164)

Again this formula is uniform for $0 \leqslant \phi_2 \leqslant \pi$ so long as r_2 is not small.

Exercise

27. Show that, when r_2 is small, the term

$$\frac{\exp\{-ik_0(r^{(1)}+r_2)\}}{2\pi(2k_0r_+)^{\frac{1}{2}}}\, g(\phi_0,\phi^{(1)}-\gamma_+)[F\{-k_0^{\frac{1}{2}}(r^{(1)}+r_2-r_+)^{\frac{1}{2}}\}-\tfrac{1}{2}ik_0^{-\frac{1}{2}}(r^{(1)}+r_2-r_+)^{-\frac{1}{2}}]$$

must be added to (8.163) and (8.164). Here r_+,γ_+ are polar coordinates with $(0,2d)$ as centre and the initial line is the line joining $(0,2d)$ to (h,d).

The above demonstrates that, when $|\phi_0-\phi^{(1)}| > \delta$, the uniform theory may be applied in a standard fashion. When $\phi_0 = \phi^{(1)}$ this is no longer true because the upper edge lies on the shadow boundary of the lower edge. The term F in (8.158) cannot be approximated near the upper diffracting edge and so the incident field does not have the structure demanded by the standard uniform theory. To circumvent this difficulty a series expansion is inserted for F.

The incident field is, from (8.158),

$$v_1(x,y) \sim -\pi^{-\frac{1}{2}}\exp(-ik_0r_1+\tfrac{1}{4}\pi i)F(-k_0^{\frac{1}{2}}\psi_1) -$$

$$-\frac{\exp(-ik_0r_1-\frac{1}{4}\pi i)}{2(2\pi k_0r_1)^{\frac{1}{2}}}\operatorname{cosec}\tfrac{1}{2}(\phi_1+\phi_0)$$

where $\psi_1 = (2r_1)^{\frac{1}{2}}\sin\frac{1}{2}(\phi_1-\phi_0)$. Replace F by

$$F(-k_0^{\frac{1}{2}}\psi_1) = \exp(ik_0\psi_1^2)\left[\tfrac{1}{2}\pi^{\frac{1}{2}} \exp(-\tfrac{1}{4}\pi i) + \right.$$

$$\left. + \sum_{q=0}^{\infty} \frac{\exp(-\frac{1}{2}q\pi i)}{q!\,(2q+1)}\, k_0^{q+\frac{1}{2}}\psi_1^{2q+1}\right].$$

This yields

$$v_1(x,y) \sim -\tfrac{1}{2}u_0(x,y) - \frac{\exp(-ik_0r_1-\frac{1}{4}\pi i)}{2(2\pi k_0r_1)^{\frac{1}{2}}} \operatorname{cosec} \tfrac{1}{2}(\phi_1+\phi_0) -$$

$$- \pi^{-\frac{1}{2}}u_0(x,y) \exp(\tfrac{1}{4}\pi i) \sum_{q=0}^{\infty} \frac{\exp(-\frac{1}{2}q\pi i)}{q!\,(2q+1)}\, k_0^{q+\frac{1}{2}}\psi_1^{2q+1}.$$

Diffraction of the plane wave $-\frac{1}{2}u_0$ furnishes the total field

$$-\tfrac{1}{2}u_0(x,y) - \tfrac{1}{2}\exp(-ik_0r^{(1)})v_1(x-h,y-d) .$$

The contribution of the cylindrical wave is

$$-\frac{\exp\{-ik_0(r^{(1)}+r_2)\}}{2\pi(2k_0r_1)^{\frac{1}{2}}} \operatorname{cosec} \tfrac{1}{2}(\phi_1+\phi_0)\{F(k_0^{\frac{1}{2}}\psi)+\tfrac{1}{2}ik_0^{-\frac{1}{2}}\psi^{-1}\}$$

$$-\frac{i\exp\{-ik_0(r^{(1)}+r_2)\}g(\phi_0,\phi_2)}{8\pi k_0(r^{(1)}r_2)^{\frac{1}{2}}\sin\phi_0} .$$

Each term in the infinite series in v_1 is now treated separately as an incident field. Let

$$V_q = \exp\{-ik_0r_1\cos(\phi_1-\phi_0)\}\psi_1^{2q+1}$$

be a typical incident field with $L^i = r_1\cos(\phi_1-\phi_0)$. In view of the factor $k_0^{q+\frac{1}{2}}$ in v_1 the full expansion (8.155) must be used rather than the approximation (8.156). However, the final field will be required only to $O(k_0^{-1})$ and so it will be sufficient to stop at $\hat{F}_{q+1}$. Now since ψ_1^{2q+1} and its first $2q$ derivatives vanish at the diffracting edge it follows from (8.148) that $\hat{F}_m$ is zero for $m = 0, \ldots, q+1$ when $q \geqslant 1$. Hence only for $q = 0$

can $\hat{F}_m$ play a role. But, when $q = 0$, (8.149) makes evident that $\hat{F}_0 = 0$, and then since $\partial\psi_1/\partial r_2 = (2r^{(1)})^{-\frac{1}{2}} \sin(\phi_2-\phi_0)$ at the edge we deduce from (8.152) that

$$\hat{F}_1 = \frac{1}{8(r^{(1)}r_2)^{\frac{1}{2}}} \frac{\cos\frac{1}{2}(\phi_2-\phi_0)}{\sin^2\frac{1}{2}(\phi_2-\phi_0)} .$$

Thus the total field U_q corresponding to V_q is

$$U_q \sim \pi^{-\frac{1}{2}} \exp\{-ik_0(r^{(1)}+r_2)+\tfrac{1}{4}\pi i\}\psi_1^{2q+1}\Big\{F(k_0^{\frac{1}{2}}\psi_2) + $$

$$+ \tfrac{1}{2}ik_0^{-\frac{1}{2}}\psi_2^{-1} \sum_{n=0}^{q+1} \left(\tfrac{1}{2}\right)_n \left(\frac{i}{k_0\psi_2^2}\right)^n\Big\} +$$

$$+ \frac{\exp\{-ik_0(r^{(1)}+r_2)+\frac{1}{4}\pi i\}}{8(\pi k_0^3 r^{(1)} r_2)^{\frac{1}{2}}} \delta_{0q} h(\phi_2,\phi_0)$$

where δ_{0q} is the usual Kronecker symbol and

$$h(\phi_2,\phi_0) = \frac{\cos\frac{1}{2}(\phi_2-\phi_0)}{\sin^2\frac{1}{2}(\phi_2-\phi_0)} + \frac{\cos\frac{1}{2}(\phi_2+\phi_0)}{\sin^2\frac{1}{2}(\phi_2+\phi_0)} . \tag{8.165}$$

The substitution

$$F(w) = \tfrac{1}{2}\pi^{\frac{1}{2}} \exp(-\tfrac{1}{4}\pi i) \sum_{m=0}^{\infty} \frac{(-)^m \exp(\frac{1}{4}m\pi i)w^m}{(\frac{1}{2}m)!} \tag{8.166}$$

leads to

$$U_q \sim \tfrac{1}{2}\mathrm{i}\ \exp\{-\mathrm{i}k_0(r^{(1)}+r_2)\ +$$

$$+\ (\tfrac{1}{4}-\tfrac{1}{2}q)\pi\mathrm{i}\}k_0^{-q-\frac{1}{2}}\eta^{2q+1}\sum_{m=0}^{\infty}\frac{(-)^m\ \exp(\frac{1}{4}m\pi\mathrm{i})}{(\frac{1}{2}m-q-\frac{1}{2})!}\ k_0^{\frac{1}{2}m}\psi_2^m\ -$$

$$-\ \frac{\exp\{-\mathrm{i}k_0(r^{(1)}+r_2)+\frac{1}{4}\pi\mathrm{i}\}}{2\pi}\left\{\frac{(q+\frac{1}{2})!\exp(\frac{1}{2}q\pi\mathrm{i})\eta^{2q+1}}{k_0^{q+3/2}\psi_2^2}\ -\right.$$

$$\left.-\ \frac{\delta_{0q}\pi^{\frac{1}{2}}}{4(k_0^3r^{(1)}r_2)^{\frac{1}{2}}}\ h(\phi_2,\phi_0)\right\}\ ,$$

where $\eta = \psi_1/\psi_2$. Thus the total field caused by the infinite series in v_1 is

$$-\ \pi^{-\frac{1}{2}}\ \exp(\tfrac{1}{4}\pi\mathrm{i})\ \sum_{q=0}^{\infty}\ \frac{\exp(-\frac{1}{2}q\pi\mathrm{i})k_0^{q+\frac{1}{2}}}{q!(2q+1)}\ U_q$$

$$=\ \tfrac{1}{2}\pi^{-\frac{1}{2}}\ \exp\{-\mathrm{i}k_0(r^{(1)}+r_2)\}\sum_{q=0}^{\infty}\frac{(-)^q\eta^{2q+1}}{q!(2q+1)}\sum_{m=0}^{\infty}\frac{(-)^m\exp(\frac{1}{4}m\pi\mathrm{i})}{(\frac{1}{2}m-q-\frac{1}{2})!}k_0^{\frac{1}{2}m}\psi_2^m$$

$$+\ \frac{\mathrm{i}\ \exp\{-\mathrm{i}k_0(r^{(1)}+r_2)\}}{4\pi k_0}\left\{\frac{\pi^{-\frac{1}{2}}}{\psi_2^2}\sum_{q=0}^{\infty}\frac{(q-\frac{1}{2})!}{q!}\eta^{2q+1}\ -\ \frac{1}{2}\,\frac{h(\phi_2,\phi_0)}{(r^{(1)}r_2)^{\frac{1}{2}}}\right\}$$

in $0 \leqslant \phi_2 \leqslant \pi$ provided that r_2 is not small. To cope with the series we note that $\psi_2^2 - \psi_1^2 = r^{(1)} + r_2 - r_1 > 0$ which implies $0 < \eta < 1$. Consequently

$$\sum_{q=0}^{\infty}\ \frac{(q-\frac{1}{2})!}{q!}\eta^{2q+1}\ =\ \pi^{\frac{1}{2}}\eta(1-\eta^2)^{-\frac{1}{2}}\ .$$

Writing

$$F(x,y) = \sum_{q=0}^{\infty} \frac{(-)^q x^{2q+1}}{q!(2q+1)} \sum_{m=0}^{\infty} \frac{(-)^m \exp(\frac{1}{4}m\pi\mathrm{i})y^m}{(\frac{1}{2}m-q-\frac{1}{2})!} \tag{8.167}$$

it may be shown (see Exercise 28) that, for real x and y with $|x| < 1$,

$$F(x,y) = \pi^{-\frac{1}{2}}G\{(1-x^2)^{\frac{1}{2}}y, xy\} + \pi^{\frac{1}{2}} \exp(\mathrm{i}y^2)H(-y) -$$

$$- 2\exp\{\mathrm{i}y^2(1-x^2)+\tfrac{1}{4}\pi\mathrm{i}\}F(-xy)H(-y) \tag{8.168}$$

where

$$\left.\begin{aligned} G(x,y) &= y\exp(\mathrm{i}x^2)\int_x^{\infty} \frac{\exp(-\mathrm{i}t^2)}{t^2+y^2}\,\mathrm{d}t \qquad (x > 0),\\ G(x,y) &= -G(-x,y) \qquad (x < 0). \end{aligned}\right\} \tag{8.169}$$

Therefore the total field due to the infinite series is

$$\frac{1}{2\pi}\exp\{-\mathrm{i}k_0(r^{(2)}+r_2)\}\Big[G\{(1-\eta^2)^{\frac{1}{2}}k_0^{\frac{1}{2}}\psi_2, k_0^{\frac{1}{2}}\eta\psi_2\} + \pi\exp(\mathrm{i}k_0\psi_2^2)H(-\psi_2) - 2\pi^{\frac{1}{2}}\mathrm{e}^{\mathrm{i}k_0\psi_2^2(1-\eta^2)+\frac{1}{4}\pi\mathrm{i}}F(-k_0^{\frac{1}{2}}\eta\psi_2)H(-\psi_2)\Big] +$$

$$+ \frac{\mathrm{i}\exp\{-\mathrm{i}k_0(r^{(1)}+r_2)\}}{4\pi k_0}\left\{\psi_2^{-2}\eta(1-\eta^2)^{-\frac{1}{2}} - \frac{1}{2}\frac{h(\phi_2,\phi_0)}{(r^{(1)}r_2)^{\frac{1}{2}}}\right\}.$$

Collecting together all the terms we find that

$$v_1(x,y)+v_2(x,y) \sim \frac{1}{2\pi}\exp\{-ik_0(r^{(1)}+r_2)\}\Big[G\{k_0^{\frac12}(r^{(1)}+r_2-r_1)^{\frac12},(2k_0r_1)^{\frac12}\sin\tfrac12(\phi_1-\phi_0)\} -$$

$$+\frac{i(2r_1)^{\frac12}}{4k_0r_2(r^{(1)}+r_2-r_1)^{\frac12}}\,\frac{\sin\frac12(\phi_1-\phi_0)}{\sin^2\frac12(\phi_2-\phi_0)}\Big] -$$

$$-\frac{\exp\{-ik_0(r^{(1)}+r_2)\}}{2\pi(2k_0r_1)^{\frac12}}\operatorname{cosec}\tfrac12(\phi_1+\phi_0)\,[F\{k_0^{\frac12}(r^{(1)}+r_2-r_1)^{\frac12}\}+\tfrac12 ik_0^{-\frac12}(r^{(1)}+r_2-r_1)^{-\frac12}] -$$

$$-\frac{i\exp\{-ik_0(r^{(1)}+r_2)\}}{8\pi k_0(r^{(1)}r_2)^{\frac12}}\left\{\frac{g(\phi_0,\phi_2)}{\sin\phi_0}+h(\phi_2,\phi_0)\right\} -$$

$$-\frac12 u_0(x,y)H(\phi_2-\phi_0) - \frac12\exp(-ik_0r^{(1)})v_1(x-h,y-d) -$$

$$-\pi^{-\frac12}\exp\{-ik_0r_1+\tfrac14\pi i\}F\{(2k_0r_1)^{\frac12}\sin\tfrac12(\phi_0-\phi_1)\}H(\phi_0-\phi_2)\,. \qquad (8.170)$$

The formula (8.170) is uniformly valid in $0 \leqslant \phi_2 \leqslant \pi$ when $\phi_0 = \phi^{(1)}$ so long as r_2 is not small. If r_2 is small extra terms must be added as in previous formulae.

On the shadow boundary $\phi_2 = \phi_0 = \phi_1$ and the dominant behaviour comes from G and the terms which have no k_0 in the denominator. Now, if both x and y are small and positive in $G(x,y)$, (8.169) supplies

$$G(x,y) = \exp(ix^2)\int_{x/y}^{\infty}\frac{\exp(-iy^2t^2)}{t^2+1}\,dt \to \tan^{-1}\frac{y}{x}\,.$$

Hence, on account of (8.98), the field on the shadow boundary $\phi_2 = \phi_0$ is

$$v_1 + v_2 \sim \exp\{-ik_0(r^{(1)}+r_2)\}\left\{\frac{1}{2\pi}\tan^{-1}\left(\frac{r_2}{r^{(1)}}\right)^{\frac12} - \frac14\right\}\,. \qquad (8.171)$$

Actually, $v_1 + v_2$ is the total field due to the diffraction by the upper half-plane of the incident field v_1. Furthermore,

v_1 is the scattered field emanating from the lower half-plane under the excitation of u_0. The diffraction of u_0 by the upper half-plane provides a total field of

$$u_0(x,y) + \exp(-ik_0r^{(1)})v_1(x-h,y-d)$$

of which the leading term on the shadow boundary is $\frac{1}{2}u_0$. We infer that the total field on the shadow boundary due to the diffraction of u_0 by the two half-planes when their interaction is ignored is

$$\left\{\frac{1}{4} + \frac{1}{2\pi}\tan^{-1}\left(\frac{r_2}{r^{(1)}}\right)^{\frac{1}{2}}\right\}u_0$$

to a first approximation, provided that r_2 is not small. It will be observed that this approaches the value of $\frac{1}{2}u_0$ for a single half-plane as $r_2 \to \infty$. In other words, when $\phi_0 = \phi^{(1)}$, the field on the shadow boundary a large distance from the two half-planes is the same as that for one half-plane. It must, however, not be forgotten that this result is based on the neglect of the interaction between the half-planes.

Exercises

28. By taking a partial derivative of (8.167) with respect to x prove that

$$\pi^{\frac{1}{2}}F(x,y) = \sin^{-1} x - 2iy\int_0^x F\{y(1-t^2)^{\frac{1}{2}}\}dt$$

for $|x| < 1$. Substitute from (8.95) in the integrand and interchange the order of integration if $y > 0$. If $y < 0$ use (8.96) first. Hence show the legitimacy of (8.168).

29. Repeat the analysis of this section for a magnetically polarized plane wave.

30. Examine the possibility of finding the diffraction by three parallel half-planes.

8.19. Emission from a waveguide

A remark at the end of the preceding section pointed out that the interaction between the two half-planes was neglected. In this section a problem is considered in which the interaction must be taken into account if accuracy is to be achieved.

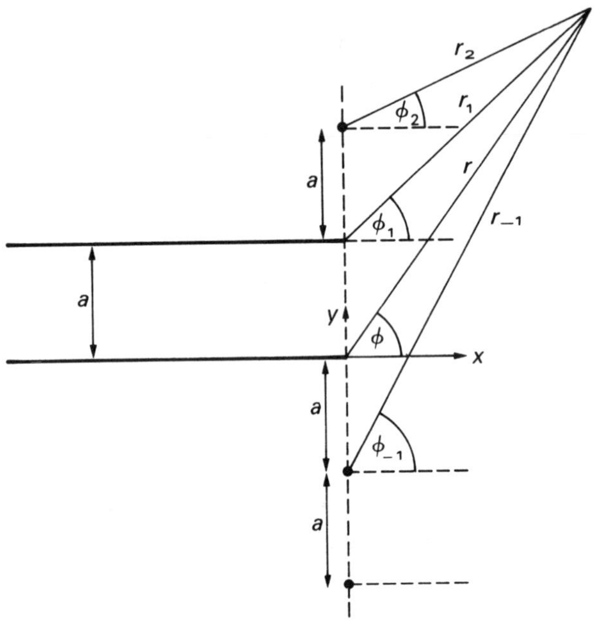

FIG. 8.21. Radiation from a parallel-plate waveguide.

Consider a parallel-plate waveguide (Fig. 8.21) which is carrying a TE mode; it is desired to determine the radiation from the open end and the reflection coefficients within the guide. The exciting mode is taken as

$$u_0(x,y) = \sin\left(\frac{N\pi y}{a}\right) \exp(-i\kappa_N x)$$

where N is a positive integer and the propagation constant κ_n is defined by

$$\kappa_n = \left(k_0^2 - \frac{n^2\pi^2}{a^2}\right)^{\frac{1}{2}}$$

being either positive real or negative imaginary. Put

$$\sin\theta_n = \frac{n\pi}{k_0 a}, \qquad \cos\theta_n = \frac{\kappa_n}{k_0}.$$

The incident mode may be decomposed as

$$u_0(x,y) = \frac{\exp\{-ik_0(x\cos\theta_N - y\sin\theta_N)\} - \exp\{-ik_0(x\cos\theta_N + y\sin\theta_N)\}}{2i}$$

which expresses it as a linear combination of two plane waves, one travelling in the direction of θ_N and the other along $-\theta_N$. The plane wave in the direction of θ_N has amplitude $-1/2i$ and, when it strikes the upper edge, generates a diffracted wave $\bar{v}_1$ according to (8.161) given by

$$\bar{v}_1(x,y) = (-)^N \frac{\exp(-ik_0r_1 - \frac{1}{4}\pi i)}{4i(2\pi k_0 r_1)^{\frac{1}{2}}}\, g(\theta_N, \phi_1) \qquad (8.172)$$

in $0 \leqslant \phi_1 \leqslant \pi$. The field $\bar{v}_1$ hits the lower half-plane and causes a scattered field v_2, which then scatters $\bar{v}_3$ from the upper half-plane. The process continues and there is a similar one due to the other plane wave in u_0. This iterative procedure was first introduced by Chester (1950, 1950*a*) in the time domain and subsequently discussed in the frequency domain in the context of rays by a number of authors (Boersma 1974, 1975*a*, Lee and Boersma 1975, Batorsky and Felsen 1973, Lee 1972, Bowman 1970, Yee and Felsen 1969, Yee, Felsen, and Keller 1969, Felsen and Yee 1968).

The interaction field v_2 can be evaluated by noting that it is due to a source at $(0,a)$ and so, by symmetry, the scattered field is the same as that at the image point when the excitation is due to the image source. Therefore, take the incident wave as

$$(-)^N \frac{\exp(-ik_0 r_{-1} - \frac{1}{4}\pi i)}{4i(2\pi k_0 r_{-1})^{\frac{1}{2}}} g(\theta_N, \phi_{-1}) \ .$$

Then, from (8.160), we deduce that for $0 \leqslant \phi \leqslant \pi$,

$$v_2(x,y) = \frac{(-)^{N+1} \exp\{-ik_0(r+a)\}}{4\pi i \ (2k_0)^{\frac{1}{2}}} \left[\frac{g(\theta_N, \phi_{-1})}{r_{-1}^{\frac{1}{2}}} \{F(-k_0^{\frac{1}{2}}\psi) - \tfrac{1}{2}ik_0^{-\frac{1}{2}}\psi^{-1}\} - \frac{i}{4(k_0 ar)^{\frac{1}{2}}} g(\theta_N, \tfrac{1}{2}\pi) g(\tfrac{1}{2}\pi, \phi) \right]$$

where $\psi = (r+a-r_{-1})^{\frac{1}{2}}$ with the same convention on sign as has been employed hitherto.

The wave v_2 strikes the upper half-plane and initiates $\bar{v}_3$. On account of the rapid variation of F near to $\phi = \frac{1}{2}\pi$ a procedure similar to that of the preceding section must be followed. Since this has to be done several times it pays to adopt a general approach.

Indeed, it is convenient to start from the representation

$$\bar{v}_1(x,y) = \frac{\exp(-ik_0 r_1)}{k_0^{\frac{1}{2}}} \sum_{q=0}^{\infty} \{\bar{a}_{1q}(r_1,\phi_1) + \bar{b}_{1q}(r_1,\phi_1) k_0^{-\frac{1}{2}}\} (k_0^{\frac{1}{2}}\psi_1)^q \tag{8.173}$$

where $\psi_1 = (r_1+a-r)^{\frac{1}{2}}$ with the same convention on sign as has been employed to date. While more complicated than (8.172), because

$$\bar{a}_{10}(r_1,\phi_1) = \frac{(-)^N \exp(-\frac{1}{4}\pi i) g(\phi_N, \phi_1)}{4i(2\pi r_1)^{\frac{1}{2}}}$$

and all other $\bar{a}_{1q}$ and $\bar{b}_{1q}$ vanish, it has the advantage of being in a form which is suitable for generalization. By replacing

r_1 and ϕ_1 by r_{-1} and ϕ_{-1} respectively we deduce that, to find v_2, we need the field scattered by the lower half-plane from the incident wave $\exp(-ik_0r_{-1})\psi_{-1}^q$ where $\psi_{-1} = (r_{-1}+a-r_{-2})^{\frac{1}{2}}$. According to (8.155), the scattered wave in $0 \leqslant \phi \leqslant \pi$ is

$$-\pi^{-\frac{1}{2}}\exp(ik_0\psi^2+\tfrac{1}{4}\pi i-ik_0r_{-1})\psi_{-1}^q\Big\{F(-k_0^{\frac{1}{2}}\psi) -$$

$$-\tfrac{1}{2}ik_0^{-\frac{1}{2}}\psi^{-1}\sum_{n=0}^{[\frac{1}{2}q]}\left(\tfrac{1}{2}\right)_n\left(\frac{i}{k_0\psi^2}\right)^n\Big\} -$$

$$-\frac{\exp\{-ik_0(a+r)-\frac{1}{4}\pi i\}}{2(2\pi k_0r)^{\frac{1}{2}}}\,g(\tfrac{1}{2}\pi,\phi)\delta_{0q}$$

with an error which is $O(k_0^{-\frac{1}{2}q-1})$. Here $[\frac{1}{2}q]$ is the largest integer which does not exceed $\frac{1}{2}q$. Consequently, in $0 \leqslant \phi \leqslant \pi$,

$$v_2(x,y) = -(\pi k_0)^{-\frac{1}{2}}\exp\{-ik_0(a+r)+\tfrac{1}{4}\pi i\}\Big[\sum_{q=0}^{\infty}\Big\{\bar{a}_{1q}(r_{-1},\phi_{-1}) +$$

$$+k_0^{-\frac{1}{2}}\bar{b}_{1q}(r_{-1},\phi_{-1})\Big\}\times(k_0^{\frac{1}{2}}\psi_{-1})^q\Big\{F(-k_0^{\frac{1}{2}}\psi)-$$

$$-\tfrac{1}{2}ik_0^{-\frac{1}{2}}\psi^{-1}\sum_{n=0}^{[\frac{1}{2}q]}\left(\tfrac{1}{2}\right)_n\left(\frac{i}{k_0\psi^2}\right)^n\Big\} + \frac{i\bar{a}_{10}(a,\frac{1}{2}\pi)}{2(2k_0r)^{\frac{1}{2}}}\,g(\tfrac{1}{2}\pi,\phi)\Big]$$

correct to $O(k_0^{-1})$. Now, on account of (8.166),

$$F(z) + \frac{1}{2}\frac{i}{z}\sum_{n=0}^{[\frac{1}{2}q]}\left(\tfrac{1}{2}\right)_n\left(\frac{i}{z^2}\right)^n = \frac{\frac{1}{2}\pi^{\frac{1}{2}}i\exp(-\frac{1}{4}q\pi i)}{(-\frac{1}{2}q-\frac{1}{2})!\,z^{q+1}} +$$

$$+\frac{1}{2}\frac{\pi^{\frac{1}{2}}}{z^q}\exp\Big\{(\tfrac{3}{4}q-\tfrac{1}{4})\pi i\Big\}\sum_{m=0}^{\infty}\frac{(-)^m\exp(\frac{1}{4}m\pi i)z^m}{(\frac{1}{2}m-\frac{1}{2}q)!}. \tag{8.174}$$

Hence

$$
\begin{aligned}
v_2(x,y) = k_0^{-\frac{1}{2}} \exp\{-\mathrm{i}k_0(a+r)\}\Big[-\sum_{q=0}^{\infty} \tfrac{1}{2}\{\bar{a}_{1q}(r_{-1},\phi_{-1}) + {} \\
+ k_0^{-\frac{1}{2}}\bar{b}_{1q}(r_{-1},\phi_{-1})\}(-)^q\eta^q \exp(\tfrac{3}{4}q\pi\mathrm{i}) \times \sum_{m=0}^{\infty} \frac{\exp(\frac{1}{4}m\pi\mathrm{i})}{(\frac{1}{2}m-\frac{1}{2}q)!}(k_0^{\frac{1}{2}}\psi)^m + {} \\
+ \frac{1}{2}\frac{\exp(\frac{3}{4}\pi\mathrm{i})}{k_0^{\frac{1}{2}}\psi} \sum_{q=0}^{\infty} \frac{\bar{a}_{1q}(r_{-1},\phi_{-1})}{(-\frac{1}{2}q-\frac{1}{2})!}(-)^q \exp(-\tfrac{1}{4}q\pi\mathrm{i})\eta^q + {} \\
+ \frac{\exp(\frac{3}{4}\pi\mathrm{i})\bar{a}_{10}(a,\frac{1}{2}\pi)}{2(2\pi k_0 r)^{\frac{1}{2}}} g(\tfrac{1}{2}\pi,\phi)\Big]
\end{aligned}
\tag{8.175}
$$

where $\eta = \psi_{-1}/\psi$. If (8.175) is written as

$$
v_2(x,y) = k_0^{-\frac{1}{2}} \exp\{-\mathrm{i}k_0(a+r)\} \sum_{q=0}^{\infty} \{a_{2q}(r,\phi)+k_0^{-\frac{1}{2}}b_{2q}(r,\phi)\}(k_0^{\frac{1}{2}}\psi)^q
\tag{8.176}
$$

it follows that

$$
a_{2q}(r,\phi) = -\frac{1}{2}\exp(\tfrac{1}{4}q\pi\mathrm{i}) \sum_{m=0}^{\infty} \frac{\bar{a}_{1m}(r_{-1},\phi_{-1})(-)^m\eta^m \exp(\frac{3}{4}m\pi\mathrm{i})}{(\frac{1}{2}q-\frac{1}{2}m)!},
\tag{8.177}
$$

$$
\begin{aligned}
b_{2q}(r,\phi) = -\frac{1}{2}\exp(\tfrac{1}{4}q\pi\mathrm{i}) \sum_{m=0}^{\infty} \frac{\bar{b}_{1m}(r_{-1},\phi_{-1})(-)^m\eta^m \exp(\frac{3}{4}m\pi\mathrm{i})}{(\frac{1}{2}q-\frac{1}{2}m)!} + {} \\
+ \tfrac{1}{2}\delta_{0q} \exp(\tfrac{3}{4}\pi\mathrm{i})\Big\{\frac{1}{\psi}\sum_{m=0}^{\infty} \frac{\bar{a}_{1m}(r_{-1},\phi_{-1})(-)^m \exp(-\frac{1}{4}m\pi\mathrm{i})\eta^m}{(-\frac{1}{2}m-\frac{1}{2})!} + {} \\
+ \frac{\bar{a}_{10}(a,\frac{1}{2}\pi)}{(2\pi r)^{\frac{1}{2}}} g(\tfrac{1}{2}\pi,\phi)\Big\}.
\end{aligned}
\tag{8.178}
$$

By commencing with (8.176) we can deduce

$$\bar{v}_3(x,y) = k_0^{-\frac{1}{2}} \exp\{-ik_0(2a+r_1)\} \sum_{q=0}^{\infty} \{\bar{a}_{3q}(r_1,\phi_1)+k_0^{-\frac{1}{2}}\bar{b}_{3q}(r_1,\phi_1)\}(k_0^{\frac{1}{2}}\psi)^q$$

where $\bar{a}_{3q}(r_1,\phi_1)$ is related to $a_{2m}(r,\phi)$ by a formula of the same type as (8.177) except that η is exchanged for $\eta_1 = \psi/\psi_1$. A similar remark is true for $\bar{b}_{3q}$.

The relationship may be expressed otherwise. By Hankel's contour integral

$$\frac{1}{(\frac{1}{2}q-\frac{1}{2}m)!} = \frac{1}{2\pi i}\int_{-\infty}^{(0+)} e^{w}\, w^{-\frac{1}{2}q+\frac{1}{2}m-1}\,dw \qquad |\mathrm{ph}\,w| \leqslant \pi$$

where the contour begins at $-\infty$, circles the origin once in the positive sense, and returns to $-\infty$. Hence, if the contour is chosen so that $|tw^{-\frac{1}{2}}| < 1$ on it,

$$\sum_{q=0}^{\infty} a_{2q}(r,\phi)t^q = -$$

$$-\frac{1}{4\pi i}\sum_{m=0}^{\infty} \bar{a}_{1m}(r_{-1},\phi_{-1})(-)^m\eta^m \exp(\tfrac{3}{4}m\pi i)\int_{-\infty}^{(0+)} e^{w} w^{\frac{1}{2}m-1}(1-\exp(\tfrac{1}{4}\pi i)\,tw^{-\frac{1}{2}})^{-1}dw\ .$$

It is now convenient to make ϕ, and hence ϕ_{-1}, equal to $\frac{1}{2}\pi$. Then $\eta = r^{\frac{1}{2}}/(r+2a)^{\frac{1}{2}}$.

Let

$$f_2(r,t) = \sum_{q=0}^{\infty} a_{2q}(r,\tfrac{1}{2}\pi)t^q\ ,$$

$$\bar{f}_1(r_{-1},t) = \sum_{q=0}^{\infty} \bar{a}_{1q}(r_{-1},\tfrac{1}{2}\pi)t^q\ .$$

It follows that

$$f_2(r,t) = -\frac{1}{4\pi i}\int_{-\infty}^{(0+)} e^{w} w^{-1}\{1-\exp(\tfrac{1}{4}\pi i)tw^{-\frac{1}{2}}\}^{-1}\,\bar{f}_1(r+a,-\exp(\tfrac{3}{4}\pi i)\eta w^{\frac{1}{2}})\,dw$$

$$= -\frac{1}{2\pi i}\int_{-\infty+i\alpha}^{\infty+i\alpha}\frac{\exp(-w^2)}{w-t\exp(-\frac{1}{4}\pi i)}\,\bar{f}_1\left(r+a,\exp(\tfrac{1}{4}\pi i)w\left(\frac{r}{r+2a}\right)^{\frac{1}{2}}\right)dw \qquad (8.179)$$

where the contour of integration passes below the pole $w = t\exp(-\frac{1}{4}\pi i)$, i.e. $\alpha < \mathscr{I}\, t\exp(-\frac{1}{4}\pi i)$. Similarly

$$\bar{f}_3(r,t) = -\frac{1}{2\pi i}\int_{-\infty}^{\infty}\frac{\exp(-w^2)}{w-t\exp(-\frac{1}{4}\pi i)}\,f_2\left(r+a,\ \exp(\tfrac{1}{4}\pi i)w\left(\frac{r}{r+2a}\right)^{\frac{1}{2}}\right)dw.$$

The application of (8.179) to the next stage of the diffraction process gives

$$\bar{f}_3(r_1,t) = \frac{1}{(2\pi i)^2}\int_{-\infty+i\alpha_1}^{\infty+i\alpha_1}\frac{\exp(-w_1^2)}{w_1-t\exp(-\frac{1}{4}\pi i)}\int_{-\infty+i\alpha_2}^{\infty+i\alpha_2}\frac{\exp(-w_2^2)}{w_2-w_1\{r_1/(r_1+2a)\}^{\frac{1}{2}}}$$

$$\times\ \bar{f}_1\left(r_1+2a,\exp(\tfrac{1}{4}\pi i)w_2\left(\frac{r_1+a}{r_1+3a}\right)^{\frac{1}{2}}\right)dw_2\,dw_1 \qquad (8.180)$$

with $\alpha_1 < \mathscr{I}\, t\exp(-\frac{1}{4}\pi i)$, $\alpha_2 < \alpha_1\{r_1/(r_1+2a)\}^{\frac{1}{2}}$. Since $\bar{a}_{1q} = 0$ if $q \neq 0$ we thus derive

$$(r_1,t) = \frac{(-)^N\exp(-\frac{1}{4}\pi i)g(\theta_N,\frac{1}{2}\pi)}{4i(2\pi)^{\frac{1}{2}}(2\pi i)^2(r_1+2a)^{\frac{1}{2}}}\int_{-\infty+i\alpha_1}^{\infty+i\alpha_1}\frac{\exp(-w_1^2)}{w_1-t\exp(-\frac{1}{4}\pi i)}\int_{-\infty+i\alpha_2}^{\infty+i\alpha_2}\frac{\exp(-w_2^2)}{w_2-w_1\{r_1/(r_1+2a)\}^{\frac{1}{2}}}dw_2\,dw_1$$

Now make the substitutions

$$(w_1-t\exp(-\tfrac{1}{4}\pi i))^{-1} = 2i(r_1+a)^{\frac{1}{2}}r_1^{-\frac{1}{2}}\int_0^{\infty}\exp\left[\frac{-2ix_1(r_1+a)^{\frac{1}{2}}\{w_1-t\exp(-\frac{1}{4}\pi i)\}}{r_1^{\frac{1}{2}}}\right],$$

$$(w_2-z)^{-1} = 2i(r_1+2a)^{\frac{1}{2}}(r_1+a)^{-\frac{1}{2}}\int_0^\infty \exp\left\{\frac{-2ix_2(r_1+2a)^{\frac{1}{2}}(w_2-z)}{(r_1+a)^{\frac{1}{2}}}\right\}$$

and carry out the integrations with respect to w_1, w_2. There results

$$\bar{f}_3(r_1,t) =$$

$$\frac{(-)^N\exp(-\frac{1}{4}\pi i)g(\theta_N,\frac{1}{2}\pi)}{4i(2\pi)^{\frac{1}{2}}\pi r_1^{\frac{1}{2}}}\int_0^\infty\int_0^\infty \exp\{2ix_1t\exp(-\tfrac{1}{4}\pi i)\frac{(r_1+a)^{\frac{1}{2}}}{r_1^{\frac{1}{2}}} - x_1^2\frac{r_1+a}{r_1} - 2x_2^2+2x_1x_2\}dx_1dx_2 \, . \tag{8.181}$$

By replacing $\bar{f}_3$, $\bar{f}_1$ in (8.180) by $\bar{f}_5$, $\bar{f}_3$ respectively and using (8.181) we find by a similar procedure that

$$\bar{f}_5(r_1,t) = \frac{(-)^N\exp(-\frac{1}{4}\pi i)g(\theta_N,\frac{1}{2}\pi)}{4i(2\pi)^{\frac{1}{2}}\pi^2 r_1^{\frac{1}{2}}}\int_0^\infty\cdots\int_0^\infty \exp\left\{2ix_1t\exp(-\tfrac{1}{4}\pi i)\left(\frac{r_1+a}{r_1}\right)^{\frac{1}{2}} - \right.$$

$$\left. - x_1^2\frac{r_1+a}{r_1} - 2(x_2^2+x_3^2+x_4^2-x_1x_2-x_2x_3-x_3x_4)\right\}dx_1 \ldots dx_4$$

and clearly

$$\bar{f}_{2n+1}(r_1,t) = \frac{(-)^{N+2n}\exp(-\frac{1}{4}\pi i)g(\theta_N,\frac{1}{2}\pi)}{4i(2\pi)^{\frac{1}{2}}\pi^n r_1^{\frac{1}{2}}}\int_0^\infty\cdots\int_0^\infty \exp\{2ix_1t\exp(-\tfrac{1}{4}\pi i)\left(\frac{r_1+a}{r_1}\right)^{\frac{1}{2}} -$$

$$- x_1^2\frac{r_1+a}{r_1} - 2\sum_{m=2}^{2n}(x_m^2-x_{m-1}x_m)\}dx_1 \ldots dx_{2n} \tag{8.182}$$

for $n = 1, 2, \ldots$. Obviously f_{2n} is given by the same formula with $n - \frac{1}{2}$ for n.

An immediate deduction from (8.182) is that

$$\bar{a}_{2n+1,q}(r_1,\tfrac{1}{2}\pi) = \frac{(-)^{N+2n}\exp(-\frac{1}{4}\pi i)g(\theta_N,\frac{1}{2}\pi)}{q!4i(2\pi)^{\frac{1}{2}}\pi^n r_1^{\frac{1}{2}}} 2^q \exp(\tfrac{1}{4}\pi i)\left(\frac{r_1+a}{r_1}\right)^{\frac{1}{2}q} \times$$
$$\int_0^\infty \cdots \int_0^\infty x_1^q \exp\{-x_1^2 \frac{r_1+a}{r_1} - 2\sum_{m=2}^{2n} (x_m^2 - x_{m-1}x_m)\}dx_1 \ldots dx_{2n}\,. \tag{8.183}$$

From (8.183) it follows that

$$\left(\frac{r+a}{a}\right)^{\frac{1}{2}}\pi^{\frac{1}{2}} \sum_{q=0}^{\infty} \frac{\bar{a}_{2n+1,q}(r+a,\frac{1}{2}\pi)}{(-\frac{1}{2}q-\frac{1}{2})!} \exp(-\tfrac{1}{4}q\pi i)\left(\frac{r}{r+2a}\right)^{\frac{1}{2}q}(-)^q$$
$$= \frac{(-)^N\exp(-\frac{1}{4}\pi i)g(\theta_N,\frac{1}{2}\pi)}{4i(2\pi)^{\frac{1}{2}}\pi^n a^{\frac{1}{2}}} \int_0^\infty \cdots \int_0^\infty \exp\{-2x_1^2 - 2\sum_{m=2}^{2n} (x_m^2 - x_{m-1}x_m)\}dx_1 \ldots dx_{2n}$$
$$= \bar{a}_{2n+1,0}(a,\tfrac{1}{2}\pi) \tag{8.184}$$

on using $(2q)! = q!(q-\frac{1}{2})!2^{2q}/\pi^{\frac{1}{2}}$ and (8.183). The conclusion (8.184) is consistent with the requirement that there be no singularity in (8.178) when $\phi = \frac{1}{2}\pi$. A similar result holds for $a_{2n,q}$.

When $\phi = \frac{1}{2}\pi$, the term involving δ_{0q} in (8.178) becomes, for general n,

$$\tfrac{1}{2}\delta_{0q} \exp(\tfrac{3}{4}\pi i)\left[\frac{2r^{\frac{1}{2}}}{\{2a(r+a)\}^{\frac{1}{2}}} \sum_{q=0}^{\infty} \frac{\exp(-\frac{1}{4}q\pi i)}{(-\frac{1}{2}q-\frac{1}{2})!}\left(\frac{r}{r+2a}\right)^{\frac{1}{2}q}\left\{\frac{\partial}{\partial\phi_{-1}}\bar{a}_{2n+1,q}(r+a,\phi_{-1})\right\}_{\phi_{-1}=\frac{1}{2}\pi} +\right.$$
$$\left. + \bar{a}_{2n+1,0}\frac{(a,\frac{1}{2}\pi)}{(2\pi r)^{\frac{1}{2}}}\right].$$

A derivative of (8.177) with respect to ϕ shows that $\partial a/\partial\phi$ satisfies the same recurrence relation as a, apart from the factor $r/(r+a)$. Hence

$$\left\{\frac{\partial}{\partial\phi_1}\bar{a}_{2n+1,q}(r_1,\phi_1)\right\}_{\phi_1=\frac{1}{2}\pi} = \frac{r_1}{r_1+2na}\frac{h(\theta_N)}{g(\theta_N,\frac{1}{2}\pi)}\bar{a}_{2n+1,q}(r_1,\tfrac{1}{2}\pi)$$

where

$$h(\theta_N) = \left\{\frac{\partial}{\partial\phi}g(\theta_N,\phi)\right\}_{\phi=\frac{1}{2}\pi} = 2^{\frac{1}{2}}(2+\cos\theta_N)\frac{\sin\frac{1}{2}\theta_N}{\cos^2\theta_N}.$$

Applying (8.184) we obtain

$$b_{2n+2,q}(r,\tfrac{1}{2}\pi) = \tfrac{1}{2}(-)^q\exp(\tfrac{1}{4}q\pi i)\sum_{m=0}^{\infty}\frac{\bar{b}_{2n+1,m}(r_{-1},\frac{1}{2}\pi)\eta^m\exp(\frac{3}{4}m\pi i)}{(\frac{1}{2}q-\frac{1}{2}m)!} +$$

$$+\tfrac{1}{2}\delta_{0q}\exp(\tfrac{3}{4}\pi i)\frac{\bar{a}_{2n+1,0}(a,\frac{1}{2}\pi)}{(2\pi r)^{\frac{1}{2}}}\left\{1+\frac{2r}{r+(2n+1)a}\frac{h(\theta_N)}{g(\theta_N,\frac{1}{2}\pi)}\right\}.$$

Evidently, because the recurrence is the same as that for $a_{n,q}$,

$$b_{n,q}(r,\tfrac{1}{2}\pi) = -\frac{2i}{(-)^N g(\theta_N,\frac{1}{2}\pi)}\sum_{m=1}^{n-1}\left\{1+2\frac{r+(n-m-1)a}{r+(n-1)a}\frac{h(\theta_N)}{g(\theta_N,\frac{1}{2}\pi)}\right\}a_{m,0}(a,\tfrac{1}{2}\pi)a_{n-m,q}(r,\tfrac{1}{2}\pi)$$ (8.185)

where, to simplify the summation, the bars on letters have been dropped.

Sufficient quantities have now been determined for a calculation of the field. Let $v_{2n}(x,y) = \exp(-ik_0r)w_{2n}(r,\phi)$. Then, according to (8.155), (8.151), and (8.152), a non-uniform expansion of $\bar{v}_{2n+1}$ for $n > 1$ is

$$\bar{v}_{2n+1} = \frac{\exp\{-ik_0(r_1+a)-\frac{1}{4}\pi i\}}{2(2\pi k_0r_1)^{\frac{1}{2}}}\left[-w_{2n}(a,\tfrac{1}{2}\pi)g(\tfrac{1}{2}\pi,\phi_1)-\frac{3i}{8k_0a}w_{2n}(a,\tfrac{1}{2}\pi)\left\{\frac{\cos^2\frac{1}{2}(\frac{1}{2}\pi-\phi_1)}{\sin^3\frac{1}{2}(\frac{1}{2}\pi-\phi_1)}+\right.\right.$$

$$\left.+\frac{\cos^2\frac{1}{2}(\frac{1}{2}\pi+\phi_1)}{\sin^3\frac{1}{2}(\frac{1}{2}\pi+\phi_1)}\right\}-\frac{i}{4k_0}\left\{\frac{w_{2n}(a,\frac{1}{2}\pi)}{r_1}+\left[\frac{\partial}{\partial r_1}w_{2n}(r,\phi)\right]_{r_1=0}\right\}\times$$

$$\left.\times\{\operatorname{cosec}^3\tfrac{1}{2}(\tfrac{1}{2}\pi-\phi_1)+\operatorname{cosec}^3\tfrac{1}{2}(\tfrac{1}{2}\pi+\phi_1)\}\right].$$

It is non-uniform because it fails in the vicinity of $\phi_1 = \pm\frac{1}{2}\pi$. Now

$$w_{2n}(a,\tfrac{1}{2}\pi) = k_0^{-\frac{1}{2}} \exp\{-ik_0(2n-1)a\}\{a_{2n,0}(a,\tfrac{1}{2}\pi)+k_0^{-\frac{1}{2}}b_{2n,0}(a,\tfrac{1}{2}\pi)\},$$

$$\left[\frac{\partial w_{2n}(r,\phi)}{\partial r_1}\right]_{r_1=0} = -\tfrac{1}{2}a^{-\frac{1}{2}}a_{2n,1}(a,\tfrac{1}{2}\pi)\cos\phi_1 \exp\{-ik_0(2n-1)a\}.$$

Hence

$$\bar{v}_{2n+1} = \frac{\exp\{-ik_0(r_1+2na)-\frac{1}{4}\pi i\}}{2(2\pi k_0 r_1)^{\frac{1}{2}}}\Big[-k_0^{-\frac{1}{2}}\{a_{2n,0}(a,\tfrac{1}{2}\pi) +$$

$$+ k_0^{-\frac{1}{2}}b_{2n,0}(a,\tfrac{1}{2}\pi)\}g(\tfrac{1}{2}\pi,\phi_1) + \frac{i}{2k_0 a^{\frac{1}{2}}}\, a_{2n,1}(a,\tfrac{1}{2}\pi)h(\pi-\phi_1)\Big]. \tag{8.186}$$

On account of (8.183)

$$a_{2n,0}(a,\tfrac{1}{2})$$

$$= \frac{(-)^{N+2n-1}\exp(-\frac{1}{4}\pi i)g(\theta_N,\frac{1}{2}\pi)}{4i(2\pi)^{\frac{1}{2}}\pi^{n-\frac{1}{2}}a^{\frac{1}{2}}}\int_0^\infty\!\!\cdots\!\int_0^\infty \exp\Big\{-2x_1^2-2\sum_{m=2}^{2n-1}(x_m^2-x_{m-1}x_m)\Big\}dx_1\,\ldots\,dx_{2n-1} \tag{8.187}$$

The substitution $x_1 = t_1$, $x_m = t_m + x_{m-1}$ $(m = 2, \ldots, n)$ gives

$$\int_0^\infty\!\cdots\int_0^\infty \exp\Big\{-2\sum_{m=1}^{n} x_m^2 + 2\sum_{m=2}^{n} x_{m-1}x_m\Big\}dx_1\,\ldots\,dx_n$$

$$= \int_0^\infty\!\int_{-t_1}^\infty\!\cdots\int_{-(t_1+\ldots+t_{n-1})}^\infty \exp\Big\{-\sum_{m=1}^{n} t_m^2 - \Big(\sum_{m=1}^{n} t_m\Big)^2\Big\}dt_1\,\ldots\,dt_n$$

$$= \int_0^\infty\!\int_{-t_1}^\infty\!\cdots\int_{-(t_1+\ldots+t_{n-1})}^\infty\int_{-\infty}^\infty \exp\Big(-\sum_{m=1}^{n+1} t_m^2\Big)\delta\Big(\sum_{m=1}^{n+1} t_m\Big)dt_1\,\ldots\,dt_{n+1}$$

where δ is the standard Dirac generalized function. If the integration with respect to t_n is carried out we obtain

$$\int_0^\infty \int_{-t_1}^\infty \cdots \int_{-(t_1+\ldots+t_{n-2})}^\infty \int_{-\infty}^\infty \exp\left\{-\sum_{m=1}^n t_m^2 - \left(\sum_{m=1}^n t_m\right)^2\right\} dt_1 \ldots dt_n .$$

Instead the integration with respect to t_{n-1} or t_{n-2} etc. may be performed. Adding together all these results gives

$$(n+1)\int_0^\infty \cdots \int_0^\infty \exp\left(-2\sum_{m=1}^n x_m^2 + 2\sum_{m=2}^n x_{m-1}x_m\right) dx_1 \ldots dx_n$$

$$= \int_{-\infty}^\infty \cdots \int_{-\infty}^\infty \exp\left\{-\sum_{m=1}^n t_m^2 - \left(\sum_{m=1}^n t_m\right)^2\right\} dt_1 \ldots dt_n$$

$$= \int_{-\infty}^\infty \cdots \int_{-\infty}^\infty \exp\left(-\sum_{m=1}^{n+1} t_m^2\right)\delta\left(\sum_{m=1}^{n+1} t_m\right) dt_1 \ldots dt_{n+1}$$

$$= \int_{-\infty}^\infty \cdots \int_{-\infty}^\infty \exp\left(-\sum_{m=1}^{n+1} t_m^2\right)\delta\left((n+1)^{\frac{1}{2}} t_{n+1}\right) dt_1 \ldots dt_{n+1}$$

$$= \int_{-\infty}^\infty \cdots \int_{-\infty}^\infty \exp\left(-\sum_{m=1}^{n} t_m^2\right) dt_1 \ldots dt_n/(n+1)^{\frac{1}{2}}$$

$$= \frac{\pi^{\frac{1}{2}n}}{(n+1)^{\frac{1}{2}}} \tag{8.188}$$

by a rotation of coordinates. Hence

$$a_{2n,0}(a,\tfrac{1}{2}\pi) = \frac{(-)^{N+2n-1}\exp(-\frac{1}{4}\pi i)g(\theta_N,\frac{1}{2}\pi)}{4i(2\pi)^{\frac{1}{2}}a^{\frac{1}{2}}(2n)^{3/2}} . \tag{8.189}$$

The same expression holds for $\bar{a}_{2n+1}$ with $n + \frac{1}{2}$ for n.

For $a_{2n,1}$ there is an extra factor x_1 in the integrand of (8.187). To obtain the formula analogous to (8.188) we note that symmetry enables us to replace x_1 by $\frac{1}{2}(x_1+x_n)$ or

$\frac{1}{4}\Sigma_{m=1}^{n}(-2x_{m-1}+4x_m-2x_{m+1})$ on the understanding that x_0 and x_{n+1} are zero. Hence

$$\int_0^\infty \cdots \int_0^\infty x_1 \exp\left(-2\sum_{m=1}^{n} x_m^2 + 2\sum_{m=2}^{n} x_{m-1}x_m\right) dx_1 \ldots dx_n$$

$$= \frac{1}{4}\sum_{m=1}^{n} \int_0^\infty \cdots \int_0^\infty \exp\left(-2\sum_{i=1}^{m-1} x_i^2 + 2\sum_{i=2}^{m-1} x_{i-1}x_i\right) dx_1 \ldots dx_{m-1} \times$$

$$\times \int_0^\infty \cdots \int_0^\infty \exp\left(-2\sum_{i=m+1}^{n} x_i^2 + 2\sum_{i=m+2}^{n} x_{i-1}x_i\right) dx_{m+1} \ldots dx_n$$

$$= \frac{1}{4}\pi^{\frac{1}{2}(n-1)} \sum_{m=1}^{n} m^{-3/2}(n-m+1)^{-3/2} \tag{8.190}$$

from (8.188). Consequently

$$a_{2n,1}(a,\tfrac{1}{2}\pi) = \frac{(-)^{N+1+2n} g(\theta_N,\frac{1}{2}\pi)}{8i\pi a^{\frac{1}{2}}} \sum_{m=1}^{2n-1} m^{-3/2}(2n-m)^{-3/2}. \tag{8.191}$$

Substitution of (8.189) in (8.185) leads to

$$b_{n,0}(a,\tfrac{1}{2}\pi) = \frac{(-)^{N+n}}{16\pi a}\{g(\theta_N,\tfrac{1}{2}\pi)+h(\theta_N)\} \sum_{m=1}^{n-1} m^{-3/2}(n-m)^{-3/2} \tag{8.192}$$

since

$$2\sum_{m=1}^{n-1} \frac{n-m}{m^{3/2}(n-m)^{3/2}} = \sum_{m=1}^{n-1} \frac{n-m}{m^{3/2}(n-m)^{3/2}} + \sum_{m=1}^{n-1} \frac{m}{(n-m)^{3/2}m^{3/2}}$$

$$= \sum_{m=1}^{n-1} \frac{n}{m^{3/2}(n-m)^{3/2}} .$$

The total diffracted field from the upper edge due to the plane wave striking it can now be calculated from (8.172), (8.186), (8.189), (8.191), and (8.192). It is

$$\bar{v}_1 + \sum_{n=1}^{\infty} \bar{v}_{2n+1} = \frac{(-)^N \exp(-ik_0r_1-\frac{1}{4}\pi i)}{2(2\pi k_0 r_1)^{\frac{1}{2}}}\Bigg(-\tfrac{1}{2}ig(\theta_N,\phi_1) +$$

$$+ \frac{\exp(-\frac{1}{4}\pi i)}{4i(2\pi k_0 a)^{\frac{1}{2}}}\, g(\tfrac{1}{2}\pi,\phi_1)g(\theta_N,\tfrac{1}{2}\pi) \sum_{n=1}^{\infty} \frac{\exp(-i2nk_0a)}{(2n)^{3/2}}$$

$$- \frac{1}{16\pi k_0 a}\Big[g(\theta_N,\tfrac{1}{2}\pi)h(\pi-\phi_1)+g(\tfrac{1}{2}\pi,\phi_1)\{g(\theta_N,\tfrac{1}{2}\pi)+h(\theta_N)\}\Big] \times$$

$$\times \sum_{m=1}^{\infty} \exp(-2ink_0a) \sum_{m=1}^{2n-1} \frac{1}{m^{3/2}(2n-m)^{3/2}}\Bigg) .$$

With regard to the plane wave striking the lower edge we have

$$v_1(x,y) = -\frac{\exp(-ik_0r-\frac{1}{4}\pi i)}{4i(2\pi k_0 r)^{\frac{1}{2}}}\, g(\theta_N,\phi) .$$

The fields $\bar{v}_2$, $\bar{v}_4$, ... successively scattered from the upper edge can be calculated by the same method as has just been described. Thus the field diffracted by the upper edge due to the incident mode is

$$\bar{v}_d(r_1,\phi_1) = \sum_{n=1}^{\infty} \bar{v}_n = \frac{(-)^N \exp(-ik_0r_1-\frac{1}{4}\pi i)}{2(2\pi k_0 r_1)^{\frac{1}{2}}}\, f(\phi_1) \tag{8.193}$$

where

$$f(\phi_1) = -\tfrac{1}{2}ig(\theta_N,\phi_1) + \frac{\exp(-\frac{1}{4}\pi i)S}{4i(2\pi k_0 a)^{\frac{1}{2}}}\, g(\tfrac{1}{2}\pi,\phi_1)g(\theta_N,\tfrac{1}{2}\pi) -$$

$$- \frac{S^2}{16\pi k_0 a}\Big[g(\theta_N,\tfrac{1}{2}\pi)h(\pi-\phi_1)+g(\tfrac{1}{2}\pi,\phi_1)\{g(\theta_N,\tfrac{1}{2}\pi)+h(\theta_N)\}\Big] , \tag{8.194}$$

$$S = \sum_{m=1}^{\infty} \frac{\exp(-2mkai)}{(2m)^{3/2}} + (-)^N \sum_{m=1}^{\infty} \frac{\exp\{-(2m-1)kai\}}{(2m-1)^{3/2}} .$$

Similarly the field diffracted by the lower edge is given by

$$v_d(r,\phi) = -\frac{\exp(-ik_0r-\frac{1}{4}\pi i)}{2(2\pi k_0 r)^{\frac{1}{2}}} f(\phi) . \quad (8.195)$$

The addition of (8.193) and (8.195) determines the radiated field in any direction apart from the exceptional directions $\phi = \pm\, \theta_N,\ \pm\, \frac{1}{2}\pi$ where the formulae are non-uniform. The field reflected back into the waveguide can be expressed in modal form via

$$u_r(x,y) = \sum_{n=1}^{\infty} R_n \sin\left(\frac{n\pi y}{a}\right) \exp(i\kappa_n x) .$$

Obviously

$$R_n \exp(i\kappa_n x) = \frac{2}{a}\int_0^a u_r(x,y)\sin\left(\frac{n\pi y}{a}\right) dy .$$

Therefore, it is necessary to consider integrals of the form

$$\int_0^a \exp\left\{-ik_0(x^2+y^2)^{\frac{1}{2}} + \frac{in\pi y}{a}\right\} \frac{F(y)}{(x^2+y^2)^{\frac{1}{4}}} dy$$

and it is sufficient for x to be large and negative. The method of stationary phase is therefore applicable. The point of stationary phase occurs at $y = -x \tan\theta_n$ and the integral has asymptotic value

$$\frac{(2\pi k_0)^{\frac{1}{2}}}{\kappa_n} \exp(-\tfrac{1}{4}\pi i + i\kappa_n x) F(-x \tan\theta_n) .$$

Use this result with (8.193) and (8.195), remembering that ϕ is near π and ϕ_1 near $-\pi$. Then

$$R_n = \{1+(-)^{N+n}\} \frac{f(\pi-\theta_n)}{2a\kappa_n} . \tag{8.196}$$

The derivation of (8.196) is strictly valid only when κ_n is real, but, in fact, (8.196) holds even when the mode is not propagating but additional argument is needed because the point of stationary phase is no longer in the interval of integration.

The formula (8.196) brings out the point that the reflection coefficient vanishes when $N + n$ is odd, consistent with there being no coupling between modes of different symmetries.

It will be noted that (8.196) breaks down at $\kappa_n = 0$ and, on account of (8.194), at $\kappa_N = 0$. In other words it fails at the cut-off frequencies of the nth and Nth modes. At other cut-off frequencies the reflection coefficient is continuous but has discontinuities in slope; as the frequency increases the modulus comes down at an angle to the vertical until cut-off is reached when it immediately starts climbing vertically while the phase descends vertically and then rises at an angle to the vertical.

The predictions of ray theory described here are in excellent agreement with calculations based on the exact solution which can be obtained for this problem (Weinstein 1948, 1969; Heins 1948).

Exercises

31. The TM mode $u_0 = \cos(N\pi y/a)\exp(-i\kappa_N x)$ is incident on the end of a parallel-plate waveguide. Show that the field diffracted by the upper plate is

$$\bar{v}_d(r_1,\phi_1) = (-)^N \frac{\exp(-ik_0r_1-\frac{1}{4}\pi i)}{2(2\pi k_0r_1)^{\frac{1}{2}}} F_1(\phi_1)$$

where

$$F_1(\phi_1) = -\frac{1}{2}G_1(\theta_N,\phi_1) + \frac{\exp(-\frac{1}{4}\pi i)}{2(2\pi k_0 a)^{\frac{1}{2}}} SG_1(\theta_N,\tfrac{1}{2}\pi)G_1(\tfrac{1}{2}\pi,\phi_1)$$

$$+ \frac{iS^2}{16\pi k_0 a}\left[G_1(\theta_N,\tfrac{1}{2}\pi)H_1(\pi-\phi_1)-G_1(\tfrac{1}{2}\pi,\phi_1)\{G_1(\theta_N,\tfrac{1}{2}\pi)-H_1(\theta_N)\}\right],$$

$$G_1(\theta,\phi) = \operatorname{cosec}\frac{1}{2}(\theta-\phi) - \operatorname{cosec}\frac{1}{2}(\theta+\phi),$$

$$H_1(\phi) = 2^{\frac{1}{2}}\cos\frac{1}{2}\phi(2-\cos\phi)\sec^2\phi.$$

Deduce that the reflection coefficient for the nth mode is $-i\{1+(-)^{N+n}\}f(\pi-\theta_n)/2a\kappa_n$ unless $n = 0$ when the coefficient is halved.

32. Check the statements made about (8.196) at cut-off frequencies.

33. The plane wave $\exp(ik_0 y)$ is incident on the parallel-plate waveguide of Fig. 8.21. Show that the distant total field produced in the direction $\phi = -\frac{1}{2}\pi$ is

$$\exp(-ik_0 r)\left[\frac{1}{2}-\frac{1}{2\pi}\left(\frac{a}{r}\right)^{\frac{1}{2}}\left\{1-\sum_{n=1}^{\infty}\left(\frac{1}{(2n)^{\frac{1}{2}}}-\frac{1}{(2n+1)^{\frac{1}{2}}}\right)\exp(-2ink_0 a)\right\}\mp\frac{\exp(-\frac{1}{4}\pi i)}{2(2\pi k_0 r)^{\frac{1}{2}}}\right]$$

the upper or lower sign being taken according as the incident wave is electrically or magnetically polarized.

34. Try the same type of problem as Exercise 33 but with the plates staggered.

35. Repeat Exercise 33 but with a line source on $x = 0$, $y > a$.

36. Examine the possibility of extending the theory of this section to a cylindrical waveguide.

37. A slit in an infinite plane occupies $|x| \leqslant a$, $y = 0$ and is irradiated from $y < 0$ by the plane wave $\exp(-ik_0 y)$. Find a uniform asymptotic expansion for the field, using polar coordinates (r_1,ϕ_1) at the left-hand edge and $(r_2,\pi-\phi_2)$ at the right-hand edge (so that the slit is in $\phi_1 = 0$ and $\phi_2 = 0$). Show that, in the electrically polarized case, the diffracted field is

$$- \frac{\exp(-ik_0r_1 - \frac{1}{4}\pi i)}{2(2\pi k_0 r_1)^{\frac{1}{2}}} g(\tfrac{1}{2}\pi, \phi_1) - \frac{\exp(-ik_0r_2 - \frac{1}{4}\pi i)}{2(2\pi k_0 r_2)^{\frac{1}{2}}} g(\tfrac{1}{2}\pi, \phi_2) -$$

$$- \frac{i\exp(-2ika)}{4\pi k_0 a^{\frac{1}{2}}} \left\{ \frac{\exp(-ik_0r_1)}{r_1^{\frac{1}{2}}} \sec \tfrac{1}{2}\phi_1 + \frac{\exp(-ik_0r_2)}{r_2^{\frac{1}{2}}} \sec \tfrac{1}{2}\phi_2 \right\} + O(k_0^{-3/2}) .$$

For magnetic polarization replace $g(\frac{1}{2}\pi,\phi)$ by $G_1(\frac{1}{2}\pi,\phi)$ (defined in Exercise 31) and drop the third term.

Compare your formulae with the analytical results of Luneburg and Westpfahl (1975) and Sologub (1972).

38. Repeat Exercise 37 with the aperture and screen interchanged. Confirm that your results are in accord with Babinet's principle.

39. In Exercise 37 the right-hand screen is such that the field vanishes whereas the left-hand screen makes the normal derivative zero. Show that the transmission coefficient is

$$1 - \frac{1}{2\pi^{\frac{1}{2}}} \frac{\cos(2k_0a - \frac{1}{4}\pi)}{(k_0a)^{3/2}} - \frac{9}{32\pi^{\frac{1}{2}}(k_0a)^{5/2}} \sin(2k_0a - \tfrac{1}{4}\pi)$$

(for non-perfectly conducting screens see Senior 1975).

40. A *splash plate* for the waveguide of Fig. 8.21 is constructed by making $y = d$ $(d > 0)$ perfectly conducting. If $k_0 d \gg 1$, calculates its effect by imaging the field found without a splash plate. If the splash plate is of finite length determine the additional field due to the edges to a first approximation.

8.20. The wedge

A canonical problem for what happens when the edge is on a surface which is not infinitesimally thin is provided by the wedge. The angle of the wedge will be taken as δ, with free space occupying $-\pi < \phi < \pi-\delta$ (Fig. 8.22). When $\delta = 0$ the wedge degenerates into the semi-infinite plane already considered. The illumination will be assumed to be due to a line source at L producing an incident field of $-\frac{1}{4}iH_0^{(2)}(k_0|\mathbf{x}-\mathbf{x}_1|)$ where $\mathbf{x}_1$ is the position of L.

It will be supposed that the wedge is perfectly conducting,

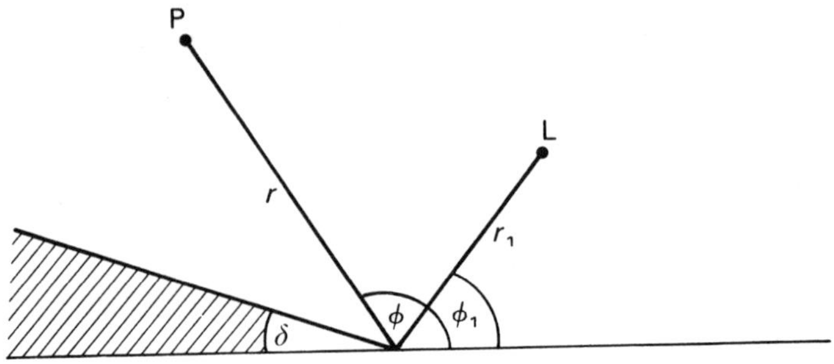

FIG. 8.22. Geometry for scattering by a wedge.

though solutions are available for impedance boundary conditions with different impedances on the two faces (Van Dantzig 1958, Lauwerier 1959, 1960, 1961, Williams 1959, 1961, Malyuzinec 1958). For electric polarization

$$E_z = w(\phi-\phi_1) - w(\phi+\phi_1+2\pi) \tag{8.197}$$

while for magnetic polarization

$$H_z = w(\phi-\phi_1) + w(\phi+\phi_1+2\pi) \tag{8.198}$$

the z axis being along the edge. Here

$$w(\phi) = -\frac{1}{8\pi}\int_{\infty-\pi i\mu}^{\infty+\pi i\mu} \frac{H_0^{(2)}[k_0\{r^2+r_1^2-2rr_1\cosh(u/\mu)\}^{\frac{1}{2}}]}{\cosh u - \cos\mu\phi} \sinh u \, du \tag{8.199}$$

where $1/\mu = 2 - \delta/\pi$ and the contour of integration passes to the left of the point where $\cosh(u/\mu) = (r^2+r_1^2)/2rr_1$ and to the right of the imaginary axis (Fig. 8.23).

Alternative ways of writing (8.192) are, when $r < r_1$,

$$w(\phi) = \frac{1}{4}\int_{-i\infty}^{i\infty} J_\nu(k_0r)H_\nu^{(2)}(k_0r_1)\,\frac{\cos\nu(|\phi|-\pi/\mu)}{\sin(\nu\pi/\mu)}\,d\nu$$

$$= -\frac{1}{2}i\mu\sum_{n=1}^{\infty} J_{n\mu}(k_0r)H_{n\mu}^{(2)}(k_0r_1)\cos n\mu\phi - \frac{1}{4}i\mu J_0(k_0r)H_0^{(2)}(kr_1)$$

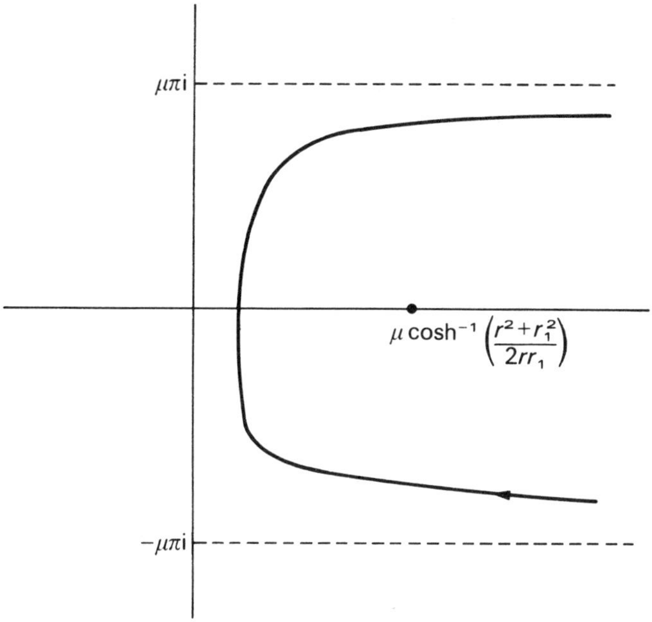

FIG. 8.23. The contour of integration for the wedge.

where the principal value of the integral is taken. For $r > r_1$, interchange r and r_1 in these formulae.

When $k_0 r \gg 1$, the integral in (8.199) can be evaluated asymptotically by the method of stationary phase, the stationary points being at $u = \pm i\pi\mu$. In deforming the contour to the points of stationary phase, poles may be passed over. The contributions of the poles represent the field of geometrical optics each corresponding to some image of the source in the faces of the wedge. Remark that it is possible for there to be no shadow and for more than one reflected wave to exist, e.g. when $\phi_1 = -\frac{1}{2}\delta$. Stationary phase provides the diffracted wave

$$w_d(\phi) \sim -\frac{\mu i \sin(\mu\pi)\exp\{-ik_0(r+r_1)\}}{4\pi k_0(rr_1)^{\frac{1}{2}}(\cos\mu\pi - \cos\mu\phi)} \tag{8.200}$$

so long as ϕ is not near a shadow boundary or the boundary of a reflected wave. Consequently

$$E_{zd} \sim -\frac{\mu i \exp\{-ik_0(r+r_1)\}}{2\pi k_0 (rr_1)^{\frac{1}{2}} l} \sin \mu\pi \sin \mu(\phi+\pi) \sin \mu(\phi_1+\pi) , \quad (8.201)$$

$$H_{zd} \sim -\frac{\mu i \exp\{-ik_0(r+r_1)\}}{2\pi k_0 (rr_1)^{\frac{1}{2}} l} \sin \mu\pi\{\cos \mu\pi - \cos \mu(\phi+\pi) \cos \mu(\phi_1+\pi)\} \quad (8.202)$$

where

$$l = \cos^2\mu\pi - 2 \cos \mu\pi \cos \mu(\phi+\pi) \cos \mu(\phi_1+\pi) + \frac{1}{2} \cos 2\mu(\phi+\pi) + \frac{1}{2} \cos 2\mu(\phi_1+\pi).$$

Notice the reciprocity displayed by (8.201) and (8.202), in that the formulae are unaltered by exchange of $\mathbf{x}$ and $\mathbf{x}_1$.

The diffraction field disappears if μ is a positive integer. In that case geometrical optics gives the exact solution, which can be obtained from a finite number of images in the faces of the wedge. The wedge angles where this takes place are all not less than π and cluster about 2π. For δ near 2π a small change in the angle of the wedge can eliminate diffraction. However, we shall be concerned mainly with situations where (8.200)–(8.202) are significant.

The extension to oblique incidence may be dealt with as in §8.15. Then, with the Ansatz that an incident ray produces a cone of edge rays similar in nature to that when $\delta = 0$, the field on an edge ray in the notation §8.15 is

$$\begin{pmatrix} E_\xi^e \\ E_\eta^e \end{pmatrix} = \begin{pmatrix} D_E & 0 \\ 0 & D_H \end{pmatrix} \begin{pmatrix} E_{\xi 0}^i \\ E_{\eta 0}^i \end{pmatrix} \frac{\exp(-ik_0 r \operatorname{cosec} \theta_0)}{(r \operatorname{cosec} \theta_0)^{\frac{1}{2}}} \quad (8.203)$$

where

$$D_E = \left(\frac{2}{\pi k_0}\right)^{\frac{1}{2}} \frac{\exp(-\frac{1}{4}\pi i)}{l} \mu \sin \mu\pi \sin \mu(\phi+\pi) \sin \mu\phi_0 \operatorname{cosec} \theta_0 , \quad (8.204)$$

$$D_H = \left(\frac{2}{\pi k_0}\right)^{\frac{1}{2}} \frac{\exp(-\frac{1}{4}\pi i)}{l} \mu\{\cos\mu\pi - \cos\mu(\phi+\pi)\cos\mu\phi_0\}\operatorname{cosec}\theta_0 \tag{8.205}$$

and now

$$l = \cos^2\mu\pi - 2\cos\mu\pi\cos\mu(\phi+\pi)\cos\mu\phi_0 + $$
$$+ \tfrac{1}{2}\cos 2\mu(\phi+\pi) + \tfrac{1}{2}\cos 2\mu\phi_0 .$$

The relation (8.203) reduces to (8.111) when $\delta = 0$ and $\mu = \frac{1}{2}$.

By means of (8.203)–(8.205) general edges of wedge type can be handled as in §8.16. The formulae so derived will be non-uniform because (8.203) is not uniformly valid. If a uniform expression is desired then (8.200) must be supplanted by one which copes with the passage of a point of stationary phase through a pole.

In view of the local approximation in GTD by a plane wave, it will be sufficient to consider an incident plane wave. The appropriate form for w can be ascertained by multiplying (8.199) by $(8\pi k_0 r_1)^{\frac{1}{2}}\exp(ik_0 r_1 + \frac{1}{4}\pi i)$ and letting $r_1 \to \infty$. This leads to

$$w_p(\phi) = \frac{1}{2\pi i}\int_{\infty-\pi i\mu}^{\infty+\pi i\mu} \frac{\exp\{k_0 r\cosh(u/\mu)\}}{\cosh u - \cos\mu\phi}\sinh u\, du \tag{8.206}$$

and the incident wave makes an angle ϕ_0 with the positive axis where $\phi_0 = \pi + \phi_1$. The integrand has simple poles which are on the imaginary axis at $u = i(2n\pi \pm \mu\phi)$, where n is any integer. Deform the contour of integration into the straight lines joining $\infty - \pi i\mu$, $-\pi i\mu$, $\pi i\mu$, $\infty + \pi i\mu$ with indentations round any poles encountered. Because of the integrand being odd, the only contribution from the imaginary axis comes from the poles so that

$$w_p(\phi) = \sum_n \exp\left\{ik_0 r\cos\left(\phi + \frac{2n\pi}{\mu}\right)\right\} + \frac{\mu}{2\pi i}\int_{-\infty}^{\infty}\frac{\exp(-ik_0 r\cosh t)\sinh\mu(t+\pi i)}{\cosh\mu(t+\pi i) - \cos\mu\phi}\, dt$$

the summation being over those values of n for which $-\pi < \phi + 2n\pi/\mu < \pi$. The residue series is obviously responsible for the incident and reflected waves as well as the shadow regions. When no pole is near $t = 0$

$$w_p(\phi) \sim \sum_n \exp\left\{ik_0 r \cos\left(\phi + \frac{2n\pi}{\mu}\right)\right\} + \frac{\mu}{(2\pi k_0 r)^{\frac{1}{2}}} \frac{\sin(\mu\pi)\exp(-ik_0 r - \frac{1}{4}\pi i)}{\cos\mu\pi - \cos\mu\phi} \tag{8.207}$$

the diffracted wave in which is consistent with (8.200).

When a pole is near $t = 0$ let it be at $t = t_N = i(\phi - \pi + 2N\pi/\mu)$. Write

$$f(t) = \frac{\sinh\mu(t+\pi i)}{\cosh\mu(t+\pi i) - \cos\mu\phi}$$

and put $\cosh t = 1 + \frac{1}{2}v^2$. Then

$$\frac{\mu}{2\pi i}\int_{-\infty}^{\infty} f(t)\exp(-ik_0 r\cosh t)\mathrm{d}t$$

$$= \frac{\mu}{2\pi i}\exp(-ik_0 r)\int_{-\infty}^{\infty}\exp(-\tfrac{1}{2}ik_0 r v^2)f(t)\,\frac{\mathrm{d}t}{\mathrm{d}v}\,\mathrm{d}v\ .$$

If $v = v_N$ when $t = t_N$

$$f(t)\frac{\mathrm{d}t}{\mathrm{d}v} = \frac{A}{v - v_N} + B$$

plus a series which is regular in a neighbourhood of the origin. Clearly $A = 1/\mu$ and

$$B = f(0)\left(\frac{\mathrm{d}t}{\mathrm{d}v}\right)_{v=0} + \frac{A}{v_N} = \frac{i\sin\mu\pi}{\cos\mu\pi - \cos\mu\phi} + \frac{1}{\mu v_N}\ .$$

Now, if $\mathscr{I}v_N < 0$,

$$\int_{-\infty}^{\infty} \frac{\exp(-\frac{1}{2}\mathrm{i}k_0 rv^2)}{v - v_N}\mathrm{d}v = -\mathrm{i}\int_{-\infty}^{\infty} \exp(-\tfrac{1}{2}\mathrm{i}k_0 rv^2)\int_0^{\infty} \exp\{\mathrm{i}t(v-v_N)\}\mathrm{d}t\ \mathrm{d}v$$

$$= -\mathrm{i}\left(\frac{2\pi}{k_0 r}\right)^{\frac{1}{2}}\exp(-\tfrac{1}{4}\pi\mathrm{i})\int_0^{\infty}\exp\left(-\mathrm{i}tv_N + \frac{\mathrm{i}t^2}{2k_0 r}\right)\mathrm{d}t$$

$$= 2\pi^{\frac{1}{2}}\exp(-\tfrac{1}{4}\pi\mathrm{i})F\{\mathrm{i}(\tfrac{1}{2}k_0 r)^{\frac{1}{2}}v_N\} .$$

Accordingly

$$\frac{\mu}{2\pi\mathrm{i}}\int_{-\infty}^{\infty} f(t)\exp(-\mathrm{i}k_0 r\cosh t)\mathrm{d}t \sim \frac{\mu A}{\pi^{\frac{1}{2}}\mathrm{i}}\exp(-\mathrm{i}k_0 r - \tfrac{1}{4}\pi\mathrm{i})F\{\mathrm{i}(\tfrac{1}{2}k_0 r)^{\frac{1}{2}}v_N\}$$

$$+ \frac{\mu B}{2\pi\mathrm{i}}\left(\frac{2\pi}{k_0 r}\right)^{\frac{1}{2}}\exp(-\mathrm{i}k_0 r - \tfrac{1}{4}\pi\mathrm{i})$$

when $\mathscr{I} v_N < 0$. If $\mathscr{I} v_N > 0$ it is only necessary to replace F by $-F\{-\mathrm{i}(\frac{1}{2}k_0 r)^{\frac{1}{2}}v_N\}$. Since $v_N = -2\mathrm{i}\cos\frac{1}{2}(\phi + 2N\pi/\mu)$ it follows that

$$w_p(\phi) \sim \sum_n{}' \exp\left\{\mathrm{i}k_0 r\cos\left(\phi + \frac{2n\pi}{\mu}\right)\right\} +$$

$$+ \frac{1}{\pi^{\frac{1}{2}}}\exp(-\mathrm{i}k_0 r + \tfrac{1}{4}\pi\mathrm{i})F\left\{-(2k_0 r)^{\frac{1}{2}}\cos\frac{1}{2}\left(\phi + \frac{2N\pi}{\mu}\right)\right\} +$$

$$+ \left\{\frac{\mu\sin\mu\pi}{\cos\mu\pi - \cos\mu\phi} + \frac{1}{2}\sec\frac{1}{2}\left(\phi + \frac{2N\pi}{\mu}\right)\right\}\frac{\exp(-\mathrm{i}k_0 r - \frac{1}{4}\pi\mathrm{i})}{(2\pi k_0 r)^{\frac{1}{2}}} \qquad (8.208)$$

where the prime on Σ indicates that $n = N$ is to be omitted. The formula (8.208) provides a smooth transition as $\phi + 2N\pi/\mu$ passes through π in contrast to the singular behaviour of (8.207). The transition as $\phi + 2N\pi/\mu$ passes through $-\pi$ is also given by (8.208).

Oblique incidence may be covered by the device of §8.15.

Exercise

41. A wedge of height h is placed symmetrically on the side of the splash plate of Exercise 40 facing the waveguide. Examine how this affects the radiation pattern for various values of h and a selection of wedge angles.

When the interior of the wedge plays a part as, for example, when it is dielectric little progress has been made with solving the canonical problem except in the special cases where $\delta = \pi$ (plane interface) or where the dielectric constants differ only slightly from the environment (Rawlins 1972).

8.21. The effect of curvature

The next problem to be examined is that of a curved obstacle (Fig. 8.24). The first matter to be disposed of is the rays which can be anticipated on the basis of Fermat's principle. If the medium is homogeneous the absolute minimum of the optical path length is provided by the straight line P_0P. It corresponds to the direct ray from P_0 to P and carries the primary radiation as if there were no obstacle present. There is also a reflected ray P_0OP, satisfying Snell's laws as in §8.5.

In addition, there is the stationary path P_0QRSP in which P_0Q and PS are tangential to the boundary of the obstacle whereas QRS is on the surface of the target. A ray of this type is called a *creeping ray* because of the way it sticks to

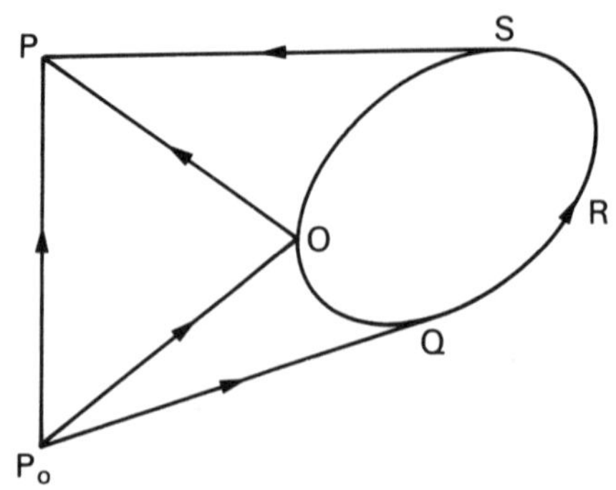

FIG. 8.24. Rays for scattering by an obstacle.

the boundary of the obstacle. Other creeping rays exist. For example, in Fig. 8.25(a) $P_0S'RQ'P$ is similar in type to that of Fig. 8.24 but going round the body in the opposite sense. In Fig. 8.25(b) $P_0QRSOQRSP$ does more than a complete circuit of the obstacle. Clearly, an infinite number of creeping rays can be constructed.

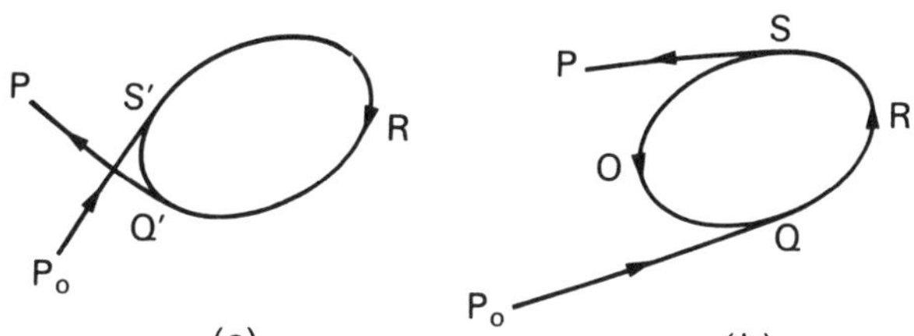

FIG. 8.25. Examples of creeping rays.

The total field at P is the sum of the contributions from the direct ray, reflected rays, and all creeping rays. If all these contributions had to be computed it would be a tremendous task but, fortunately, it turns out that most of the creeping rays can be neglected because they are heavily damped. The above picture has omitted any rays in the interior of the body; should there be interior rays there may be contributions at P from rays which have emerged from the body after refraction and internal reflection.

The canonical problem for creeping rays is the circular cylinder of radius a illuminated by a line source at (r_1, ϕ_1) or $\mathbf{x}_1$. The case of electric polarization will be discussed first and then results stated for other boundary conditions. Let the incident wave be $-\frac{1}{4}iH_0^{(2)}(k_0|\mathbf{x}-\mathbf{x}_1|)$. Now

$$H_0^{(2)}(k_0|\mathbf{x}-\mathbf{x}_1|) = \sum_{m=-\infty}^{\infty} \exp\{im(\phi-\phi_1)\}J_m(k_0r_1)H_m^{(2)}(k_0r)$$

for $r > r_1$. Hence the total electric intensity in the field is given by

$$E_z = -\tfrac{1}{4}\mathrm{i}H_0^{(2)}(k_0|\mathbf{x}-\mathbf{x}_1|) +$$

$$+\tfrac{1}{4}\mathrm{i}\sum_{m=-\infty}^{\infty}\exp\{\mathrm{i}m(\phi-\phi_1)\}\frac{J_m(k_0a)}{H_m^{(2)}(k_0a)}H_m^{(2)}(k_0r_1)H_m^{(2)}(k_0r)$$

in order to satisfy the boundary and radiation conditions. The series representation is admirable at low frequencies but as k_0a increases the convergence becomes too slow for computational convenience. Efforts have therefore been made to convert it to a more suitable form at high frequencies (White 1922, Fok 1946, 1965, Franz and Deppermann 1952, Imai 1954, Franz 1954, Friedlander 1954, Wu and Rubinow 1956, Wu 1956, Beckmann and Franz 1957, Goriainov 1958, Logan 1969).

Noting that $J_m = \frac{1}{2}(H_m^{(1)}+H_m^{(2)})$ we see that, when $r < r_1$,

$$E_z = -\tfrac{1}{8}\mathrm{i}\sum_{m=-\infty}^{\infty}\Big\{H_m^{(1)}(k_0r) -$$

$$-\frac{H_m^{(1)}(k_0a)}{H_m^{(2)}(k_0a)}H_m^{(2)}(k_0r)\Big\}H_m^{(2)}(k_0r_1)\exp\{\mathrm{i}m(\phi-\phi_1)\}\ .$$

The Poisson summation formula is a device for transforming a series rapidly convergent for low values of a parameter into one which converges well for large values of the parameter. By means of it

$$E_z = -\tfrac{1}{8}\mathrm{i}\sum_{n=-\infty}^{\infty}\int_{-\infty}^{\infty}\Big\{H_\nu^{(1)}(k_0r) -$$

$$-\frac{H_\nu^{(1)}(k_0a)}{H_\nu^{(2)}(k_0a)}H_\nu^{(2)}(k_0r)\Big\}H_\nu^{(2)}(k_0r_1)\exp\{\mathrm{i}\nu(\phi-\phi_1)-2\mathrm{i}n\nu\pi\}\mathrm{d}\nu\ . \qquad (8.209)$$

The singularities of the integrand are simple poles at the zeros of the denominator. They are located in the second and fourth quadrants. Denote those in the fourth quadrant by ν_s, i.e.

$$H^{(2)}_{\nu_s}(k_0 a) = 0 \quad . \tag{8.210}$$

The poles in the second quadrant are then at $\nu = -\nu_s$ since $H^{(2)}_{-\nu}(z) = \exp(-\nu\pi i)H^{(2)}_{\nu}(z)$. The aim is now to deform the contour over the poles. By virtue of

$$H^{(1)}_{\nu}(x) \sim -i\left(\frac{2}{\pi\nu}\right)^{\frac{1}{2}} \exp\left[\nu\left\{\ln\left(\frac{2\nu}{x}\right) - 1\right\}\right] ,$$

$$H^{(2)}_{\nu}(x) \sim i\left(\frac{2}{\nu\pi}\right)^{\frac{1}{2}} \exp\left[\nu\left\{\ln\left(\frac{2\nu}{x}\right) - 1\right\}\right] ,$$

$$J_{\nu}(x) \sim \frac{\exp(\frac{1}{4}\pi i)}{(2\pi\nu)^{\frac{1}{2}}} \exp\left[-\nu\left\{\ln\left(\frac{2\nu}{x}\right) - 1\right\}\right]$$

valid for $|\nu| \gg x$, $|\mathrm{ph}\nu| \leqslant \frac{1}{2}\pi - \delta$ $(\delta > 0)$ and $H^{(1)}_{-\nu}(z) = \exp(\nu\pi i)H^{(1)}_{\nu}(z)$ it may be verified that deformation in the lower half-plane is permissible when $\phi - \phi_1 < 2n\pi$ and in the upper half-plane when $\phi - \phi_1 > 2n\pi$. Consequently, for $0 < \phi-\phi_1 < 2\pi$,

$$E_z = \frac{1}{4}\pi \sum_{s=1}^{\infty} \frac{H^{(1)}_{\nu_s}(k_0 a)H^{(2)}_{\nu_s}(k_0 r)H^{(2)}_{\nu}(k_0 r_1)}{\tilde{H}^{(2)}_{\nu_s}(k_0 a)\{1-\exp(-2\pi i\nu_s)\}}\Big[\exp\{i\nu_s(\phi-\phi_1-2\pi)\} +$$
$$+ \exp\{-i\nu_s(\phi-\phi_1)\}\Big]$$

where $\tilde{H}^{(2)}_{\nu}(z) = \partial H^{(2)}_{\nu}(z)/\partial\nu$. Another way of writing the series comes from the Wronskian

$$H^{(1)}_{\nu}(z)H^{(2)\prime}_{\nu}(z) - H^{(1)\prime}_{\nu}(z)H^{(2)}_{\nu}(z) = -\frac{4i}{\pi z}$$

so that

$$H^{(1)}_{\nu_s}(k_0a)H^{(2)\prime}_{\nu_s}(k_0a) = -\frac{4\mathrm{i}}{\pi k_0 a} .$$

Applying the recurrence relation

$$zH^{(2)\prime}_{\nu}(z) - H^{(2)}_{\nu}(z) = -zH^{(2)}_{\nu+1}(z)$$

we derive

$$H^{(1)}_{\nu_s}(k_0a) = \frac{4\mathrm{i}}{\pi k_0 a H^{(2)}_{\nu_s+1}(k_0a)} .$$

Also, if B_ν and C_ν are any pair of Bessel functions of order ν,

$$\int^{x} B_\nu(kr)C_\nu(kr)\,\frac{dr}{r} = \frac{kx}{2\nu}\{B_{\nu+1}(kx)\tilde{C}_\nu(kx)-B_\nu(kx)\tilde{C}_{\nu+1}(kx)\} +$$

$$+ \frac{B_\nu(kx)C_\nu(kx)}{2\nu} .$$

Hence

$$N_s^2 = \nu_s\int_a^{\infty}\{H^{(2)}_{\nu_s}(k_0t)\}^2\,\frac{dt}{t} = -\tfrac{1}{2}k_0aH^{(2)}_{\nu_s+1}(k_0a)\tilde{H}^{(2)}_{\nu_s}(k_0a) .$$

Accordingly,

$$E_z = -\tfrac{1}{2}\mathrm{i}\sum_{s=1}^{\infty}\frac{H^{(2)}_{\nu_s}(k_0r)H^{(2)}_{\nu_s}(k_0r_1)}{\{1-\exp(-2\pi\mathrm{i}\nu_s)\}N_s^2}\Big[\exp\{\mathrm{i}\nu_s(\phi-\phi_1-2\pi)\} +$$

$$+ \exp\{-\mathrm{i}\nu_s(\phi-\phi_1)\}\Big] . \qquad (8.211)$$

Eqn (8.211) was derived for $r < r_1$. When $r > r_1$ the only alteration to (8.209) is to interchange r and r_1; therefore, the same applies to (8.211). Hence (8.211) holds without restriction on r so long as $0 < \phi - \phi_1 < 2\pi$.

Since $k_0 a$ is large the standard asymptotic expansion of a Hankel function of large argument shows that $H_\nu^{(2)} = 0$ cannot have any solutions for small and moderate values of $|\nu|$. Thus $|\nu_s|$ must be at least of the same order of magnitude as $k_0 a$. Now, a uniformly valid asymptotic expression for a Hankel function of large argument and order is

$$H_\nu^{(2)}(x) \sim 2^{3/2}\,\nu^{-1/3}\exp(\tfrac{1}{3}\pi \mathrm{i})\left\{\frac{\xi}{(x/\nu)^2-1}\right\}^{\frac{1}{4}}\mathrm{Ai}\{\nu^{2/3}\xi\,\exp(\tfrac{1}{3}\pi \mathrm{i})\} \tag{8.212}$$

where Ai is the Airy function and

$$\tfrac{2}{3}\xi^{3/2} = \left\{\left(\frac{x}{\nu}\right)^2 - 1\right\}^{\frac{1}{2}} - \sec^{-1}\left(\frac{x}{\nu}\right).$$

If x/ν is near unity $\xi \approx 2^{1/3}(1-\nu/x)$ and

$$H_\nu^{(2)}(x) \sim 2^{4/3}\nu^{-1/3}\,\exp(\tfrac{1}{3}\pi \mathrm{i})\mathrm{Ai}(y) \tag{8.213}$$

where $y = 2^{1/3}\exp(\frac{1}{3}\pi \mathrm{i})x^{-1/3}(x-\nu)$. Consequently, when ν_s is near $k_0 a$ it is related to a zero of the Airy function. It is well known that $\mathrm{Ai}(\alpha_s e^{\pi \mathrm{i}}) = 0$ where α_s is real and positive; the first few are

$$\left.\begin{array}{lll} \alpha_1 = 2.338, & \alpha_2 = 4.088, & \alpha_3 = 5.520\,, \\ \alpha_4 = 6.787, & \alpha_5 = 7.944, & \alpha_6 = 9.023\,. \end{array}\right\} \tag{8.214}$$

Hence the poles near $k_0 a$ are given by

$$\nu_s = k_0 a + (k_0 a)^{1/3} 2^{-1/3}\,\exp(-\tfrac{1}{3}\pi \mathrm{i})\alpha_s\,. \tag{8.215}$$

As s increases and ν deviates appreciably from $k_0 a$ this formula fails and (8.212) shows that the poles approach $-\mathrm{i}\infty$.

It is evident from (8.215) that the first term in the [ ] of (8.211) becomes more exponentially damped as ϕ decreases, whereas the second term does so when ϕ increases. Thus the first term corresponds to a creeping wave travelling in the clockwise direction while the second term is an anticlockwise creeping wave. In view of the size of the imaginary part of ν_s the factor $1 - \exp(-2\pi i\nu_s)$ in (8.211) can be replaced by unity without significant loss.

When $x - \nu \gg x^{1/3}$, the asymptotic formula

$$\mathrm{Ai}(z) \sim \frac{\exp(-\frac{2}{3}z^{3/2})}{2\pi^{\frac{1}{2}}z^{\frac{1}{4}}} \qquad (|\mathrm{ph}\ z| < \pi) \tag{8.216}$$

converts (8.212) to

$$H_\nu^{(2)}(x) \sim \left(\frac{2}{\pi}\right)^{\frac{1}{2}}(x^2-\nu^2)^{-\frac{1}{4}} \exp i\left\{\frac{1}{4}\pi-(x^2-\nu^2)^{\frac{1}{2}} + \nu \sec^{-1}\left(\frac{x}{\nu}\right)\right\} .$$

Hence, if k_0r/ν_s is bounded away from 1,

$$E_z \sim -\frac{1}{2}i\left(\frac{2}{\pi}\right)^{\frac{1}{2}}\exp(\tfrac{1}{4}\pi i) \sum_{s=1}^{\infty} \frac{H_{\nu_s}^{(2)}(k_0r_1)}{N_s^2(k_0^2r^2-\nu_s^2)^{\frac{1}{4}}} \exp\{i\nu_s \sec^{-1}\frac{k_0r}{\nu_s} - i(k_0^2r^2-\nu_s^2)^{\frac{1}{2}}\} \times$$

$$\times\ [\exp\{i\nu_s(\phi-\phi_1-2\pi)\} + \exp\{-i\nu_s(\phi-\phi_1)\}] . \tag{8.217}$$

For a first approximation (8.215) indicates that ν_s is k_0a. If also r_1 is not too near to a

$$E_z \sim \frac{1}{\pi}\sum_s \frac{\exp\{-ik_0(r_0+r')\}}{N_s^2k_0(r_0r')^{\frac{1}{2}}}\left\{\exp\left(-\frac{i\nu_sr''}{a}\right) + \exp\left(-\frac{i\nu_sr'''}{a}\right)\right\} \tag{8.218}$$

where

$$r_0 = (r_1^2-a^2)^{\frac{1}{2}}, \qquad r' = (r^2-a^2)^{\frac{1}{2}} ,$$

$$\frac{r''}{a} = \phi - \phi_1 - \sec^{-1}\left(\frac{r}{a}\right) - \sec^{-1}\left(\frac{r_1}{a}\right) ,$$

$$\frac{r'''}{a} = 2\pi - \phi + \phi_1 - \sec^{-1}\left(\frac{r}{a}\right) - \sec^{-1}\left(\frac{r_1}{a}\right) .$$

This result may be interpreted in the following way. The upper incident ray tangent to the circle (Fig. 8.26) hits it at the point of *glancing incidence* or *upper penumbral* point and then initiates a creeping ray in the anticlockwise direction. The creeping ray travels along an arc of length r'' and then covers a distance r' to the point of observation along the tangent at the other end of the arc. In other words, the incident wave at the penumbral point is multiplied by

$$\frac{4 \exp(\frac{1}{4}\pi i - i\nu_s r''/a)}{(2\pi k_0)^{\frac{1}{2}} N_s^2} \frac{\exp(-ik_0 r')}{r'^{\frac{1}{2}}}$$

in order to produce the cylindrical wave at the point of observation. The exponential decay associated with the arc of boundary traversed is due to the steady shedding of energy in a tangential direction as the creeping ray goes round. Via (8.215) and (8.213) we may derive the approximation

$$N_s^2 = -2^{7/3} \exp(-\tfrac{2}{3}\pi i)(k_0 a)^{-1/3}\{\mathrm{Ai}'(-\alpha_s)\}^2 . \qquad (8.219)$$

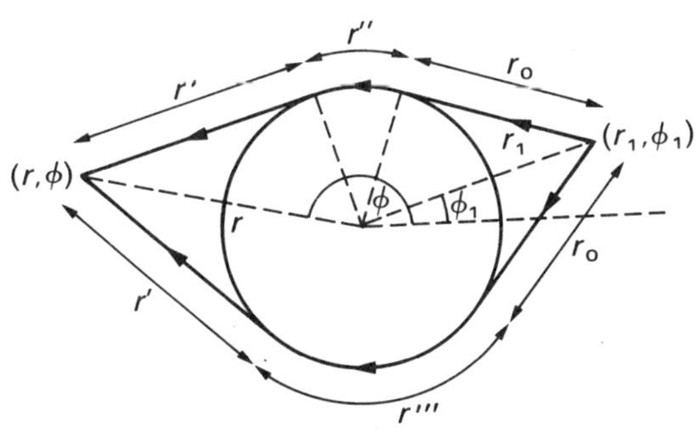

FIG. 8.26. The ray-theory lengths for a circular cylinder.

There is a similar creeping ray from the lower penumbral point and it is launched with a diffraction coefficient calculated by the same rules. If the source is on the boundary put $r_1 = a$ in (8.217) and a picture similar to that of Fig. 8.26 emerges.

Creeping waves which have made a complete circuit of the boundary are multiplied by the factor $\exp(-2\pi i \nu_s)$ and so are rarely large enough to be of any significance. Therefore, the first one or two terms of (8.217) or (8.218) should be sufficient for a good approximation provided that the arc traversed on the cylinder is non-zero. This means that it is in the *deep shadow* of Fig. 8.27 that a term or two will suffice. As the diffracted field is exponentially small, *the shadow behind a large obstacle is darker when the boundary is curved than when it has a straight edge*. In the penumbra or region of deep illumination the residue series (8.211) converges so slowly that it is not helpful. The dominant terms in the deep illumination will be the direct and reflected rays, though it is quite difficult to prove rigorously (Ursell, 1957, 1968, Babich 1962, Buslaev 1964, Grimshaw 1966, Brown 1966, Ludwig 1967, Morawetz and Ludwig 1968, Leppington 1968, Bloom and Matkowsky 1969). The penumbral region needs a formula which makes the

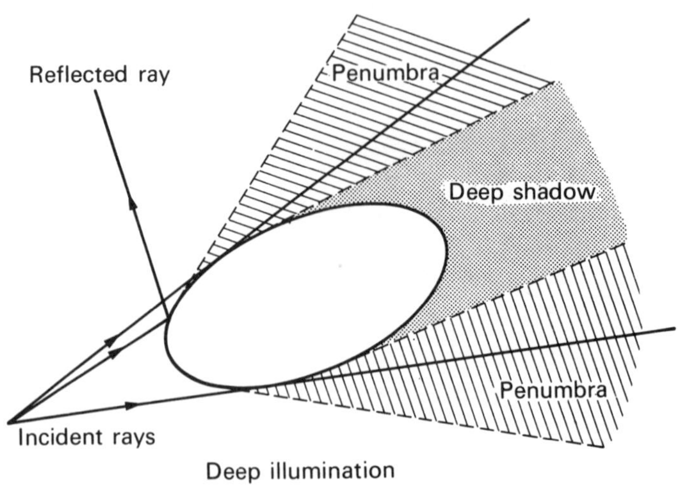

FIG. 8.27. Regions associated with creeping rays.

transition between deep shadow and deep illumination. The penumbra occupies the space limited by tangents drawn at arc lengths of approximately $a(k_0a)^{-1/3}$ from the penumbral point and is shown shaded in Fig. 8.27.

An alternative representation is derived from (8.209). Attention will first be concentrated on the range $\nu \geqslant 0$. Replacing $H_\nu^{(1)}$ by J_ν, as is legitimate, we see from (8.212), (8.216), and

$$J_\nu(x) \sim 2^{\frac{1}{2}}\nu^{-\frac{1}{3}}\left\{\frac{\xi}{(x/\nu)^2-1}\right\}^{\frac{1}{4}}\mathrm{Ai}(\nu^{\frac{2}{3}}\xi e^{\pi i}) \tag{8.220}$$

that the integrand diminishes exponentially for $\nu > k_0r$ (note that (8.209) holds for $r < r_1$). Therefore, the integral from k_0r to infinity is discarded for a first approximation. Also only dominant terms will be retained in the following; a rigorous treatment (Jones 1963) demonstrates that this is justified. Both r and r_1 will be kept bounded away from a.

Between 0 and k_0r, (8.212) and (8.216) may be adopted for $H_\nu^{(2)}(k_0r_1)$. If we do likewise for $H_\nu^{(1)}(k_0r)$ we obtain for the first part of (8.209)

$$-\frac{i}{4\pi}\int_0^{k_0r}\frac{\exp i\{(k_0^2r^2-\nu^2)^{\frac{1}{2}}-\nu\sec^{-1}(k_0r/\nu)-(k_0^2r_1^2-\nu^2)^{\frac{1}{2}}+\nu\sec^{-1}(k_0r_1/\nu)+\nu\alpha\}}{(k_0^2r^2-\nu^2)^{\frac{1}{4}}(k_0^2r_1^2-\nu^2)^{\frac{1}{4}}}\,d\nu$$

where $\alpha = \phi-\phi_1 - 2n\pi$. The exponent has a point of stationary phase where

$$\cos^{-1}\frac{\nu}{k_0r_1} - \cos^{-1}\frac{\nu}{k_0r} = -\alpha$$

provided that α is negative and greater than $-\frac{1}{2}\pi$; otherwise there is no stationary point. Since the left-hand side is an increasing function of ν there is no solution unless $r_1\cos\alpha > r$.

The contribution of this stationary point is

$$- \frac{i \exp(-ik_0|\mathbf{x}-\mathbf{x}_1|+\frac{1}{4}\pi i)}{(8\pi k_0|\mathbf{x}-\mathbf{x}_1|)^{\frac{1}{2}}} .$$

It therefore reproduces the incident field. Since the stationary value of ν/k_0 may be interpreted as the perpendicular distance from the centre of the circle to the incident ray, the point of observation must lie in the region of deep illumination. In particular, if $\nu < k_0 a$, it corresponds to a point on a direct ray which will strike the cylinder but before the point of impact.

If we do exactly the same with the second part of (8.209) we obtain

$$\frac{i}{4\pi} \int \exp i\{-(k_0^2r^2-\nu^2)^{\frac{1}{2}}+\nu \sec^{-1}(k_0r/\nu)-(k_0^2r_1^2-\nu^2)^{\frac{1}{2}}+\nu \sec^{-1}(k_0r_1/\nu) + $$
$$+ 2(k_0^2a^2-\nu^2)^{\frac{1}{2}}-2\nu \sec^{-1}(k_0a/\nu)+\nu\alpha\}/(k_0^2r^2-\nu^2)^{\frac{1}{4}}(k_0^2r_1^2-\nu^2)^{\frac{1}{4}}\, d\nu .$$

In this case the stationary point satisfies

$$\cos^{-1}\left(\frac{\nu}{k_0r_1}\right) + \cos^{-1}\left(\frac{\nu}{k_0r}\right) - 2\cos^{-1}\left(\frac{\nu}{k_0a}\right) = -\alpha. \tag{8.221}$$

It can exist only if $\nu < k_0a$ and $\alpha < 0$. It is responsible for a reflected ray which is in harmony with Snell's laws. However, the asymptotic behaviour of the Hankel functions of argument k_0a alters as ν passes through k_0a and for $\nu > k_0a$ the appropriate integral is

$$\frac{1}{4\pi} \int \frac{\exp i\{\nu \sec^{-1}(k_0r/\nu)-(k_0^2r^2-\nu^2)^{\frac{1}{2}}+\nu \sec^{-1}(k_0r_1/\nu)-(k_0^2r_1^2-\nu^2)^{\frac{1}{2}}+\nu\alpha\}}{(k_0^2r^2-\nu^2)^{\frac{1}{4}}(k_0^2r_1^2-\nu^2)^{\frac{1}{4}}}\, d\nu$$

The equation for the stationary point is now

$$\cos^{-1}\left(\frac{\nu}{k_0r_1}\right) + \cos^{-1}\left(\frac{\nu}{k_0r}\right) = -\alpha . \tag{8.222}$$

For its existence it is necessary that $-\pi < \alpha < 0$ and $r_1 \cos\alpha < r$. This time the incident wave is provided in that part in $\pi < \phi - \phi_1 < 2\pi$ of the region of deep illumination not already dealt with.

The field of geometrical optics has therefore been verified in the deep illumination of Fig. 8.27. It does not extend to the penumbra because both stationary points in (8.221) and (8.222) are at $\nu = k_0 a$ on the shadow boundary. But, when $\nu = k_0 a$ it is no longer legitimate to replace the Airy functions of the Hankel functions of argument $k_0 a$ by their asymptotic formulae. Accordingly, it is necessary to reconsider the integral in this neighbourhood.

In the penumbra the main contribution comes from near $\nu = k_0 a$ and so now (8.213) is relevant. Thus

$$-\frac{\exp(-\frac{2}{3}\pi i)}{4\pi}\int \frac{\mathrm{Ai}\{2^{\frac{1}{3}}\exp(-\frac{1}{3}\pi i)(k_0 a-\nu)/(k_0 a)^{\frac{1}{3}}\}}{\mathrm{Ai}\{2^{\frac{1}{3}}\exp(\frac{1}{3}\pi i)(k_0 a-\nu)/(k_0 a)^{\frac{1}{3}}\}} \times$$

$$\times \frac{\exp i\{\nu\sec^{-1}(k_0 r/\nu)-(k_0^2 r^2-\nu^2)^{\frac{1}{2}}+\nu\sec^{-1}(k_0 r_1/\nu)-(k_0^2 r_1^2-\nu^2)^{\frac{1}{2}}+\nu\alpha\}}{(k_0^2 r^2-\nu^2)^{\frac{1}{4}}(k_0 r_1^2-\nu^2)^{\frac{1}{4}}}\,d\nu$$

has to be discussed. Since ν is near $k_0 a$ we may expand the exponent as far as quadratic powers and ignore the variation of ν in the denominator. After the substitution $\nu = k_0 a+(k_0 a)^{\frac{1}{3}}\beta/2^{\frac{1}{3}}$ the result is

$$-\frac{\exp(-\frac{2}{3}\pi i)}{4\pi}\,\frac{(\frac{1}{2}k_0 a)^{\frac{1}{3}}\exp(-ik_0 L)}{k_0(r^2-a^2)^{\frac{1}{4}}(r_1^2-a^2)^{\frac{1}{4}}}\int\frac{\mathrm{Ai}(-\beta\exp(-\frac{1}{3}\pi i))}{\mathrm{Ai}(-\beta\exp(\frac{1}{3}\pi i))}\times$$

$$\times\exp\left\{-i(\tfrac{1}{2}k_0 a)^{\frac{1}{3}}\frac{L_b\beta}{a}-\tfrac{1}{2}i(\tfrac{1}{2}k_0 a)^{\frac{2}{3}}\frac{L_1\beta^2}{k_0}\right\}d\beta$$

where

$$L = (r_1^2-a^2)^{\frac{1}{2}} + (r^2-a^2)^{\frac{1}{2}} - a\cos^{-1}\frac{a}{r} - a\cos^{-1}\frac{a}{r_1} - a\alpha ,$$

$$-\frac{L_b}{a} = \cos^{-1}\frac{a}{r} + \cos^{-1}\frac{a}{r_1} + \alpha ,$$

$$L_1 = (r^2-a^2)^{-\frac{1}{2}} + (r_1^2-a^2)^{-\frac{1}{2}} .$$

As with the usual asymptotic approach we wish to make the limits of integration $\pm\infty$ and drop the quadratic terms as far as possible. Lack of convergence at ∞ prevents the immediate implementation of this process. So, for $\beta > 0$, take advantage of the fact that

$$\mathrm{Ai}(z\exp(-\tfrac{1}{3}\pi\mathrm{i})) = \exp(\tfrac{1}{3}\pi\mathrm{i})\mathrm{Ai}(z\exp(\pi\mathrm{i})) - \exp(\tfrac{2}{3}\pi\mathrm{i})\mathrm{Ai}(z\exp(\tfrac{1}{3}\pi\mathrm{i})). \tag{8.223}$$

The first term causes no trouble at infinity and the second may be dealt with separately. Hence, we obtain

$$E_z \sim \frac{\mathrm{i}(\frac{1}{2}k_0a)^{\frac{1}{3}}\exp(-\mathrm{i}k_0L)}{2\pi^{\frac{1}{2}}k_0(r^2-a^2)^{\frac{1}{4}}(r_1^2-a^2)^{\frac{1}{4}}}\, p\left\{\tfrac{1}{2}k_0L_b(\tfrac{1}{2}k_0a)^{-\frac{2}{3}}\right\}+$$

$$+\frac{\exp(-\mathrm{i}k_0L)F\{k_0^{\frac{1}{2}}L_b/a(2L_1)^{\frac{1}{2}}\}}{2^{3/2}\pi k_0^{\frac{1}{2}}(r^2-a^2)^{\frac{1}{4}}(r_1^2-a^2)^{\frac{1}{4}}L_1^{\frac{1}{2}}} \tag{8.224}$$

where

$$p(\tau) = \frac{\exp(\frac{1}{6}\pi\mathrm{i})}{2\pi^{\frac{1}{2}}}\int_0^\infty \frac{\mathrm{Ai}(\mu)\exp(-\mathrm{i}\mu\tau)}{\mathrm{Ai}(\mu\exp(-\frac{2}{3}\pi\mathrm{i}))}\,\mathrm{d}\mu +$$

$$+\frac{\exp(-\frac{1}{6}\pi\mathrm{i})}{2\pi^{\frac{1}{2}}}\int_{-\infty}^0 \frac{\mathrm{Ai}(-\mu\exp(-\frac{1}{3}\pi\mathrm{i}))}{\mathrm{Ai}(-\mu\exp(\frac{1}{3}\pi\mathrm{i}))}\exp(-\mathrm{i}\mu\tau)\,\mathrm{d}\mu . \tag{8.225}$$

So far only that part of (8.209) has been examined in which $\nu \geqslant 0$. A contribution has been found when $0 \leqslant \phi-\phi_1 \leqslant 2\pi$ only if $n = 1$ and it represents the effect of the lower penumbral point in Fig. 8.26 by putting $\alpha = \phi-\phi_1-2\pi$. For $\nu \leqslant 0$, change the sign of ν in the integrand; the sole effect is to reverse the sign in front of α. There is now a contribution only when $n = 0$ and it is given by the foregoing with $\alpha = -(\phi-\phi_1)$. Therefore (8.224) gives the field in the upper and lower penumbra with a correct choice of α in the two domains.

The formula (8.224) will now be investigated a little more closely, sticking to the upper penumbral zone in which $\alpha = -(\phi-\phi_1)$. In the second integral of (8.225) deform the contour into the radius vector through $\exp(-\frac{2}{3}\pi i)$. For $\tau > 0$ use (8.223) in the numerator and then

$$p(\tau) = \frac{\exp(\frac{1}{6}\pi i)}{2\pi^{\frac{1}{2}}} \int_{\infty \exp(-\frac{2}{3}\pi i)}^{\infty} \frac{\mathrm{Ai}(\mu)\exp(-i\mu\tau)}{\mathrm{Ai}(\mu \exp(-\frac{2}{3}\pi i))}\, d\mu + \frac{1}{2\pi^{\frac{1}{2}}\tau} \quad (\tau > 0). \tag{8.226}$$

The contour can now be deformed over the poles of the integrand to give

$$p(\tau) = \frac{1}{2\pi^{\frac{1}{2}}\tau} - \sum_{s=1} \frac{\exp(\frac{1}{6}\pi i - i\alpha_s \tau \exp(-\frac{1}{3}\pi i))}{2\pi^{\frac{1}{2}}\{\mathrm{Ai}'(-\alpha_s)\}^2} \quad (\tau > 0). \tag{8.227}$$

At the shadow boundary $L_b = 0$ and L_b becomes positive as one moves towards the deep shadow. Substitute (8.227) in (8.224) and then if L_b is sufficiently positive for (8.97) to be applicable (8.218) is recovered for the upper penumbral point when account is taken of (8.215) and (8.219). Thus (8.224) supplies a continuous transition from the shadow boundary to the deep shadow. Moreover, since $p(0)$ is finite from (8.225), (8.227) shows that (8.218) blows up like $1/L_b$ as the shadow boundary is approached.

When $\tau < 0$, deform the contour of integration of the first integral of (8.225) into the radius vector through $\exp(\frac{1}{3}\pi i)$. Reverse the sign of μ and employ (8.223). The consequence is that

$$p(\tau) = \frac{\exp(-\frac{1}{6}\pi i)}{2\pi^{\frac{1}{2}}} \int_{\infty \exp(-\frac{2}{3}\pi i)}^{\infty} \frac{\mathrm{Ai}(\mu \exp(-\frac{1}{3}\pi i))\exp(i\mu\tau)}{\mathrm{Ai}(\mu \exp(\frac{1}{3}\pi i))} d\mu + \frac{1}{2\pi^{\frac{1}{2}}\tau}$$

$$(\tau < 0) . \tag{8.228}$$

For $|\mu| \gg 1$, the asymptotic formulae can be substituted for the Airy function and a point of stationary phase appears. In fact, it can be shown that

$$\begin{aligned} p(\tau) &\sim \frac{1}{2\pi^{\frac{1}{2}}} \int_{-\infty}^{\infty} \exp(\tfrac{4}{3}i\mu^{3/2}+i\mu\tau)d\mu + \frac{1}{2\pi^{\frac{1}{2}}\tau} \\ &\sim \tfrac{1}{2}(-\tau)^{\frac{1}{2}} \exp(\tfrac{1}{4}\pi i + \tfrac{1}{12}i\tau^3) + \frac{1}{2\pi^{\frac{1}{2}}\tau} \end{aligned} \tag{8.229}$$

as $\tau \to -\infty$. If L_b is small but large enough in magnitude for (8.229), (8.96), and (8.97) to be valid we regain from (8.224) the incident and reflected waves on observing that

$$\begin{aligned} L^i &= L - \frac{1}{2}\frac{L_b^2}{a^2 L_1} , \\ L^r &= L - \frac{1}{24}\frac{L_b^3}{a^2} \end{aligned} \tag{8.230}$$

in the vicinity of the shadow boundary. However, as L_b increases further in magnitude the matching is unsuccessful because of the miscarriage of the approximations for L^i and L^r.

The correct way to use the transition formula (8.224) therefore is as follows. Start at the shadow boundary $L_b = 0$. As L_b becomes more negative a situation will be reached where the incident and reflected fields are reproduced. As soon as that occurs use the true incident and reflected waves for any larger deviation from the shadow boundary. On the other hand, for positive L_b, the residue series should be used from the point where it first becomes applicable. Values on the shadow boundary require $p(0)$ which is given by

$$p(0) = 0.354 \exp(\tfrac{1}{6}\pi i) . \tag{8.231}$$

For later purposes it is worth remarking the following identification

$$L = L_1^i + L^c + L_b , \tag{8.232}$$

$$L_1 = 1/L_1^i + 1/L^c \tag{8.233}$$

where L_1^i is the length of the incident ray to the penumbral point P_1 and L^c is the length of a ray from the point of observation to a point of tangency. L_b is the distance along the boundary between the two points of contact, being counted negative for observation above the shadow boundary as in Fig. 8.28 and positive for the points on the shadow side.

The method has been described for the case of electric polarization but it works equally well for an impedance boundary

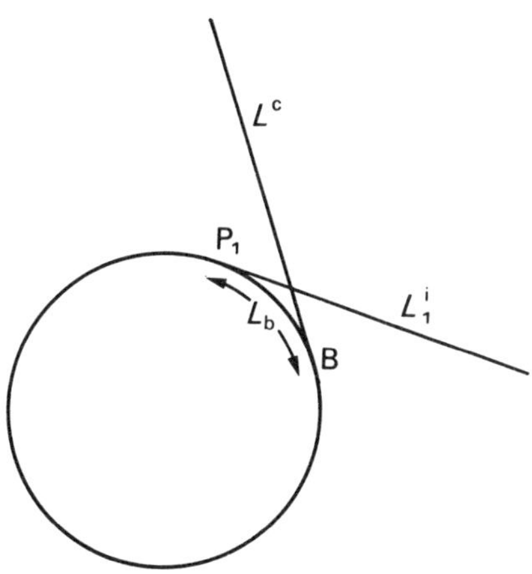

FIG. 8.28. Parameters for points on the illuminated side.

condition

$$\frac{\partial E_z}{\partial n} + k_0 Z(\tfrac{1}{2}k_0 a)^{-\frac{1}{3}} E_z = 0 \tag{8.234}$$

the factor $(\frac{1}{2}k_0 a)^{-\frac{1}{3}}$ being inserted to simplify some later formulae. The expression (8.211) still has veracity with the understanding that

$$N_s^2 = \nu_s \int_a^\infty \frac{\{H_{\nu_s}^{(2)}(k_0 t)\}^2 dt}{t} \tag{8.235}$$

with ν_s a solution of

$$H_{\nu_s}^{(2)\prime}(k_0 a) + Z(\tfrac{1}{2}k_0 a)^{-\frac{1}{3}} H_{\nu_s}^{(2)}(k_0 a) = 0 \tag{8.236}$$

instead of (8.210). The early solutions are given by (8.215) with $\alpha_s e^{\pi i}$ replaced by δ_s where

$$\mathrm{Ai}'(\delta_s) + Z\exp(-\tfrac{1}{3}\pi i)\mathrm{Ai}(\delta_s) = 0.$$

With this change (8.218) can still be used but now

$$N_s^2 = -\frac{2^{7/3}\exp(-\frac{2}{3}\pi i)(k_0 a)^{-1/3}\{\mathrm{Ai}'(\delta_s)\}^2\{Z^2 - \delta_s \exp(\frac{2}{3}\pi i)\}}{Z^2} \tag{8.237}$$

instead of (8.219). If $Z = 0$ exchange $Z/\mathrm{Ai}'(\delta_s)$ and $-\exp(\frac{1}{3}\pi i)/\mathrm{Ai}(\delta_s)$. The transition formula (8.224) can also be used if $p_Z(\tau)$ is put in place of $p(\tau)$ where

$$p_Z(\tau) = \frac{\exp(\frac{1}{6}\pi i)}{2\pi^{\frac{1}{2}}} \int_0^\infty \frac{Z\,\mathrm{Ai}(\mu) - \mathrm{Ai}'(\mu)}{\exp(\frac{1}{3}\pi i)\mathrm{Ai}'(\mu\exp(-\frac{2}{3}\pi i)) + Z\mathrm{Ai}(\mu\exp(-\frac{2}{3}\pi i))} \exp(-i\mu\tau)d\mu \tag{8.238}$$

$$+ \frac{\exp(-\frac{1}{6}\pi i)}{2\pi^{\frac{1}{2}}} \int_{-\infty}^0 \frac{\exp(-\frac{1}{3}\pi i)\mathrm{Ai}'(-\mu\exp(-\frac{1}{3}\pi i)) + Z\mathrm{Ai}(-\mu\exp(-\frac{1}{3}\pi i))}{\exp(\frac{1}{3}\pi i)\mathrm{Ai}'(-\mu\exp(\frac{1}{3}\pi i)) + Z\mathrm{Ai}(-\mu\exp(\frac{1}{3}\pi i))} \exp(-i\mu\tau)d\mu\ .$$

Observe that $p_\infty(\tau) = p(\tau)$. The formulae analogous to (8.227) and (8.229) are

$$p_Z(\tau) = \frac{1}{2\pi^{\frac{1}{2}}\tau} + \sum_{s=1} \frac{Z^2\exp\{\frac{1}{6}\pi i + i\delta_s\tau \ \ \exp(-\frac{1}{3}\pi i)\}}{\{\delta_s \ \exp(\frac{2}{3}\pi i) - Z^2\}2\pi^{\frac{1}{2}}\{Ai'(\delta_s)\}^2} \qquad (\tau > 0) \tag{8.239}$$

$$\sim \frac{1}{2}(-\tau)^{\frac{1}{2}} \frac{Z - \frac{1}{2}i\tau}{Z + \frac{1}{2}i\tau} \exp(\frac{1}{4}\pi i + \frac{1}{12}i\tau 3) + \frac{1}{2\pi^{\frac{1}{2}}\tau} \qquad (\tau \to -\infty) . \tag{8.240}$$

Some values of p_0 and p_∞ are shown in Table 8.1; in particular,

$$p_0(0) = -0.308 \exp(\frac{1}{6}\pi i) . \tag{8.241}$$

Values of $p_Z(0)$ when $Z \exp(\frac{1}{4}\pi i)$ is a real constant are also available (Wait and Conda 1959).

TABLE 8.1

	$p_\infty(\tau)$		$p_0(\tau)$	
τ	Mod	Phase (deg)	Mod	Phase (deg)
-3	0.883	-93.5	0.891	104.3
-2	0.623	- 1.8	0.790	192.4
-1.5	0.549	+21.6	0.688	211.3
-1.0	0.484	32.5	0.560	219.5
-0.5	0.419	34.2	0.429	218.6
0	0.354	30.0	0.308	210.0
0.5	0.290	22.9	0.204	193.7
1.0	0.232	15.2	0.129	167.8
1.5	0.183	8.48	0.0894	130.6
2.0	0.144	+ 3.79	0.0793	92.0
3.0	0.0953	+ 0.12	0.0812	45.1
4.0	0.0705	- 0.12	0.0745	22.3
5.0	0.0564	- 0.02	0.0623	9.71

The field on the boundary may also be of interest. The preceding results then need some modification because $r = a$, but the treatment follows the same lines. It is found that near the penumbral point

$$E_z = -\frac{\mathrm{i}\ \exp(\frac{1}{4}\pi\mathrm{i}-\mathrm{i}k_0L)}{(8\pi k_0)^{\frac{1}{2}}(r_1^2-a^2)^{\frac{1}{4}}}\, q_Z\{\tfrac{1}{2}k_0L_\mathrm{b}(\tfrac{1}{2}k_0a)^{-\frac{2}{3}}\}, \tag{8.242}$$

$$\frac{\partial E_z}{\partial n} = \mathrm{i}k_0\left(\frac{2}{k_0a}\right)^{\frac{1}{3}} Z\, \frac{\exp(\frac{1}{4}\pi\mathrm{i}-\mathrm{i}k_0L)}{(8\pi k_0)^{\frac{1}{2}}(r_1^2-a^2)^{\frac{1}{4}}}\, q_Z\{\tfrac{1}{2}k_0L_\mathrm{b}(\tfrac{1}{2}k_0a)^{-\frac{2}{3}}\} \tag{8.243}$$

where

$$q_Z(\tau) = \frac{\exp(-\frac{1}{3}\pi\mathrm{i})}{2\pi\mathrm{i}}\int_{-\infty}^{\infty} \frac{\exp(-\mathrm{i}\mu\tau)}{\exp(\frac{1}{3}\pi\mathrm{i})\mathrm{Ai}'(\mu\ \exp(-\frac{2}{3}\pi\mathrm{i}))+Z\mathrm{Ai}(\mu\ \exp(-\frac{2}{3}\pi\mathrm{i}))}\mathrm{d}\mu. \tag{8.244}$$

Moving away from the penumbral point soon leads to geometrical optics or a residue series in the shadow because

$$q_Z(\tau) = \sum_{s=1} \frac{\exp\{\mathrm{i}\delta_s\tau\ \exp(-\frac{1}{3}\pi\mathrm{i})\}}{(Z^2\mathrm{e}^{-\frac{2}{3}\pi\mathrm{i}}-\delta_s)\mathrm{Ai}(\delta_s)} \qquad (\tau > 0) \tag{8.245}$$

$$\sim \frac{2\tau\ \exp(\frac{1}{3}\mathrm{i}\tau^3)}{\tau - \mathrm{i}Z} \qquad (\tau \to -\infty)\ . \tag{8.246}$$

Therefore (8.242) and (8.243) may be employed in the same manner as (8.224). A selection of values for q_Z appears in Table 8.2. Additional computations for various values of Z including complex values have also been made (Rice 1954, Fok 1946, 1965, Wait and Conda 1958).

By reciprocity these results will also predict the field off the cylinder due to a line source on the cylinder.

Although the circular cylinder is the easiest to analyse theoretically, it is not completely suitable as a canonical problem because of its invariable radius of curvature. A parabolic cylinder is better from this point of view; it has been studied (Jones 1964, Lowdon 1970) but the detail will not be presented here. There is concord between the conclusions and the approach to be delineated in the next section.

TABLE 8.2

	$Zq_Z(\tau)$ $(Z = \infty)$		$q_0(\tau)$	
τ	Modulus	Phase (deg)	Modulus	Phase (deg)
-1.0	2.16	-115.9	1.861	- 15.43
-0.5	1.38	-106.8	1.682	+ 1.52
0.0	0.77	-120	1.399	0
0.3	0.515	-134.8	1.197	- 6.06
0.6	0.327	-152.9	0.991	- 14.23
1.0	0.167	-180.1	0.738	- 26.63
1.5	0.066	-215.9	0.488	- 42.57
2.0	0.025	-251.6	0.315	- 57.98
3.0	0.0033	-320.7	0.130	- 87.57
4.0	0.00043	-358.0	0.054	-116.75

8.22. Generalization

The analysis of the preceding section has shown that (8.211) is valid for a variety of boundary conditions (8.234) with a pertinent choice of ν_s. The aim is now to broaden its range of validity (Ludwig 1975). First suppose that Z is a function of ϕ which varies continuously but only slightly within a wavelength. Persist in making ν_s satisfy (8.236); ν_s will then be a function of ϕ which may be indicated by writing $\nu_s(\phi)$. Evidently, $\nu_s(\phi)$ will be large but it will be assumed, on account of (8.215) that its derivative with respect to ϕ is of a lower order of magnitude. Now consider

$$E_z = -\frac{1}{2}i\sum_{s=1}^{\infty}\frac{H_{\nu_s}^{(2)}(k_0r)}{N_s^2}H_{\nu_s}^{(2)}(k_0r_1)\exp\{-iP_s(\phi)\} \qquad (8.247)$$

as a candidate for the contribution from an upper penumbral point, the more general P_s superseding the $\nu_s(\phi-\phi_1)$ of (8.211) but N_s still being given by (8.235). P_s will be large and so will its derivatives. If (8.247) is entered in the governing partial differential equation and the derivatives of ν_s and N_s with respect to ϕ neglected by virtue of their smallness, the disappearance of the largest terms demands

$$P'_s = \pm \nu_s .$$

The upper sign is the proper choice for the penumbral point under consideration and so

$$P_s(\phi) = \int_{\phi_1}^{\phi} \nu_s(t)\mathrm{d}t$$

which reduces to the right value when Z is a constant. When r,ϕ are very near to r_1,ϕ_1 respectively, ν_s and N_s are effectively constant so that (8.247) displays the correct singularity for a source. Consequently, (8.247) can be regarded as a satisfactory solution when Z is variable.

Next, suppose that the boundary is not circular, but is convex. To adapt (8.247) to this situation take a in (8.236) as the radius of curvature of the boundary at the point under consideration. The polar coordinates r,ϕ are no longer adequate. Instead use the arc length σ of the boundary (increasing in the same sense as ϕ does) and a transverse coordinate R. On the boundary $R = a$ and $\partial R/\partial n = 1$, so that R does not change significantly with a wavelength alteration in σ. The choice of these coordinates is reasonable if we do not stray too far from the boundary. Now, ν_s is a function of σ, as determined by (8.236) and

$$E_z = -\frac{1}{2}\mathrm{i}\sum_{s=1}^{\infty} \frac{H_{\nu_s}^{(2)}(k_0R)}{N_s^2} H_{\nu_s}^{(2)}(k_0R_1)\exp\{-\mathrm{i}P_s(\sigma)\} \tag{8.248}$$

with N_s defined by (8.235). In view of the choice of R the boundary condition (8.234) is satisfied. Therefore (8.248) is appropriate to an upper penumbral point if

$$\frac{\mathrm{d}P_s}{\mathrm{d}\sigma} = \frac{\nu_s}{a} . \tag{8.249}$$

The expression (8.248) can be expected to be valid only when both the source and point of observation are near the boundary because R and σ are coordinates local to the boundary.

For a distant source, the plane-wave approximation can be employed. In this case, r_1 is taken to infinity before the generalizations of (8.247) and (8.248) are attempted. It is found that for an incident plane wave of unit amplitude

$$E_z = 2 \sum_{s=1}^{\infty} \exp\{-\mathrm{i}P_s(\sigma)\} \frac{H_{\nu s}^{(2)}(k_0 R)}{N_s^2} \tag{8.250}$$

where now P_s is zero at the penumbral point where the incident ray grazes the body. Although (8.250) is valid only near the body, the Hankel function can be approximated as soon as R is somewhat different from a and (8.250) has a cylindrical wave structure similar to (8.218). By matching to a suitable cylindrical wave E_z can be continued to points well away from the body.

When Z is infinite and the Hankel functions are replaced by their asymptotic formulae the basic difference between (8.218) and (8.249) is that $\nu_s(\phi-\phi_1)$ becomes $\int \nu_s \, \mathrm{d}\sigma/a$. On account of (8.215) this is

$$k_0\sigma + (\tfrac{1}{2}k_0)^{\frac{1}{3}}\alpha_s \exp(-\tfrac{1}{3}\pi\mathrm{i}) \int \frac{\mathrm{d}\sigma}{a^{\frac{2}{3}}} \,.$$

It may thus be surmised that (8.224) will be relevant to a uniformly valid representation if $L_\mathrm{b}/a^{\frac{2}{3}}$ is replaced by $\int a^{-\frac{2}{3}}\mathrm{d}\sigma$. Some rewriting of (8.224) will improve its generality. Suppose that not too close to the source the incident field is $\gamma^\mathrm{i} \exp(-\mathrm{i}k_0 L^\mathrm{i})$ and that a line source $-\frac{1}{4}\mathrm{i}H_0^{(2)}$ at the point of observation would generate nearby a field $\Gamma^\mathrm{c} \exp(-\mathrm{i}k_0 L')$, L' being the optical path length from the point of observation. Then, by virtue of (8.230), (8.224) may be expressed as

$$E_z \sim \frac{\gamma^{\mathrm{i}}}{\pi^{\frac{1}{2}}} \exp(\tfrac{1}{4}\pi\mathrm{i}-\mathrm{i}k_0 L)F[\{k_0(L-L^{\mathrm{i}})^{\frac{1}{2}}\}]$$

$$- 4\pi^{\frac{1}{2}}(\tfrac{1}{2}k_0 a_1 \tfrac{1}{2}k_0 a_{\mathrm{B}})^{\frac{1}{6}}\gamma_1^{\mathrm{i}}\Gamma_{\mathrm{B}}^{\mathrm{c}} \exp(-\mathrm{i}k_0 L)p\{(\tfrac{1}{2}k_0)^{\frac{1}{3}} \int_{P_1}^{\mathrm{B}} a^{-\frac{2}{3}}\, \mathrm{d}\sigma\} \qquad (8.251)$$

where P_1 is the penumbral point and B is the point of departure from the boundary of the tangential ray to the point of observation. The values of a and γ^{i} at P_1 are denoted by a_1 and γ_1^{i} respectively, whereas a_{B} and $\Gamma_{\mathrm{B}}^{\mathrm{c}}$ are the values of a and Γ^{c} at B. L is the optical path length from source to point of observation via the boundary as defined by (8.232) (cf. Fig. 8.28). The square root $(L-L^{\mathrm{i}})^{\frac{1}{2}}$ is chosen to be positive in the shadow and negative in the illuminated region. The field (8.251) is entirely consistent with (8.224) when the boundary is circular and is employed in a similar fashion in predicting the field, but is applicable for arbitrary convex boundaries. It exhibits reciprocity in that the field is unaltered by interchange of the source point and point of observation.

Clearly, impedance boundary conditions are also within grasp by the introduction of p_Z as defined in (8.238). In particular, the analogues of (8.242) and (8.243) are

$$E_z = \gamma_1^{\mathrm{i}} \exp(-\mathrm{i}k_0 L)q_Z\{(\tfrac{1}{2}k_0)^{\frac{1}{3}} \int_{P_1}^{\mathrm{C}} a^{-\frac{2}{3}}\, \mathrm{d}\sigma\}, \qquad (8.252)$$

$$\frac{\partial E_z}{\partial n} = - k_0(\tfrac{1}{2}k_0 a)^{-\frac{1}{3}}Z\gamma_1^{\mathrm{i}} \exp(-\mathrm{i}k_0 L)q_Z\{(\tfrac{1}{2}k_0)^{\frac{1}{3}} \int_{P_1}^{\mathrm{C}} a^{-\frac{2}{3}}\, \mathrm{d}\sigma\} \qquad (8.253)$$

where C is the point of the boundary under consideration.

When the source is very close to the boundary but not actually on it further investigation has to be undertaken (Jones 1967).

The medium outside the cylinder may not be homogeneous and its refractive index N may vary from point to point. The laws (8.251)–(8.253) are still serviceable provided that L, L^{i} are optical path lengths and $1/a$ is replaced by (Jones 1963)

$$\frac{1}{N}\left(\frac{1}{a} + \frac{1}{\rho_0}\right)$$

where ρ_0 is the radius of curvature of the tangential ray, being counted positive when the centre of curvature is on the same side of the boundary tangent as the inhomogeneous medium. Also $d\sigma$ becomes $N d\sigma$ and, in (8.253), the right-hand side is multiplied by N_C. The boundary condition now takes the form

$$\frac{\partial E_z}{\partial n} + k_0 N \left\{\frac{2}{k_0 N}\left(\frac{1}{a} + \frac{1}{\rho_0}\right)\right\}^{\frac{1}{3}} Z E_z = 0 \ .$$

The formulae (8.248) and (8.250) are also applicable so long as $k_0 N$ is put for k_0 and a,σ are modified as already indicated.

Exercises

42. Determine the field in the shadow region when the obstacle is a perfectly conducting (*a*) parabola, (*b*) ellipse. How would your results be modified if the boundary possessed variable impedance?

43. For perfect conductivity in Exercise 42 find the field in the penumbra for both electric and magnetic polarization.

44. Show that the scattering coefficient of a perfectly conducting convex cylinder in a homogeneous medium under an electrically polarized plane wave is

$$2 + \frac{2^{5/3}\pi^{\frac{1}{2}}}{k_0^{2/3} D}\left(a_1^{1/3} + a_2^{1/3}\right) \mathscr{I} p(0)$$

where D is the length of wavefront intercepted by the obstacle and a_1, a_2 are the radii of curvature at the penumbral points.

Is it sufficient to replace $p(0)$ by $p_0(0)$ for magnetic polarization?

45. The medium above the perfectly conducting plane $y = 0$ is stratified so that the refractive index is $(1 + qy)^{\frac{1}{2}}$ where q is a positive constant. Determine the field in the penumbra and deep shadow for both types of polarization.

46. A circular cylinder of radius a is surrounded by a medium in which the refractive index N increases steadily with radial distance but is independent of angle. Determine the field in the penumbra due to a line source

parallel to the axis of the cylinder.

The source and point of observation are at radial distances $a + h_1$ and $a + h$ respectively where $1 \gg h/a$, $h_0/a \gg (k_0 a)^{-\frac{2}{3}}$. Show that propagation takes place as if it were in a homogeneous medium but the cylinder had an effective radius a_e so long as the actual distance $a(\phi-\phi_1)$ along the surface of the cylinder is employed. The effective radius is given by

$$a_e = \left\{1 + a\left(\frac{\mathrm{d}N}{\mathrm{d}r}\right)_{r=a}\right\}^{-1}.$$

Compare your results with those of Exercise 45 when $q = 2/a_e$.

8.23. Optimal curvature

Sometimes obstacles are designed to shield regions from high-frequency radiation. It may be conjectured that such obstacles should have curved boundaries because the shadow is deeper than when edges are present. The damping which occurs is governed by the integral along the boundary in (8.251). It has been suggested (Leppington 1970) that the shadow would be darkest if the curvature were optimized so that the integral became as large as possible. However, the practical effects do not seem to be large.

The attenuation of the optimal design has been compared (Butler 1972) with that due to an ellipse with the same penumbral point and it was found that the optimal boundary gave less than $2dB$ extra shielding over the elliptic cylinder. Indeed, for the smaller angular deviations into the shadow the elliptic cylinder offered somewhat greater attenuation than the optimal design.

Both shapes provided appreciably better shielding than a sharp edge. It therefore appears that it is sufficient to ensure that the boundary has a large radius of curvature. The degree of shielding afforded will be relatively insensitive to the precise shape of the boundary and little will be gained by attempting to make it fit the optimal design.

8.24. The diffraction matrix for a curved boundary

Creeping rays in the deep shadow of a perfectly conducting cylinder in a homogeneous medium make a convenient jumping off point for eliciting the polarization connections in diffraction by a curved boundary. The coordinates based on a ray which were fixed upon in §8.15 permit a satisfactory descriptive mechanism here too. From (8.251) and (8.227), on a creeping ray

$$\begin{pmatrix} E_\xi^{\mathrm{c}} \\ E_\eta^{\mathrm{c}} \end{pmatrix} = \begin{pmatrix} D_E^{\mathrm{c}} & 0 \\ 0 & D_H^{\mathrm{c}} \end{pmatrix} \begin{pmatrix} E_\xi^{\mathrm{i}}(P_1) \\ E_\eta^{\mathrm{i}}(P_1) \end{pmatrix} \frac{\exp\{-ik_0(L_{\mathrm{b}}+L^{\mathrm{c}})\}}{(L^{\mathrm{c}})^{\frac{1}{2}}} \tag{8.254}$$

where P_1 is the penumbral point, L_{b} and L^{c} are as already defined and

$$D_E^{\mathrm{c}} = \frac{(a_1 a_{\mathrm{B}})^{\frac{1}{6}}\exp(-\frac{1}{12}\pi \mathrm{i})}{2\pi^{\frac{1}{2}}(\frac{1}{2}k_0)^{\frac{1}{6}}} \sum_{s=1} \frac{\exp\{-\alpha_s \exp(\frac{1}{6}\pi\mathrm{i})(\frac{1}{2}k_0)^{\frac{1}{3}}\int_{P_1}^{\mathrm{B}} a^{-\frac{2}{3}}\mathrm{d}\sigma\}}{\{\mathrm{Ai}'(-\alpha_s)\}^2}, \tag{8.255}$$

$$D_H^{\mathrm{c}} = \frac{(a_1 a_{\mathrm{B}})^{\frac{1}{6}}\exp(-\frac{1}{12}\pi \mathrm{i})}{2\pi^{\frac{1}{2}}(\frac{1}{2}k_0)^{\frac{1}{6}}} \sum_{s=1} \frac{\exp\{-\beta_s \exp(\frac{1}{6}\pi\mathrm{i})(\frac{1}{2}k_0)^{\frac{1}{3}}\int_{P_1}^{\mathrm{B}} a^{-\frac{2}{3}}\mathrm{d}\sigma\}}{\beta_s\{\mathrm{Ai}(-\beta_s)\}^2}, \tag{8.256}$$

β_s being a positive number such that $\mathrm{Ai}'(-\beta_s) = 0$. Some typical values of β_s are

$$\begin{aligned} \beta_1 &= 1.019, \quad \beta_2 = 3.248, \quad \beta_3 = 4.820, \\ \beta_4 &= 6.163, \quad \beta_5 = 7.372, \quad \beta_6 = 8.488 \end{aligned} \tag{8.257}$$

and values of α_s appear in (8.214); other relevant constants are given in Table 8.3. The magnetic intensity on a creeping ray is given by

$$\mathbf{H}^{\mathrm{c}} = \left(\frac{\varepsilon_0}{\mu_0}\right)^{\frac{1}{2}} \hat{\mathbf{L}}^{\mathrm{c}} \wedge \mathbf{E}^{\mathrm{c}} \tag{8.258}$$

TABLE 8.3

s	$\mathrm{Ai}(-\beta_s)$	$\mathrm{Ai}'(-\alpha_s)$
1	0.536	0.701
2	-0.419	-0.803
3	0.380	0.865
4	-0.358	-0.911
5	0.342	0.947
6	-0.330	-0.978

where $\hat{\mathbf{L}}^c$ is a unit vector along the creeping ray towards the point of observation.

As in §8.15, (8.254) may be applied immediately to oblique incidence on a cylinder though now the creeping ray will follow a helical-type path on the cylinder and a must be taken as the radius of curvature in the osculating plane of this path. For general convex targets the creeping ray will follow a geodesic on the surface in accordance with Fermat's principle. Analytic expressions for geodesics are not easily derivable except for the simplest surfaces and it will usually be necessary to solve the governing differential equation numerically. Furthermore, each ray of an incident tube will traverse its own geodesic and so the cross-section of the tube will, in general, be different at the point of departure from the surface from that at incidence. Therefore, the field will be altered by the factor C where C^2 is the ratio of the cross-sections at the beginning and end of the geodesic on the obstacle. In general, C will have to be calculated numerically.

Again, when the creeping rays leave the surface the spreading of energy due to variation of the cross-section of a tube must be allowed for by a factor $\{\rho_1^c/(\rho_1^c+L^c)\}^{\frac{1}{2}}$ as for the edge rays of §8.16. Here ρ_1^c is the principal radius of curvature associated with the tube (the surface is a caustic), but it cannot be specified further, in contrast to the situation for edge rays, and it must usually be evaluated numerically.

On this basis the field on a creeping ray in the deep shadow of a convex perfect conductor satisfies

$$\begin{pmatrix} E^{c}_{\xi} \\ E^{c}_{\eta} \end{pmatrix} = C \begin{pmatrix} D^{c}_{E} & 0 \\ 0 & D^{c}_{H} \end{pmatrix} \begin{pmatrix} E^{i}_{\xi}(P_1) \\ E^{i}_{\eta}(P_1) \end{pmatrix} \left\{ \frac{\rho^{c}_{1}}{L^{c}(\rho^{c}_{1}+L^{c})} \right\}^{\frac{1}{2}} \exp\{-ik_0(L_b+L^c)\} \quad (8.259)$$

with a the radius of curvature of the geodesic and the magnetic intensity given by (8.258).

The prediction (8.259) is founded on the approximation (8.215). More accurate expansions have been discussed (Franz and Klante 1959, Hong 1967) but details are omitted here.

The penumbra requires the transition formula (8.251) in its entirety. It leads to

$$\mathbf{E}^{c} = \exp(\tfrac{1}{4}\pi i)\pi^{-\frac{1}{2}} \mathbf{E}^{i} \exp\{-ik_0(L-L^{i})\}F[\{k_0(L-L^{i})\}^{\frac{1}{2}}] + \\ + \begin{pmatrix} D^{c}_{E} & 0 \\ 0 & D^{c}_{H} \end{pmatrix} \begin{pmatrix} E^{i}_{\xi}(P_1) \\ E^{i}_{\eta}(P_1) \end{pmatrix} \left\{ \frac{\rho^{c}_{1}}{L^{c}(\rho^{c}_{1}+L^{c})} \right\}^{\frac{1}{2}} \exp\{-ik_0(L_b+L^c)\} \quad (8.260)$$

where now

$$D^{c}_{E} = -\frac{(a_1 a_B)^{\frac{1}{6}} \exp(-\frac{1}{4}\pi i)}{(\frac{1}{2}k_0)^{\frac{1}{6}}} p\{(\tfrac{1}{2}k_0)^{\frac{1}{3}} \int_{P_1}^{B} a^{-\frac{2}{3}} \,d\sigma\}, \quad (8.261)$$

$$D^{c}_{H} = -\frac{(a_1 a_B)^{\frac{1}{6}} \exp(-\frac{1}{4}\pi i)}{(\frac{1}{2}k_0)^{\frac{1}{6}}} p_0\{(\tfrac{1}{2}k_0)^{\frac{1}{3}} \int_{P_1}^{B} a^{-\frac{2}{3}} \,d\sigma\}. \quad (8.262)$$

Comparison with §8.4 indicates that the substitution

$$L - L^{i} = \frac{L^{c}\rho_1\rho_2(\rho^{c}_{1}+L^{c})}{2\rho^{c}_{1}(\rho_1+L^{c})(\rho_2+L^{c})}$$

where ρ_1, ρ_2 are the principal radii of curvature of the incident tube at P_1 can be helpful near the shadow boundary. The formula (8.260) must be utilized according to the same conventions as (8.224) and (8.251), i.e. starting from the shadow boundary and then transferring as soon as possible to either

(8.259) or the geometric optics field in the illuminated zone.

Radiation from sources on a convex object can be calculated from (8.252) and (8.253) by invoking reciprocity (Pathak and Kouyoumjian 1974).

8.25. Diffraction by a discontinuity in curvature

The preceding exploration has tacitly assumed that there is no discontinuity in the radius of curvature. When a discontinuity does occur another canonical problem needs to be probed (Weston 1962, 1965, Senior 1971, 1972, Boyd 1977, Sharples 1962*a*,*b*). Primary concern is not with the influence on a passing creeping wave (which is expected to be weak) nor with the creeping waves generated by the sources induced at the discontinuity but with the effect on geometrical optics in the illuminated region. The model selected is a cylinder tangential to the x axis and with the discontinuity at the origin (Fig. 8.29). The radii of curvature to the right and left of the discontinuity will be designated a_1 and a_2 respectively. The incident wave will be assumed plane and electrically polarized. Thus the primary field is

$$E^{\mathrm{i}} = \exp\{-\mathrm{i}k_0 r \cos(\phi-\phi_0)\}$$

the subscript z being dropped for convenience. The angle ϕ_0 is subject to $0 < \phi_0 < \pi$ in order to prevent grazing incidence. The equation of the boundary curve is $y = f(x)$ where f and f' are continuous. Both $f(0)$ and $f'(0)$ are zero but $f''(+0) = 1/a_1$ and $f''(-0) = 1/a_2$.

The boundary condition for the total field $E(x,y)$ is

$$E(x,f(x)) = 0. \qquad (8.263)$$

With the intent of concentrating on the neighbourhood of the origin introduce *stretched coordinates* X,Y where

$$X = k_0 x, \qquad Y = k_0 y .$$

Assume that for k_0 large

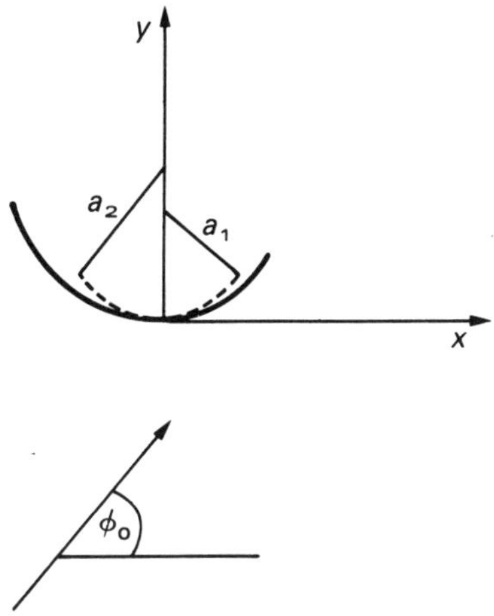

FIG. 8.29. Parameters for a discontinuity in curvature.

$$E(x,y) = E(k_0^{-1}X, k_0^{-1}Y) \sim \sum_{m=0} k_0^{-m} E_m(X,Y)$$

which will be called an *inner expansion* of E. The partial differential equation to be satisfied by E_m is

$$\left(\frac{\partial^2}{\partial X^2} + \frac{\partial^2}{\partial Y^2} + 1\right)E_m = 0.$$

The boundary condition (8.263) transforms to

$$\sum_{m=0} k_0^{-m} E_m\left\{X, k_0 f\left(\frac{X}{k_0}\right)\right\} = 0.$$

Expanding f about the origin we obtain to a first approximation

$$f\left(\frac{X}{k_0}\right) = \frac{1}{2}\,\frac{X^2}{k_0^2 a_1}$$

for $X > 0$ and the same formula with a_2 for $X < 0$. Hence the boundary condition may be written

$$E_0(X,0) + k_0^{-1}\left\{E_1(X,0) + \frac{1}{2}\frac{X^2}{a_{1,2}}\frac{\partial}{\partial y}E_0(X,0)\right\} + \ldots = 0$$

where a_1 or a_2 transpires according to $X > 0$ or < 0. This will be applied as

$$E_0(X,0) = 0, \quad E_1(X,0) = -\frac{1}{2}\frac{X^2}{a_{1,2}}\frac{\partial}{\partial y}E_0(X,0) \; . \qquad (8.264)$$

Similar boundary conditions can be derived for the higher-order coefficients but will not be needed for our purposes.

To complete the picture the conditions at infinity need to be specified. Now

$$E^i = \exp\{-i(X\cos\phi_0 + Y\sin\phi_0)\}$$

and so it can be regarded as the incident field for E_0. Had terms in inverse powers of k_0 appeared they would have had to be treated as incident fields for E_1, ... At any rate, E_0 has to be a radiating field after the subtraction of E^i and E_1 consists entirely of outgoing waves. Thus

$$E_0(X,Y) = \exp\{-i(X\cos\phi_0 + Y\sin\phi_0)\} - \exp\{-i(X\cos\phi_0 - Y\sin\phi_0)\}$$

and so, from (8.264),

$$E_1(X,0) = \frac{iX^2}{a_{1,2}}\sin\phi_0 \exp(-iX\cos\phi_0) \; .$$

By means of the Green's function which vanishes on the plane we obtain for $Y < 0$

$$-E_1(X,Y) = \frac{Y\sin\phi_0}{2a_2}\int_{-\infty}^{0}\frac{X'^2\exp(-iX'\cos\phi_0)}{\{(X'-X)^2+Y^2\}^{\frac{1}{2}}}H_1^{(2)}[\{(X'-X)^2+Y^2\}^{\frac{1}{2}}]\,dX'$$

$$+\frac{Y\sin\phi_0}{2a_1}\int_{0}^{\infty}\frac{X'^2\exp(-iX'\cos\phi_0)}{\{(X'-X)^2+Y^2\}^{\frac{1}{2}}}H_1^{(2)}[\{(X'-X)^2+Y^2\}^{\frac{1}{2}}]\,dX' .$$

Strictly, these integrals do not converge at infinity. This is because (8.264) has been applied for large values of $|X|$ where it is not truly valid. In essence, the largeness of X is limited by the requirement to keep x small. Therefore, tolerable approximations may be expected so long as the intervals of integration are visualized as finite with significant contributions from small X'.

When $X^2 + Y^2 \gg 1$, insert the asymptotic approximation for the Hankel function and then

$$\int_0 \frac{X'^2\exp(-iX'\cos\phi_0)}{\{(X'-X)^2+Y^2\}^{\frac{1}{2}}}H_1^{(2)}[\{(X'-X)^2+Y^2\}^{\frac{1}{2}}]\,dX'$$

$$\sim\left(\frac{2}{\pi}\right)^{\frac{1}{2}}\exp(\tfrac{3}{4}\pi i)\int_0 \frac{X'^2\exp[-iX'\cos\phi_0 - i\{(X'-X)^2+Y^2\}^{\frac{1}{2}}]}{\{(X'-X)^2+Y^2\}^{\frac{3}{4}}}\,dX' .$$

Restricting attention to $Y < 0$, we see that the exponent is a monotonic function of X' if $X\sin\phi_0 + Y\cos\phi_0 < 0$ or if $\phi < -\phi_0$ on putting $X = R\cos\phi$, $Y = R\sin\phi$. The main contribution will then spring from the origin. If $\phi > -\phi_0$ there is a stationary point at $X' = R\operatorname{cosec}\phi_0\sin(\phi+\phi_0)$ and it is straightforward to certify that this corresponds to a reflected wave emanating from the right portion of the cylinder, so once again the diffracted field stems from the origin. An expansion in powers of X' supplies a contribution from the origin of

$$\left(\frac{8}{\pi}\right)^{\frac{1}{2}}\frac{\exp(\frac{1}{4}\pi i - iR)}{R^{3/2}(\cos\phi-\cos\phi_0)^3} .$$

Consequently the diffracted part of E_1 is given by

$$E_1^d(X,Y) = \left(\frac{2}{\pi R}\right)^{\frac{1}{2}} \frac{\sin\phi \sin\phi_0}{(\cos\phi - \cos\phi_0)^3}\left(\frac{1}{a_2} - \frac{1}{a_1}\right)\exp(\tfrac{1}{4}\pi i - iR). \quad (8.265)$$

Not too far from the cylinder the field on a ray should have standard form and contain factors such as $\{\rho/(\rho+s)\}^{\frac{1}{2}}\exp(-ik_0 s)$ where ρ is the radius of curvature on a pencil at the cylinder and s is the distance along a ray from the cylinder. As s diminishes this field should merge with the field in stretched coordinates as R increases. As far as reflection is concerned there is no new feature and so the calculation will be omitted. The diffraction pattern can be matched with (8.265) if

$$E^d(x,y) = \frac{1}{k_0}\left(\frac{2}{\pi k_0 r}\right)^{\frac{1}{2}} \frac{\sin\phi \sin\phi_0}{(\cos\phi - \cos\phi_0)^3}\left(\frac{1}{a_2} - \frac{1}{a_1}\right)\exp(\tfrac{1}{4}\pi i - ik_0 r). \quad (8.266)$$

Obviously, failure of (8.266) must be accepted in the vicinity of $\phi = -\phi_0$ where there is a transition region for the reflected wave. Uniform expansions in this domain will require a more sophisticated Ansatz such as that of §8.17 to match the inner expansion.

The same method applied to magnetic polarization provides

$$H^d(x,y) = \frac{1}{k_0}\left(\frac{2}{\pi k_0 r}\right)^{\frac{1}{2}} \frac{1 - \cos\phi \cos\phi_0}{(\cos\phi - \cos\phi_0)^3}\left(\frac{1}{a_2} - \frac{1}{a_1}\right)\exp(\tfrac{1}{4}\pi i - ik_0 r). \quad (8.267)$$

The effect of a discontinuity in curvature is more subdued than that of an edge. Nevertheless it may be important in determining the polarization characteristics of a reflecting surface. Slight flaws in manufacture might not be sufficient to raise a surface kink but might be large enough to cause the curvature to change discontinuously and there could then be untoward cross-polarization.

There is no reason why a_1 should not be negative corresponding to a concavity on one side of the discontinuity. However, the reflected pencil will now be concentrated because of focusing and the diffracted field is unlikely to be significant in comparison except possibly at very great distances from the obstacle.

Oblique incidence can be dealt with as in §8.15. The diffracted rays from the discontinuity now lie on a cone. The analogue of (8.117) is

$$\mathbf{E}^{d} = \left\{\frac{\rho_1^{d}}{(\rho_1^{d}+s)s}\right\}^{\frac{1}{2}} \exp(-ik_0 s)D'\mathbf{E}^{i}(0) \tag{8.268}$$

where

$$D' = \begin{pmatrix} D_E' & 0 \\ 0 & D_H' \end{pmatrix}$$

and

$$D_E' = \frac{\exp(\frac{1}{4}\pi i)}{k_0}\left(\frac{2}{\pi k_0}\right)^{\frac{1}{2}} \frac{\sin\phi\,\sin\phi_0}{(\cos\phi-\cos\phi_0)^3}\left(\frac{1}{a_2}-\frac{1}{a_1}\right)\operatorname{cosec}^2\theta_0, \tag{8.269}$$

$$D_H' = \frac{\exp(\frac{1}{4}\pi i)}{k_0}\left(\frac{2}{\pi k_0}\right)^{\frac{1}{2}} \frac{1-\cos\phi\,\cos\phi_0}{(\cos\phi-\cos\phi_0)^3}\left(\frac{1}{a_2}-\frac{1}{a_1}\right)\operatorname{cosec}^2\theta_0. \tag{8.270}$$

The method delineated in this section will handle impedance boundary conditions, curves of discontinuity on three-dimensional surfaces or discontinuities at a point such as the vertex of a cone (Kaminetzky and Keller 1972). There is no necessity for the medium to be homogeneous. Discontinuities on interfaces between two media are also within the scope of this technique. The discontinuities may, of course, be in higher derivatives than the second.

Exercises

47. If the boundary condition (8.263) is replaced by $\partial E/\partial n + ik_0 Z(x)E = 0$

where $\partial/\partial n$ is a normal derivative directed out of the obstacle and discontinuities due to Z are ignored, show that

$$E^{\mathrm{d}} = \frac{1}{k_0}\left(\frac{2}{\pi k_0 r}\right)^{\frac{1}{2}} E^{\mathrm{i}}(0)\ \frac{\sin\phi\sin\phi_0\{1-\cos\phi\cos\phi_0-Z^2(0)\}}{\{\sin\phi_0-Z(0)\}\{\sin\phi+Z(0)\}(\cos\phi-\cos\phi_0)^3}\left(\frac{1}{a_2}-\frac{1}{a_1}\right)$$

$$\times\exp(\tfrac{1}{4}\pi\mathrm{i}-\mathrm{i}k_0 r)\ .$$

48. If in Exercise 47 Z is continuous but its derivative jumps by an amount Z' in going from negative to positive values of x show that the result of Exercise 47 still holds provided that $1/a_2 - 1/a_1$ is replaced by

$$\frac{1}{a_2}-\frac{1}{a_1}-Z'(\cos\phi-\cos\phi_0)\{1-\cos\phi\cos\phi_0-Z^2(0)\}^{-1}\ .$$

49. The medium above the boundary in Fig. 8.29 is homogeneous with refractive index N. The conditions on the interface are $E^{(1)} = a(x)E^{(2)}$, $\partial E^{(1)}/\partial n = b(x)\partial E^{(2)}/\partial n$ where (1) and (2) refer to the lower and upper media respectively. Find the diffracted ray in the lower medium assuming that a, b, and their derivatives are continuous.

50. In Fig. 8.29 f and its first $r - 1$ derivatives are continuous. The boundary condition (8.263) is applied. Prove that (8.266) may still be used provided that $(\cos\phi-\cos\phi_0)^3$ is replaced by $-\mathrm{i}^r k_0^{r-2}(\cos\phi-\cos\phi_0)^{r+1}$ and the corresponding discontinuity of $f^{(r)}$ is substituted for $1/a_2 - 1/a_1$.

Demonstrate that the same substitutions work for (8.267) and Exercise 47.

Other canonical problems (sometimes called *etalon problems* in the Russian literature) have been studied. For example, a curved surface which also possesses an edge has been investigated (Borovikov 1973), as has the scattering from a cone (Goryainov 1961, Borovikov 1962, Senior 1971*a*, Senior and Uslenghi 1971, 1973). The effect of a surface impedance when the boundary is curved has also been examined (Bahar 1971).

Creeping waves generally follow geodesics on the surface but it is not necessary that both radii of curvature be finite. A disc, when illuminated by a plane wave at grazing incidence with the electric vector in the plane of the disc, exhibits scattering behaviour which is compatible with a wave reflected from the front edge together with one which has circumscribed the edge in shadow (Hong 1967, Ryan and Peters 1968, Senior 1969). Although creeping rays are customarily regarded as determined by the field at the shadow boundary some care is required if the creeping rays come to a focus on the diffracting surface since more complicated effects may then arise (Kazarinoff and Senior 1962).

Once rays have left their generating surface they are governed by the general considerations of §§8.2, 8.3. Often the propagation will be in an inhomogeneous medium which may have curved boundaries (Lewis 1965, Ahluwalia, Keller and Matkowsky 1974) and frequently the rays may form caustic surfaces within the medium when another canonical problem must be solved (Kravstsov 1964, Ludwig 1966, Lewis, Bleistein and Ludwig 1967, Tokatly and Kinber 1971, Gazazian and Kinber 1971). Conditions under which rays do not produce caustics have been formulated (Bloom and Kazarinoff 1976).

The tracing of rays is, in many applications, a matter of integrating numerically ordinary differential equations as already indicated. However, the determination of the field also involves the solution of the transport equations in order to fix the amplitudes of the field components. In effect, this is much the same as calculating the cross-section of a ray tube (see (8.19)). Sometimes this can be carried out analytically by evaluating the Jacobian

$$J_1 = \frac{\partial(x_1, x_2, x_3)}{\partial(s, \sigma_2, \sigma_3)}$$

where (x_1, x_2, x_3) is the point of the ray where the arc length is s and the parameters characterizing the ray are σ_2, σ_3. If the equations of the ray are

$$f(x_1, x_2, x_3, \theta, \phi) = 0, \qquad g(x_1, x_2, x_3, \theta, \phi) = 0$$

where θ,ϕ are perhaps angles showing the initial direction of launching of the ray

$$J_1 = \left(\frac{\partial g}{\partial \theta}\frac{\partial f}{\partial \phi} - \frac{\partial g}{\partial \phi}\frac{\partial f}{\partial \theta}\right)\left\{\left(\frac{\partial f}{\partial x_2}\frac{\partial g}{\partial x_3} - \frac{\partial f}{\partial x_3}\frac{\partial g}{\partial x_2}\right)^2 + \left(\frac{\partial f}{\partial x_3}\frac{\partial g}{\partial x_1} - \frac{\partial f}{\partial x_1}\frac{\partial g}{\partial x_3}\right)^2 + \left(\frac{\partial f}{\partial x_1}\frac{\partial g}{\partial x_2} - \frac{\partial f}{\partial x_2}\frac{\partial g}{\partial x_1}\right)^2\right\}^{-\frac{1}{2}} \tag{8.271}$$

and other formulae are given in §8.3.

When f and g are known only numerically, use of (8.271) or other analytical expressions can necessitate awkward numerical derivatives. One way of overcoming this is to trace a large number of trajectories starting in uniformly distributed directions; then the local density of rays (and hence the field amplitude) can be estimated by measuring the distance between adjacent rays for instance. Such a procedure is uneconomical in computer time and its precision diminishes as the range along a ray increases.

A more efficient method has been proposed (Hayes 1970) though it is in no way cheap in computer time. It consists in solving the differential equations for the element of area of the phase front. One advantage is that it is also applicable to anisotropic media.

Let the eikonal be L and put $\mathbf{p}$ = grad L so that the eikonal equation is

$$p^2 = N^2. \tag{8.272}$$

The equations of the rays, when L is selected as the parameter to vary along a ray, are

$$\frac{d\mathbf{x}}{dL} = \frac{\mathbf{p}}{N^2}, \quad \frac{d\mathbf{p}}{dL} = \frac{1}{N}\,\text{grad}\,N \tag{8.273}$$

where L is related to the arc length s by $(ds/dL)^2 = 1/N^2$. Suppose that the solution of (8.273) starting from a specified point in the direction characterized by θ,ϕ is

$$\mathbf{x} = \mathbf{x}(L,\theta,\phi), \quad \mathbf{p} = \mathbf{p}(L,\theta,\phi) .$$

Write

$$\mathbf{Q}_\theta = \frac{\partial \mathbf{x}}{\partial \theta}, \quad \mathbf{Q}_\phi = \frac{\partial \mathbf{x}}{\partial \phi} .$$

Then an element of area of the phase front L = constant is given by

$$\delta\boldsymbol{\Sigma} = \mathbf{Q}_\theta \wedge \mathbf{Q}_\phi \; \delta\theta\delta\phi . \tag{8.274}$$

The cross-section δA of a ray tube now follows from

$$\delta A = \delta\Sigma \, \frac{\mathbf{p}}{p} \cdot \frac{\mathrm{d}\mathbf{x}}{\mathrm{d}s} = \delta\Sigma \tag{8.275}$$

though it should be noted that the factor of $\delta\Sigma$ is not unity in anisotropic media in general.

Now, by applying $\partial/\partial\theta$ (which implies that L is constant) to (8.273)

$$\frac{\mathrm{d}\mathbf{Q}_\theta}{\mathrm{d}L} = \frac{1}{N^2}\mathbf{R}_\theta - \frac{2}{N^3}\,\mathbf{p}(\mathbf{Q}_\theta \cdot \mathrm{grad}\, N) \tag{8.276}$$

where $\mathbf{R}_\theta = \partial\mathbf{p}/\partial\theta$. Also

$$\frac{\mathrm{d}\mathbf{R}_\theta}{\mathrm{d}L} = \frac{1}{N}(\mathbf{Q}_\theta \cdot \mathrm{grad})\mathrm{grad}\, N - \frac{1}{N^2}(\mathbf{Q}_\theta \cdot \mathrm{grad}\, N)\mathrm{grad}\, N. \tag{8.277}$$

Similarly, if $\mathbf{R}_\phi = \partial\mathbf{p}/\partial\phi$, both (8.276) and (8.277) hold if θ is replaced by ϕ.

Thus **Q** and **R** can be found by integrating the differential equations (8.276) and (8.277) instead of by taking derivatives of **x** and **p**. Then (8.274) and (8.275) supply the area of the ray tube.

Initial conditions are needed for the differential equations. If the rays originate from a point source where $L = 0$, $\mathbf{x}$ is a linear function of L initially and so $\mathbf{Q}_\theta = \mathbf{Q}_\phi = 0$ at $L = 0$. On the other hand, $\mathbf{R}_\theta$ and $\mathbf{R}_\phi$ take the initial values of $\partial\mathbf{p}/\partial\theta$ and $\partial\mathbf{p}/\partial\phi$ which are usually not zero.

The calculation of the field has now been converted to the numerical integration of the 18 ordinary differential equations (8.273), (8.276) and (8.277) (for both θ and ϕ). The process is started by stating the position of the source and the initial angles θ,ϕ of the ray. The inaugural value of p then follows from the eikonal equation. These six conditions, coupled with the 12 on $\mathbf{Q}$ and $\mathbf{R}$, are sufficient for the integration to proceed. Any of the methods described earlier, such as the fourth-order Runge–Kutta or predictor–corrector, may be employed. Accuracy may be checked by comparing the value of p obtained by numerical integration with that provided by the eikonal equation. Furthermore, the behaviour of the phase front area can be measured and, if

$$\left|1 - \frac{\mathbf{p}.\delta\boldsymbol{\Sigma}}{p|\delta\boldsymbol{\Sigma}|}\right|$$

is not sufficiently small (say, less than 10^{-3}), the accuracy could be regarded as inadequate.

If a perfectly reflecting body is encountered the method will still cope so long as Snell's laws are appropriate, i.e. creeping and edge rays can be ignored. New starting values will be wanted at the body for the differential equations. Denote these by the subscript 2, while values on the ray just before the body is reached are signified by the subscript 1. Then, if $\mathbf{n}$ is the unit normal from the body into the medium, $p_2 = p_1$ and

$$\mathbf{p}_2 = \mathbf{p}_1 - 2(\mathbf{p}_1.\mathbf{n})\mathbf{n} \tag{8.278}$$

and the same relationship (8.278) is valid for $\mathbf{x}/|\mathbf{x}|$ and $\mathbf{Q}$. However,

$$R_{2i} = R_{1i} - 2(\mathbf{R}_1.\mathbf{n})\mathbf{n} - 2(\mathbf{p}_1.\mathbf{n})\left(\frac{\partial \mathbf{n}}{\partial x_i}\right) . \frac{\mathbf{n} \wedge (\mathbf{Q}_1 \wedge \mathbf{x}_1)}{\mathbf{n} . \mathbf{x}_1} \quad (8.279)$$

in which the last term is negligible unless the curvature is large.

Exercises

51. Prove (8.271).

52. Show that $p\partial p/\partial\theta = \mathbf{p} . \mathbf{R}_\theta$.

53. In a certain medium the eikonal equation is $p^2 = (N\text{-}\mathbf{M}.\mathbf{p})^2$ where $\mathbf{M}$ is a known vector. Show that the differential equations of the rays are

$$\frac{d\mathbf{x}}{dL} = \frac{1}{N}\left(\mathbf{M} + \frac{\mathbf{p}}{p}\right),$$

$$\frac{d\mathbf{p}}{dL} = \frac{1}{N}\{\text{grad } N - (\mathbf{p}.\text{grad})\mathbf{M}\} .$$

Prove also that $\delta A = (p+\mathbf{M}.\mathbf{p})\delta\Sigma|\mathbf{p}+p\mathbf{M}|^{-1}$ and that

$$\frac{d\mathbf{Q}}{dL} = \frac{1}{pN}\left\{\mathbf{R} - \frac{(\mathbf{p}.\mathbf{R})\mathbf{p}}{p^2}\right\} - \frac{p\mathbf{M} + \mathbf{p}}{pN^2}(\mathbf{Q}.\text{grad } N) + \frac{1}{N}(\mathbf{Q}.\text{grad})\mathbf{M} .$$

54. Justify (8.278) and (8.279).

8.26. Reflector antennas

Communication links at microwave frequencies often place severe demands on the designers of antennas. Usually the specification calls for a radiation or reception pattern which forms a *pencil beam*, i.e. the radiation is concentrated in a small angle about a certain direction, often known as the *boresight*. The sidelobes into which radiation outside the main beam goes are commonly restricted in their maxima or location and there may be limitations on the total radiation outside the main beam. The main reason for these constraints is the reduction of interference with other antennas. The probability of such interference grows with the expansion of microwave telecommunications which is occurring steadily because of its involvement in point-to-point communication, passing information via satellites, microwave broadcasting, and radioastronomy. Similar considerations can also be relevant in the context of radar where a highly directional beam may be essential to secure an accurate fix on a target. The designer is consequently faced with the twin objectives of high efficiency and low sidelobes; not infrequently, these have to be met at the least possible cost and with the best possible match at the feed.

A solution of many years standing is to employ a reflector, there being two main types according as the reflector is fed from a source symmetrically or unsymmetrically placed relative to the reflector (Fig. 8.30). The source is usually a horn at the end of a waveguide. Many aspects of the design are unaffected by whether the antenna is for transmission or for reception; there are in any case close connections between the two on account of reciprocity theorems.

Most microwave antennas operate with a single direction of linear polarization. However, to augment the possible traffic, future antennas are likely to be pressed to perform with two perpendicular linear polarizations at the same frequency simultaneously. Such *frequency re-use antennas* will push design to its limits in order to achieve the desired polarization discrimination.

The reflector for on-axis feed must, in effect, be a paraboloid with uniform phase over the aperture of the reflector.

(a) (b)

FIG. 8.30. (a) On-axis front-fed reflector; (b) offset front-fed reflector.

With the feed at the focus a good pencil beam with quite low sidelobes will be produced. If the beam has to be moved about in order to scan a certain region either the whole structure has to be steered mechanically or the feed has to be shifted. In the latter case a spherical reflector may be worth consideration though the sidelobes may then rise to too high a level. Beam scanning may raise the problem of blockage of the antenna aperture by the feed; an offset feed may then be viewed favourably. However, offset designs are usually more complicated and costly than those with on-axis feed.

The overall effectiveness of an antenna system is highly dependent upon the efficiency with which the feed couples to the nearby field. Early designs were based on conventional circular waveguide operating in a single mode. Better efficiency at narrow bandwidths can be attained by multi-mode feeds (Ludwig 1969) but nowadays the tendency is to press into service single hybrid-mode excitation from a corrugated waveguide (Clarricoats and Saha 1971, Clarricoats, Olver and Chong 1975, Parini, Clarricoats, and Olver 1975).

The detailed design of an antenna system is clearly a matter for the specialist so that here we shall be content with a description of the principles involved in this task, leaving the reader to find complete calculations elsewhere (see for example James 1976). Assuming a point source the first duty is to determine the trajectories of the incident and reflected rays. For a given point of observation this means discovering the point or points on the reflector profile which make the optical path length stationary. Two simultaneous equations will have to be solved and, in general, this must be undertaken numerically. Once the points of reflection have been settled the curvature matrices C^{i} and C^{r} of §8.5 are evaluated. The formulae of §8.5 will then supply the field of geometrical optics. Thus a first approximation to the radiation pattern is derived and a similar approach may be adopted for general excitation.

Geometrical optics neglects the diffraction due to the rim of the reflector. To furnish a correction for diffraction the trajectories of edge rays are necessary. These require the points on the rim where the optical path length is stationary;

again a numerical procedure such as Newton's method (§§ 4.5, 1.8, 1.9) will be involved. The angles ϕ, ϕ_0, and θ_0 needed for (8.117) must then be determined, followed by the curvature matrix C^i and the rim curvature $1/\rho_1^e$. The diffracted field can then be obtained from (8.117) unless the point of observation is near a shadow or reflection boundary when recourse to the more recondite formulae of §8.17 is enforced.

It may happen that the edge rays form a caustic. When the point of observation is nowhere near the caustic all that is necessary is to allow for a possible change of phase of $\frac{1}{2}\pi$ as described in §8.13. If the point of observation is close to a caustic, take advantage of the fact that an edge ray appears to be generated by a line source at the rim. Thus equivalent electric and magnetic currents flowing in the rim can be postulated and the field produced computed from the integrals of Chapter 6 for wire antennas. One misfortune is that the equivalent currents depend upon the direction of observation so that the numerical integration may be slow. Should this procedure not resolve the singularity of the field it may be desirable to resort to physical optics since the situation is now one in which the quadratic approximation of §8.13 is no longer adequate.

The physical optics of §8.13 constitutes another way of enumerating the pattern. Its value lies in estimating the field at caustics where GTD fails; in such regions the numerical computation of the double integral is feasible because the phase is not varying over rapidly.

Most high-gain reflector antennas are constructed with dual reflectors rather than a single reflector. They are typed as *Cassegrain*, the prototype of which consists of a paraboloidal main reflector and a hyperboloidal subreflector (Fig. 8.31). More uniform illumination of the main reflector and higher efficiency than a front-fed paraboloid can be accomplished by this arrangement. A common feed for a Cassegrain antenna is a conical hybrid mode (Clarricoats and Salema 1973).

One aim of the original Cassegrain design was to contrive a uniform phase over the aperture of the main reflector. One can commence with geometrical optics despite the subreflector being habitually not many wavelengths in dimension so that

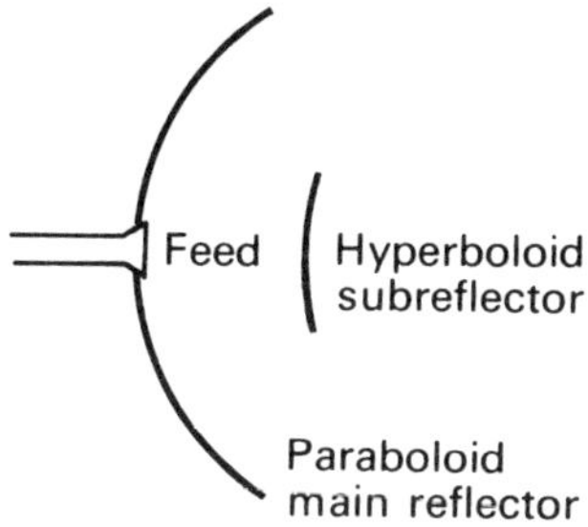

FIG. 8.31. Typical Cassegrain antenna.

diffraction effects are not negligible. The contingency of deviations from predicted patterns due to the struts keeping the subreflector in position must not be forgotten. They are likely to cause deterioration of cross-polarization discrimination as well as scattering energy into the far sidelobes. The far sidelobes will also be influenced by diffraction at the rims of the two reflectors and energy from the feed going past the subreflector (what might be termed *spillover*); there may be spillover of the energy from the subreflector to the main reflector as well. Blockage of the aperture of the main reflector will spring from the subreflector and struts, exerting an effect on the nearer sidelobes.

An extra degree of freedom is available in dual reflectors in that the position and profile of the subreflector can be modified. One possibility is to create a specified phase and amplitude distribution in the aperture of the main reflector via geometrical optics (Galindo 1964, Williams 1965, Potter 1967), so long as the main reflector is in the far field of the subreflector.

Optimization of a different nature can be fulfilled by means of reciprocity theory (Wood 1970, 1971, 1972). According to a standard reciprocity theorem (Jones 1964) if $\mathbf{E}_1$ is the electric intensity due to a current source $\mathbf{a}$ at $\mathbf{x}_1$ and $\mathbf{E}_2$ is that due to $\mathbf{b}$ at $\mathbf{x}_2$, the same conductors being present in both cases,

$$\mathbf{b} \cdot \mathbf{E}_1(\mathbf{x}_2) = \mathbf{a} \cdot \mathbf{E}_2(\mathbf{x}_1) .$$

Let now $\mathbf{E}_1$ be the field when a plane wave of unit power is incident along the axis of the main reflector and the subreflector is ignored. Suppose that when the feed is transmitting unit power it induces a current on the subreflector in harmony with physical optics. In other words, a current density $\mathbf{n} \wedge \mathbf{H}_t$ is produced on the subreflector where $\mathbf{n}$ is the unit normal and $\mathbf{E}_t$, $\mathbf{H}_t$ is the field due to the feed at the subreflector. Regard this current density as responsible for the field $\mathbf{E}_2$ via the main reflector. Then the reciprocity theorem tells us that the field radiated along the boresight is effectively

$$I = \int_S \mathbf{n} \wedge \mathbf{H}_t \cdot \mathbf{E}_1 \mathrm{d}S \qquad (8.280)$$

the integration being over the front of the subreflector. Hence a measure of the *overall efficiency* of the dual reflector system is

$$\eta = II^* \qquad (8.281)$$

provided that physical optics is acceptable on the subreflector. The field $\mathbf{E}_1$ can be estimated by GTD or any other convenient method and so if the field from the feed is known either by computation or measurement the integral in (8.280) can be evaluated. The efficiency is thereby determined.

An arbitrary profile can be assumed for the subreflector and the efficiency maximized by any of the optimization procedures, such as conjugate gradients, described in Chapter 4. (In practice, shaping of the subreflector is much more important than modification of the main reflector.) The effect of small deviations of profile may also be predicted (Poulton 1975). One may also maximize η at the same time as minimizing

$$\left| \int_S \mathbf{n} \wedge \mathbf{H}_t \cdot \mathbf{E}_t^* \mathrm{d}S \right|^2$$

the power delivered to the subreflector by the feed. Moreover, by integrating over a range of frequencies and then optimizing, we have the possibility of arriving at a subreflector profile which will be optimally efficient over a band of frequencies.

LEAKY RAYS

8.27. Gaussian beams and complex sources

The function $\exp(-ikR)/R$ where $R^2 = (x-x_0)^2 + (y-y_0)^2 + (z-z_0)^2$ is a solution of Helmholtz's equation in (x,y,z) for any fixed (x_0,y_0,z_0). Even if x_0, y_0, and z_0 are complex this statement remains true. Suppose, in particular, that $x_0 = y_0 = 0$ and $z = -ib$ with b positive real so that

$$R^2 = \{\rho^2+(z+ib)^2\}^{\frac{1}{2}}$$

where $\rho = (x^2+y^2)^{\frac{1}{2}}$. Then the real values of (x,y,z) at which R vanishes are given by $z = 0$, $\rho = b$. Thus R is a multiple-valued function in (x,y,z) space. To render it single valued introduce a cut in $z = 0$, $\rho \leqslant b$ with R specified to be $z + ib$ on $\rho = 0$ when $z > 0$ and to be $-(z+ib)$ when $z < 0$. Values elsewhere are determined by continuity. Such a definition is consistent with the customary one for R when $b = 0$.

If $\rho^2 \ll z^2 + b^2$ and $z > 0$

$$\frac{\exp(-ikR)}{R} \approx \frac{1}{z + ib} \exp\left\{-ik\left(z+ib + \frac{1}{2}\,\frac{\rho^2}{z+ib}\right)\right\} \tag{8.282}$$

$$\approx \frac{1}{(z^2+b^2)^{\frac{1}{2}}} \exp\left\{-ikz\left(1 + \frac{1}{2}\,\frac{\rho^2}{z^2+b^2}\right) - \frac{1}{2}\,\frac{kb\rho^2}{z^2+b^2} - kb - i\tan^{-1}\frac{b}{z}\right\}. \tag{8.283}$$

Let the range of observation points which satisfy $\rho^2 \ll z^2 + b^2$ be called the *paraxial region*. Then these formulae show that in that part of the paraxial region where $z^2 \ll b^2$ the wave propagates parallel to the z axis with little distortion of the phase front but subject to an exponential decay perpendicular to the z axis proportional to ρ^2. In fact, the field falls to $1/e$ of its value at $\rho = 0$ when $\rho = (2b/k)^{\frac{1}{2}}$. Thus, for $z^2 \ll b^2$,

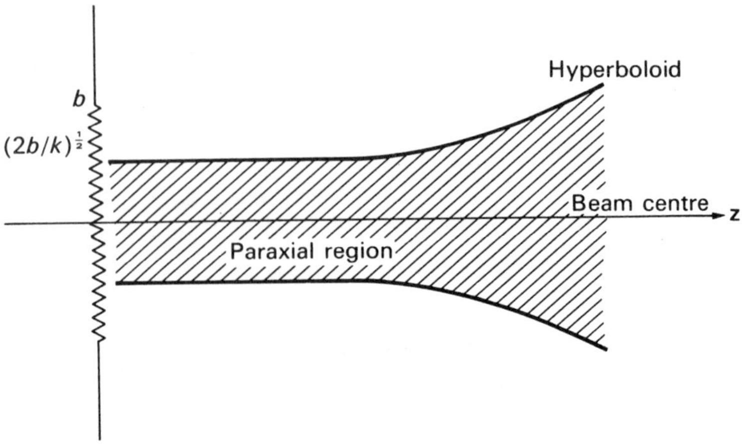

FIG. 8.32. The paraxial region of a Gaussian beam.

a well-collimated beam is formed in the paraxial region (Fig. 8.32). For larger values of z the 1/e points lie on a hyperboloid which is asymptotic to the hyperbolic cone $\frac{1}{2}kb\rho^2 = z^2$. Fields of this type are known as *Gaussian beams*. They are important in beam waveguides and lasers (Goubau and Schwering 1961, Kogelnik and Li 1966). Often the plane $z = 0$ is known as the *beam waist* and $(2b/k)^{\frac{1}{2}}$ is reckoned as the *spot size* at the beam waist.

The above derivation demonstrates that a Gaussian beam is an approximate representation in the paraxial region of an exact solution of Helmholtz's equation due to a source at a complex point (Deschamps 1971). Outside the paraxial region the Gaussian approximation is no longer valid. Nevertheless, it is clear that any known solutions for real point sources which can be continued analytically to sources at complex points will provide solutions for excitation by a Gaussian beam. For example, by employing the standard formulae for a magnetic dipole we may obtain the electromagnetic beam field whose components are, if $G = \exp(-\mathrm{i}kR)/4\pi R$,

$$E_x = \frac{m\omega^2}{cR}(z+\mathrm{i}b)G, \quad E_y = 0, \quad E_z = -\frac{m\omega^2}{cR}xG ,$$

$$H_x = -\frac{mk^2}{\mu_0 R^2}\,xyG, \quad H_y = \frac{mk^2}{\mu_0 R^2}\{x^2+(z+ib)^2\}G, \quad H_z = -\frac{mk^2}{\mu_0 R^2}(z+ib)yG$$

at a distant point. At the centre this beam is transverse electromagnetic. Beams with other polarizations can be produced by giving the magnetic dipole other orientations and by utilizing the electric dipole representation.

Two-dimensional Gaussian beams can be generated in the same way by starting from the line source

$$G_2 = -\tfrac{1}{4}\,iH_0^{(2)}(kr_2)$$

where $r_2 = \{y^2+(z+ib)^2\}^{\frac{1}{2}}$. For large kr_2

$$G_2 \sim \left(\frac{1}{8\pi kr_2}\right)^{\frac{1}{2}} \exp(-ikr_2-\tfrac{1}{4}\pi i)$$

and in the paraxial region $y^2 \ll z^2 + b^2$

$$r_2 \approx z + ib + \tfrac{1}{2}y^2(z+ib)^{-1} \,.$$

By this device problems such as the reflection of a Gaussian beam by a dielectric can be resolved (Ra, Bertoni, and Felsen 1973).

Allowing a complex source to go off to infinity gives rise to a plane wave propagating along a complex angle of incidence so that the wave is evanescent. If such a wave is electrically polarized and strikes a perfectly conducting half-plane the solution is still given by (8.99) but now ϕ_0 is the complex angle $\phi_r - i\phi_i$ with ϕ_r and ϕ_i real. When $k_0 r \gg 1$, asymptotic expressions may be substituted for F unless its argument is small. In contrast to the parabolas of Fig. 8.12 when ϕ_0 is real, failure now occurs within two ellipses which become smaller as ϕ_i increases (Green 1976). The relatively sharp switch across the shadow boundary from the plane wave to the diffraction field (8.100) also does not materialize. Indeed, a little care is necessary in dealing with the shadow boundary because of the evanescent nature of the incident wave.

The asymptotic formula (8.97) ignores exponentially small contributions. If they are included there is no need to alter (8.97) for $-\frac{3}{4}\pi < \text{ph}\ w < \frac{1}{4}\pi$ but for other ranges we have

$$F(w) = \tfrac{1}{2}\pi^{\frac{1}{2}} \exp(iw^2 - \tfrac{1}{4}\pi i) - \frac{i}{2w} + \frac{1}{4w^3} + \ldots \qquad (0 < \text{ph}\ w < \tfrac{1}{2}\pi), \tag{8.284}$$

$$F(w) = \pi^{\frac{1}{2}} \exp(iw^2 - \tfrac{1}{4}\pi i) - \frac{i}{2w} + \frac{1}{4w^3} + \ldots \qquad (\tfrac{1}{4}\pi < \text{ph}\ w < \tfrac{3}{4}\pi). \tag{8.285}$$

It is evident from (8.284), (8.285), and (8.99) that a term of the same order of magnitude as the incident wave is present for $\pi \geqslant \phi \geqslant \phi_r$ though it is negligible compared with the diffracted field as soon as ϕ is sufficiently greater than ϕ_r. If it be required that the term have the precise form of the incident wave the transition to the shadow can be conceived of as complete when $\text{ph}\ w = \frac{1}{4}\pi$, i.e. when $\phi = \phi_c$ where $\tan \frac{1}{2}(\phi_c - \phi_r) = \tanh \frac{1}{2}\phi_i$ or

$$\phi_c = \phi_r + \tan^{-1}(\sinh \phi_i).$$

There is thus a diffuse region stretching from $\phi = \phi_r$ to $\phi = \phi_c$ in which the incident wave disappears and there is no clear-cut shadow boundary. If one wishes to suggest a definite line as the shadow boundary it would appear that any line between $\phi = \phi_r$ and $\phi = \phi_c$ could be chosen (Otis 1974) according to taste.

Exercises

55. Derive the beams for magnetic dipoles parallel to the x and z axes, as well as those for electric dipoles.

56. A line source parallel to the x axis produces a Gaussian beam field $E_x = -\frac{1}{4}\omega\mu_0 H_0^{(2)}(kr_2)$ in a medium $z < 0$ with constants μ_0, ε_0, where

$$r_2^2 = (y + ib \sin\alpha)^2 + (z + d + ib \cos\alpha)^2.$$

Show that in the paraxial region

$$\{(z+d)\sin\alpha - y\cos\alpha\}^2 \ll \{(z+d)\cos\alpha + y\sin\alpha\}^2 + b^2$$

the beam can be visualized as travelling at an angle α to the z axis from $(0, -d)$.

In $z > 0$, ε_0 is changed to ε_1. Show that, in the reflected wave,

$$E_x \sim -\frac{\omega\mu_0 R(\theta)}{2(2\pi kD)^{\frac{1}{2}}}\exp(-ikD+\tfrac{1}{4}\pi i)$$

where

$$D^2 = (y+ib\sin\alpha)^2 + (z-d-ib\cos\alpha)^2,$$

$$\sin\theta = \frac{y+ib\sin\alpha}{D},$$

$$R(\theta) = \frac{\cos\theta - (\varepsilon_1/\varepsilon_0 - \sin^2\theta)^{\frac{1}{2}}}{\cos\theta + (\varepsilon_1/\varepsilon_0 - \sin^2\theta)^{\frac{1}{2}}}.$$

57. In Exercise 56 put $y = \rho_0\sin\theta_0$, $z - d = -\rho_0\cos\theta_0$. Prove that, when θ_0 is near α, $R(\theta) = R(\alpha) + \delta R$ where

$$\frac{\delta R}{R(\alpha)} = \frac{\rho_0}{\rho_0 + ib}\,\frac{2\sin\alpha}{(\varepsilon_1/\varepsilon_0-\sin^2\alpha)^{\frac{1}{2}}}(\theta_0-\alpha).$$

Show that the centre of the reflected beam, defined by those θ_0 for which $\mathcal{R}\{-ikD+\delta R/R(\alpha)\}$ is a maximum, occurs at

$$\theta_0 = \alpha + \frac{2\sin\alpha}{kb\rho_0}\,\mathcal{R}\,\frac{\rho_0 - ib}{(\varepsilon_1/\varepsilon_0-\sin^2\alpha)^{\frac{1}{2}}}$$

when D is replaced by its paraxial approximation. Deduce that, when $\alpha > \sin^{-1}(\varepsilon_1/\varepsilon_0)^{\frac{1}{2}}$, the reflected beam has suffered a lateral displacement of $(2\sin\alpha)k/(\sin^2\alpha - \varepsilon_1/\varepsilon_0)^{\frac{1}{2}}$ parallel to the y axis (Horowitz and Tamir 1971).

If $\alpha < \sin^{-1}(\varepsilon_1/\varepsilon_0)^{\frac{1}{2}}$ prove that the beam shift depends upon the incident beam width and the observation distance ρ_0 from the image source.

8.28. Complex rays

The notion of a source at a complex point invites the proposal that energy propagates from it to the point of observation by means of rays which must, of course, lie in a complex space in order to pass through the source. Examination of the theory of §§8.1–8.3 reveals that most of it is still valid if L is taken to be complex instead of real, though some features such as the curvature of rays cannot be accepted. Complex L implies that the local plane wave of geometrical optics is inhomogeneous. If L is complex the eikonal equation (8.9) and the ray equations (8.25) and (8.27) are still valid provided that one does not reject the possibility that x, y, and z are complex. Suppose, for example, that $N = 1$. Then, from (8.27),

$$\frac{d^2x}{d\sigma^2} = 0 = \frac{d^2y}{d\sigma^2} = \frac{d^2z}{d\sigma^2}$$

whence

$$\frac{x - x_0}{A} = \frac{y - y_0}{B} = \frac{z - z_0}{C} \qquad (8.286)$$

where $A^2 + B^2 + C^2 = 1$ on account of (8.9) and (8.25). In (8.286) any of the quantities x, y, z, x_0, y_0, z_0, A, B, C may be complex and so (8.286) is the equation of a *complex ray*. However, at the point of observation x, y, and z must be real so that the complex initial point (x_0,y_0,x_0) and complex slope (A,B,C) must be chosen to ensure that this is so. In the Gaussian beam x_0 and y_0 are both zero whereas z_0 is pure imaginary.

Another aspect is that it will not be possible to trace L unless its initial values at complex points are known. It is usually feasible to surmount this difficulty by means of analytic continuation.

It is not always necessary to make all the variables complex. To illustrate this point we consider the two-dimensional problem in which

$$L(y,0) = (y^2-b^2)^{\frac{1}{2}} \tag{8.287}$$

on $z = 0$. In view of the preceding section this is equivalent to specifying the spot size of a Gaussian beam at the beam waist. The theory of complex rays should therefore be competent to reconstruct the Gaussian beam in $z > 0$.

Allow y to be complex but keep z real. On $z = 0$ let L satisfy (8.287) even when y is complex, there being no difficulty about the analytic continuation in the paraxial region. On the complex ray $y - y_0 = \lambda z$ the value of L is given by

$$L = \{z^2+(y-y_0)^2\}^{\frac{1}{2}} + L(y_0,0) = \{z^2+(y-y_0)^2\}^{\frac{1}{2}} + (y_0^2-b^2)^{\frac{1}{2}} . \tag{8.288}$$

The relation between λ and y_0 has to be determined. Now, from (8.278), $\partial L(y,0)/\partial y = y/(y^2-b^2)^{\frac{1}{2}}$ and hence the eikonal equation (8.9) supplies $\partial L(y,0)/\partial z = \mathrm{i}b/(y^2-b^2)^{\frac{1}{2}}$. It follows from (8.25) that the slope of the ray must satisfy $\mathrm{d}y/\mathrm{d}z = y_0/\mathrm{i}b$ on $z = 0$. Thus $\lambda = y_0/\mathrm{i}b$ and the equation of the ray is $y - y_0 = y_0 z/\mathrm{i}b$. Eliminating y_0 from (8.288) by means of this equation, we obtain

$$L = \{y^2+(z+\mathrm{i}b)^2\}^{\frac{1}{2}}$$

in agreement with the earlier formula for the Gaussian beam.

There are consequently two alternative viewpoints. Either the Gaussian beam may be visualized as composed of complex rays passing through a single point in complex (y,z) space or as consisting of rays in complex y and real z space whose sources are distributed over $z = 0$.

Equations which are entirely real can be achieved by separating the complex equations into their real and imaginary parts (Felsen 1976). Let $L = L_r + \mathrm{i}L_i$ where L_r and L_i are real. Then (8.9) becomes

$$\text{grad}^2 L_r - \text{grad}^2 L_i = N^2, \qquad \text{grad } L_r \cdot \text{grad } L_i = 0. \quad (8.289)$$

Let $\mathbf{s}_r, \mathbf{s}_i$ be the unit vectors

$$\mathbf{s}_r = \frac{\text{grad } L_r}{|\text{grad } L_r|}, \qquad \mathbf{s}_i = \frac{\text{grad } L_i}{|\text{grad } L_i|}$$

and s_r, s_i the corresponding arc lengths on the curves to which they are tangential. Analogous to (8.28) we have

$$\frac{d}{ds_r}(|\text{grad } L_r|\mathbf{s}_r) = \text{grad}|\text{grad } L_r|,$$

$$\frac{d}{ds_i}(|\text{grad } L_i|\mathbf{s}_i) = \text{grad}|\text{grad } L_i|. \quad (8.290)$$

From the second part of (8.289), L_i is constant on a trajectory parallel to $\mathbf{s}_r$ and such a trajectory is therefore known as a *phase path*. L_r is constant on trajectories parallel to $\mathbf{s}_i$, and so they are called *equiphase curves*. Phase paths and equiphase curves are orthogonal to one another because of (8.289).

Corresponding to (8.18) the curvatures of phase paths and equiphase curves are given by

$$\kappa_r = \boldsymbol{\nu}_r \cdot \text{grad ln}|\text{grad } L_r|, \qquad \kappa_i = \boldsymbol{\nu}_i \cdot \text{grad ln}|\text{grad } L_i|.$$

Thus a phase path bends whenever $|\text{grad } L_r|$ varies along an equiphase curve and, *mutatis mutandis*, there is similar bending of equiphase curves. In an inhomogeneous plane wave the phase paths and equiphase curves are orthogonal straight lines.

If the spacing between equiphase curves contracts $|\text{grad } L_r|$ must increase and then (8.289) implies that $|\text{grad } L_i|$ also grows when N is constant. Thus the field becomes steadily more damped and evanescent in this direction. Conversely, when the spacing between equiphase curves expands $|\text{grad } L_i|$ diminishes. Should the situation be reached where $|\text{grad } L_i| \to 0$ the possibility of a non-evanescent field arises. Thus there can be a conversion from a zone of evanescence to non-evanescence with consequent leakage of energy. For this reason complex rays are sometimes called *leaky rays*. However, it is important to realize that phase paths and equiphase curves are real curves which should not be identified with complex or leaky rays, these being straight lines in a complex space when N is constant.

Amplitudes can be dealt with via (8.8), (8.16), and (8.30) which may also be split into their real and imaginary parts if desired.

Exercises

58. If (8.287) is replaced by $L(y,0) = f(y)$ where f is analytic show that the slope dy/dz of the complex ray through $(y_0,0)$ is $f'(y_0)[1-\{f'(y_0)\}^2]^{-\frac{1}{2}}$.

59. Suppose that in two dimensions the phase paths are concentric circles with centre at the origin and that, on $r = a$, $|\text{grad } L_r| = \beta\ (> 1)$. If $N = 1$ prove that, for $r \geqslant a$, $|\text{grad } L_r| = \beta a/r$ and $|\text{grad } L_i| = (\beta^2 a^2/r^2 - 1)^{\frac{1}{2}}$. Deduce that $L_r = \beta s_r$ and

$$L_i = L_{ia} + \beta a(\tanh \tau - \tanh \tau_a - \tau + \tau_a)$$

where $\cosh \tau = \beta a/r$ and the subscript a indicates that $r = a$.

Examine what happens as $r \to \beta a$.

8.29. Optical fibres

The simplest kind of optical fibre consists of a circular cylinder of dielectric material surrounded by another dielectric. In more sophisticated versions there may be additional concentric annular regions of dielectric and the refractive index may vary in a radial direction transverse to the cylindrical axis. Such structures may be tackled in terms of modes by employing separation of variables (§6.23). Modal expansions are not, however, as straightforward as for metal waveguides because an appreciable part of the energy can be transported outside the guiding cylinder and because a contribution from a continuous spectrum may have to be allowed for. A full modal investigation of an optical fibre can therefore be highly complicated. For this reason the discussion of this section will be limited to those circumstances in which the propagation can be regarded as taking place along rays. This simplification has the advantage of avoiding the global difficulties of modes at the price of concentrating on frequencies for which the ray picture is adequate. A further benefit is that, because rays depend only upon local properties, generalization to fibres which are not strictly circular or whose refractive index defies exact enumeration of the modes becomes a possibility.

It will be sufficient for our purposes to consider what happens at a curved interface where the dielectric constants change discontinuously but are uniform everywhere else. Such an interface may be able to support a surface wave whose energy is confined to an immediate neighbourhood of the boundary. A two-dimensional model of the situation is that in which the phase paths are concentric circles, one of the phase paths coinciding with the dielectric interface (Fig. 8.33). Then, if $|\text{grad } L_r| = \beta$ on $r = a$,

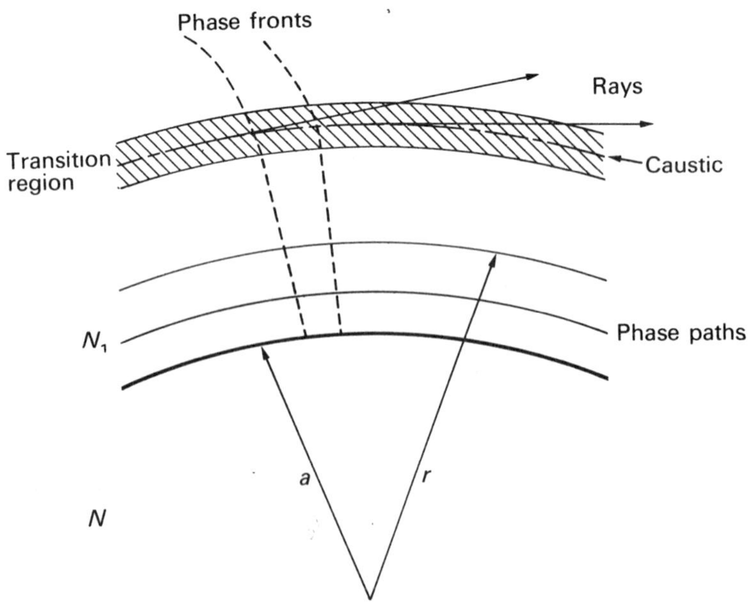

FIG. 8.33. Model of evanescent field outside a curved dielectric.

$$|\text{grad } L_r| = \frac{\beta a}{r}, \qquad |\text{grad } L_i| = \left(\frac{\beta^2 a^2}{r^2} - N_1^2\right)^{\frac{1}{2}}$$

where N_1 is the refractive index of the outer medium. So long as $\beta > N_1$ there is a region of evanescence adjacent to the interface. As r increases to $\beta a/N_1$, $|\text{grad } L_i|$ decreases to zero. Thus the evanescent wave is converted to a non-evanescent wave which carries energy in $r > \beta a/N_1$ along conventional rays tangential to the caustic $r = \beta a/N_1$. In other words, after a period of exponential decay the complex ray is transformed into a real ray whose diminution of amplitude is due solely to spreading by radiation. Rays which reappear as conventional real ones after traversing a region of evanescence are sometimes known as *tunnelling rays*. Although the rays in the state of propagation lose amplitude only by radiation, they will be exponentially smaller than rays which have not tunnelled because of the dissipation of energy in reaching the caustic. Nevertheless there may be an appreciable leakage of energy from the caustic.

It is important to observe that tunnelling rays do not occur for a plane interface since the caustic moves off to infinity as $a \to \infty$. Hence tunnelling rays represent a surface wave which is peculiar to a curved dielectric.

The amplitude changes which take place at a dielectric interface may be calculated from Fresnel's laws for a plane boundary to a first approximation (§8.5). To obtain the appropriate coefficients in *three dimensions*, recall that if the plane wave

$$\mathbf{E}^i = (l\mathbf{i}+m\mathbf{j}+n\mathbf{k})\exp\{-ikN(x \sin\theta_0 \cos\phi_0 + y \sin\theta_0 \sin\phi_0 + z \cos\theta_0)\}$$

with

$$l \sin\theta_0 \cos\phi_0 + m \sin\theta_0 \sin\phi_0 + n \cos\theta_0 = 0$$

is incident on $y = 0$, the reflected wave has components (the exponential factor being omitted)

$$E_x^{\mathrm{r}} = R^{\mathrm{E}}l + \frac{(R^{\mathrm{E}}+R^{\mathrm{M}})m \cos\theta_{\mathrm{i}} \sin\theta_0 \cos\phi_0}{\sin^2\theta_{\mathrm{i}}}, \tag{8.291}$$

$$E_y^{\mathrm{r}} = R^{\mathrm{M}}m, \tag{8.292}$$

$$E_z^{\mathrm{r}} = R^{\mathrm{E}}n + \frac{(R^{\mathrm{E}}+R^{\mathrm{M}})m \cos\theta_{\mathrm{i}} \cos\theta_0}{\sin^2\theta_{\mathrm{i}}} \tag{8.293}$$

with R^{E} and R^{M} given by (8.55) and (8.57) respectively, θ being replaced by the angle of incidence θ_{i} where $\cos\theta_{\mathrm{i}} = \sin\theta_0 \sin\phi_0$. Similarly, for the transmitted field

$$E_x^{\mathrm{t}} = T^{\mathrm{E}}l + \frac{(R^{\mathrm{E}}+R^{\mathrm{M}})m \cos\theta_{\mathrm{i}} \sin\theta_0 \cos\phi_0}{\sin^2\theta_{\mathrm{i}}}, \tag{8.294}$$

$$E_y^{\mathrm{t}} = \frac{NT^{\mathrm{M}}m}{N_1}, \tag{8.295}$$

$$E_z^{\mathrm{t}} = T^{\mathrm{E}}n + \frac{(R^{\mathrm{E}}+R^{\mathrm{M}})m \cos\theta_{\mathrm{i}} \cos\theta_0}{\sin^2\theta_{\mathrm{i}}} \tag{8.296}$$

where T^{E} and T^{M} are given by (8.56) and (8.58) respectively (θ_{i} being substituted for θ). Noteworthy is the fact that components of the reflected and transmitted waves tangential to the interface can be created by the normal components of the incident field. Also, when dealing with rays in general, the twisting of the plane of polarization as described in §8.3 must be borne in mind.

The formulae for the reflection and transmission coefficients are defined for all angles of incidence θ_{i} but there are important differences of interpretation if $\sin\theta_{\mathrm{i}} > N_1/N$; such an eventuality can arise only when $N_1/N < 1$. If $N_1/N < 1$ define the *critical angle of incidence* θ_{c} by $\sin\theta_{\mathrm{c}} = N_1/N$. For $\theta_{\mathrm{i}} > \theta_{\mathrm{c}}$, $(N_1^2 - N^2 \sin^2\theta_{\mathrm{i}})^{\frac{1}{2}}$ is taken to be negative imaginary and the transmitted wave with coefficients (8.294)–(8.296) will be exponentially attenuated as it moves along. Thus, unless the coefficients are modified, the phenomenon of tunnelling which occurs for curved interfaces will be concealed.

By consideration of the canonical problem it is elicited (Snyder and Love 1975, Jones 1978) that the appropriate reflection coefficients are those of (8.55) and (8.57) with $\{(N_1/N)^2-\sin\theta_i\}^{\frac{1}{2}}$ replaced by

$$-\frac{\mathrm{Ai}'(u_0\exp(\frac{1}{3}\pi\mathrm{i}))}{\{\frac{1}{2}kN\rho_0\exp(\frac{1}{2}\pi\mathrm{i})\}^{\frac{1}{3}}\mathrm{Ai}(u_0\exp(\frac{1}{3}\pi\mathrm{i}))}$$

where Ai is the standard Airy function. The quantity ρ_0 is the radius of curvature of the section of the interface by the plane of incidence, i.e. the plane containing the incident ray and surface normal, whereas

$$u_0 = \frac{(\frac{1}{2}kN\rho_0)^{\frac{2}{3}}(\sin^2\theta_c-\sin^2\theta_i)}{\{1+\rho_0(\sin^2\theta_c-\sin^2\theta_i)/b\}^{\frac{2}{3}}} \tag{8.297}$$

b being the larger of the principal radii of curvature. It is understood that $\sin^2\theta_c = N_1^2/N^2$ even when $N_1 > N$. Corresponding results for transmission are more recondite because of the presence of a caustic when $\theta_i > \theta_c$. These formulae are relevant when the incident ray is on the concave side of the interface; no change to R^E and R^M is necessary for incidence from the convex side because there is then no tunnelling.

Denote the reflection coefficients just delineated by $\tilde{R}^E$ and $\tilde{R}^M$. Unless $\theta_i \approx \theta_c$, $|u_0|$ will be large since ray theory demands radii of curvature which are not small. Hence the asymptotic formulae

$$\mathrm{Ai}(w) \sim \frac{\exp(-\frac{2}{3}w^{3/2})}{2\pi^{\frac{1}{2}}w^{\frac{1}{4}}} \qquad (|\mathrm{ph}\ w| < \pi)$$

$$\mathrm{Ai}'(w) \sim -\frac{w^{\frac{1}{4}}}{2\pi^{\frac{1}{2}}}\exp(-\frac{2}{3}w^{3/2}) \qquad (|\mathrm{ph}\ w| < \pi) \tag{8.298}$$

$$\mathrm{Ai}(w) \sim \frac{\exp(\frac{1}{4}\pi\mathrm{i})}{\pi^{\frac{1}{2}}w^{\frac{1}{4}}}\cos(\tfrac{2}{3}\mathrm{i}w^{3/2}-\tfrac{1}{4}\pi) \qquad (\tfrac{1}{3}\pi < \mathrm{ph}\, w < \tfrac{5}{3}\pi) \tag{8.299}$$

$$\mathrm{Ai}'(w) \sim \frac{w^{\frac{1}{4}}}{\pi^{\frac{1}{2}}}\exp(-\tfrac{1}{4}\pi\mathrm{i})\sin(\tfrac{2}{3}\mathrm{i}w^{3/2}-\tfrac{1}{4}\pi) \qquad (\tfrac{1}{3}\pi < \mathrm{ph}\, w < \tfrac{5}{3}\pi)$$

may be employed. It follows that $\tilde{R}^{\mathrm{E}} \sim R^{\mathrm{E}}$, $\tilde{R}^{\mathrm{M}} \sim R^{\mathrm{M}}$ when θ_{i} is not near θ_{c}. Consequently, the reflection coefficients are, to a first approximation, those of a plane interface when the angle of incidence is not approaching critical.

For a particular polarization a guide to the energy which is transmitted is $1-|\tilde{R}^{\mathrm{E}}|^2$ or $1-|\tilde{R}^{\mathrm{M}}|^2$. From

$$\exp(\tfrac{1}{3}\pi\mathrm{i})\mathrm{Ai}'(x\exp(\tfrac{1}{3}\pi\mathrm{i}))\mathrm{Ai}(x\exp(\pi\mathrm{i})) - $$

$$- \exp(\pi\mathrm{i})\mathrm{Ai}(x\exp(\tfrac{1}{3}\pi\mathrm{i}))\mathrm{Ai}'(x\exp(\pi\mathrm{i})) = -\frac{\exp(\frac{1}{6}\pi\mathrm{i})}{2\pi} \tag{8.300}$$

and

$$\mathrm{Ai}(x\exp(\pi\mathrm{i})) = \exp(\tfrac{1}{3}\pi\mathrm{i})\mathrm{Ai}(x\exp(\tfrac{1}{3}\pi\mathrm{i})) + \exp(-\tfrac{1}{3}\pi\mathrm{i})\mathrm{Ai}(x\exp(-\tfrac{1}{3}\pi\mathrm{i})) \tag{8.301}$$

it is discovered that

$$1 - |\tilde{R}^{\mathrm{E}}|^2 = \frac{4\mu\mu_1 NN_1\cos\theta_{\mathrm{i}}|\cos\theta_{\mathrm{r}}|H(u_0)}{N^2\mu_1^2\cos^2\theta_{\mathrm{i}}+2\mu\mu_1 NN_1\cos\theta_{\mathrm{i}}|\cos\theta_{\mathrm{r}}|H+\mu^2N_1^2|\cos\theta_{\mathrm{r}}|^2HJ}, \tag{8.302}$$

$$1 - |\tilde{R}^{\mathrm{M}}|^2 = \frac{4\varepsilon\varepsilon_1 NN_1\cos\theta_{\mathrm{i}}|\cos\theta_{\mathrm{r}}|H(u_0)}{N^2\varepsilon_1^2\cos^2\theta_{\mathrm{i}}+2\varepsilon\varepsilon_1 NN_1\cos\theta_{\mathrm{i}}|\cos\theta_{\mathrm{r}}|H+\varepsilon^2N_1^2|\cos\theta_{\mathrm{r}}|^2HJ} \tag{8.303}$$

where $N_1\cos\theta_{\mathrm{r}} = N(\sin^2\theta_{\mathrm{c}}-\sin^2\theta_{\mathrm{i}})^{\frac{1}{2}}$, θ_{r} being negative imaginary when $\theta_{\mathrm{i}} > \theta_{\mathrm{c}}$, and

$$\frac{1}{H(u_0)} = 4\pi|u_0|^{\frac{1}{2}}|\mathrm{Ai}(u_0\ \exp(\tfrac{1}{3}\pi i))|^2 , \tag{8.304}$$

$$J(u_0) = \frac{4\pi|\mathrm{Ai}'(u_0\ \exp(\frac{1}{3}\pi i))|^2}{|u_0|^{\frac{1}{2}}} \tag{8.305}$$

with $u_0 = |u_0|\exp(-\pi i)$ when negative.

When $|u_0| \gg 1$ it may be inferred from (8.289) and (8.299) that

$$\begin{aligned} H(u_0) &\sim 1 \quad (u_0 > 0) \\ &\sim \exp(-\tfrac{4}{3}|u_0|^{3/2}) \quad (u_0 < 0), \end{aligned} \tag{8.306}$$

$$\begin{aligned} J(u_0) &\sim 1 \quad (u_0 > 0) \\ &\sim \exp\ (\tfrac{4}{3}|u_0|^{3/2}) \quad (u_0 < 0). \end{aligned} \tag{8.307}$$

Hence, for $|u_0| \gg 1$,

$$1 - |\tilde{R}^E|^2 \sim \frac{4\mu\mu_1 NN_1 \cos\theta_i \cos\theta_r}{(N\mu_1\cos\theta_i + N_1\mu\cos\theta_r)^2} \quad (u_0 > 0) \tag{8.308}$$

$$\sim \frac{4\mu\mu_1 NN_1\cos\theta_i|\cos\theta_r|\exp(-\frac{4}{3}|u_0|^{3/2})}{N^2\mu_1^2\cos^2\theta_i + N_1^2\mu^2|\cos\theta_r|^2} \quad (u_0 < 0) \tag{8.309}$$

When $|u_0|$ is not large, $\theta_i \approx \theta_c$ because of the huge factor $kN\rho_0$ in (8.297). Furthermore, the second and third terms of the denominator of (8.302) can be viewed as products of $(\frac{1}{2}kN\rho_0)^{-\frac{1}{3}}$ and a bounded quantity. They are therefore negligible provided that

$$\tfrac{1}{2}kN\rho_0\mu_1 \cos^3\theta_c \gg \mu . \tag{8.310}$$

Accordingly, for moderate values of u_0

$$1 - |\tilde{R}^E|^2 \approx \frac{4N_1\mu\mu_1\cos\theta_i|\cos\theta_r|H(u_0)}{N\mu_1^2\cos^2\theta_i}. \tag{8.311}$$

Since $\theta_i \approx \theta_c$, the denominator does not differ by much from $|N\mu_1\cos\theta_i+N_1\mu\cos\theta_r|^2/N$. On this basis (8.308), (8.309) and (8.311) may be combined into the single formula

$$1 - |\tilde{R}^E|^2 = \frac{4NN_1\mu\mu_1\cos\theta_i|\cos\theta_r|H(u_0)}{|N\mu_1\cos\theta_i+N_1\mu\cos\theta_r|^2} \tag{8.312}$$

with tolerable accuracy over the whole range of u_0.

There is a corresponding formula for $\tilde{R}^M$ deduced from (8.303) by exchanging μ,μ_1 and $\varepsilon,\varepsilon_1$ in (8.312).

From (8.297), there is the simplification

$$u_0 \approx (\tfrac{1}{2}kN\rho_0)^{2/3}(\sin^2\theta_c-\sin^2\theta_i) \tag{8.313}$$

when $\theta_i \approx \theta_c$. However, (8.313) must be dropped in favour of (8.297) when $\theta_i - \theta_c$ becomes increasingly positive otherwise (8.312) will overestimate the amount of transmitted energy.

While (8.312) is satisfactory for an incident wave that is E polarized some care is necessary when both polarizations exist because then cross-multiplication may produce algebraic decay rather than exponential for $u_0 \ll -1$ and additional terms in (8.306) and (8.307) may be necessary.

For the special case of a circular optical fibre, one principal radius of curvature b is infinite and the other is the radius a of the cylinder. Let the z axis be the axis of the cylinder and suppose that a typical ray in the cylinder travelling in the direction of z increasing makes an angle θ_z with the z axis. At its point of impact with the cylindrical boundary let the projection of the ray on the cross-section of the cylinder make an angle θ_s with the radius vector from the centre of the cross-section to the point of impact (Fig. 8.34). Then

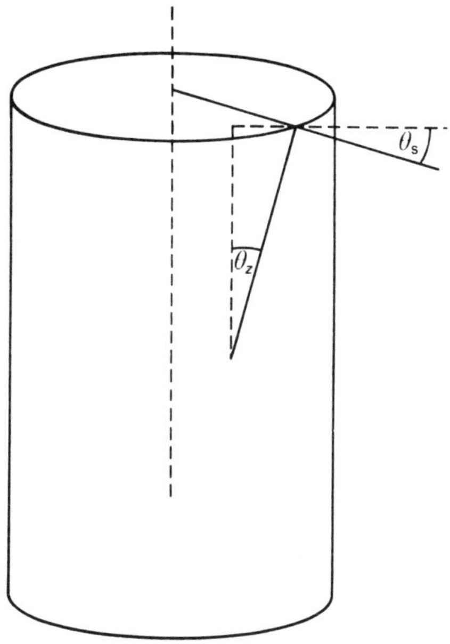

FIG. 8.34. Angles for propagation along a fibre.

$$\cos\theta_i = \sin\theta_z \cos\theta_s. \tag{8.314}$$

Meridional rays pass through the axis of the cylinder and are such that $\theta_s = 0$, $\cos\theta_i = \sin\theta_z$. Rays in which $\theta_s \neq 0$ are called *skew rays*.

To illustrate the procedure we shall concentrate solely on rays which are polarized so that (8.312) is applicable. For meridional rays ρ_0/b is unity and ρ_0 is infinite. Thus u_0 will be positive or negative infinity according as θ_i is less or greater than θ_c. Consequently, a meridional ray will either radiate energy by refraction as if at a plane interface $(\theta_i < \theta_c)$ or radiate no energy at all $(\theta_i > \theta_c)$, there being a virtually instantaneous transition when θ_i passes through θ_c. If the initiating energy in a hollow cone between the angles θ_z and $\theta_z + d\theta_z$ is $2\pi G(\theta_z)\sin\theta_z\, d\theta_z$ the energy which leaves the fibre by refraction in a distance z is

$$2\pi \int_{\frac{1}{2}\pi-\theta_c}^{\frac{1}{2}\pi} G(\theta)\{1-|R^E|^{2n}\}\sin\theta\, d\theta$$

where n is the number of reflections in a distance z given by

$$n = \frac{z}{2a} \tan \theta .$$

If $N_1 > N$, the lower limit of integration is 0.

Sources off the axis will be responsible for skew rays and maybe meridional rays. If a ray starts from $z = 0$ at the point whose polar coordinates are (c, χ) and goes in a direction making an angle θ_z with the cylindrical axis with its projection on $z = 0$ having the inclination of the polar angle χ_0

$$a \sin \theta_s = c \sin(\chi - \chi_0) \tag{8.315}$$

at the point of reflection. Suppose that $G(\theta_z, c, \chi, \chi_0)$ is a measure of the incident energy along the ray so that

$$G(\theta_z, c, \chi, \chi_0) c\,dc\,d\chi \sin \theta_z \, d\theta_z \, d\chi_0$$

is the energy launched from a small area about (c, χ) into an elemental solid angle about the specified direction of propagation. Then the energy which radiates from a length z of the cylinder due to the reflection of this elementary beam is

$$G(\theta_z, c, \chi, \chi_0)\{1 - |\tilde{R}^{\mathrm{E}}|^{2\nu}\} c\,dc\,d\chi \sin \theta_z \, d\theta_z \, d\chi_0$$

where ν is the number of reflections given by

$$\nu = \frac{1}{2} \frac{z \tan \theta_z}{a \cos \theta_s}$$

and θ_i is defined by (8.314). Moreover, from (8.297),

$$u_0 = (\tfrac{1}{2}ka \operatorname{cosec}^2\theta_s \operatorname{cosec}^2\theta_z)^{\frac{2}{3}} \left\{ \left(\frac{N_1}{N}\right)^2 - 1 + \sin^2\theta_z \cos^2\theta_s \right\} . \tag{8.316}$$

The total radiation due to all rays is obtained by integration, in which it is advantageous to convert χ_0 to θ_s via (8.315). Thus, if τ is the total radiation,

$$\tau = 4\int_0^a\int_0^{2\pi}\int_0^{\frac{1}{2}\pi}\int_0^{\sin^{-1}(c/a)} G(\theta_z,c,\chi,\chi_0)\left\{1 - |\tilde{R}^E|^{2\nu}\right\}\frac{a\cos\theta_s}{(c^2-a^2\sin^2\theta_s)^{\frac{1}{2}}}\, c\,dc\,d\chi\,\sin\theta_z\,d\theta_z\,d\theta_s\,. \tag{8.317}$$

If, in particular, G is independent of c and χ_0 put

$$\int_0^{2\pi} G(\theta_z,c,\chi,\chi_0)d\chi = \frac{2G_0(\theta_z)}{a^2}$$

and then, because of the structure of $\tilde{R}^E$ and (8.316),

$$\tau = 8\int_0^{\frac{1}{2}\pi}\int_0^{\frac{1}{2}\pi} G_0(\theta_z)\{1-|\tilde{R}^E|^{2\nu}\}\sin\theta_z\,d\theta_z\,d\theta_s\,. \tag{8.318}$$

The value $\theta_s = 0$ is admitted in (8.317) and (8.318) so that both include the radiation due to meridional rays. Also tunnelling is allowed for. The separate contributions from tunnelling and refraction can be assessed by limiting θ_s to a value less than that for which $N\cos\theta_s = (N^2-N_1^2)^{\frac{1}{2}}\text{cosec}\,\theta_z$ (this limit being $\frac{1}{2}\pi$ if $N_1 > N$) in refraction and letting θ_s range above this limit for tunnelling. Obviously, as soon as θ_s exceeds the limiting value by a significant amount the contribution will be exponentially small on account of (8.309). Whether the rays near critical incidence will supply an appreciable effect depends on the relative sizes of various parameters but they should not be ignored without calculation. The integrand of (8.318) decreases as θ_s increases so that the rays which are not far off meridional tend to radiate more than the rays which are more skew. On the other hand, the integrand (apart from G_0) is an increasing function of θ_z with the consequence that isotropic sources will produce more radiated energy from their rays with larger values of θ_z.

Exercise

60. By assuming that the z dependence of all fields is $\exp(-ikN\cos\theta_s)$ in the fibre problem, deduce that when two-dimensional rays are refracted by a curve of radius of curvature a

$$1 - |\tilde{R}^E|^2 = \frac{4\mu\mu_1 N\cos\theta_s |N_1^2 - N^2\sin^2\theta_s|^{\frac{1}{2}}}{|N\mu_1\cos\theta_s + \mu(N_1^2 - N^2\sin^2\theta_s)^{\frac{1}{2}}|^2} H(u_0)$$

with

$$u_0 = \left(\frac{\frac{1}{2}kNa}{\sin^2\theta_s}\right)^{\frac{2}{3}} \left\{\left(\frac{N_1}{N}\right)^2 - \sin^2\theta_s\right\}$$

and θ_s the (two-dimensional) angle of incidence.

What is the analogous result for the other polarization?

9

SOURCE DETECTION

9.1. General considerations

The location of the origins of an electromagnetic disturbance can be of considerable importance, either because one wishes to know where a source of scattering or radiation is situated or because one wishes to eliminate its influence on other signals. A related problem of practical significance is the determination of the properties of a medium from its effect on a wave; there are obvious applications in the detection of minerals, in the investigation of materials, and in assessing the state of the atmosphere for propagation purposes.

It is vital to get straight right away that problems of this type do not usually have a unique solution and that assertions about the location of sources are highly dependent upon the assumptions made in analysing the signals. A common supposition is that there is a closed surface S containing all the sources and that a homogeneous isotropic medium lies between S and the observer O (Fig. 9.1). The general representations (6.75) and (6.76), which are *exact*, demonstrate that the observed field is equivalent to that generated by a suitable distribution of sources on S. One can infer that these are *apparent* (i.e. not *real*) sources if one has *a priori* knowledge

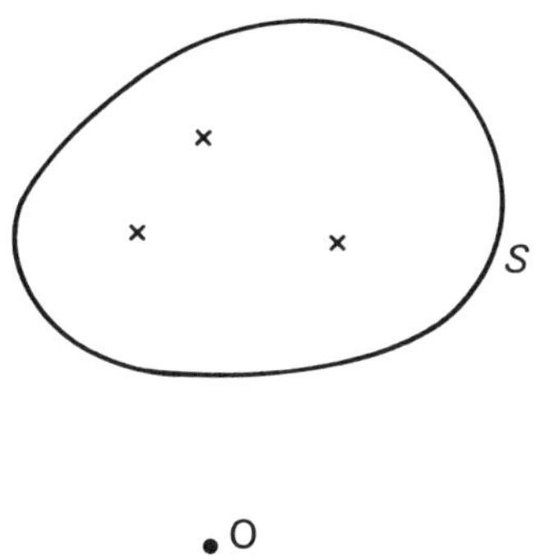

FIG. 9.1. Surface containing sources.

that the sources are not actually on S. However, a similar representation in terms of apparent sources exists for any closed surface surrounding S but excluding O. Therefore, even the determination of apparent sources is subject to a fair degree of ambiguity. The most that can be asserted is that, given S, the field can be accounted for by a certain distribution of sources on S.

One may also have to be equivocal about the source distribution because there are sources which do not radiate energy at infinity. To see this (Friedlander 1973) consider Maxwell's equations in a homogeneous isotropic dielectric

$$\operatorname{curl} \mathbf{E} + i\omega\mu \mathrm{H} = \mathbf{0}, \quad \operatorname{curl} \mathrm{H} - i\omega\varepsilon \mathbf{E} = \mathbf{j} \tag{9.1}$$

where the current density $\mathbf{j}$ is reasonably smooth and vanishes outside a finite region. Apply the operator (curl curl $-k^2$), where $k^2 = \omega^2\mu\varepsilon$, to (9.1). Then a current density is produced which supplies an electromagnetic field in which the electric intensity is $-i\omega\mu\mathbf{j}$ and the magnetic intensity is curl $\mathbf{j}$. Since $\mathbf{j}$ vanishes outside a finite region so does this electromagnetic field and so a source distribution which is responsible for a non-radiating field has been created.

It has been proved (Bleistein and Cohen 1977) that, if $\mathbf{j}$ is zero outside a finite domain, eqns (9.1) possess a non-radiating solution if and only if

$$|\kappa|^2\tilde{\mathbf{j}}(\kappa, c|\kappa|) - \kappa\{\kappa.\tilde{\mathbf{j}}(\kappa, c|\kappa|)\} = \mathbf{0} \tag{9.2}$$

where $c = 1/(\mu\varepsilon)^{\frac{1}{2}}$,

$$\tilde{\mathbf{j}}(\kappa,\omega) = \int \mathbf{j}(x,\omega)\exp(i\kappa.\mathbf{x})\,d\mathbf{x} \tag{9.3}$$

and the integration in (9.3) is over the whole space. Convergence of (9.3) is immediate since $\mathbf{j}$ is non-zero only in a finite domain. Current densities for which (9.2) is satisfied need to be treated with care.

It must therefore be recognized that the formulation for source detection may often not yield a unique solution and that, even when additional information is available, the difficulty may not be resolved. For instance, suppose it is alleged that the sources are to be regarded as radiating in free space and their positions are to be determined from distant observations by tracing rays backwards. Then, consider the situation in which there is a point source M in a dielectric of higher refractive index than where the observer is located, the interface between the two being a plane (Fig. 9.2).

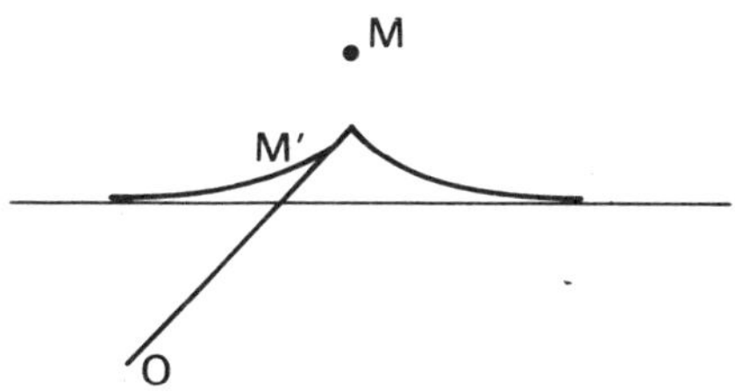

FIG. 9.2. An observer at O places M at M′.

The observer will place the apparent source at the point where nearby rays focus, i.e. on the caustic of the refracted rays. An observer at O therefore puts the source at M′ where OM′ is tangent to the caustic. Thus the apparent location of the source depends upon the position of the observer and its apparent direction from the observer will not be that of the true source except at normal incidence. Only detailed information about the medium containing the source would allow any likelihood of detecting the real source from the caustic traced by the apparent sources.

Additional knowledge may be forthcoming from other approaches. For example, by probing at other frequencies, say optical or infrared, it may be possible to deduce properties of the medium, or again a source might give out a variety of disturbances some of which were more easily analysed than others. However, one does need to be convinced that it is the same source which emits the different kinds of radiation before this latter avenue is open.

The lack of uniqueness may appear, at first sight, to contradict everyday experience because telescopes and eyes seem to supply unambiguous images (except maybe in optical illusions).

However, objects which convey sharp images usually do so by radiation which is incoherent in space. Highly coherent radiation from lasers is often afflicted by *speckle* and good visual holograms usually demand that the illumination be sufficiently diffuse. Similarly, polished objects viewed by specularly reflected light are often difficult to recognize. The knowledge that the radiation from a source is spatially incoherent can be a valuable piece of information.

Even if the theory predicted a unique solution, enough data, as uncontaminated by noise as practicable, has to be collected for the application of the theory in a numerically sound fashion.

It is clear therefore that the detection of sources can be fraught with anxiety for the experimenter and a considerable body of experience and skill may have to be built up before interpretations of physical significance can be accepted with confidence.

INVERSE SCATTERING

9.2. Low frequencies

The first problem to be investigated is that in which it is known that a far field has been scattered by a perfectly conducting obstacle in an otherwise homogeneous isotropic dielectric but the shape and location of the scatterer are to be found. It will be sufficient for illustrative purposes to consider the two-dimensional case; the discussion can easily be extended in principle to three-dimensional problems (Imbriale and Mittra 1970).

Let the scatterer have generators parallel to the z axis and be irradiated by a plane wave with its electric intensity perpendicular to the (x,y) plane. Denoting the electric intensity by E and choosing the x axis in the direction of propagation of the plane wave we have $E^{\mathrm{i}} = \exp(-ikx)$ in the incident wave, k being the wavenumber of the medium. A scattered field E^{s} will be produced such that $E^{\mathrm{i}} + E^{\mathrm{s}}$ is zero on the obstacle and

$$E^{s} \sim \left(\frac{2}{\pi kr}\right)^{\frac{1}{2}} A(\phi)\exp\{-i(kr-\tfrac{1}{4}\pi)\} \tag{9.4}$$

as the distance r from the origin increases, ϕ being the polar angle which is zero on the x axis. The pattern function $A(\phi)$ is supposed to be known and from it the position of the scatterer is to be determined.

Write $A(\phi) = \Sigma a_n \exp(in\phi)$. The coefficients a_n can be calculated by standard Fourier analysis (cf. §§1.5, 2.14) and, in practice, the expansion is limited to a finite number of terms.

Outside the minimum radius $r = a$ of the circle enclosing the obstacle (Fig. 9.3) it is known that (9.4) is the asymptotic expansion of

$$E^{s} = \Sigma a_n H_n^{(2)}(kr)\exp\{in(\phi-\tfrac{1}{2}\pi)\} \tag{9.5}$$

where $H_n^{(2)}$ is the Hankel function of the second kind. Actually, a is not given because the location of the scatterer is unknown. However, if $E^{i} + E^{s}$ is calculated from (9.5) for decreasing r a value of r will be reached at which $E^{i} + E^{s}$ is zero at a point on the circumference of the circle. This value of r will be adjudged to be a and the point as the point of contact between the minimum circle and perimeter of the obstacle. In practice, a zero will rarely be achieved precisely and it will be necessary to be satisfied when $E^{s} + E^{i}$ is less than some pre-assigned tolerance (which must not be tighter than the data can attain).

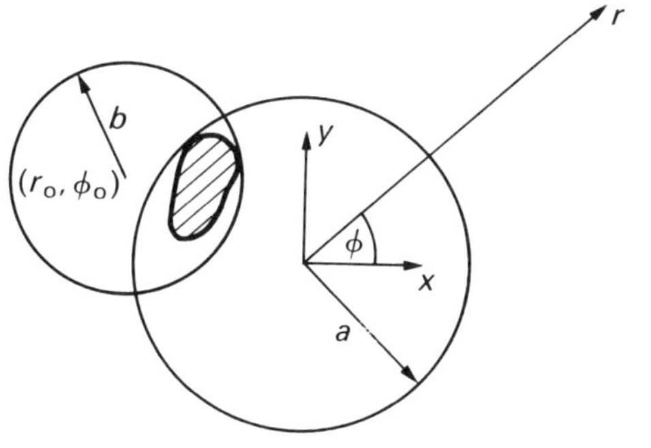

FIG. 9.3. Geometry for inverse scattering.

To discover another point on the perimeter of the body select a new origin (r_0,ϕ_0) and repeat the process. The new pattern function $A'(\phi)$ is given by

$$A'(\phi) = A(\phi)\exp\{-ikr_0 \cos(\phi-\phi_0)\} = \Sigma a'_n \exp(in\phi)$$

where

$$a'_n = \sum_m a_m J_{n-m}(kr_0)\exp\{i(n-m)(\tfrac{1}{2}\pi-\phi_0)\} .$$

Then the minimum radius $r = b$ (Fig. 9.3) is obtained via (9.5) with a'_n for a_n and r,ϕ measured from the new origin.

The above procedure is generally satisfactory for single convex scatterers. For more complicated obstacles it may be supplemented by drawing a circle of radius d and centre (r_0,ϕ_0) whose interior lies entirely outside $r = a$. Inside this new circle

$$E^{\mathrm{s}} = \Sigma c_n J_n(kr')\exp(in\phi)$$

where r' is measured from (r_0,ϕ_0) and

$$c_n = \sum_m a_m H^{(2)}_{n-m}(kr_0)\exp\{i(n-m)(\pi-\phi_0)-\tfrac{1}{2}m\pi i\} .$$

Now increase r' until a zero of $E^{\mathrm{i}} + E^{\mathrm{s}}$ is discovered; this corresponds to another point on the perimeter.

If $A(\phi)$ is not given for the whole range $|\phi| \leqslant \pi$ but only for $\phi_1 \leqslant \phi \leqslant \phi_2$ say, then an approximation A_n to a_n is found by minimizing (§1.5)

$$\int_{\phi_1}^{\phi_2} |A(\phi) - \sum_n A_n \exp(in\phi)|^2 d\phi .$$

Thereafter, the procedure is followed as before with A_n substituting for a_n.

It must be realized that it is rarely possible to make

observations as complete as required by theory. Measurements can usually be made at only a discrete number of angles and have to be extended by interpolation or other devices so as to cover a continuous range. Variations in time and wavenumber will be similarly incomplete. Therefore estimates deduced from real data are bound to be in error as well as suffering from the uncertainties (which cannot be avoided) from the noise inherent in any measurement. So predictions in inverse scattering theory are always highly idealized.

The procedure seems to work tolerably well when a typical dimension of the scatterer is somewhat less than a wavelength but at higher frequencies the computation time becomes excessive and the method of the next section should be tried.

Exercises

1. From the known analytical solution for a circular cylinder of radius $1/k$ use the above procedure to locate 8–12 points on the boundary. Repeat the calculation when the far-field data are restricted to $|\phi| \leqslant 120°$.

2. By means of one of the methods described in earlier chapters calculate the far field scattered by an elliptic cylinder with semi-axes $1/k$ and $1/2k$. Use this as data for the inverse problem and examine how much inaccuracy is introduced if the data are restricted to $|\phi| \leqslant \frac{1}{2}\pi$.

Does the orientation of the cylinder have much effect?

3. Repeat Exercise 2 for two circular cylinders, each of radius $1/k$, whose centres are separated by $6/k$, the line of centres being perpendicular to the direction of propagation of the incident plane wave.

4. Formulate the procedure for three dimensions and test it on a sphere and a spheroid.

9.3. High frequencies

At high frequencies the location of a perfectly conducting convex obstacle or target is assisted by the approximation of physical optics. The representation of the field is

$$\mathbf{E} = \mathbf{E}^i + \mathbf{E}^s$$

where

$$\mathbf{E}^s(P) = \frac{1}{i\omega\varepsilon}(\text{grad div} + k^2)\int_S \mathbf{n}_q \wedge \mathbf{H}\psi(P,q)\,\mathrm{d}S_q ,$$

$$\psi(P,Q) = \frac{\exp(-ik|\mathbf{x}_P-\mathbf{x}_Q|)}{4\pi|\mathbf{x}_P-\mathbf{x}_Q|}$$

and $\mathbf{n}_q$ is the unit normal from the interior of S to the exterior.

H is not known exactly on S but, at high frequencies, the reasonable approximation can be made that each small portion of the surface behaves as if it were plane. Each small portion will be subject to the incident wave and a secondary field generated by the rest of the surface. To a first approximation the secondary field may be neglected. Thus we may suppose that an incident plane wave illuminates a section of S called the lit side and denoted by L in Fig. 9.4. The remainder of S is dark and denoted by D. On L, $\mathbf{n} \cdot \mathbf{n}_q < 0$ where $\mathbf{n}$ is the unit vector in the direction of propagation of the plane wave. In the shadow region D the current distribution is assumed to be zero and on the lit side L the approximation

$$\mathbf{n}_q \wedge \mathbf{H} = 2\mathbf{n}_q \wedge \mathbf{H}^i$$

is introduced. Thus

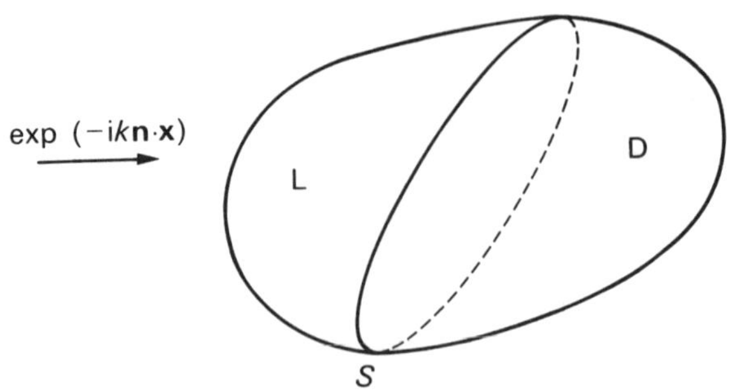

FIG. 9.4. The lit and dark portions of the target.

$$\mathbf{E}^s(P) = \frac{2}{i\omega\varepsilon}(\text{grad div} + k^2)\int_L \mathbf{n}_q \wedge \mathbf{H}^i\psi(P,q)\,dS_q$$

with the result that at distant points

$$\mathbf{E}^s(\mathbf{x}) \sim \frac{k^2\exp(-ik|\mathbf{x}|)}{2\pi\varepsilon|\mathbf{x}|}\int_L \{\mathbf{P}-(\mathbf{P}.\hat{\mathbf{x}})\hat{\mathbf{x}}\}\exp(ik\hat{\mathbf{x}}.\mathbf{x}_q)\,dS_q$$

where $\hat{\mathbf{x}}$ is a unit vector in the direction of $\mathbf{x}$ and $i\omega\mathbf{P} = \mathbf{n}_q \wedge \mathbf{H}^i$.
Write

$$\mathbf{E}^s(\mathbf{x}) \sim \mathbf{A}(\mathbf{n},\hat{\mathbf{x}})\,\frac{\exp(-ik|\mathbf{x}|)}{|\mathbf{x}|}$$

so that

$$\mathbf{A}(\mathbf{n},\mathbf{x}) = \frac{\omega k}{2\pi}\left(\frac{\mu}{\varepsilon}\right)^{\frac{1}{2}}\int_L \{\mathbf{P}-(\mathbf{P}.\hat{\mathbf{x}})\hat{\mathbf{x}}\}\exp(ik\hat{\mathbf{x}}.\mathbf{x}_q)\,dS_q \ .$$

Let the incident plane wave be

$$\mathbf{E}^i = \mathbf{e}_0\exp(-ik\mathbf{n}.\mathbf{x}) \ ,$$

$$\mathbf{H}^i = \left(\frac{\varepsilon}{\mu}\right)^{\frac{1}{2}} n \wedge \mathbf{e}_0\exp(-ik\mathbf{n}.\mathbf{x})$$

where $\mathbf{e}_0$ is a fixed unit vector such that $\mathbf{n}\,.\,\mathbf{e}_0 = 0$. Then

$$\mathbf{A}(\mathbf{n},\hat{\mathbf{x}}) = \frac{k}{2\pi i}\int_L [(\mathbf{n}_q.\mathbf{e}_0)\mathbf{n}-(\mathbf{n}_q.\mathbf{n})\mathbf{e}_0-\{(\mathbf{n}_q.\mathbf{e}_0)(\mathbf{n}.\hat{\mathbf{x}}) - (\mathbf{n}_q.\mathbf{n})(\mathbf{e}_0.\hat{\mathbf{x}})\}\hat{\mathbf{x}}]\exp\{ik\mathbf{x}_q.(\hat{\mathbf{x}}-\mathbf{n})\}\,dS_q \ .$$

Now irradiate the target with the plane wave $\mathbf{e}_0\exp(ik\mathbf{n}.\mathbf{x})$, travelling in the opposite direction to the previous one. Then

L and D interchange roles so that

$$\mathbf{A}(-\mathbf{n},\hat{\mathbf{x}}) = \frac{-k}{2\pi i}\int_D [(\mathbf{n}_q.\mathbf{e}_0)\mathbf{n}-(\mathbf{n}_q.\mathbf{n})\mathbf{e}_0-\{(\mathbf{n}_q.\mathbf{e}_0)(\mathbf{n}.\hat{\mathbf{x}}) - (\mathbf{n}_q.\mathbf{n})(\mathbf{e}_0.\hat{\mathbf{x}})\}\hat{\mathbf{x}}]\exp\{ik\mathbf{x}_q.(\hat{\mathbf{x}}+\mathbf{n})\}dS_q \quad .$$

Hence

$$\mathbf{A}(\mathbf{n},\hat{\mathbf{x}}) + \{\mathbf{A}(-\mathbf{n},-\hat{\mathbf{x}})\}^*$$

$$= \frac{k}{2\pi i}\int_S [(\mathbf{n}_q.\mathbf{e}_0)\mathbf{n}-(\mathbf{n}_q.\mathbf{n})\mathbf{e}_0-\{(\mathbf{n}_q.\mathbf{e}_0)(\mathbf{n}.\hat{\mathbf{x}}) - (\mathbf{n}_q.\mathbf{n})(\mathbf{e}_0.\hat{\mathbf{x}})\}\hat{\mathbf{x}}]\exp\{ik\mathbf{x}_q.(\hat{\mathbf{x}}-\mathbf{n})\}dS_q \qquad (9.6)$$

where the asterisk indicates a complex conjugate. Therefore

$$\mathbf{n} \,.\, [\mathbf{A}(\mathbf{n},\hat{\mathbf{x}}) + \{\mathbf{A}(-\mathbf{n},-\hat{\mathbf{x}})\}^*]$$

$$= \frac{k}{2\pi i}\int_S \mathbf{n}_q \,.\, [\{1-(\mathbf{n}.\hat{\mathbf{x}})^2\}\mathbf{e}_0+(\mathbf{e}_0.\hat{\mathbf{x}})(\mathbf{n}.\hat{\mathbf{x}})\mathbf{n}]\exp\{ik\mathbf{x}_q.(\hat{\mathbf{x}}-\mathbf{n})\}dS_q \quad .$$

By the divergence theorem

$$\mathbf{n} \,.\, [\mathbf{A}(\mathbf{n},\hat{\mathbf{x}})+\{\mathbf{A}(-\mathbf{n},-\hat{\mathbf{x}})\}^*]$$

$$= \frac{k^2}{2\pi}\,\mathbf{e}_0 \,.\, \hat{\mathbf{x}}(1-\mathbf{n}.\hat{\mathbf{x}})\int_V \exp\{ik\mathbf{x}_q.(\hat{\mathbf{x}}-\mathbf{n})\}dV \qquad (9.7)$$

where V is the volume enclosed by S. Eqn (9.7) constitutes a basic identity (Bojarski 1967, Lewis 1969, Perry 1974, Bleistein and Bojarski 1975) from which the shape of the obstacle is obtained.

Let $H_V(\mathbf{x})$ be unity when $\mathbf{x}$ is in V and zero elsewhere. Then (9.7) can be written as

$$B = \int_{-\infty}^{\infty} H_V(\mathbf{x})\exp(-i\boldsymbol{\alpha}.\mathbf{x}_q)\,d\mathbf{x}_q \tag{9.8}$$

where

$$B = \frac{2\pi\mathbf{n}\ .\ [\mathbf{A}(\mathbf{n},\hat{\mathbf{x}})+\{\mathbf{A}(-\mathbf{n},-\hat{\mathbf{x}})\}^*]}{k^2\mathbf{e}_0.\hat{\mathbf{x}}(1-\mathbf{n}.\hat{\mathbf{x}})},$$

$$\boldsymbol{\alpha} = k(\mathbf{n}-\hat{\mathbf{x}})$$

and the integration in (9.8) is over all space. Regarding B as known from far-field measurements we can view (9.8) as giving the Fourier transform of H_V, i.e. H_V can be found from

$$H_V(\mathbf{x}) = \frac{1}{8\pi^3}\int_{-\infty}^{\infty} B\ \exp(i\boldsymbol{\alpha}.\mathbf{x})\,d\boldsymbol{\alpha}\ . \tag{9.9}$$

Consequently H_V can be determined and the location of the target thereby discovered provided that B can be calculated for all $\boldsymbol{\alpha}$. All values of $\boldsymbol{\alpha}$ can be covered by picking $\hat{\mathbf{x}}$ to differ from $\mathbf{n}$ by a small vector in a chosen direction and then allowing k to range over large values. The advantage of this choice is that physical optics can be expected to be a good approximation at high frequencies for observations within a small cone about the direction of propagation of the incident plane wave. Alternatively, we can set $\hat{\mathbf{x}} = -\mathbf{n}$ so that only observations of backscattering are involved and again physical optics may be anticipated to furnish good results. Unfortunately, (9.7) is then nugatory since both sides vanish identically. However, by scalar multiplication of (9.6) by $\mathbf{e}_0$ we can still retain (9.8) by putting

$$B = \frac{2\pi\mathbf{e}_0\ .\ [\mathbf{A}(\mathbf{n},-\mathbf{n})\ +\ \{\mathbf{A}(-\mathbf{n},\mathbf{n})\}^*]}{k^2}\ . \tag{9.10}$$

Nevertheless, a large number of measurements of the amplitude and phase of the scattered wave will still be required.

A check on accuracy is provided by the fact that strictly H_V should take only the values 0 or 1. If, therefore, (9.9) supplies an H_V which differs significantly from either of these values it is clear that errors have been committed. In particular, this will be true if the evaluation of the integral in (9.8) leads to an appreciable imaginary part. Actually, it is easier to compute $\mathbf{n} \cdot \operatorname{grad} H_V$ in which the integrand of (9.8) is multiplied by $i\boldsymbol{\alpha} \cdot \mathbf{n}$, eliminating a small denominator in B when $\mathbf{n}$ is near $\hat{\mathbf{x}}$. Since grad H_V vanishes except on S a trace of the boundary is obtained. Usually, real data will be band limited so that the resolution of the boundary will be no better than a half wavelength.

An obvious disadvantage of the above technique is the requirement to make observations for all values of $\boldsymbol{\alpha}$. The procedure can be adapted to obviate this difficulty at the cost of solving an integral equation. Suppose that measurements are possible only for $\boldsymbol{\alpha}$ ranging over a volume T. Then, instead of B being known for all $\boldsymbol{\alpha}$, only $BH_T(\boldsymbol{\alpha})$ is available where H_T is unity for $\boldsymbol{\alpha}$ in T and zero elsewhere. Since B satisfies (9.8), the Fourier convolution theorem gives

$$\frac{1}{8\pi^3}\int_{-\infty}^{\infty} BH_T(\boldsymbol{\alpha})\exp(i\boldsymbol{\alpha}.\mathbf{x})\,d\boldsymbol{\alpha} = \int_{-\infty}^{\infty} h_T(\mathbf{x}-\mathbf{x}_q)H_V(\mathbf{x}_q)\,d\mathbf{x}_q \qquad (9.11)$$

where

$$h_T(\mathbf{x}) = \frac{1}{8\pi^3}\int_{-\infty}^{\infty} H_T(\boldsymbol{\alpha})\exp(i\boldsymbol{\alpha}.\mathbf{x})\,d\boldsymbol{\alpha} .$$

Since the left-hand side of (9.11) and h_T are both known, (9.11) is an integral equation to determine H_V. It is an integral equation of the first kind with kernel h_T. Such integral equations are generally ill posed and there is no guarantee that a solution exists. Even if a solution exists it may not be unique and may not depend continuously on the data. Notwithstanding, it can be tackled by methods of preceding chapters and some success has been reported (Bojarski 1974) in obtaining numerical solutions.

The method can be modified to deal with the detection of protrusions on plane sheets when observations are limited to one side (Bleistein 1976).

Exercises

5. From the known solution for a circular cylinder of radius $20/k$ use the above method to estimate the boundary. Repeat the calculation if the data are restricted to a wedge of 120°.

6. Repeat Exercise 2 with semi-axes of $20/k$ and $10/k$.

7. Repeat Exercise 3 with the radii of the cylinders $10/k$ and their centres separated by $50/k$.

8. Repeat Exercise 5 for a sphere of radius $10/k$.

9. Test the above procedure on (*a*) a spheroid, (*b*) a parallelepiped.

9.4. Scattering in the time domain

The technique of the previous section is also capable of coping with pulse problems (Bleistein 1976*a*). Define the time transform $F(\omega)$ of $f(t)$ by

$$F(\omega) = \int_{-\infty}^{\infty} f(t)\exp(-i\omega t)\,dt \; .$$

For real f this implies that

$$\{F(\omega)\}^* = F(-\omega) \tag{9.12}$$

which enables F to be determined for negative values of ω from its behaviour for positive ω. Further

$$f(t) = \frac{1}{2\pi}\,\mathcal{R}\int_{0}^{\infty} F(\omega)\exp(i\omega t)\,d\omega \; . \tag{9.13}$$

When backscattering data are utilized, $\hat{\mathbf{x}} = -\mathbf{n}$ and the functional dependence of B may be denoted by $B(\mathbf{n},\omega)$. From the definition of $\mathbf{A}$ it follows that B satisfies (9.12) whether given by (9.8) or (9.10). Therefore, introduce a real function b such that

$$\omega^2 B(\mathbf{n},\omega) = \int_{-\infty}^{\infty} b(\mathbf{n},t)\exp(-i\omega t)\,dt \quad . \tag{9.14}$$

Now, the left-hand side of (9.9) is real so that we can write

$$H_V(\mathbf{x}) = \frac{1}{8\pi^3}\mathcal{R}\int_{-\infty}^{\infty} B(\mathbf{n},\omega)\exp\left(2i\omega\mathbf{n}.\frac{\mathbf{x}}{c}\right)d\boldsymbol{\alpha}$$

where $c = 1/(\mu\varepsilon)^{\frac{1}{2}}$. By carrying out the integration with respect to ω, employing (9.13), we have

$$H_V(\mathbf{x}) = \frac{1}{2\pi^2 c^2}\int_{\Omega} b\left(\mathbf{n},\frac{2\mathbf{n}.\mathbf{x}}{c}\right)d\Omega \tag{9.15}$$

where $d\Omega$ is the surface element of the unit sphere. Eqn (9.15) offers the possibility of making observations in the time domain if b can be related to the time-varying field. Otherwise, a Fourier transform will be necessary so that the results of the preceding section can be applied in the frequency domain.

It is convenient to deduce the relation for b from an exact result which will now be derived. Suppose the excitation in the time domain is such that the incident electric intensity near the target is the plane wave

$$\mathbf{e}^{i} = \mathbf{e}_0 f\left(t-\frac{\mathbf{n}.\mathbf{x}}{c}\right) \quad .$$

Then in the frequency domain the incident electric intensity is $\mathbf{E}^{i} = F(\omega)\mathbf{e}_0 \exp(-ik\mathbf{n}.\mathbf{x})$ and

$$\mathbf{E}^{s}(\mathbf{x}) \sim F(\omega)\mathbf{A}(\mathbf{n},\hat{\mathbf{x}},\omega)\frac{\exp(-ik|\mathbf{x}|)}{|\mathbf{x}|}$$

the dependence of $\mathbf{A}$ on ω now being explicitly indicated. This holds for positive ω and is extended to negative values by means of (9.12). Hence, in the time domain the distant scattered field is given by

$$\mathbf{e}^{s}(\mathbf{x},t) \sim \frac{\mathbf{e}(\mathbf{n},\hat{\mathbf{x}},t-|\mathbf{x}|/c)}{|\mathbf{x}|}$$

where

$$\mathbf{e}(\mathbf{n},\hat{\mathbf{x}},t) = \frac{1}{2\pi}\int_{-\infty}^{\infty} F(\omega)\mathbf{A}(\mathbf{n},\hat{\mathbf{x}},\omega)\exp(i\omega t)\mathrm{d}\omega \tag{9.16}$$

or, in terms of positive ω only,

$$\mathbf{e}(\mathbf{n},\hat{\mathbf{x}},t) = \Re\frac{1}{\pi}\int_{0}^{\infty} F(\omega)\mathbf{A}(\mathbf{n},\hat{\mathbf{x}},\omega)\exp(i\omega t)\mathrm{d}\omega\ . \tag{9.17}$$

If

$$\mathbf{A}(\mathbf{n},\hat{\mathbf{x}},\omega) = \int_{-\infty}^{\infty} \mathbf{a}(\mathbf{n},\hat{\mathbf{x}},t)\exp(-i\omega t)\mathrm{d}t$$

(9.16) can be expressed in the convolution form

$$\mathbf{e}(\mathbf{n},\hat{\mathbf{x}},t) = \int_{-\infty}^{\infty} f(t-\tau)\mathbf{a}(\mathbf{n},\hat{\mathbf{x}},\tau)\mathrm{d}\tau\ . \tag{9.18}$$

Eqns (9.16)–(9.18) are equivalent forms of the exact relation referred to above. When $f(t) = \delta(t)$ or $F(\omega) = 1$, $\mathbf{e}(\mathbf{n},\hat{\mathbf{x}},t) = \mathbf{a}(\mathbf{n},\hat{\mathbf{x}},t)$. Thus, in general, (9.18) shows that *the scattered wave in the time domain is the convolution of the incident signal waveform f and the scattered waveform for an impulsive source*. We also conclude that $\mathbf{a}(\mathbf{n},\hat{\mathbf{x}},t) = \mathbf{0}$ for $t < 0$ to be consistent with the finite rate of spreading of the disturbance caused by the impulsive source.

When the approximation of physical optics is made, (9.14)

and (9.10) imply that

$$\left.\begin{aligned} b(\mathbf{n},t) &= 2\pi c^2\mathbf{e}_0 \,.\, \mathbf{a}(\mathbf{n},-\mathbf{n},t) \qquad (t>0) \\ &= 2\pi c^2\mathbf{e}_0 \,.\, \mathbf{a}(-\mathbf{n},\mathbf{n},-t) \qquad (t<0)\,. \end{aligned}\right\} \tag{9.19}$$

Substitution in (9.18) supplies

$$\mathbf{e}_0 \,.\, \mathbf{e}(\mathbf{n},\hat{\mathbf{x}},t) = \frac{1}{2\pi c^2}\int_0^\infty \{f(t-\tau)b(\mathbf{n},\tau)+f(t+\tau)b(-\mathbf{n},\tau)\}\mathrm{d}\tau\,. \tag{9.20}$$

The left-hand side is known by measurement and f is prescribed by the incident wave so that (9.20) constitutes an integral equation to determine b. Once solved the position of the obstacle is provided by (9.15). Should the incident source be impulsive (9.20) does not have to be solved because $\mathbf{a}$ can be measured directly and b found from (9.19).

Exercises

10. If $f(t) = \delta'(t)$ find b in terms of $\mathbf{e}$ and see what formula results for grad $H_V(\mathbf{x})$.

11. Locate a sphere by examining the scattering in the time domain of (*a*) an impulsive plane wave, (*b*) a Gaussian plane wave pulse.

9.5. Moving targets

The exact formula (9.16) refers to a target which is stationary with respect to the source of the irradiating field. If the source or target is moving it needs modification. The case of an obstacle travelling with constant velocity while the source is fixed will be considered here. Formulae when the roles of the obstacle and source are interchanged can be dealt with by a simple transformation. If, however, different parts of the system possess different velocities a more elaborate attack has to be mounted (Jones 1964, 1977, Ffowcs Williams and Hawkings 1969).

Let $\mathbf{x},t$ be the coordinate system in which the sources are at rest and $\mathbf{x}',t'$ a system in which the target is stationary.

The two coordinate systems are related by a Lorentz transformation. It will be assumed that the spatial axes of the two systems are aligned and that the target moves along the positive x_1 axis with speed v. By writing formulae in terms of the velocity $\mathbf{v}$, however, the special choice of the axes can be eliminated.

The Lorentz transformation can be expressed as

$$x_1' = \beta(x_1 - vt), \quad x_2' = x_2, \quad x_3' = x_3, \quad t' = \beta\left(t - \frac{vx_1}{c^2}\right), \tag{9.21}$$

$$x_1 = \beta(x_1' + vt'), \quad x_2 = x_2', \quad x_3 = x_3', \quad t = \beta\left(t' + \frac{vx_1'}{c^2}\right) \tag{9.22}$$

where $\beta = (1-v^2/c^2)^{-\frac{1}{2}}$. In the $\mathbf{x},t$ system the incident wave is

$$\mathbf{e}^{\mathrm{i}} = \mathbf{e}_0 f\left(t - \frac{\mathbf{n}.\mathbf{x}}{c}\right), \quad \mathbf{h}^{\mathrm{i}} = \left(\frac{\varepsilon}{\mu}\right)^{\frac{1}{2}} \mathbf{n} \wedge \mathbf{e}^{\mathrm{i}} \, .$$

On substitution from (9.22)

$$f\left(t - \frac{\mathbf{n}.\mathbf{x}}{c}\right) = f\left\{\gamma\left(t' - \frac{\mathbf{n}'.\mathbf{x}'}{c}\right)\right\} = g\left(t' - \frac{\mathbf{n}'.\mathbf{x}'}{c}\right)$$

where

$$\gamma = \beta\left(1 - \frac{n_1 v}{c}\right) = \beta\left(1 - \frac{\mathbf{n}.\mathbf{v}}{c}\right),$$

$$n_1' = \frac{\beta(n_1 - v/c)}{\gamma}, \quad n_2' = \frac{n_2}{\gamma}, \quad n_3' = \frac{n_3}{\gamma}. \tag{9.23}$$

In the $\mathbf{x}',t'$ system the incident wave is

$$\begin{aligned}\mathbf{e}^{\mathrm{i}\prime} &= \beta\mathbf{e}^{\mathrm{i}} + \frac{(1-\beta)(\mathbf{e}^{\mathrm{i}}.\mathbf{v})\mathbf{v}}{v^2} + \beta\mu\mathbf{v} \wedge \mathbf{h}^{\mathrm{i}} \\ &= \mathbf{e}_0' g\left(t' - \frac{\mathbf{n}'.\mathbf{x}'}{c}\right)\end{aligned}$$

where

$$\mathbf{e}_0' = \gamma\mathbf{e}_0 + (\mathbf{e}_0.\mathbf{v})\left\{(1-\beta)\frac{\mathbf{v}}{v^2} + \frac{\beta\mathbf{n}}{c}\right\}. \tag{9.24}$$

Thus if the incident plane wave has angular frequency ω in the $\mathbf{x},t$ space it appears to have angular frequency $\gamma\omega$ in the system in which the target is stationary. This change in frequency represents the *Doppler shift* due to motion. In addition, the apparent direction of propagation is $\mathbf{n}'$ instead of $\mathbf{n}$. As is clear from (9.24) the polarization of the wave is also altered (unless $\mathbf{e}_0$ is perpendicular to $\mathbf{v}$); this is to accommodate not only the change in direction of propagation but also the interdependence of electric and magnetic fields.

Indicate explicitly the dependence of $\mathbf{A}$ on the polarization by writing $\mathbf{A}(\mathbf{e},\mathbf{n},\hat{\mathbf{x}},\omega)$. In the $\mathbf{x}',t'$ system the obstacle is motionless so that (9.16) can be applied. Hence

$$\mathbf{e}^{s\prime}(\mathbf{x}',t') = \frac{\mathbf{e}'(\mathbf{n}',\hat{\mathbf{x}}',t'-|\mathbf{x}'|/c)}{|\mathbf{x}'|}$$

where

$$\mathbf{e}'(\mathbf{n}',\hat{\mathbf{x}},t') = \frac{1}{2\pi}\int_{-\infty}^{\infty} G(\omega)\mathbf{A}(\mathbf{e}_0',\mathbf{n}',\hat{\mathbf{x}}',\omega)\exp(\mathrm{i}\omega t')\,\mathrm{d}\omega \;.$$

Now

$$\gamma G(\omega) = F\left(\frac{\omega}{\gamma}\right)$$

and, from (9.21),

$$t' - \frac{|\mathbf{x}'|}{c} = t' - \frac{\mathbf{x}'.\hat{\mathbf{x}}'}{c} = \left(t - \frac{\mathbf{x}.\hat{\mathbf{x}}}{c}\right)\Big/\gamma_1$$

where $\gamma_1 = \beta(1-\hat{\mathbf{x}}.\mathbf{v}/c)$,

$$\hat{x}_1' = \frac{\beta(\hat{x}_1-v/c)}{\gamma_1}, \qquad \hat{x}_2' = \frac{\hat{x}_2}{\gamma_1}, \qquad \hat{x}_3' = \frac{\hat{x}_3}{\gamma_1} \tag{9.25}$$

in analogy with (9.23). Hence

$$\mathbf{e}'\left(\mathbf{n}',\hat{\mathbf{x}}',t'-\frac{|\mathbf{x}'|}{c}\right) = \mathbf{e}_1\left(\mathbf{n}',\hat{\mathbf{x}}',t-\frac{|\mathbf{x}|}{c}\right)$$

with

$$\mathbf{e}_1(\mathbf{n}',\hat{\mathbf{x}}',t) = \frac{\gamma_1}{2\pi\gamma}\int_{-\infty}^{\infty} F\left(\frac{\gamma_1}{\gamma}\omega\right)\mathbf{A}(\mathbf{e}',\mathbf{n}',\hat{\mathbf{x}}',\gamma_1\omega)\exp(i\omega t)\,d\omega \ . \quad (9.26)$$

The distant field observed in the $\mathbf{x},t$ system is now obtained from

$$\mathbf{e}^s(\mathbf{x},t) = \beta\mathbf{e}^{s\prime} + (1-\beta)(\mathbf{e}^{s\prime}.\mathbf{v})\frac{\mathbf{v}}{v^2} - \frac{\beta}{c}\mathbf{v}\wedge(\mathbf{x}'\wedge\mathbf{e}^{s\prime}) \ . \quad (9.27)$$

Comparison of (9.26) with (9.16) reveals that (9.26) reduces to (9.16) when the target is immobile so that $v = 0$, $\beta = 1 = \gamma = \gamma_1$, $\hat{\mathbf{x}} = \hat{\mathbf{x}}'$, $\mathbf{n} = \mathbf{n}'$. Also (9.26) displays evidence of the effect of motion in four places. There is a frequency shift in the incident signal waveform due to the factor γ_1/γ. The scattering amplitude **A** has a frequency shift caused by the presence of ω, an angular distortion due to $\mathbf{n}',\hat{\mathbf{x}}'$ replacing $\mathbf{n},\mathbf{x}$ and a polarization warping because $\mathbf{e}_0'$ substitues for $\mathbf{e}_0$. An additional polarization twist also occurs in the conversion back to source coordinates through (9.27). The whole of (9.27) can be expressed in terms of $\mathbf{x},t$ by means of (9.23)–(9.26).

The integral in (9.26) can be written in forms analogous to (9.17) and (9.18) if desired.

THE INVERSE SOURCE PROBLEM

9.6. Harmonic sources

The assumption made in this section is that the observed radiation is to be accounted for by a distribution of sources placed in a homogeneous isotropic medium. Then a solution of (9.1) is sought such that **j** is confined to some part of the interior T of a closed surface S. The tangential components of both **E** and **H** are to be measured on S and the aim is to determine **j** from these data (Fig. 9.5).

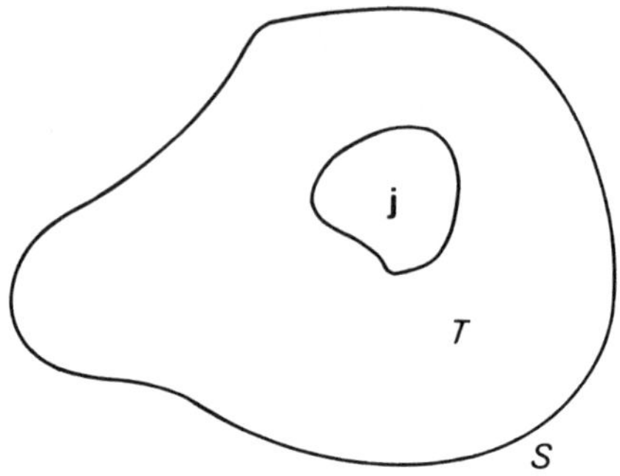

FIG. 9.5. The inverse-source problem.

Now

$$\mathbf{E}(\mathbf{x}) = (\text{grad div} + k^2)\frac{1}{\text{i}\omega\varepsilon}\int_T \mathbf{j}(\mathbf{x}_q)\psi(P,q)\,\mathrm{d}\mathbf{x}_q\ , \tag{9.28}$$

$$\mathbf{H}(\mathbf{x}) = \text{curl}\int_T \mathbf{j}(\mathbf{x}_q)\psi(P,q)\,\mathrm{d}\mathbf{x}_q \tag{9.29}$$

where $\mathbf{x}$ or P denotes the point of observation and $\psi(P,q) = \exp(-\text{i}kr)/4\pi r$ with r the distance between P and q.

Since $\mathbf{j}$ is non-zero only in a portion of T there is no loss in making the volume of integration the whole of T.

The fields inside S can be exposed by means of volume and surface integrals. Although it is customary to base such a representation on retarded potentials, it will prove to be advantageous here to employ advanced potentials. The appropriate relations are

$$\mathbf{E}(\mathbf{x}) = (\text{grad div} + k^2)\left\{\frac{1}{\text{i}\omega\varepsilon}\int_T \mathbf{j}(\mathbf{x}_q)\psi^*(P,q)\,\mathrm{d}\mathbf{x}_q - \right.$$

$$\left. - \frac{1}{\text{i}\omega\varepsilon}\int_S \mathbf{n}_q \wedge \mathbf{H}\psi^*(P,q)\,\mathrm{d}S_q\right\} - \text{curl}\int_S \mathbf{n}_q \wedge \mathbf{E}\psi^*(P,q)\,\mathrm{d}S_q\ , \tag{9.30}$$

$$\mathbf{H}(\mathbf{x}) = \operatorname{curl}\left\{\int_T \mathbf{j}(\mathbf{x}_q)\psi^*(P,q)\,d\mathbf{x}_q - \int_S \mathbf{n}_q \wedge \mathbf{H}\psi^*(P,q)\,dS_q\right\} +$$

$$+ \frac{1}{i\omega\mu}(\text{grad div} + k^2)\int_S \mathbf{n}_q \wedge \mathbf{E}\psi^*(P,q)\,dS_q \qquad (9.31)$$

where $\mathbf{n}_q$ is the unit normal out of S. The surface integrals in (9.30) and (9.31) cannot be reckoned to vanish because advanced potentials have been utilized. They are, however, known from the measurements of the tangential components made on S. Hence

$$\mathbf{E}(\mathbf{x}) = (\text{grad div} + k^2)\frac{1}{i\omega\varepsilon}\int_T \mathbf{j}(\mathbf{x}_q)\psi^*(P,q)\,d\mathbf{x}_q + \mathbf{g}(\mathbf{x}), \qquad (9.32)$$

$$\mathbf{H}(\mathbf{x}) = \operatorname{curl}\int_T \mathbf{j}(\mathbf{x}_q)\psi^*(P,q)\,d\mathbf{x}_q + \mathbf{h}(\mathbf{x}) \qquad (9.33)$$

where $\mathbf{g}$ and $\mathbf{h}$ are known. Combining (9.32) with (9.28) and (9.33) with (9.29) we obtain

$$(\text{grad div} + k^2)\frac{1}{i\omega\varepsilon}\int_T \mathbf{j}(\mathbf{x}_q)\{\psi^*(P,q)-\psi(P,q)\}\,d\mathbf{x}_q + \mathbf{g}(\mathbf{x}) = \mathbf{0}, \qquad (9.34)$$

$$\operatorname{curl}\int_T \mathbf{j}(\mathbf{x}_q)\{\psi^*(P,q)-\psi(P,q)\}\,d\mathbf{x}_q + \mathbf{h}(\mathbf{x}) = \mathbf{0}. \qquad (9.35)$$

These two integral equations determine $\mathbf{j}$ from the given data. Notice that they are exact and do not require measurements to be made in the far field or the approximation of physical optics to be adopted. The integral equations are not independent because (9.35) is the curl of (9.34) since $\operatorname{curl}\mathbf{g} + i\omega\mu\mathbf{h} = \mathbf{0}$. Nevertheless, they serve as useful verifiers of one another in numerical work where their relationship will not be exact.

The kernel of (9.34) (or (9.35)) is non-singular; indeed

$$\psi^*(P,q) - \psi(P,q) = \frac{ik}{2\pi}\, j_0(kr)$$

where j_0 is the spherical Bessel function. Nevertheless, (9.34) is an ill-posed and ill-conditioned integral equation. Despite this fact, experiments with synthetic data (i.e. generated analytically) suggest that a tolerable if not very precise indication of the source distribution can be achieved (Bojarski 1974).

Exercises

12. Place an electric dipole at the origin and then calculate **g** and **h** if S is a sphere of radius a with centre at the origin. Then solve (9.34) and (9.35) numerically to see how well they predict the source.

13. Repeat Exercise 12 for a magnetic dipole.

14. Repeat Exercise 12 with two parallel electric dipoles on the z axis and equidistant from the origin.

15. If $\mathbf{h} = \mathbf{0}$ and T is a sphere of radius a find the solutions of (9.35) by using the addition theorem to expand j_0 in a series of spherical harmonics. (This provides confirmation of the non-uniqueness referred to already.)

16. If T is a sphere of radius a, find the values of λ for which

$$\int_T j(\mathbf{x}_q) j_0(kr)\,\mathrm{d}\mathbf{x}_q = \lambda j(\mathbf{x})$$

inside T and show that they can be expressed as

$$\tfrac{1}{2}a^3[\{j_m(ka)\}^2 - j_{m-1}(ka)j_{m+1}(ka)]$$

for $m = 0, 1, 2, \ldots$. As $m \to \infty$ show that these eigenvalues tend to zero rapidly (like m^{-2m-3}), thereby verifying that (9.34) and (9.35) are ill conditioned and ill posed.

If the tangential components of **E** and **H** can be observed only over part of S, say S_1, and not over the remainder S_2 put

$$\mathbf{g} = \mathbf{g}_1 + \mathbf{g}_2 = \int_{S_1} + \int_{S_2} .$$

Now $\mathbf{g}_2$ is unknown. However, it can be expressed in terms of the unknown **j** by finding **E** and **H** on S_2 from (9.28) and (9.29).

Then (9.34) becomes an integral equation of the first kind for **j** in which the kernel also contains an integral over S_2. While the problem can be tackled in principle the integral equation is likely to become more ill behaved as S_1 shrinks in size.

9.7. Inhomogeneities

The method of the preceding section is also adaptable to the investigation of finite regions of inhomogeneity in an otherwise homogeneous medium. Suppose, in particular, that the permittivity is $\varepsilon + \varepsilon'$ where ε' is non-zero only over a finite region. Excite the medium with a known wave $\mathbf{E}^i$, e.g. a plane wave. Then the scattered wave satisfies (9.1) with $\mathbf{j} = i\omega\varepsilon'(\mathbf{E}^i+\mathbf{E}^s)$. The current density **j** is determined from (9.34) or (9.35) and then $\mathbf{E}^s$ can be calculated from (9.28). Since $\mathbf{E}^i$ is known it follows that ε' can be computed.

Variations in μ can also be coped with by using representations of the electromagnetic field in terms of magnetic current (cf. §7.2).

The method does not seem to be very satisfactory from a numerical point of view and other modes of tackling the problem have been put forward (Cohen and Bleistein 1977). For scalar problems one suggestion (Bates, Boerner, and Dunlop 1976) has been the introduction of the *Rytov approximation* in which $E = E^i E'$ and the equation for E' handled. For stratified media where the variations are one dimensional more intricate analysis is feasible (Meyer 1975, 1975*a*, 1976, Bates and Wall 1976).

Another technique, which is applicable to the general inverse source problem, is to expand **j** in a finite series of known functions such as polynomials but with coefficients to be discovered. The resulting field on S can then be calculated from (9.28) and (9.29). The coefficients are now found by minimizing the sum of the squares of the differences between the predicted and measured tangential components of **E** and **H** on S. In general, this will involve an optimization technique such as conjugate gradients as described in Chapter 4.

It is likely that this method will be superior numerically to the integral equation approach if the expansion functions are well chosen. Simultaneous probing at several frequencies is also possible with this approach.

Exercises

17. Repeat Exercises 12–14 using the optimization technique with the expansion functions chosen as polynomials. Do you think some other choice would give better results?

18. Choose a simple ε' and compute the field scattered from an incident plane wave. Use the values as data to determine ε' by (*a*) integral equations, (*b*) the optimization technique. Which do you think is the better method? What happens if you vary the frequency?

19. In fields which depend only on z, μ and ε have the constant values μ_0 and ε_0 except for $0 \leqslant z \leqslant 1$ where $\varepsilon = \varepsilon_0 e^z$. Use the known form of ε to generate data for the inverse problem and see how accurate your solution to the inverse problem is. Consider a number of frequencies.

Repeat the investigation for (i) $\varepsilon = \varepsilon_0(1+z)$, (ii) $\varepsilon = \varepsilon_0(1+\sin \pi z)$, (iii) $\varepsilon = \varepsilon_0(1+\sinh 2z)$.

9.8. Statistical considerations

Many natural sources change their positions and frequencies with time. The assumption of monochromatic radiation is then inappropriate. While it would be possible to contemplate the formulation of the integral equation (9.34) in the time domain it is doubtful whether sufficient measurements could be undertaken to make the calculation of the corresponding **g** a practical realizability.

For the inverse source problem an outgoing solution of

$$\text{curl } \mathbf{E} + \mu \frac{\partial \mathbf{H}}{\partial t} = \mathbf{0}, \qquad \text{curl } \mathbf{H} - \varepsilon \frac{\partial \mathbf{E}}{\partial t} = \mathbf{J}(\mathbf{x}, t)$$

is required. Assuming that **J** is zero outside some finite region, we have from eqn (2) of Chapter 7

$$\mathbf{E}(\mathbf{x}, t) = -\frac{\partial}{\partial t} \frac{\mu}{4\pi} \int_{-\infty}^{\infty} \frac{\mathbf{J}(\mathbf{x}', t-|\mathbf{x}-\mathbf{x}'|/c)}{|\mathbf{x}-\mathbf{x}'|} \, d\mathbf{x}' -$$

cont.

$$- \operatorname{grad} \frac{1}{4\pi\varepsilon} \int_{-\infty}^{\infty} \frac{\rho(\mathbf{x}';t-|\mathbf{x}-\mathbf{x}'|/c)}{|\mathbf{x}-\mathbf{x}'|} \, d\mathbf{x}' \qquad (9.36)$$

where the integration is over all space and

$$\operatorname{div} \mathbf{J} + \frac{\partial\rho}{\partial t} = 0 \ .$$

If random processes are operating it will usually be necessary to consider averages for comparison with experimental results. It is customary to regard **E** as a typical member of an ensemble of functions which characterize the statistical properties of the process. Common practice is to suppose that the ensembles which are encountered in electromagnetism are *stationary* and *ergodic*. Roughly speaking, an ensemble is stationary if all ensemble averages are independent of the time origin. Likewise, an ergodic process is one in which each ensemble average agrees with the corresponding time average for a typical member of the ensemble. Thus, in stationary ergodic random processes attention can be restricted to time averages and the absolute position of the time origin is irrelevant.

Any practical disturbance will exist only during some finite time interval, say $-T \leqslant t \leqslant T$. It is mathematically desirable to allow $T \to \infty$ so that the assumption of stationarity can be invoked. For this reason the time average is defined by

$$\langle h \rangle = \lim_{T\to\infty} \frac{1}{2T} \int_{-T}^{T} h(t)\,dt \ .$$

It is clear at once that, if this limit is non-zero, $\int_{-\infty}^{\infty} h(t)\,dt$ must be divergent. Since we expect to deal with non-zero averages this difficulty is bound to arise and there will be technical problems with Fourier analysis. To surmount these problems we work with truncated functions such as h_T where

$$h_T = h(t) \quad (|t| \leqslant T)$$
$$= 0 \quad (|t| > T)$$

carry out the desired procedures with h_T, and then proceed to the limit as $T \to \infty$. Provided that suitable smoothing is accomplished by ensemble averaging, the limit customarily exists and is the same as would be achieved by direct formal Fourier analysis. Accordingly, the technicalities will be ignored in the following and Fourier transforms will be manipulated as if there were no difficulties.

On this understanding the average

$$\langle \mathbf{E}(\mathbf{x}_1, t_1+t) \,.\, \mathbf{E}(\mathbf{x}_2, t_2+t)\rangle$$

can be introduced. In view of the assumption that the random process is stationary, only the time difference $t_2 - t_1$ will occur. It is therefore sufficient to concentrate on

$$\langle \mathbf{E}(\mathbf{x}_1, t) \,.\, \mathbf{E}(\mathbf{x}_2, t+\tau)\rangle$$

which is known as the *cross-correlation function* at $\mathbf{x}_1$ and $\mathbf{x}_2$. When $\mathbf{x}_1 = \mathbf{x}_2$ this becomes

$$\langle \mathbf{E}(\mathbf{x}_1, t) \,.\, \mathbf{E}(\mathbf{x}_1, t+\tau)\rangle$$

which is often called the *autocorrelation function*.

An associated normalized quantity is

$$\gamma_{12}(\tau) = \frac{\langle \mathbf{E}(\mathbf{x}_1, t) . \mathbf{E}(\mathbf{x}_2, t+\tau)\rangle}{\{\langle \mathbf{E}^2(\mathbf{x}_1, t)\rangle\langle \mathbf{E}^2(\mathbf{x}_2, t)\rangle\}^{\frac{1}{2}}}$$

which may be deemed a measure of the degree of *coherence*. As a consequence of the Schwarz inequality (§1.5)

$$0 \leqslant |\gamma_{12}(\tau)| \leqslant 1 \; .$$

If $|\gamma_{12}(\tau)| = 1$ the disturbance is often said to be *coherent*, whereas if $|\gamma_{12}(\tau)| = 0$ is called *incoherent*; for other values of $|\gamma_{12}(\tau)|$ there is *partial coherence*.

Write

$$<\mathbf{E}(\mathbf{x}_1,t)\ .\ \mathbf{E}(\mathbf{x}_2,t+\tau)> = \int_{-\infty}^{\infty} G_{12}(\omega)\exp(-i\omega\tau)d\omega\ , \qquad (9.37)$$

$$<\mathbf{E}(\mathbf{x}_1,t)\ .\ \mathbf{E}(\mathbf{x}_1,t+\tau)> = \int_{-\infty}^{\infty} G_1(\omega)\exp(-i\omega\tau)d\omega\ . \qquad (9.38)$$

Then the cross-correlation and autocorrelation functions are being effectively analysed into their Fourier components. That is why G_1 is termed the *power spectral density* and G_{12} the *cross-power spectral density*. Since most random processes are not periodic with respect to τ, G_{12} and G_1 are liable to vary continuously with ω rather than contain a number of discrete frequencies. They are, obviously, functions of $\mathbf{x}_1$ and $\mathbf{x}_2$ as well as ω in general.

By Fourier inversion

$$G_{12}(\omega) = \frac{1}{2\pi}\int_{-\infty}^{\infty} <\mathbf{E}(\mathbf{x}_1,t)\ .\ \mathbf{E}(\mathbf{x}_2,t+\tau)> \exp(i\omega\tau)d\tau\ , \qquad (9.39)$$

$$G_1(\omega) = \frac{1}{2\pi}\int_{-\infty}^{\infty} <\mathbf{E}(\mathbf{x}_1,t)\ .\ \mathbf{E}(\mathbf{x}_1,t+\tau)> \exp(i\omega\tau)d\tau\ . \qquad (9.40)$$

Since the correlation functions are real, the spectral densities must satisfy

$$\{G_{12}(\omega)\}^* = G_{12}(-\omega), \qquad \{G_1(\omega)\}^* = G_1(-\omega) \qquad (9.41)$$

as may be seen directly from either (9.37) or (9.39) (cf. (9.12)).

On account of (9.41) the spectral density is known at negative frequencies as soon as its values at positive frequencies are available. Therefore, the mean $\bar{\omega}$ is defined by

$$\bar{\omega} = \frac{\int_0^\infty \omega |G_1(\omega)|^2 d\omega}{\int_0^\infty |G_1(\omega)|^2 d\omega}$$

and the *effective spectral width* $\Delta\omega$ by

$$(\Delta\omega)^2 = \frac{\int_0^\infty (\omega-\bar{\omega})^2 |G_1(\omega)|^2 d\omega}{\int_0^\infty |G_1(\omega)|^2 d\omega} . \tag{9.42}$$

The *coherence time* $\Delta\tau$ is specified by

$$(\Delta\tau)^2 = \frac{\int_{-\infty}^\infty \tau^2 |g_1(\tau)|^2 d\tau}{\int_{-\infty}^\infty |g_1(\tau)|^2 d\tau} \tag{9.43}$$

where

$$g_1(\tau) = \langle \mathbf{E}(\mathbf{x}_1, t) \cdot \mathbf{E}(\mathbf{x}_1, t+\tau) \rangle . \tag{9.44}$$

The mean value of τ is zero because (9.41) implies, via (9.38) and (9.44), that $|g_1(\tau)|^2$ is an even function of τ.

Now

$$\int_{-\infty}^\infty |g_1(\tau)|^2 d\tau = 2\pi \int_{-\infty}^\infty |G_1(\omega)|^2 d\omega = 4\pi \int_0^\infty |G_1(\omega)|^2 d\omega$$

because the integrand is even by virtue of (9.41). Similarly

$$\int_{-\infty}^\infty \tau^2 |g_1(\tau)|^2 d\tau = 4\pi \int_0^\infty \left| \frac{\partial G_1(\omega)}{\partial \omega} \right|^2 d\omega .$$

Hence, from (9.43),

$$(\Delta\tau)^2 = \frac{\int_0^\infty |\partial G_1(\omega)/\partial\omega|^2 d\omega}{\int_0^\infty |G_1(\omega)|^2 d\omega} . \tag{9.45}$$

If α is real

$$\int_0^\infty |h_1(\omega)+\alpha h_2^*(\omega)|^2 d\omega \geqslant 0 .$$

Choosing

$$\alpha \int_0^\infty |h_2|^2 d\omega = -\frac{1}{2}\int_0^\infty (h_1 h_2 + h_1^* h_2^*) d\omega$$

we derive

$$\left\{\int_0^\infty (h_1 h_2 + h_1^* h_2^*) d\omega\right\}^2 \leqslant 4\int_0^\infty |h_1|^2 d\omega \int_0^\infty |h_2|^2 d\omega . \tag{9.46}$$

Make the selection

$$h_1(\omega) = (\omega-\bar{\omega})G_1(\omega), \qquad h_2(\omega) = \frac{\partial G_1^*(\omega)}{\partial\omega} .$$

Then

$$\int_0^\infty (h_1 h_2 + h_1^* h_2^*) d\omega = \Big[(\omega-\bar{\omega})|G_1(\omega)|^2\Big]_0^\infty - \int_0^\infty |G_1(\omega)|^2 d\omega$$

after an integration by parts. The contribution from the upper limit must vanish if (9.42) is to have a meaning. The lower limit does not contribute if $G_1(0) = 0$, which will now be assumed. Hence, from (9.46),

$$\left(\int_0^\infty |G_1(\omega)|^2 d\omega\right)^2 \leqslant 4 \int_0^\infty (\omega-\bar{\omega})^2 |G_1(\omega)|^2 d\omega \int_0^\infty \left|\frac{\partial G_1}{\partial \omega}\right|^2 d\omega \ .$$

It follows from (9.42) and (9.45) that

$$\Delta\tau \ . \ \Delta\omega \geqslant \frac{1}{2} \tag{9.47}$$

when $G_1(0) = 0$. Thus the effective frequency range is related to the reciprocal of the time over which the signal remains coherent. The smaller the time of coherence the greater the bandwidth involved.

The definitions of spectral width and coherence time are satisfactory in the sense that they are based on the measurable correlation g_1 (or its Fourier transform) rather than the electric intensity itself which may be rapidly varying.

There is another feature to be noted. In any region which does not contain sources

$$\left(\nabla_2^2 - \frac{1}{c^2}\frac{\partial^2}{\partial\tau^2}\right)\langle \mathbf{E}(\mathbf{x}_1, t) . \mathbf{E}(\mathbf{x}_2, t+\tau)\rangle = 0$$

where ∇_2 refers to the point $\mathbf{x}_2$, because each component of $\mathbf{E}$ satisfies such an equation. This permits the cross-correlation function to be represented by Kirchhoff's integral and the way in which it propagates can be determined. In particular, if it and its normal derivative are known on a closed surface it can be found elsewhere provided that no source intervenes.

Correspondingly

$$\nabla_2^2 G_{12} + k^2 G_{12} = 0$$

in any region where there are no sources. If G_{12} and its normal derivative were measured on a closed surface an integral equation analogous to (9.34) and (9.35) could be set up for the source term appropriate to the cross-power spectral density partial differential equation. Alternatively, the technique described at the end of the preceding section could be adopted.

In either case we arrive at some indication of the average source distribution. However, this direct attack does not seem to have been explored to any extent yet.

Although the correlations have been defined in terms of $\mathbf{E}.\mathbf{E}$ it is straightforward to handle other quantities. For instance, the cross-correlation of a single component of the field

$$\langle E_x(\mathbf{x}_1,t)E_x(\mathbf{x}_2,t+\tau)\rangle$$

or that of two components

$$\langle E_x(\mathbf{x}_1,t)E_y(\mathbf{x}_2,t+\tau)\rangle$$

(which might be relevant to cross-polarization effects) or between the electric and magnetic intensities might be considered. Only minor modifications to the foregoing analysis would be needed.

9.9. Correlation techniques

The first scheme to be discussed is based on measurements of a single component of the electric field such as E_x. This satisfies

$$\nabla^2 E_x - \frac{1}{c^2}\frac{\partial^2 E_x}{\partial t^2} = \mu\frac{\partial J_x}{\partial t} + \frac{1}{\varepsilon}\frac{\partial\rho}{\partial x} = -q(\mathbf{x},t)$$

and q furnishes a guide to the sources.

At large distances from the source distribution

$$E_x(\mathbf{x},t) \sim \frac{1}{4\pi|\mathbf{x}|}\int q\left(\mathbf{x}',t-\frac{|\mathbf{x}-\mathbf{x}'|}{c}\right)\mathrm{d}\mathbf{x}' . \tag{9.48}$$

If now correlations are defined in terms of this sole component, as explained at the end of §9.8, at large distances

$$g_1(0) = \langle\{E_x(\mathbf{x},t)\}^2\rangle = \frac{1}{16\pi^2 c^4|\mathbf{x}|^2}\int S(\mathbf{y})\,d\mathbf{y} \tag{9.49}$$

$$= \frac{1}{16\pi^2 c^4|\mathbf{x}|^2}\int R(\mathbf{y})\,d\mathbf{y} \tag{9.50}$$

where

$$S(\mathbf{y}) = \int \left\langle c^4 q\left(\mathbf{x}',t-\frac{|\mathbf{x}-\mathbf{x}'|}{c}\right)q\left(\mathbf{y},t-\frac{|\mathbf{x}-\mathbf{y}|}{c}\right)\right\rangle \,d\mathbf{x}' ,$$

$$R(\mathbf{y}) = \left\langle 4\pi c^4|\mathbf{x}|E_x(\mathbf{x},t)q\left(\mathbf{y},t-\frac{|\mathbf{x}-\mathbf{y}|}{c}\right)\right\rangle .$$

In view of the similarity of (9.49) and (9.50) to the corresponding result for a simple point source both R and S can be thought of as the source strength per unit volume. Since they involve averages measurement of them may be feasible. However, the insertion of probes close to the sources can scarcely be avoided and the noise generated thereby may pollute the readings. Even if this were not so the correlations to be observed may well be small with consequent doubt as to their accuracy and reliability. For example, should the right-hand side of (9.49) or (9.50) turn out to be negative the observations would have to be rejected.

In any case there is no certainty that R and S are being correctly interpreted. Any function, whose integral over space was zero, could be added to R or S without invalidating (9.49) or (9.50). So there is extensive scope for alternative interpretation and there seems to be no convincing argument why the additional function should be zero rather than some other value.

Correlation can also arise in connection with harmonic problems. An illustration is the finding of a local deviation in a medium which is not varying with time. Suppose that ε has the constant value ε_0 except in some finite region. Then a plane wave, whose wavelength is much larger than the inhomogeneity, can be fired at the medium. The Rayleigh–Gans or Born approximation is then reasonable and the second part of (9.1) can be written as

$$\operatorname{curl} \mathbf{H} - i\omega\varepsilon_0\mathbf{E} = i\omega(\varepsilon-\varepsilon_0)\mathbf{e}_0 \exp(-ik\mathbf{x}.\mathbf{i})$$

if the x axis is picked to be along the direction of propagation of the incident wave. The distant scattered field is therefore

$$\mathbf{E}^s(\mathbf{x}) \sim \frac{k^2\exp(-ik|\mathbf{x}|)}{4\pi\varepsilon_0|\mathbf{x}|}\{\mathbf{e}_0-(\mathbf{e}_0.\hat{\mathbf{x}})\hat{\mathbf{x}}\} \int_{-\infty}^{\infty} n(\mathbf{y})\exp\{ik\mathbf{y}.(\hat{\mathbf{x}}-\mathbf{i})\}\mathrm{d}\mathbf{y}$$

where $n = \varepsilon-\varepsilon_0$ and k^2 is now $\omega^2\mu\varepsilon_0$. Clearly n is zero outside the blob of deviating matter. Hence

$$|\mathbf{E}^s(\mathbf{x})|^2 \sim \frac{k^4}{16\pi^2\varepsilon_0^2|\mathbf{x}|^2}|\mathbf{e}_0-(\mathbf{e}_0.\hat{\mathbf{x}})\hat{\mathbf{x}}|^2 \int_{-\infty}^{\infty}\int_{-\infty}^{\infty} n(\mathbf{y})n(\mathbf{z})\exp\{ik(\mathbf{y}-\mathbf{z}).(\hat{\mathbf{x}}-\mathbf{i})\}\mathrm{d}\mathbf{y}\mathrm{d}\mathbf{z}$$

$$\sim \frac{k^4|\mathbf{e}_0-(\mathbf{e}_0.\hat{\mathbf{x}})\hat{\mathbf{x}}|^2}{16\pi^2\varepsilon_0^2|\mathbf{x}|^2} \int_{-\infty}^{\infty}\exp\{ik\mathbf{w}.(\hat{\mathbf{x}}-\mathbf{i})\}\int_{-\infty}^{\infty} n(\mathbf{x})n(\mathbf{w}+\mathbf{z})\mathrm{d}\mathbf{z}\mathrm{d}\mathbf{w}\,. \qquad (9.51)$$

The inner integral in (9.51) can be perceived as the autocorrelation function of n (though now in space rather than time) so that measurements of the far field y yield some information about this function. It is perhaps most profitable to make some simple assumption about the form of the autocorrelation function, such as that it is Gaussian with adjustable parameters, with the intent of evaluating the right-hand side of (9.51). Comparison with the measured left-hand side should then sanction the determination of the parameters, optimization techniques being utilized if convenient. If tolerable agreement can be secured by adjustment of the parameters some idea can be obtained of the rate at which the autocorrelation functions falls to zero. That enables one to get some notion of the correlation between deviations from the average and gain some indication of the extent of the inhomogeneity.

This procedure is likely to be especially valuable in eliciting the properties of biological specimens which are usually subject to statistical variability.

9.10. Far-field cross-correlation technique

There is another way in which correlation measurements can be related to source distributions. Let the observations be made at points sufficiently distant from the sources for (9.48) to hold. Then, if $|\mathbf{x}_1| = |\mathbf{x}_2|$, the cross-correlation function g_{12} is given by

$$g_{12}(\tau) = \langle E_1(\mathbf{x}_1,t)E_1(\mathbf{x}_2,t+\tau)\rangle$$

$$= \frac{1}{16\pi^2|\mathbf{x}_1|^2}\int_{-\infty}^{\infty}\int_{-\infty}^{\infty}\left\langle q\left(\mathbf{y},t-\frac{|\mathbf{x}_1-\mathbf{y}|}{c}\right)q\left(\mathbf{z},t+\tau-\frac{|\mathbf{x}_2-\mathbf{z}|}{c}\right)\right\rangle \mathrm{d}\mathbf{y}\,\mathrm{d}\mathbf{z}$$

with the slight generalization that E_1 is an arbitrary, but fixed, component of the field and q its associated effective source. The assumption of statistical stationarity enables this to be written as

$$g_{12}(\tau) = \frac{1}{16\pi^2|\mathbf{x}_1|^2}\int_{-\infty}^{\infty}\int_{-\infty}^{\infty}\left\langle q(\mathbf{y},t)q\left(\mathbf{z},t+\tau+\frac{|\mathbf{x}_1-\mathbf{y}|}{c}-\frac{|\mathbf{x}_2-\mathbf{z}|}{c}\right)\right\rangle \mathrm{d}\mathbf{y}\,\mathrm{d}\mathbf{z}.$$

The cross-power spectral density is, from (9.39),

$$G_{12}(\omega) = \frac{1}{32\pi^3|\mathbf{x}_1|^2}\int_{-\infty}^{\infty}\int_{-\infty}^{\infty}\exp\{-ik(|\mathbf{x}_1-\mathbf{y}|-|\mathbf{x}_2-\mathbf{z}|)\}\int_{-\infty}^{\infty}\langle q(\mathbf{y},t)q(\mathbf{z},t+\tau)\rangle\exp(i\omega\tau)\,\mathrm{d}\tau\,\mathrm{d}\mathbf{y}\,\mathrm{d}\mathbf{z}.$$

The quantity

$$\frac{1}{2\pi}\int_{-\infty}^{\infty}\langle q(\mathbf{y},t)q(\mathbf{z},t+\tau)\rangle\exp(i\omega\tau)\,\mathrm{d}\tau$$

is the cross-power spectral density of the source distribution; denote it by $\bar{q}(\mathbf{y},\mathbf{z},\omega)$. Then

$$G_{12}(\omega) = \frac{1}{16\pi^2|\mathbf{x}_1|^2}\int_{-\infty}^{\infty}\int_{-\infty}^{\infty}\bar{q}(\mathbf{y},\mathbf{z},\omega)\exp\{-ik(|\mathbf{x}_1-\mathbf{y}|-|\mathbf{x}_2-\mathbf{z}|)\}\mathrm{d}\mathbf{y}\mathrm{d}\mathbf{z}. \tag{9.52}$$

Likewise, the power spectral density at $\mathbf{x}_1$ is given by

$$G_1(\omega) = \frac{1}{16\pi^2|\mathbf{x}_1|^2}\int_{-\infty}^{\infty}\int_{-\infty}^{\infty}\bar{q}(\mathbf{y},\mathbf{z},\omega)\exp\{-ik(|\mathbf{x}_1-\mathbf{y}|-|\mathbf{x}_1-\mathbf{z}|)\}\mathrm{d}\mathbf{y}\mathrm{d}\mathbf{z}. \tag{9.53}$$

Define

$$Q_1(\mathbf{z},\omega) = \int_{-\infty}^{\infty}\bar{q}(\mathbf{y},\mathbf{z},\omega)\exp\{-ik(|\mathbf{x}_1-\mathbf{y}|-|\mathbf{x}_1-\mathbf{z}|)\}\mathrm{d}\mathbf{y}\ . \tag{9.54}$$

Then

$$G_1(\omega) = \frac{1}{16\pi^2|\mathbf{x}_1|^2}\int_{-\infty}^{\infty}Q_1(\mathbf{z},\omega)\mathrm{d}\mathbf{z}\ , \tag{9.55}$$

$$G_{12}(\omega) = \frac{1}{16\pi^2|\mathbf{x}_1|^2}\int_{-\infty}^{\infty}Q_1(\mathbf{z},\omega)\exp\{ik(|\mathbf{x}_2-\mathbf{z}|-|\mathbf{x}_1-\mathbf{z}|)\}\mathrm{d}\mathbf{z}\ . \tag{9.56}$$

The resemblance between (9.55) and (9.49) suggests that $Q_1(\mathbf{z},\omega)$ might be called the apparent source strength per unit volume at the point $\mathbf{z}$, radiating at the angular frequence ω. From (9.54), Q_1 relies on observations at $\mathbf{x}_1$ and so it is the apparent source strength as viewed from that point. In general, changes in the reference point $\mathbf{x}_1$ will entail alterations in the apparent source distribution.

Determining Q_1 from (9.55) suffers from many of the objections levelled against (9.49) and (9.50). However, there is more opportunity for manoeuvre with (9.56) because of the variation with $\mathbf{x}_2$. Suppose that there is a fixed sensor at $\mathbf{x}_1$

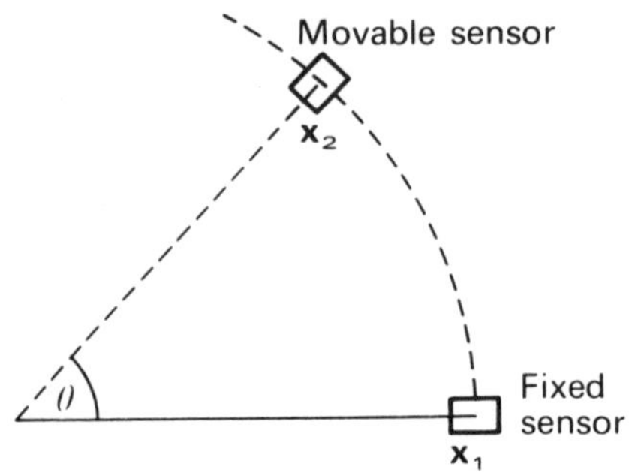

FIG. 9.6. Determination of cross-power spectral density.

and that there is a traversing sensor at $\mathbf{x}_2$ which is always maintained at the same distance from the origin as the fixed sensor (Fig. 9.6). The record of simultaneous signals at the two sensors is Fourier analysed at a given angular separation θ by passing it through narrow-band filters which will transmit only a small range of frequencies centred on the $\omega/2\pi$ at which information about Q_1 is desired. The cross-correlation function of the filtered signal then leads to $G_{12}(\omega)$ as a function of θ. In ideal circumstances, values for all θ are feasible but in practical situations the magnitude of G_{12} frequently diminishes with increasing angular separation so that the correlation may not be measurable when θ exceeds, say, 60°. Also the observations will probably be made at discrete rather than continuous values of θ. Notwithstanding, we shall assume that G_{12} is known as a continuous function of θ on a circle with centre at the origin and whose circumference passes through $\mathbf{x}_1$.

To find Q_1 from G_{12} the relationship (9.56) has to be inverted. Unfortunately, G_{12} is known only on a circle so that there is no hope of recovering full knowledge of Q_1. It therefore pays to make simplifying assumptions about the source distribution. To put it another way, the measured cross-power spectral density is accounted for by a simple, but artificial, distribution of sources.

Take the plane of the circle as the (x,y) plane with $\mathbf{x}_1$ on the x axis. It will now be assumed that the sources are not very far from the y axis. Then (9.56) simplifies to

$$G_{12}(\omega) = \frac{1}{16\pi^2|\mathbf{x}_1|^2}\int_{-\infty}^{\infty} q_1(y,\omega)\exp\{ik(|\mathbf{x}_2-y\mathbf{j}|-|x_1\mathbf{i}-y\mathbf{j}|)\}dy$$

where

$$q_1(z_2,\omega) = \int_{-\infty}^{\infty}\int_{-\infty}^{\infty} Q_1(\mathbf{z},\omega)\,dz_1 dz_3$$

and **i**, **j** are unit vectors parallel to the x and y axes respectively.

It will now be postulated that the points of observation are many wavelengths from the places where q_1 is non-zero. This is a non-trivial assumption because the positions of the apparent sources are unknown and forcing them to lie on the y axis may oblige them to be some distance from the genuine sources about whose rough location a fair degree of confidence may have been felt. Actually, the postulate can only be truly checked *a posteriori* by examining whether (*a*) the resulting apparent sources are, in fact, nowhere near the circle of measurement and (*b*) increasing $|\mathbf{x}_1|$ leaves the apparent sources substantially unchanged.

At any rate, squares and higher powers of y will now be neglected so that $|x_1\mathbf{i}-y\mathbf{j}| \sim x_1$ and

$$|\mathbf{x}_2-y\mathbf{j}| \sim x_1 - y\sin\theta\ .$$

Then

$$G_{12}(\omega) = \frac{1}{16\pi^2|\mathbf{x}_1|^2}\int_{-\infty}^{\infty} q_1(y,\omega)\exp(-iky\sin\theta)\,dy\ . \qquad (9.57)$$

Eqn (9.57) can be regarded as a Fourier transform in which $k\sin\theta$ is the transform variable. But $k\sin\theta$ ranges only from $-k$ to k so that G_{12} is constrained to that interval. By defining G_{12} to be zero outside the interval (which is plausible when $|G_{12}|$ drops sufficiently rapidly with increasing θ) we can take a Fourier inverse which will give a q_1' which will satisfy (9.57) on the interval of measurement. The pertinent formula is

$$q_1'(y,\omega) = 8\pi|\mathbf{x}_1|^2 k \int_{-\frac{1}{2}\pi}^{\frac{1}{2}\pi} G_{12}(\omega)\exp(iky \sin \theta)\cos \theta \, d\theta \; . \tag{9.58}$$

In essence q_1' is a diffraction-limited image of the apparent source distribution q_1.

Assuming that the far-field approximation is legitimate we can be sure that the resolving power is independent of $|\mathbf{x}_1|$ because it is necessary only to maintain the same range of θ for differing $\mathbf{x}_1$. On the other hand, a disadvantage is that the method prejudges the issue by insisting that the apparent sources be located on the y axis.

It is advisable to see how well (9.58) performs when the real sources are actually on the y axis. To this end let

$$q(\mathbf{x},t) = Q(y)\delta(x)\delta(z)\cos\{\omega_0 t+\phi(y)\} \tag{9.59}$$

where ω_0 is positive and the real function $Q(y)$ is assumed to be zero outside some finite interval. The phase $\phi(y)$ is chosen so that $\phi(0) = 0$ which may necessitate a change of time origin but that is of no significance here. From (9.59)

$$\bar{q}(\mathbf{x}',\mathbf{x}'',\omega) = \tfrac{1}{4}Q(y')\delta(x')\delta(z')Q(y'')\delta(x'')\delta(z'') \times$$

$$\times \; (\delta(\omega+\omega_0)\exp[i\{\phi(y'')-\phi(y')\}]+\delta(\omega-\omega_0)\exp[-i\{\phi(y'')-\phi(y')\}]).$$

Substitute in (9.57) via (9.54). Because of (9.41) only positive frequencies need be considered so that the term involving $\delta(\omega+\omega_0)$ can be omitted without loss. Hence

$$G_{12}(\omega) = \frac{\delta(\omega-\omega_0)}{64\pi^2|\mathbf{x}_1|^2}\int_{-\infty}^{\infty}\int_{-\infty}^{\infty} Q(y')Q(y'')\exp[-i\{\phi(y'')-\phi(y')+k(|\mathbf{x}_1-y'\mathbf{j}|-|\mathbf{x}_2-y''\mathbf{j}|]\}dy'd$$

$$\sim \frac{\delta(\omega-\omega_0)}{64\pi^2|\mathbf{x}_1|^2}\int_{-\infty}^{\infty}\int_{-\infty}^{\infty} Q(y')Q(y'')\exp[-i\{\phi(y'')-\phi(y')+ky''\sin \theta\}]dy'dy''$$

subject to the same conditions as imposed in (9.57). From (9.58)

$$q_1'(y,\omega) = \frac{\delta(\omega-\omega_0)}{4\pi} \int_{-\infty}^{\infty}\int_{-\infty}^{\infty} Q(y')Q(y'')\frac{\sin k(y-y'')}{y-y''}\exp[-i\{\phi(y'')-\phi(y')\}]dy'dy''.$$

For large values of k, $(\sin kx)/x$ is approximately $\pi\delta(x)$ as far as integration is concerned (Jones 1966). Hence, for high frequencies

$$q_1'(y,\omega) = \tfrac{1}{4}\delta(\omega-\omega_0)Q(y)\exp\{-i\phi(y)\}\int_{-\infty}^{\infty} Q(y')\exp\{i\phi(y')\}dy' . \tag{9.60}$$

Thus q_1' reproduces the amplitude and phase of the genuine source apart from a normalization factor.

To get rid of the normalization factor put

$$q_1'(y,\omega) = q_1''(y,\omega)\delta(\omega-\omega_0)$$

and suppose that

$$\int_{-\infty}^{\infty} Q(y')\exp\{i\phi(y')\}dy' = A\ \exp(i\alpha)$$

where A is real. Then

$$q_1''(0,\omega) = \tfrac{1}{4}Q(0)A\ \exp(i\alpha) .$$

Since q_1'' is presumed to be known from the measurements, α can be found to within a multiple of π. The indeterminacy springs from the fact that we do not know whether $Q(0)A$ is positive or negative. Also

$$\int_{-\infty}^{\infty} q_1''(y,\omega)dy = \tfrac{1}{4}A^2$$

since Q is a real function. Therefore, if we define

$$q_1'''(y,\omega) = \frac{2q_1''(y,\omega)\exp(-i\alpha)}{\left(\int_{-\infty}^{\infty} q_1''(y,\omega)\,dy\right)^{\frac{1}{2}}} \tag{9.61}$$

we obtain

$$q_1'''(y,\omega) = \pm Q(y)\exp\{-i\phi(y)\} \tag{9.62}$$

and all quantities on the right-hand side of (9.61) can be obtained from measurements.

The ambiguity of sign in (9.62) is not surprising because if the sign of q is changed, G_{12} remains unaltered. Consequently, the absolute sign of a source distribution can never be found by correlation techniques. Of course, as y varies, the sign in (9.62) stays invariable, be it plus or minus.

From (9.62) $Q(y)$ and $\phi(y)$ can be determined, apart from a possible sign error, and hence the actual sources found instead of the apparent sources. This leads to the suggestion that q_1''' as defined by (9.61) should be employed as the indicator for apparent sources in general. It gives the actual sources when they are on a line and can be expected to be a good predictor if they do not depart too widely from a straight line.

The proposal to adopt q_1''' rests on the high-frequency approximation deployed in deriving (9.60). Whether q_1''' would be at all reliable at low frequencies is an open question. The necessity for high frequencies also enforces measurements at sufficiently many values of θ to resolve the structure of the field.

Even if the sources are not nearly rectilinear q_1''' can perform a helpful function as a standardizing mechanism, for it says that the cross-power spectral density agrees with that produced by a certain (fictitious) rectilinear distribution. Therefore, relations between actual source distributions may be clarified by comparing their (fictitious) rectilinear equivalents. It is, of course, true that the real sources may not be perfectly correlated as in (9.59) but that is irrelevant. The idea of standardization is that polar correlation measurements should always be interpreted by means of a rectilinear perfectly correlated distribution but it is not argued that this will coincide with the actual sources in general.

HOLOGRAPHIC TECHNIQUES

9.11. Basic principles of holography

In holography (Gabor 1948, 1949, 1951, 1956) an object is illuminated by a coherent light wave and the diffraction pattern or *hologram* recorded on a photographic plate. After suitable processing the plate, when re-illuminated conveniently, conveys an image of the object. Holography therefore represents a means of solving the inverse scattering problem from a photographic copy of the far field. It has the obvious benefit of providing a permanent record of the solution.

The basic theory which validates the reconstruction of the image will now be derived. Since the primary concern is with optical frequencies it will be sufficient to consider scalar fields only (Born and Wolf 1965, Wolf and Shewell 1970). A monochromatic wave u^i is incident on an object and produces the field u_1 in the aperture A of the plane $z = z_1$ with $z_1 < 0$ (Fig. 9.7). The aperture is designed so that light which misses

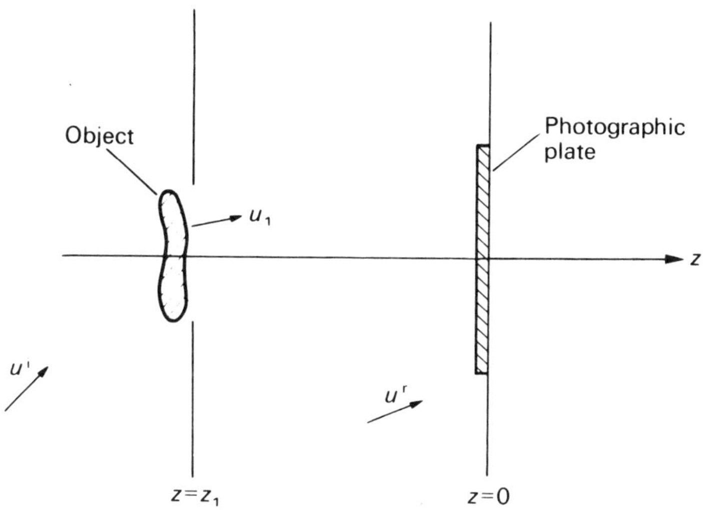

FIG. 9.7. Formation of a hologram.

the object does not pass beyond $z = z_1$. The field u_1 spreads in $z > z_1$ and has added to it a *reference field* u^r (of the same frequency as u^i) so that the total field is $u_1 + u^r$. This total field is recorded on a photographic plate at $z = 0$. The plate responds to the intensity I of the total field where

$$I = (u_1+u^r)(u_1+u^r)^* = I_1 + I^r + u_1 u^{r*} + u_1^* u^r , \qquad (9.63)$$

I_1 and I^r being the intensities of u_1 and the reference wave respectively at the plate.

The plate is now processed so that its effect is to make a wave incident on it emerge multiplied by the factor $a_0 + a_1 I$ where a_0 and a_1 are real constants. Usually, a_0 is positive and a_1 negative. The hologram so formed is placed in the plane $z = 0$ and illuminated by the monochromatic *reconstruction field* v^c (Fig. 9.8). We shall discuss only the situation in which v^c is the same frequency as u^r but the case when they are of different frequencies can be considered. The field appearing on the far side of the hologram is then

$$v(0) = a_1 v^c u_1 u^{r*} + a_1 v^c u_1^* u^r + v^c\{a_0 + a_1(I_1 + I^r)\} . \qquad (9.64)$$

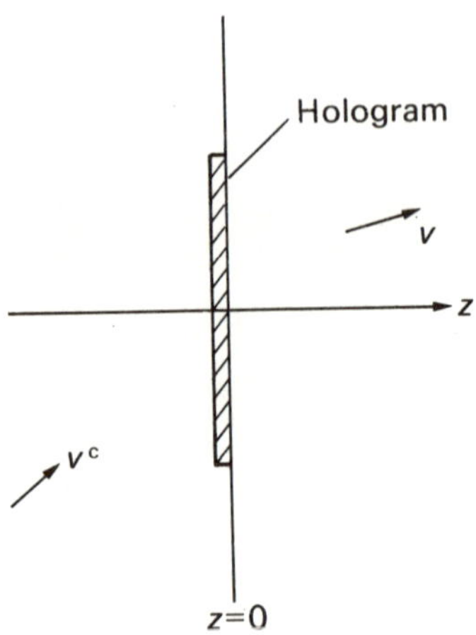

FIG. 9.8. Reconstruction of object.

If matters have been arranged aright v will reproduce an image of the object somewhere in $z > 0$. Circumstances under which this is possible (Leith and Upatnieks 1963) will now be investigated.

Let u^{r} and v^{c} be plane waves so that

$$u^{\mathrm{r}} = B\exp(-ik\mathbf{n}^{\mathrm{r}}.\mathbf{x}) ,$$

$$v^{\mathrm{c}} = C\exp(-ik\mathbf{n}^{\mathrm{c}}.\mathbf{x}) .$$

Then (9.64) gives

$$v(0) = v_1(0) + v_2(0) + v_3(0)$$

where

$$v_1(0) = a_1 CB^* u_1 \exp\{ik(\mathbf{n}^{\mathrm{r}}-\mathbf{n}^{\mathrm{c}}).\mathbf{x}\} , \qquad (9.65)$$

$$v_2(0) = a_1 CB u_1^* \exp\{-ik(\mathbf{n}^{\mathrm{r}}+\mathbf{n}^{\mathrm{c}}).\mathbf{x}\} , \qquad (9.66)$$

$$v_3(0) = \{a_0 + a_1(I_1 + BB^*)\}\exp\{-ik\mathbf{n}^{\mathrm{c}}.\mathbf{x}\} \qquad (9.67)$$

with $\mathbf{x} \equiv (x,y,0)$.

Suppose now that $\mathbf{n}^{\mathrm{r}} = \mathbf{n}^{\mathrm{c}}$ so that the reference and reconstruction waves travel in the same direction. Then, from (9.65), $v_1(0) = a_1 CB^* u_1$ and so v_1 is proportional to the field created at $z = 0$ by the light coming from the original obstacle in the absence of any reference wave. Thus, in $z > 0$, v_1 will be a constant multiple of u_1 and will appear to be propagating from a virtual image of the original object, this virtual image being in precisely the same position relative to the hologram as was the original object relative to the photographic plate.

Define $U(l,m)$ by

$$U(l,m) = \frac{k^2}{4\pi^2}\int_{-\infty}^{\infty}\int_{-\infty}^{\infty} u_1(x,y,0)\exp\{ik(lx+my)\}\mathrm{d}x\mathrm{d}y ; \qquad (9.68)$$

then a representation for u_1 valid for all z is

$$u_1(z) = \int_{-\infty}^{\infty}\int_{-\infty}^{\infty} U(l,m)\, \exp\{-ik(lx+my+nz)\}dl\,dm \tag{9.69}$$

where $n^2 = 1 - l^2 - m^2$, n being positive when $l^2 + m^2 < 1$ and negative imaginary when $l^2 + m^2 > 1$. In fact, we shall suppose that $U(l,m)$ is effectively zero when $l^2 + m^2 > 1$ so that only real n need be considered in (9.69). Eqn (9.69) is a representation in plane waves, homogeneous if $l^2 + m^2 < 1$ and inhomogeneous otherwise. Since evanescent waves do not radiate the waves with $l^2 + m^2 < 1$ are often described as the *visible* part of the spectrum while the inhomogeneous waves are the *invisible* part.

Let v_2' be the field which coincides with u_1^* on $z = 0$. Then

$$v_2'(z) = \int_{-\infty}^{\infty}\int_{-\infty}^{\infty} \{U(-l,-m)\}^*\exp\{-ik(lx+my+nz)\}dl\,dm$$

$$=\left\{\int_{-\infty}^{\infty}\int_{-\infty}^{\infty} U(l,m)\exp\{-ik(lx+my-nz)\}dl\,dm\right\}^* \tag{9.70}$$

on account of our assumption about U. It follows from (9.69) and (9.70) that

$$v_2'(z) = \{u_1(-z)\}^* \tag{9.71}$$

which is a sort of *reciprocity theorem*. That (9.69) is valid for $z > 0$ and $z < 0$ is a consequence of u_1 travelling to the right near $z = 0$.

Having established (9.71) in general we now turn to (9.66) when $\mathbf{n}^r = \mathbf{n}^c$. If the components of $\mathbf{n}^r$ are (n_1^r, n_2^r, n_3^r)

$$\frac{k^2}{4\pi^2}\int_{-\infty}^{\infty}\int_{-\infty}^{\infty} u_1^*(x,y,0)\exp\{-2ik\mathbf{n}^r.\mathbf{x}+ik(lx+my)\}dx\,dy$$

$$= \{U(2n_1^r-l, 2n_2^r-m)\}^*$$

and hence

$$v_2(z) = a_1 CB\left\{\int_{-\infty}^{\infty}\int_{-\infty}^{\infty} U(l+2n_1^{\mathrm{r}}, m+2n_2^{\mathrm{r}})\exp\{-ik(lx+my-nz)\}\mathrm{d}l\mathrm{d}m\right\}^* . \tag{9.72}$$

We shall now make the additional assumption that $U(l,m)$ is negligible unless $l^2 + m^2 \ll 1$. Then squares of l and m can be ignored and (9.69) becomes

$$u_1(x,y,z) = \exp(-ikz)\int_{-\infty}^{\infty}\int_{-\infty}^{\infty} U(l,m)\exp\{-ik(lx+my)\}\mathrm{d}l\mathrm{d}m . \tag{9.73}$$

If, in (9.72), $l + 2n_1^{\mathrm{r}}$ and $m + 2n_2^{\mathrm{r}}$ are taken as new variables

$$n = \{1-(l-2n_1^{\mathrm{r}})^2-(m-2n_2^{\mathrm{r}})^2\}^{\frac{1}{2}}$$

$$\approx \{\nu^2+4(ln_1^{\mathrm{r}}+mn_2^{\mathrm{r}})\}^{\frac{1}{2}}$$

on rejecting squares of l, m and putting $\nu^2 = 1-4(n_1^{\mathrm{r}})^2-4(n_2^{\mathrm{r}})^2$. Whenever ν is not near zero, the approximation

$$n \approx \nu + \frac{2(ln_1^{\mathrm{r}}+mn_2^{\mathrm{r}})}{\nu}$$

is legitimate. Consequently, in these circumstances

$$v_2(z) = a_1 CB\left\{u_1\left(x-2n_1^{\mathrm{r}}\frac{z}{\nu},\ y-2n_2^{\mathrm{r}}\frac{z}{\nu}, -\frac{z}{\nu}\right)\right\}^* \times$$

$$\times \exp\left\{-2ik(n_1^{\mathrm{r}}x+n_2^{\mathrm{r}}y) + \frac{ikz(1-\nu^2)}{\nu}\right\} . \tag{9.74}$$

Thus v_2 carries much the same information for positive z that u_1 does for negative z. Since u_1 must account for the presence of the obstacle in negative z, v_2 must produce a *real* image of the object to the right of the hologram. If most of the object lies pretty well on $z = z_1$, the image will occur on $z = -\nu z_1$.

Apart from a factor of proportionality and a phase factor, the complex amplitudes of corresponding points of the object and image are complex conjugates of one another.

However, whereas the object is disposed about the z axis, the image is displaced from it. In fact, if the object point $(x_0, y_0, z_1+\xi)$, ξ being small, corresponds to the image point (x_i, y_i, z_i)

$$x_i = x_0 - 2n_1^r z_1, \quad y_i = y_0 - 2n_2^r z_1, \quad z_i = -\nu(z_1+\xi) \ . \tag{9.75}$$

These relations show firstly that the image is displaced by a distance $(1-\nu^2)^{\frac{1}{2}}|z_1|$ from the z axis. Secondly, transverse changes in which only x_0 and y_0 vary are reproduced exactly but longitudinal displacements in which ξ alters are subject to a magnification ν. Thus the image is a distorted replica of the object, being squashed lengthwise (Fig. 9.9).

The analysis requires that ν be not small. Since ν is zero when the reference wave makes an angle of 30° with the z axis, we want the direction of propagation of the reference wave to be inclined at somewhat less than 30° to the z axis. However, for a good image the angle must not be too small. The reason is that v_3 essentially reproduces the reconstruction wave and a radiating wave. The image will therefore be easily distinguished from v_1 (another radiating wave) and v_3 provided that the image is not in the same direction as the reconstruction wave. The formulae (9.75) indicate that this is true provided that n_1^2 and n_2^r are not too small.

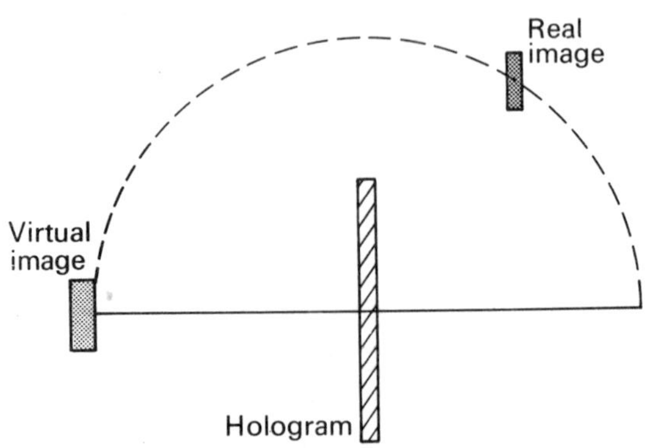

FIG. 9.9. Positions of real and virtual images.

Exercises

20. If $\mathbf{n}^r.\mathbf{k} = \mathbf{n}^c.\mathbf{k}$ and $\mathbf{n}^r \wedge \mathbf{k} = \mathbf{n}^c \wedge \mathbf{k}$, where $\mathbf{k}$ is a unit vector along the z axis, prove that v_2 is responsible for a real image on the z axis. (Hint: eqn (9.71).) Is this image a distorted version of the object? Prove also that v_1 has a virtual image which is displaced from the z axis and subject to longitudinal deformation.

21. If $\mathbf{n}^c$ is parallel to the z axis but $\mathbf{n}^r$ is not show that v_1 entails a virtual image and v_2 a real image, both images being displaced from the z axis and being deformed versions of the object.

22. In the situation of Exercise 21 prove that $v(0)$ is real. (Because of the photographic processing it is non-negative.) If $U(l,m)$ is non-zero only where $|l| \leqslant L$, $|m| \leqslant L$ where $18L^2 \leqslant (n_1^r)^2 + (n_2^r)^2$ prove that

$$u_1(x,y) = \frac{1}{\pi^2 a_1 CB^*}\int_{-\infty}^{\infty}\int_{-\infty}^{\infty} v(0)\,\frac{\sin kL(x-x')\sin kL(y-y')}{(x-x')(y-y')}\exp\{-ik(n_1^r x'+n_2^r y')\}\mathrm{d}x'\mathrm{d}y'.$$

This demonstrates that the *amplitude and phase* of the scattered field can be determined from measurements of the *intensity* of the field emerging from the hologram.

Deduce that when the reference wave makes an angle of about 30° with the z axis details down to about nine wavelengths of the illuminating light can be distinguished (Wolf 1970).

9.12. Location of an inhomogeneity

In the detection of an inhomogeneity in an otherwise uniform medium by holography the blob of deviant material is treated as the object. The incident wave u^i is chosen to be a plane wave moving in the direction of the unit vector $\mathbf{n}^i$. If the blob is sufficiently small for the Rayleigh–Gans or Born approximation to be applicable then, at large distances,

$$u_1(\mathbf{x}) \sim \frac{A^i \exp(-ik|\mathbf{X}|)}{4\pi|\mathbf{X}|}\int_{-\infty}^{\infty} n(\mathbf{y})\exp\{ik\mathbf{y}.(\hat{\mathbf{X}}-\mathbf{n}^i)\}\mathrm{d}\mathbf{y}$$

where $u^i = A^i\exp(-ik\mathbf{n}^i.\mathbf{x})$ at the blob, $\mathbf{X} = \mathbf{x} - z_1\mathbf{k}$, and n represents the difference between the refractive index in the blob and that in the surrounding medium. According to Exercise 22 measurements of the intensity emergent from the hologram will give u_1 and thereby supply the angular Fourier transform of n. By varying the angle of incidence the three-dimensional Fourier transform becomes available and so n follows by Fourier inversion (Wolf 1969). Although this is theoretically true the fact that the resolution cannot be brought below nine wavelengths imposes severe practical limitations.

9.13. Field in the aperture of an antenna

The theory of holography developed above has concentrated on optical aspects, but the principles have been applied in other areas. One such is the calculation of the aperture distribution of an antenna from measurements of the radiation pattern (Bates 1971, Napier and Bates 1971, 1973, Ransom and Mittra 1971).

To avoid the complication of vector fields, which intensify the detail without affecting the principles, the analysis is restricted to scalar fields. The field u is generated from the aperture A in the antenna at $z = z_1$ and the aperture distribution is to be inferred from measurements on $z = 0$ (Fig. 9.10). There is therefore a distinct similarity to the problem treated in §9.11.

Assuming that u is non-zero on the antenna outside the aperture

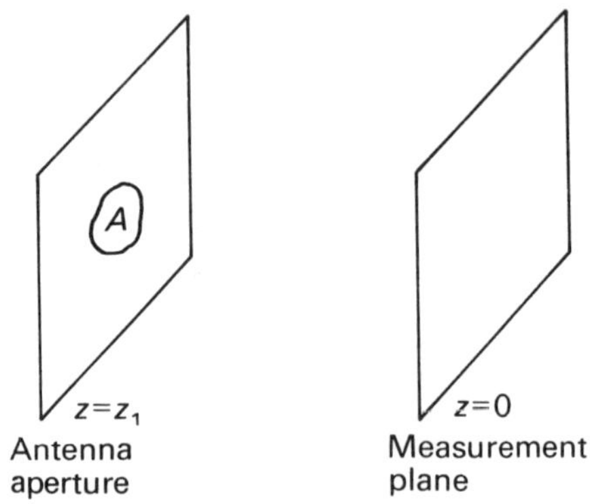

FIG. 9.10. Geometry for aperture calculation.

$$u(x,y,0) = -\frac{1}{2\pi}\frac{\partial}{\partial z_1}\int_A u(x',y',z_1)\,\frac{\exp[-ik\{(x-x')^2+(y-y')^2+z_1^2\}^{\frac{1}{2}}]}{\{(x-x')^2+(y-y')^2+z_1^2\}^{\frac{1}{2}}}\,dx'dy' \quad . \tag{9.76}$$

Because of the postulate on u, the integral in (9.76) can be extended over the whole plane $z = z_1$. Then the measured u is the convolution of the required distribution and a known function.

A direct solution of (9.76) by discretization or Galerkin methods is notoriously awkward. Instead, it is more remunerative to invert the convolution equation which is, in effect, what holography does.

The inversion of a convolution is a common occurrence because the effect of many linear devices on a signal is to form a convolution. For example, this happens when the reception pattern of an antenna, together with any linear filtering and detection, is taken into account. Again, correlation functions, transmission through a fluctuating medium, X-ray diffraction, crysto-holography, and the restoration of images blurred by motion are all examples where convolution plays an important part (Jones and Misell 1970, Sondhi 1972, Frieden and Burke 1972, Bates, Gough, and Peters 1974, Bates and Gough 1975, Bates and Lewitt 1975). A few remarks on the difficulties involved are therefore in order.

It is sufficient to discuss convolution in one dimension, so we consider the problem of finding g given f and h in

$$f(x) = \int_{-\infty}^{\infty} g(t)h(x-t)\,dt \quad . \tag{9.77}$$

Define Fourier transforms by capital letters so that, for example,

$$F(\alpha) = \int_{-\infty}^{\infty} f(x)\exp(-i\alpha x)\,dx \quad .$$

Then (9.77) leads to

$$F(\alpha) = G(\alpha)H(\alpha)$$

whence

$$G(\alpha) = \frac{F(\alpha)}{H(\alpha)} \quad . \tag{9.78}$$

Thus g is the convolution of f and the inverse transform of the reciprocal of H.

Conditions for the validity of this process depend upon the properties one is prepared to ascribe to f, g, and h. For example, (9.77) can be formulated in terms of generalized functions (Jones 1966) but for our purposes ample scope is available by sticking to conventional functions. In that case F/H and F would need to be square integrable when h is integrable. At any rate, whenever H vanishes for a real α, F must do likewise if an unacceptable singularity in G is to be evaded.

We might attempt to find g by applying a fast (inverse) Fourier transform to (9.78). This should work satisfactorily so long as $|H|$ is never small compared with $|F|$. It has already been remarked that F and H should vanish together but, regrettably, F is obtained from measurements and so suffers noise contamination. For that reason there is likely to be some discrepancy between a zero of F and the corresponding one of H. The result is that the inversion will be very noise sensitive. The complaint will be aggravated if H is also derived from observations. In other words, the inverse transform should be employed if $|H|$ is always large enough to swamp the noise in F but when this is not so an alternative technique is apposite.

One possibility is to deform the contour off the real axis to one where $|H|$ is never small. F must then be calculated for complex α from the data for f. Unfortunately, the net result is to amplify the noise from the data so the inversion will probably be as intolerably noise sensitive as before.

Nevertheless, it has been suggested (Bates, Napier, McKinnon, and McDonnell 1976) that it is advantageous to study (9.78) in the complex α plane with particular reference to zeros in the left half-plane where $\mathcal{R}(\alpha) < 0$. Wherever F has a zero either G or H must have one also. By identifying those zeros of F which agree with those of H we can deduce the positions of the zeros of G. In practice, when F and H are obtained from measured data, there will not be precise coincidence and it will be necessary to resort to determining by inspection which pairs are closest together, supplemented by minimizing the sum of the squares of distances between pairs where it is not clear which are correspondents. There may be some zeros of H left over by this process. The data are then inconsistent but just ignoring the surplus zeros of H may be successful at the expense of increasing the uncertainty in g. Any zeros of F now remaining are attributed to G. Assuming that properties of g can be deduced from the zeros of G, a matter which will be turned to shortly (§9.14), a method has been devised for carrying out the inverse convolution.

It might happen that, instead of knowing H, we were given $|H|$ for real α. Since $|H|^2$ is the Fourier transform of the autocorrelation of h for real α we can find $|H|^2$ in the complex plane from this autocorrelation. In such circumstances h is usually real and the autocorrelation is then an even function. As a consequence the complex conjugate of a zero of H is also a zero of $|H|^2$. Despite the zeros of $|H|^2$ occurring in complex conjugate couples the ones belonging to H can be identified as those common to F. Thus the zeros to be associated with G can be determined. Consequently we can hope to find g even though information on the phase of H is lacking.

Another case in which progress can be made is that in which $g(t)$ and $h(t)$ are known to be zero outside the intervals (t_0, t_0+l_g) and $(-t_1-l_h, -t_1)$ respectively (l_g, l_h being positive) and the modulus of the Fourier transform of $g(t) + h(-t)$ is available. Effectively, this is what transpires in holography if we make the further supposition that $t_0 - t_1 - l_h$ exceeds the greater of l_g and l_h. The convolution of $g(t)$ and $g(-t)$ or autocorrelation of g vanishes outside $(-l_g, l_g)$ while that of h disappears outside $(-l_h, l_h)$. On the other hand, f is zero outside $(t_0-t_1-l_h, t_0-t_1+l_g)$ which is disjoint from $(-l_g, l_g)$ and $(-l_h, l_h)$ because of our supposition. Thus, when x surpasses the greater of l_g and l_h, the autocorrelation of $g(t) + h(-t)$ reproduces that of f in the whole interval where it is non-zero and nothing else. But this autocorrelation is the inverse of the square of the modulus of the Fourier transform. Consequently, f can be discovered from the information supplied by an inverse transform. If, now, $|H|$ is known the method of the last paragraph provides g.

9.14. Zeros of entire functions

In this section the relation between f and the zeros of its transform will be discussed. It will be assumed, mainly on physical grounds, that f is non-zero only on the interval $(-a,a)$; a simple shift places this interval anywhere on the real axis. Then

$$F(\alpha) = \int_{-a}^{a} f(x)\exp(-i\alpha x)dx \ . \tag{9.79}$$

We shall have in mind that f is continuous though this is by no means essential. The integral in (9.79) converges uniformly for every bounded (complex) α and so do its derivatives with respect to α. The integrand is a continuous function of α and x for real x and complex α. Furthermore, the integrand is a regular function of α for the real x involved. Hence $F(\alpha)$ is a regular function of α for all finite complex α and is, in fact, an entire (or integral) function. Moreover

$$|F(\alpha)| < C \exp(a|\mathscr{I}\alpha|) \tag{9.80}$$

for some finite constant C as $|\alpha| \to \infty$. Thus $F(\alpha)$ is actually *an entire function of exponential type*. As a matter of fact, a theorem of Paley and Wiener (1934) demonstrates the converse, namely that, if F is an entire function satisfying (9.80), f is zero outside $(-a,a)$.

It is transparent from (9.80) that F is of less than exponential growth on the real α axis. An application of Carleman's theorem (Titchmarsh 1939) reveals that the zeros of F must cluster about the real α axis. More specific information is available from consideration of the behaviour of f near its end-points. Suppose that

$$f(x) \approx (x+a)^{e_1} B_1 \qquad (x \approx -a)$$

$$\approx (a-x)^{e_2} B_2 \qquad (x \approx a)$$

with $e_1, e_2 > -1$. Then, as $|\alpha| \to \infty$ with $\mathcal{R}(\alpha) > 0$,

$$F(\alpha) \sim \frac{e_1! B_1}{\alpha^{e_1+1}} \exp\{i a\alpha - \tfrac{1}{2}\pi i(e_1+1)\} + \\ + \frac{e_2! B_2}{\alpha^{e_2+1}} \exp\{-i a\alpha + \tfrac{1}{2}\pi i(e_2+1)\} . \tag{9.81}$$

Actually, (9.81) holds if both e_1 and e_2 are non-positive. If either is positive there may be more important contributions from interior points where the derivation of f is discontinuous (Jones 1966*a*, 1969). If the derivative of f is continuous (9.81) is valid if neither e_1 nor e_2 exceeds unity. We shall assume that the conditions are such that it is legitimate to employ (9.81).

The case of most interest is when $e_1 = e_2$. Then (9.81) shows that if F vanishes for some particular α it does so when α is increased by an integer multiple of π/a. So there is an infinite string of zeros running roughly parallel to the real axis. They cannot be far from the real axis otherwise one of the exponentials would dominate and F could not be zero; this is in accordance with Carleman's theorem.

In the particular case $e_1 = e_2$ and $B_1 = B_2$ the zeros of (9.81) occur at $\alpha = \pi(\frac{1}{2}e_1 + n)/a$, n being a large positive integer. If, instead, $B_1 = -B_2$ the large zeros of F are at $\alpha = \pi(\frac{1}{2}e_1 + \frac{1}{2} + n)/a$. A similar investigation on the negative real axis places zeros at $\alpha = -\pi(\frac{1}{2}e_1 + n)/a$ and $\alpha = -\pi(\frac{1}{2}e_1 + \frac{1}{2} + n)/a$ respectively in the two cases.

Of course, these locations of the zeros are only a first approximation because lower-order terms have been neglected in (9.81). More refined calculations may push the zeros somewhat off the real axis.

The device for fabricating F from its zeros consists in taking care of the large zeros analytically whereas those at a finite distance from the origin are handled by inspection. However, some additional information about f is essential because the large zeros when $e_1 = \frac{1}{2}$ and $B_1 = B_2$ occupy the same positions as those when $e_1 = -\frac{1}{2}$ and $B_1 = -B_2$. Suppose therefore that $B_1 = B_2$. Now

$$\int_{-a}^{a}\left(1-\frac{x^2}{a^2}\right)^{\nu-\frac{1}{2}}\exp(-i\alpha x)\frac{dx}{a} = \left(\nu-\frac{1}{2}\right)!\,\pi^{\frac{1}{2}}\,\frac{J_\nu(\alpha a)}{(\frac{1}{2}\alpha a)^\nu}. \qquad (9.82)$$

Since the integrand of (9.82) has the same end-point behaviour as f when $\nu = e_1+\frac{1}{2}$ the large zeros of $J_\nu(\alpha a)/\alpha^\nu$ will agree with those of F. Let α_1, α_2, ... be the zeros of F arranged so that $|\alpha_1| \leqslant |\alpha_2| \leqslant \ldots$ and suppose that those for $n > 2N$ coincide with those of the Bessel function (note that an entire function can have only a finite number of zeros within a finite distance of the origin). Consider

$$\tilde{F}(\alpha) = \frac{J_\nu(\alpha a)}{\alpha^\nu}\,\frac{\Pi_{n=1}^{2N}(\alpha_n-\alpha)}{\Pi_{n=1}^{N}(j_{\nu n}^2-\alpha^2a^2)} \qquad (9.83)$$

where $\nu = e_1+\frac{1}{2}$ and $j_{\nu n}$ is the nth *positive* zero of J_ν so that $J_\nu(j_{\nu n}) = 0$ and $j_{\nu n} > j_{\nu,n-1}$. The zeros of the denominator remove those of the Bessel function which may not be in agreement with ones in F. Thus $\tilde{F}$ has the same zeros as F. Also $\tilde{F}$ is entire because $J_\nu(\alpha a)/\alpha^\nu$ is entire and the zeros of the denominator are cancelled, as just explained, so that there is no trouble from that quarter. Hence an entire function has been manufactured with the same zeros as F.

Now $F/\tilde{F}$ is an entire function without zeros on account of the coincidence of the zeros of F and $\tilde{F}$. Hence $F = A\exp(b\alpha)\tilde{F}$ for some complex constants A and b. By allowing $|\alpha| \to \infty$ we see from (9.83) that (9.80) will be violated unless $b = 0$. Thus $\tilde{F}$ is, in fact, a constant multiple of F. Therefore, by inversion of $\tilde{F}$, f is found to within a constant multiple. This constant multiple cannot be resolved unless a further piece of information, such as the specification of a non-zero value of f or F at some particular point, is available.

If $B_1 = -B_2$, the contender to replace J_ν is the Struve function $\mathbf{H}_\nu$. If $|B_1| \neq B_2$ a linear combination of J_ν and $\mathbf{H}_\nu$ would be appropriate.

The preceding approach is based on the assumption of data uncontaminated by noise. Procedures when noise is a vital feature of the scene have been proposed (Bates and Napier 1972, Bates 1969) but the cardinal principle that only a few zeros not far from the origin necessitate special treatment remains unaltered.

SYNTHESIS OF RADIATION PATTERNS

9.15. General considerations

The problems that have been considered hitherto in this chapter are, in essence, all examples of solving an equation of the type

$$Tf = g \tag{9.84}$$

where T is an operator. Measurements of g have been assumed to be available and we have to determine f, the originator of these measurements. Often, it is handy to be able to design antennas which radiate a prescribed pattern. Then we have a recipe for g and require the antenna or f that is responsible for it. We are still faced with the resolution of (9.84) but there are important differences of emphasis.

If g is chosen arbitrarily, it may not be a possible radiation pattern and exact synthesis is then out of the question. None the less, an approximate synthesis may be accessible and be accurate enough for design purposes in the sense that Tf is sufficiently close to g. If g is a realizable pattern then (9.84) has a solution; whether it is unique is of less significance than in inverse probing because it is enough to find a solution acceptable to the designer. The designer will, of course, wish to know if large changes in the radiation will arise if he does not construct f precisely; this will depend upon the properties of T. Conversely, he will also wish to be aware if small changes in the prescription of g will enforce large alterations to f; a dominant role will be played by T^{-1} in this context.

Some idea of what happens can be gained from a discussion of a one-dimensional array which occupies the interval $(-a,a)$ of the x axis. The excitation $f(x)$ is taken to be continuous on this interval and zero outside. In other words, the excitation is *aperture limited*. If the angle θ is measured from the broadside direction the radiation pattern is given by

$$g(\alpha) = e(\alpha)\int_{-a}^{a}\exp(\mathrm{i}\alpha x)f(x)\mathrm{d}x \tag{9.85}$$

where $\alpha = k\sin\theta$ and $e(\alpha)$ is the pattern due to an element of the aperture.

Evidently, g is defined only for $-k \leqslant \alpha \leqslant k$ but the discussion of (9.79) indicates that the values in this interval must be those attained by a product of $e(\alpha)$ and an entire function of exponential type which satisfies (9.80). Thus g cannot be chosen arbitrarily for (9.85) to be soluble. On the other hand, if g is such that (9.85) can be solved the solution is unique, for the difference between any two solutions for the same g would have zero Fourier transform in $-k \leqslant \alpha \leqslant k$ and the only entire function with this property is itself zero.

If g is continuous without being a member of the class for which (9.85) possesses a solution it has been proved (Bouwkamp and De Bruijn 1946) that there are functions f_n with corresponding pattern functions g_n such that g_n can be made as close to g as desired by increasing n. However, as n grows, f_n becomes highly oscillatory and it becomes steadily more difficult to create a design incorporating it. The problem is related to *supergain* and *superdirectivity*; it may be regarded as an unwelcome growth in the norm of the aperture function. To put it another way superdirectivity demands aperture currents which are prohibitively large for a given radiated power and which are highly susceptible to errors of manufacture (Woodward and Lawson 1948). Consequently, discussion of the solution of

(9.85) is incomplete if an eye is not kept on the norm of f. Succeeding sections will examine how this might be done. Despite the concentration on (9.85) the fundamental ideas carry over to general two-dimensional finite arrays.

9.16. Synthesis by series expansion

One method of attacking (9.85) is to expand f via a series of known functions through

$$f(x) = \sum_{n=0}^{\infty} f_n h_n(x) \tag{9.86}$$

and determine the coefficients f_n by sampling g at a number of points $\alpha = \lambda_m$. By this means we arrive at the equations

$$h(\lambda_m) = \sum_{n=0}^{\infty} f_n \int_{-a}^{a} h_n(x)\exp(\mathrm{i}\lambda_m x)\,\mathrm{d}x \tag{9.87}$$

if $h(\alpha)$ is written for $g(\alpha)/e(\alpha)$.

The equations (9.87) simplify considerably if it is possible to choose the functions h_n so that

$$\begin{aligned}\int_{-a}^{a} h_n(x)\exp(\mathrm{i}\lambda_m x)\,\mathrm{d}x &= 0 \qquad (m \neq n)\\ &= 1 \qquad (m = n)\end{aligned} \tag{9.88}$$

i.e. the sets $\{h_n\}$ and $\{\exp(-\mathrm{i}\lambda_m x)\}$ are biorthogonal (§5.2), for then (9.87) reduces to $f_n = h(\lambda_n)$ and the coefficients of (9.86) are determined immediately from pattern values at the sampling points.

To find the correspondence between h_n and λ_m let

$$K(\alpha) = \prod_{m=0}^{\infty}\left(1 - \frac{\alpha}{\lambda_m}\right)$$

on the understanding that $1 - \alpha/\lambda_m$ is replaced by $-\alpha$ if $\lambda_m = 0$. Under proper conditions on the incrementation of λ_m it can be demonstrated (Levinson 1940) that K is effectively bounded by $\exp(a|\mathscr{I}\alpha|)$ as $|\alpha| \to \infty$. Therefore, if we define

$$h_n(x) = \frac{1}{2\pi}\int_{-\infty}^{\infty} \frac{K(\alpha)\exp(-i\alpha x)}{(\alpha-\lambda_n)K'(\lambda_n)}\,d\alpha \,, \tag{9.89}$$

h_n is zero for $x < -a$ and $x > a$ as can be seen by deforming the contour to infinity. Furthermore, by Fourier inversion

$$\frac{K(\alpha)}{(\alpha-\lambda_n)K'(\lambda_n)} = \int_{-a}^{a} h_n(x)\exp(i\alpha x)\,dx \,.$$

The substitution $\alpha = \lambda_m$ leads instantly to (9.88).

Consequently, (9.89) determines h_n once the sampling points have been chosen. The advantage of this is that the sampling points can be placed conveniently for measurement and can be arranged so that they fit in with any specified end-point behaviour consistent with §9.14. The theory holds strictly if there is one and precisely one sampling point within a distance of $\pi/4a$ of $n\pi/a$ for every integer n and no other sampling points. However, this is not likely to be of any significance in practical synthesis when the expansion in (9.86) will probably be limited to a finite number of terms and g itemized at a modest number of points.

Indeed, values of g are at hand only for $-k \leqslant \alpha \leqslant k$. Therefore, if f were expanded in a Fourier series ($h_n(x) = \exp\{-in\pi x/a\}$, $\lambda_n = n\pi/a$) the number of sampling points available would be $2[ka/\pi] + 1$ where $[x]$ is the largest integer which does not exceed x. The number of determinable coefficients in f would be the same and could be increased (if this were felt to improve accuracy) only at the expense of making ka larger, i.e. putting more wavelengths into the aperture. More sampling points with fixed ka can be achieved by defining λ_n suitably and finding the associated h_n. There is always the danger, however, that if the number of coefficients is much larger than twice the number of wavelengths in the aperture f will be oscillatory and of high norm, as described in the previous section.

More freedom of action can be secured by dropping the condition (9.88) on the basis functions h_n. Also it is well to recognize that, in practice, (9.86) will be constrained to a finite number of terms. Therefore, put

$$f(x) = \sum_{n=1}^{N} f_n h_n(x)$$

whence

$$g(\alpha) = \sum_{n=1}^{N} f_n e(\alpha) \int_{-a}^{a} h_n(x) \exp(\mathrm{i}\alpha x)\,\mathrm{d}x \ . \tag{9.90}$$

Obviously, (9.90) is exactly soluble only when g is in the space spanned by the elements of the right-hand side. Whether this is a severe limitation will depend upon the pattern to be synthesized and the basis functions employed.

If g is laid down at $\alpha = \lambda_m$ then (9.90) implies

$$g(\lambda_m) = \sum_{n=1}^{N} f_n e(\lambda_m) \int_{-a}^{a} h_n(x) \exp(\mathrm{i}\lambda_m x)\,\mathrm{d}x \tag{9.91}$$

for $m = 1, \ldots, M$. Define the column vectors $\mathbf{f}$, $\mathbf{g}$ to have the components $f_1, f_2, \ldots$ and $g(\lambda_1), g(\lambda_2), \ldots$ respectively. Then, if B is the $M \times N$ matrix with elements

$$B_{mn} = e(\lambda_m) \int_{-a}^{a} h_n(x) \exp(\mathrm{i}\lambda_m x)\,\mathrm{d}x \ , \tag{9.92}$$

(9.91) can be expressed in matrix form as

$$\mathbf{g} = B\mathbf{f} \ . \tag{9.93}$$

The problem has therefore been converted to the solution of the algebraic system (9.93) subject to whatever constraints are imposed on the norms. The norms that we shall adopt are

$$\|\mathbf{f}\|^2 = \mathbf{f}^{\mathrm{H}}C\mathbf{f} = \sum_{i,j}^{N} f_i^* C_{ij} f_j , \tag{9.94}$$

$$\|\mathbf{g}\|^2 = \mathbf{g}^{\mathrm{H}}D\mathbf{g} = \sum_{i,j}^{M} g_i^* D_{ij} g_j$$

where $\mathbf{f}^{\mathrm{H}}$ is the conjugate transpose of $\mathbf{f}$ (§1.10), and C and D are positive definite $N \times N$ and $M \times M$ matrices respectively. Frequently, it will be opportune to make C and D diagonal or even unit matrices. We shall assume that C is such that $\Sigma_i^N f_i^* C_{ii} f_i \leqslant \|\mathbf{f}\|^2$.

Suppose that a solution $\tilde{\mathbf{f}}$ of (9.93) has been obtained by some method. Then often $B\tilde{\mathbf{f}}$ will not reproduce $\mathbf{g}$ exactly but generate instead $\tilde{\mathbf{g}}$. The *synthesis error* may then be defined by

$$E = \frac{\|\mathbf{g}-\tilde{\mathbf{g}}\|^2}{\|\mathbf{g}\|^2} . \tag{9.95}$$

Another measure of the merit of $\tilde{\mathbf{f}}$ is the *quality factor*

$$Q = \frac{M\|\tilde{\mathbf{f}}\|^2}{\|\tilde{\mathbf{g}}\|^2} . \tag{9.96}$$

9.17. Construction errors

It has to be realized that aberrations in fabrication will involve departures from $\tilde{\mathbf{f}}$ in any actual construction. So, before investigating methods of finding $\tilde{\mathbf{f}}$, we try to estimate the effect of small errors.

Assume that $\tilde{\mathbf{f}}$ becomes $\tilde{\tilde{\mathbf{f}}}$ owing to the introduction of errors where

$$\tilde{\tilde{f}}_n = \tilde{f}_n(1+r_n) \ .$$

Here r_n is a random variable with zero mean and variance v^2. It will also be supposed that the data points are sufficiently separated for the r_n to be uncorrelated with one another. In terms of expectations

$$E(r_n) = 0, \quad E(r_n^* r_m) = v^2 \delta_{nm} \ .$$

The restriction $v^2 \ll 1$ will also be imposed since it is anticipated that the errors are small.

From (9.94),

$$E(\|\tilde{\tilde{\mathbf{f}}}\|^2) = \|\tilde{\mathbf{f}}\|^2 + v^2 \sum_{n=1}^{N} c_{nn} |\tilde{f}_n|^2$$

implying $\|\tilde{\mathbf{f}}\|^2 \leqslant E(\|\tilde{\tilde{\mathbf{f}}}\|^2) \leqslant (1+v^2)\|\tilde{\mathbf{f}}\|^2$ by our assumption on C. As would be expected, the norm of $\tilde{\tilde{\mathbf{f}}}$ does not differ appreciably on average from that of $\tilde{\mathbf{f}}$.

Analogous to (9.95) the average error in the radiated pattern is taken as

$$\frac{E(\|\tilde{\tilde{\mathbf{g}}}-\tilde{\mathbf{g}}\|^2)}{\|\tilde{\mathbf{g}}\|^2} = \frac{v^2 \sum_{n=1}^{N} P_{nn} |\tilde{f}_n|^2}{\|\tilde{\mathbf{g}}\|^2} \tag{9.97}$$

where P_{ij} is a typical element in the matrix $P = B^{\mathrm{H}}DB$, and $\tilde{\tilde{\mathbf{g}}} = B\tilde{\tilde{\mathbf{f}}}$. We cannot assert that the right-hand side of (9.97) is small even though v^2 is much less than unity because we cannot guarantee that the sum in the numerator is not much greater than the denominator.

To estimate the quality put

$$\tilde{Q} = \frac{ME(\|\tilde{\tilde{\mathbf{f}}}\|^2)}{E(\|\tilde{\tilde{\mathbf{g}}}\|^2)}$$

by analogy with (9.96). Then

$$\frac{1}{\tilde{Q}} = \frac{1}{Q} + \frac{v^2}{\|\tilde{\mathbf{f}}\|^2} \sum_{n=1}^{N} \left(\frac{P_{nn}}{M} - \frac{C_{nn}}{Q}\right)|\tilde{f}_n|^2 \tag{9.98}$$

when terms of the order of v^4 are neglected.

Another figure of merit which can be relevant arises from $\tilde{\tilde{\mathbf{G}}} = b\tilde{\tilde{\mathbf{g}}}$ where $b^2 = \|\tilde{\mathbf{g}}\|^2/E(\|\tilde{\tilde{\mathbf{g}}}\|^2)$. Then $E(\|\tilde{\tilde{\mathbf{G}}}\|^2) = \|\tilde{\mathbf{g}}\|^2$ which may be interpreted as saying that $\tilde{\tilde{\mathbf{G}}}$ is a normalized pattern whose average 'power' is the same as that of $\tilde{\mathbf{g}}$. Hence

$$E_n = \frac{E(\|\tilde{\tilde{\mathbf{G}}}-\tilde{\mathbf{g}}\|^2)}{\|\tilde{\mathbf{g}}\|^2} = 2(1-b) \;. \tag{9.99}$$

It can be forecast that b will be near unity if there is little pattern deterioration so that the smaller E_n is the more likely it is that the average design pattern does not deviate far from that required. So, to a certain extent, a small E_n will warrant a design being robust against minor constructional errors.

If $\tilde{\tilde{\mathbf{f}}},\tilde{\mathbf{f}}$ are exchanged for $\tilde{\mathbf{f}},\mathbf{f}$ respectively the above analysis can be used to provide estimates of the errors that will result from imperfections in the solution of (9.93) by numerical methods.

9.18. Constrained aperture norm

A solution of (9.93) can be obtained by means of the generalized inverse (§1.15) of a matrix. According to Exercise 80 of Chapter 1 the solution of (9.93) which minimizes $\|\mathbf{g}-B\mathbf{f}\|^2$ when $B^{\mathrm{H}}B$ is positive definite is

$$\mathbf{f} = (B^{\mathrm{H}}DB)^{-1}B^{\mathrm{H}}D\mathbf{g} \;. \tag{9.100}$$

As already explained this is sensitive to small errors and leads to unduly high values of the norm of $\mathbf{f}$. To prevent this the concept of *regularization* (Tihonov 1944, 1964, Deschamps and Cabayan 1972) is introduced; however, if it is desired to use (9.100) a possible numerical method is discussed in §9.21.

We seek to minimize

$$J = \|\mathbf{g}-B\mathbf{f}\|^2 + \mu\|\mathbf{f}\|^2 \qquad (9.101)$$

where $\mu \geqslant 0$. The extra term $\mu\|\mathbf{f}\|^2$ is a *penalty function* whose aim is to impose an increasing penalty on J as the norm of f gets too big. The parameter μ controls the level of the penalty. If $\mu = 0$ we recover (9.100) so we shall have in mind that $\mu > 0$. However, μ must not be too great otherwise there will be too much of a discrepancy between the **f** found from (9.101) and that sought.

Keep μ fixed. The minimization will depend upon μ, which will be denoted by writing $\mathbf{f}_\mu$. Variations of $\mathbf{f}^H$ alone in J indicate that the minimizing solution must satisfy

$$(B^H DB+\mu C)\mathbf{f}_\mu = B^H D\mathbf{g} \; . \qquad (9.102)$$

Since the matrix operating on $\mathbf{f}_\mu$ is positive definite it possesses an inverse and

$$\mathbf{f}_\mu = (B^H DB+\mu C)^{-1} B^H D\mathbf{g} \; . \qquad (9.103)$$

Useful theoretical results can be adduced by importing the eigenfunctions $\boldsymbol{\phi}_i$ and eigenvalues μ_i satisfying

$$B^H DB\boldsymbol{\phi}_i = \mu_i C\boldsymbol{\phi}_i \; . \qquad (9.104)$$

Since C is positive definite and $B^H DB$ positive semi-definite $\mu_i \geqslant 0$ and the $\boldsymbol{\phi}_i$ can be normalized so that (§1.10)

$$\boldsymbol{\phi}_j^H C\boldsymbol{\phi}_i = \delta_{ij}, \qquad \boldsymbol{\phi}_j^H B^H DB\boldsymbol{\phi}_i = \mu_i\delta_{ij} \; . \qquad (9.105)$$

Also any vector can be expressed as a linear combination of the $\boldsymbol{\phi}_i$ so that

$$\mathbf{f}_\mu = \sum_i a_{i\mu}\boldsymbol{\phi}_i \; .$$

Since **f** itself can be expanded similarly it follows that **g** is representable by

$$\mathbf{g} = \sum_i b_i B\boldsymbol{\phi}_i .$$

Substitution in (9.102) and application of (9.105) leads immediately to $a_{i\mu} = \mu_i b_i(\mu+\mu_i)^{-1}$ so that

$$\mathbf{f}_\mu = \sum_i \mu_i b_i(\mu+\mu_i)^{-1}\boldsymbol{\phi}_i . \tag{9.106}$$

From (9.106) we deduce that

$$\|\mathbf{f}_\mu\|^2 = \sum_i \mu_i^2|b_i|^2(\mu+\mu_i)^{-2}$$

and that the synthesis error is

$$E = \sum_i \frac{\mu^2|b_i|^2(\mu+\mu_i)^{-2}}{\|\mathbf{g}\|^2} .$$

Consequently, as μ increases the norm of $\|\mathbf{f}_\mu\|$ decreases monotonically whereas E increases steadily. The price to be paid for holding down the norm of $\mathbf{f}$ is larger error in the radiated pattern.

If $\mathbf{g}$ is corrupted by noise, the change to $\mathbf{f}_\mu$ is that b_i becomes $b_i(1+r_i)$. Then, if $\mathbf{f}$ is the solution of (9.93) when there is no contamination by noise, a measure of the effect of μ is

$$\frac{E(\|\mathbf{f}_\mu-\mathbf{f}\|^2}{\|\mathbf{f}\|^2} = \sum_i \frac{(v^2+\mu^2)|b_i|^2}{(\mu+\mu_i)^2\|\mathbf{f}\|^2} .$$

As μ increases this quantity first decreases steadily and then grows monotonically, the minimum occurring at the value of μ which satisfies

$$\sum_i \frac{(\mu\mu_i - v^2)|b_i|^2}{(\mu+\mu_i)^2} = 0 \quad . \tag{9.107}$$

Thus, by choosing μ from (9.107), we can ensure that $\mathbf{f}_\mu$ is in its most stable regime against noise disturbances. The smallness of v suggests that the best choice of μ will be round about v^2. Hence, when information is available about the noise characteristics, an optimal design can be attempted, though it involves the solution of (9.104). The corresponding values of $\|\mathbf{f}_\mu\|$ and the synthesis error can be derived from formulae already given.

An alternative approach (Mautz and Harrington 1973) is to require that $\|\mathbf{f}_\mu\|^2 \leqslant K$ where K is some constant prescribed in advance. In view of the monotonic decrease of $\|\mathbf{f}_\mu\|$ with increasing μ, the least value of μ which will be successful satisfies

$$\sum_i \mu_i^2 |b_i|^2 (\mu+\mu_i)^{-2} = K \quad . \tag{9.108}$$

Observe that there is no solution to (9.108) if K exceeds $\Sigma_i |b_i|^2$. When there is a solution it may deviate widely from that of (9.107) so that substantial noise corruption may occur.

So far it is implicit that g is known in both magnitude and phase. Should it transpire that the magnitude alone of the pattern has to be matched (as is frequently the eventuality in practice) the procedure needs modification. Let $\mathbf{h}$ be the vector whose mth component is $|g(\lambda_m)|$. Then a minimum of

$$J_1 = \sum_{m,n}^{M} D_{mn}\left(\left|\sum_p B_{mp} f_p\right| - h_m\right)\left(\left|\sum_q B_{nq} f_q\right| - h_n\right) + \mu\|\mathbf{f}\|^2 \tag{9.109}$$

is pursued.

When D is diagonal, as will now be assumed, the awkward modulus in (9.109) can be avoided. Consider

$$J_2 = \sum_m^M D_{mm}\left\{\sum_p B^*_{mp} f^*_p - \exp(i\beta_m)h_m\right\}\left\{\sum_q B_{mq} f_q - \exp(-i\beta_m)h_m\right\} + \mu\|\mathbf{f}\|^2$$

which we aim to minimize with respect to f and the parameters β_m simultaneously. The variation of J_2 with respect to β_m shows that

$$\exp(-i\beta_m) = \frac{\sum_q B_{mq} f_q}{|\sum_q B_{mq} f_q|} \quad . \tag{9.110}$$

On substitution in J_2, J_2 resumes the form of J_1. Thus the minimization of J_2 will lead to the same results as working with J_1.

The variation of J_2 with respect to $\mathbf{f}^H$ supplies (9.102) again with $\mathbf{g}$ replaced by a vector with components $\exp(-i\beta_m)h_m$. The function $\mathbf{f}_\mu$ may still be expanded in terms of the $\boldsymbol{\phi}_i$ but, on account of (9.110), the coefficients now satisfy non-linear equations. Theoretical treatment can therefore be involved but numerical work will proceed with the minimization of J_2 directly.

Exercises

23. If $e(\alpha) = (1-\alpha^2/k^2)^{\frac{1}{2}}$ and $h_n(x) = \delta(x-x_n)$ synthesize the radiation pattern $g(\alpha) = 1$ when $ka = 2\pi$. Space the x_n equally throughout the interval with one at the origin and take N to be (i) 3, (ii) 5, (iii) 7. Consider various values of M and μ (including zero), computing the norm of $\mathbf{f}$ and the synthesis error. Take C and D to be unit matrices.

Add a small amount of random noise to g and repeat your calculations, examining the robustness of the synthesis. Is your best value of μ comparable with v^2?

24. Repeat Exercise 23 with $g(\alpha)$ equal to (*a*) $(1-\alpha^2/k^2)^2$, (*b*) $(1-\alpha^2/k^2)^2 \exp(i \sin \alpha)$, (*c*) $\frac{1}{2} - \alpha^2/k^2$.

25. Try the synthesis of Exercise 24 when only $|g(\alpha)|$ is given.

26. Consider some of the previous questions for wider apertures.

27. Examine the effect of changing C and D from unit matrices to diagonal matrices with unequal diagonal elements.

28. Study the problem of synthesizing $g(\alpha) = (1-\alpha^2/k^2)^{-\frac{1}{2}}$ and decide

whether weighting D in favour of observations away from $\alpha = \pm k$ is helpful.

29. By minimizing

$$\|B\mathbf{f}-\mathbf{g}\|^2 + \frac{\mu M\|\mathbf{f}\|^2}{\|B\mathbf{f}\|^2}$$

carry out a synthesis in which the quality factor is controlled. Show that

$$\left\{\left(1 - \frac{\mu M\|\mathbf{f}\|^2}{\|B\mathbf{f}\|^4}\right)B^{\mathrm{H}}DB + \frac{\mu MC}{\|B\mathbf{f}\|^2}\right\}\mathbf{f} = B^{\mathrm{H}}D\mathbf{g} \quad .$$

See how your answers with this synthesis compare with those in Exercise 23.

9.19. Directivity

Hitherto, consideration has centred on trying to make the synthesized pattern close to specification on average, but when an antenna has to discriminate against signals arriving from sources other than in some specified direction it may be desirable to ask that it has a main beam which falls to zero within a certain angle and insignificant sidelobes. Remark that it is pointless to attempt a design in which there are no sidelobes because an entire function which vanishes on a stretch of the interval $(-k,k)$ is zero everywhere.

We are concerned only with patterns which are symmetrical about $\alpha = 0$ which is assumed to coincide with the maximum of the main beam. However, it is convenient to work first with a variable y which takes values from -1 to 1. Consider the Chebyshev polynomial $T_{2m}(by)$ where $b > 1$ (§1.5). As by increases from -1, $T_{2m}(by)$ passes through maxima of 1 and minima of -1 until by reaches 1 (there being $2m + 1$ of these stationary points); thereafter it steadily increases. Hence its maximum for $-1/b \leqslant y \leqslant 1$ is $T_{2m}(b)$. The null nearest to $y = 1$ is at $y_1 = (1/b)\cos(\pi/4m)$. Suppose now that $p(y)$ is a polynomial of degree m in y^2 such that $p(1) = T_{2m}(b)$ and that $|p| \leqslant 1$ at any stationary points (excluding $y = 1$). Suppose further that p vanishes at $y_2 > y_1$. Then, for $y < y_1$, $T_{2m}(by)-p(y)$ is non-negative at maxima of T_{2m} and non-positive at minima. The same assertion cannot be made at the maximum $y = 1$, but positivity is assured at $y = y_2$. Since $T_{2m}(by)-p(y)$ vanishes at $y = 1$ and is a function of y^2 it must have $2m + 2$ zeros. Being a polynomial of degree $2m$ it must be identically zero. Thus $T_{2m}(by)$ gives the null nearest to $y = 1$ of all polynomials with the stated properties.

Now make the change of variable $y = 1-b_1\alpha^2/b$. Then the preceding result can be interpreted as stating that $T_{2m}(b-b_1\alpha^2)$ is a radiation pattern which has a main beam to sidelobe ratio of $T_{2m}(b)$ on $\alpha^2 < (1+b)/b_1$ and, of all polynomials of degree $2m$ in α^2 with the same main beam to sidelobe ratio, it gives the narrowest main beam.

Let $b_1 = a^2/8m^2$ and $b = 1 + C^2/8m^2$. Allow $m \to \infty$. We infer that $\cos\{(a\alpha)^2-C^2\}^{\frac{1}{2}}$ is the pattern with the narrowest main beam of those with a main beam to sidelobe ratio of $\cosh C$.

The large zeros of this pattern occur at $\pm(n+\frac{1}{2})\pi/a$. These may not be consistent with the behaviour enforced by conditions at the aperture end-points as described in §9.14. However, it is a simple matter to shift the large zeros, say those for $n \geqslant N$, by replacing the factors in the infinite product for the cosine. The result is

$$\frac{\{(\frac{1}{2}e_1)!\}^2\left(\frac{a\alpha}{\pi}-\frac{1}{2}e_1\right)!}{\left(\frac{a\alpha}{\pi}+\frac{1}{2}e_1\right)!}\;\frac{\sin(a\alpha-\frac{1}{2}e_1\pi)}{a\alpha-\frac{1}{2}e_1\pi}\;\frac{\prod_{n=1}^{N-1}\left[1-\dfrac{a^2\alpha^2}{\rho^2\{C^2+(n-\frac{1}{2})^2\pi^2\}}\right]}{\prod_{n=1}^{N-1}\left[1-\dfrac{a^2\alpha^2}{(n+\frac{1}{2}e_1)^2\pi^2}\right]}$$

where

$$\rho = \frac{(N+\frac{1}{2}e_1)\pi}{\{C^2+(N-\frac{1}{2})^2\pi^2\}^{\frac{1}{2}}}\,.$$

The main beam width in this pattern is larger than in the ideal pattern, the factor (Taylor 1955) being ρ. This factor is usually only a few per cent greater than unity.

The question of directivity can also be approached by planning to maximize the radiation in a prescribed direction while holding the total radiated power fixed. If $\alpha = \lambda_0$ is the special direction this comes down to maximizing $|g(\lambda_0)|^2/\|B\mathbf{f}\|^2$ where $g(\lambda_0)$ is obtained from (9.91). Experimentally, it is not easy to produce a directivity that is much in excess of that generated by an aperture distribution which is uniform in both magnitude and phase. It is normally preferable to recognize this fact and avoid the theoretical superdirectivity feasible under ideal circumstances by constraints on the source norm or quality factor. A better quantity to maximize is therefore

$$\frac{|g(\lambda_0)|^2}{\|B\mathbf{f}\|^2} + \mu\|\mathbf{f}\|^2$$

or a similar expression based on the quality factor. Apart from matters of detail this problem is similar to those already considered for minimization and so will not be discussed further.

9.20. Penalty functions

The expressions (9.101) and (9.109) are not by any means the only ones which could be adopted in order to impose contraints. For example, if it is wished to make $\|\mathbf{f}\|^2 \leqslant K$ a possible penalty function is

$$J_3 = \|\mathbf{g}-B\mathbf{f}\|^2 + \tfrac{1}{2}\sigma(K-\|\mathbf{f}\|^2-\nu)_-^2$$

where x_- is x if $x \leqslant 0$ and 0 if $x \geqslant 0$. The constraint is taken into account only when it is violated, as it were. The extra parameter ν permits the avoidance of unpleasant behaviour of the second derivatives of J_3 at $\nu = 0$.

Indeed the method can be generalized to cover more than one constraint (see also §4.7). Suppose that the constraining relations are $c_m(\mathbf{f}) \geqslant 0$. Then consider

$$J_4 = \|\mathbf{g}-B\mathbf{f}\|^2 + \frac{1}{2}\sum_m \sigma_m\{c_m(f)-\nu_m\}_-^2 \quad .$$

A strategy for dealing with J_4 is based on iteration (Fletcher 1973). Let $\boldsymbol{\sigma}$ and $\boldsymbol{\nu}$ be vectors with components σ_m and ν_m respectively. Select some starting values $\boldsymbol{\sigma}^{(1)}$ and $\boldsymbol{\nu}^{(1)}$ (often $\boldsymbol{\nu}^{(1)} = \mathbf{0}$ will be satisfactory). Calculate the minimizer $\mathbf{f}^{(1)}$ of J_4. If

$$\|\mathbf{c}(\mathbf{f}^{(1)})\|_\infty = \max_m |\min\{c_m(\mathbf{f}^{(1)}),\nu_m\}| \leqslant \varepsilon$$

where ε is a pre-assigned tolerance, stop. Otherwise set $K^{(1)} = \|\mathbf{c}(\mathbf{f}^{(1)})\|_\infty$ and define

$$\boldsymbol{\nu}^{(2)} = \boldsymbol{\nu}^{(1)} - (N^{\mathrm{T}}G^{-1}N)^{-1}\mathbf{c}(\mathbf{f}^{(1)})$$

where G is the Hessian matrix of J_4 and N is the matrix (grad c_1, grad c_2, ...). Increase each σ_m if necessary so that

$$\sigma_m^{(2)} \geqslant 4\sigma_m^{(1)} \left| \frac{\{\nu_m^{(1)}-\nu_m^{(2)}\}}{c_m(\mathbf{f}^{(1)})} \right| \quad .$$

With these new values recalculate the minimizer $\mathbf{f}^{(2)}$ of J_4. Either $\|\mathbf{c}(\mathbf{f}^{(2)})\|_\infty \geqslant K^{(1)}$ or $\|\mathbf{c}(\mathbf{f}^{(2)})\|_\infty < K^{(1)}$. In the latter case repeat the procedure just described. In the former put $\sigma_m^{(3)} = 10\sigma_m^{(2)}$ for all those m for which $|c_m(\mathbf{f}^{(2)})| \geqslant K^{(1)}$ and return to finding the minimizer of J_4.

The evidence of numerical experiments is that this strategy is attractive in practice.

ARRAY SIGNAL PROCESSING

9.21. Adaptive beam forming

Radar surveillance in which existing targets are tracked and new targets observed places heavy demands on antenna facilities. Because of its importance in connection with satellites, military applications, and air traffic control (to mention a few examples) efforts have been devoted to making antenna structures which are economical by being multi-purpose. These purposes are sometimes in conflict. For instance, transmission may call for maximum power in given directions, whereas reception will require information to be absorbed from directions (perhaps unknown in advance) with maximum efficiency. It will therefore frequently be necessary to scan a portion of space either mechanically or electronically. Mechanical scanning will usually be sufficiently slow that the pattern can be estimated instantaneously from the position of the array as if it were not moving, but electronic sweeping may be rapid enough for time to enter as a new factor in the calculations. In any case the beam of the antenna is being regularly adapted as a result of the information available.

The information itself is normally obtained by samples which need only be taken at the rate of $1/2B$ where B is the communication bandwidth. While there is no lower limit to the duration of the sample, the information may be corrupted in the absence of suitable precautions. *Aliasing* arises from the introduction of additional unwanted information if the bandwidth being sampled is greater than that required for signal communication.

To obtain some idea of the considerations involved in beam steering, let a plane wave irradiate a linear array of equispaced receptors (Fig. 9.11). The difference in phase between two adjacent receptors is $kd \sin \theta$ so that if all the receptors are alike the signal in the nth is

$$s_n = \exp\{\mathrm{i}(n-1)kd \sin \theta\}$$

taking the phase of the first as zero. If these signals are given a phase delay δ_n and added, the net signal is

$$S(\theta) = \sum_{n=1}^{N} \exp\{\mathrm{i}(n-1)kd \sin \theta + \mathrm{i}\delta_n\} . \qquad (9.111)$$

If $\delta_n = (n-1)\delta$ for all n, corresponding to an inserted constant

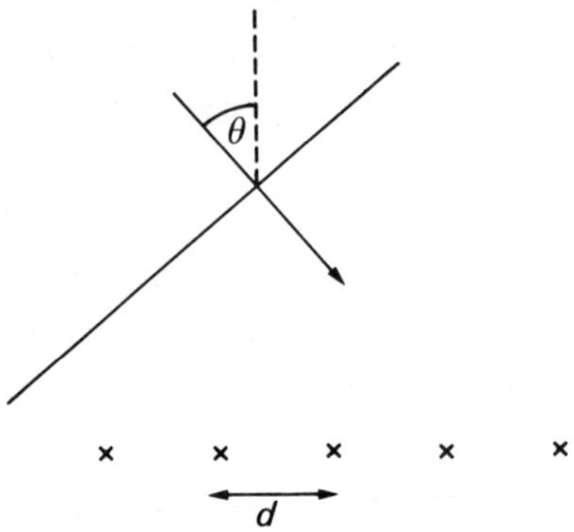

FIG. 9.11. Geometry of beam steering.

phase delay between adjacent elements,

$$S(\theta) = \frac{\exp\{\frac{1}{2}i(N-1)u\}\sin\frac{1}{2}Nu}{\sin\frac{1}{2}u}$$

where $u = kd \sin\theta + \delta$. The normalized power pattern

$$\frac{|S(\theta)|^2}{|S_{u=0}|^2} = \left(\frac{\sin\frac{1}{2}Nu}{N\sin\frac{1}{2}u}\right)^2 . \qquad (9.112)$$

From (9.112) it can be seen that, when $\delta = 0$, the maximum occurs at $\theta = 0$, i.e. the main beam is the broadside direction. If $\delta \neq 0$, the maximum is at $\theta = -\sin^{-1}(\delta/kd)$ and the main beam has been shifted. Thus, weighting the signals by phase permits beam shifting and if the magnitudes are also weighted it may be possible to introduce some of the effects discussed under synthesis.

Therefore we consider beam forming in which

$$S = \sum_{n=1}^{N} w_n^* s_n = \mathbf{w}^{\mathrm{H}}\mathbf{s}$$

where **w** and **s** are the weight and signal column vectors with components w_m and s_m respectively. Now suppose that a plane wave of unit amplitude from a certain direction, called the *look direction*, produces a signal $\mathbf{1}$. The magnitude is correctly reproduced if

$$\mathbf{w}^{\mathrm{H}}\mathbf{1} = 1 . \tag{9.113}$$

To ensure that the look direction is always preferentially treated, (9.113) is applied as a constraint. Then the weights are adjusted so that the power output $|S|^2$ is a minimum.

If λ is a Lagrange multiplier the objective function to be minimized is

$$\mathbf{w}^{\mathrm{H}}\mathbf{S}\mathbf{w} + \lambda(\mathbf{w}^{\mathrm{H}}\mathbf{1}-1) + \lambda^*(\mathbf{1}^{\mathrm{H}}\mathbf{w}-1)$$

where $\mathbf{S} = \mathbf{s}\mathbf{s}^{\mathrm{H}}$. Variations in $\mathbf{w}^{\mathrm{H}}$ alone lead immediately to $\mathbf{S}\mathbf{w} + \lambda\mathbf{1} = \mathbf{0}$ and then (9.113) implies that $\lambda^* = -1/\mathbf{1}^{\mathrm{H}}\mathbf{S}^{-1}\mathbf{1}$. Hence the optimal choice of $\mathbf{w}$ is given by

$$\mathbf{w}_{\mathrm{opt}} = \frac{\mathbf{S}^{-1}\mathbf{1}}{\mathbf{1}^{\mathrm{H}}\mathbf{S}^{-1}\mathbf{1}} \tag{9.114}$$

and the best beam former is

$$S_{\mathrm{opt}} = \mathbf{w}_{\mathrm{opt}}^{\mathrm{H}}\mathbf{s} .$$

The method has affinities with those in §9.18 but is somewhat simpler in character. In both substantial computational effort may have to be expended in the evaluation of inverses of matrices unless the following iterative algorithm (Frost 1972) is adopted.

If $\mathbf{w}_j$ is the result of some iteration, define

$$\mathbf{w}_{j+1} = \left\{I - \beta\left(I - \frac{\mathbf{1}\mathbf{1}^{\mathrm{H}}}{\mathbf{1}^{\mathrm{H}}\mathbf{1}}\right)\mathbf{S}\right\}\mathbf{w}_j$$

where I is the unit matrix and β is a non-zero feedback factor which controls the rate of convergence of the iteration. Then $\mathbf{w}_{j+1}^{H}\mathbf{1} = \mathbf{w}_{j}^{H}\mathbf{1} = 1$ so long as (9.113) was satisfied at the start of the iteration. If $\mathbf{w}_j \to \mathbf{w}_0$ as $j \to \infty$, we have $\mathbf{w}_0^H\mathbf{1} = 1$ and

$$\left(I - \frac{\mathbf{1}\mathbf{1}^H}{\mathbf{1}^H\mathbf{1}}\right)\mathbf{S}\mathbf{w} = \mathbf{0}$$

whence

$$\mathbf{w}_0 = \frac{\mathbf{S}^{-1}\mathbf{1}\mathbf{1}^H\mathbf{S}\mathbf{w}_0}{\mathbf{1}^H\mathbf{1}} . \tag{9.115}$$

Substituting for $\mathbf{w}_0$ from (9.115) in $\mathbf{w}_0^H\mathbf{1} = 1$ we obtain $\mathbf{1}^H\mathbf{S}\mathbf{w}_0 = \mathbf{1}^H\mathbf{1}/\mathbf{1}^H\mathbf{S}^{-1}\mathbf{1}$ which, when inserted in the right-hand side of (9.115), shows that $\mathbf{w}_0$ coincides with $\mathbf{w}_{opt}$ as given by (9.114).

To discuss the rate of convergence let $\mathbf{v}_j = \mathbf{w}_j - \mathbf{w}_0$. Then

$$\mathbf{v}_{j+1} = \left\{I - \beta\left(I - \frac{\mathbf{1}\mathbf{1}^H}{\mathbf{1}^H\mathbf{1}}\right)\mathbf{S}\right\}\mathbf{v}_j .$$

Since $\mathbf{1}^H\mathbf{v}_j = 0$

$$\left(I - \frac{\mathbf{1}\mathbf{1}^H}{\mathbf{1}^H\mathbf{1}}\right)\mathbf{v}_j = \mathbf{v}_j$$

so that $\mathbf{v}_{j+1} = (I - \beta Q)\mathbf{v}_j$ where

$$Q = \left(I - \frac{\mathbf{1}\mathbf{1}^H}{\mathbf{1}^H\mathbf{1}}\right)\mathbf{S}\left(I - \frac{\mathbf{1}\mathbf{1}^H}{\mathbf{1}^H\mathbf{1}}\right) .$$

Because Q is Hermitian there is a matrix P such that $P^HQP = \Lambda$ and $P^{-1} = P^H$, Λ being diagonal with elements the eigenvalues of Q. Hence

$$P^{H}\mathbf{v}_{j+1} = (I-\beta\Lambda)P^{H}\mathbf{v}_{j} .$$

Thus the system is stable and the error decays rapidly to zero provided that

$$0 < \beta < 2/\lambda_{max} . \qquad (9.116)$$

The restriction (9.116) can be severe in practice and convergence mav be slow. There may also be instability due to noisy data as already explained. For that reason it is probably advantageous to employ the theory for linear constraints set out in §4.7 or to use the following *method of elimination* (Gill and Murray 1974). It is based on a Givens–Householder transformation (§1.14).

Let $\mathbf{L}$ be the vector with zero elements except for the first which is $l_1(\mathbf{1}^H\mathbf{1})^{\frac{1}{2}}/|l_1|\{(\mathbf{1}^H\mathbf{1})^{\frac{1}{2}}$ if $l_1 = 0\}$. Choose $\mathbf{u}$ so that

$$\mathbf{1} - 2\mathbf{u}\mathbf{u}^H\mathbf{1} = \mathbf{L}$$

and $\mathbf{u}^H\mathbf{u} = 1$. This is always feasible and makes the matrix unitary (§1.10). If $\mathbf{y} = \mathbf{w} - 2\mathbf{u}\mathbf{u}^H\mathbf{w}$,

$$\mathbf{y}^H\mathbf{L} = \mathbf{w}^H\mathbf{1} .$$

Thus the constraint (9.113) is met if and only if

$$y_1^* = \frac{|l_1|}{(\mathbf{1}^H\mathbf{1})^{\frac{1}{2}}l_1} . \qquad (9.117)$$

Hence an equivalent minimization problem is that of $\mathbf{y}^*(I-2\mathbf{u}\mathbf{u}^H)\mathbf{S}(I-2\mathbf{u}\mathbf{u}^H)\mathbf{y}$ subject to (9.117). Let $\mathbf{b}$ be the first column of the matrix after the first element has been removed and $\mathbf{B}$ the submatrix after the deletion of the first row and column. Then, if $\mathbf{y}_2$ is y after the extraction of y_1,

$$\mathbf{B}\mathbf{y}_2 = -y_1\mathbf{b} . \qquad (9.118)$$

The matrix $\mathbf{B}$, in view of the unitary character of $I - 2\mathbf{u}\mathbf{u}^H$, can be written in its Cholesky decomposition (§1.12) LL^{H} and so $\mathbf{y}_2$ can be found from (9.118) by solving two sets of linear equations. Since y_1 is known from (9.117), $\mathbf{w}_{\text{opt}}$ can be calculated without difficulty. This method can be efficient especially when combined with the penalty function technique described earlier.

A possible disadvantage is that the optimization is carried out afresh with each set of signals so that any earlier information about the values of the weights is wasted.

9.22. Simultaneous multiple beams

The observation that different values of δ in (9.112) cause shifts in the main beam prompts the suggestion that by attaching a number of independent feeding systems each incorporating a different phase shift, the array could be operated as if its maximum beam were in various directions simultaneously, i.e. a multiple-beam mode would be achieved by means of multiple feeders. Such a mode could be attractive so it is vital to establish the point that if the antenna and feeds are lossless and passive there are strict limitations on the patterns which are permitted (White 1962, Allen 1961, Shelton and Kelleher 1961).

It will be convenient to discuss the antenna when transmitting since reciprocity will then cope with reception. Suppose that for a unit drive in the mth feed the radiated electric intensity is $\mathbf{E}_m(\theta',\phi)$ in the direction indicated by the spherical polar angles θ',ϕ. Then if the drive is v_m the electric intensity is $v_m\mathbf{E}_m$. Consequently, if all feeds are operated simultaneously, the output is $\Sigma_m v_m \mathbf{E}_m$ and the input power is of the form $\Sigma_m R_m |v_m|^2$ since the feeds are independent. On the other hand, the output power is proportional to

$$\sum_m \sum_n v_m v_n^* \int_0^{\pi} \int_0^{2\pi} \mathbf{E}_m(\theta',\phi) \,.\, \mathbf{E}_n^*(\theta',\phi)\sin\theta' d\theta' d\phi \,.$$

In the lossless passive case the two powers must be the same for arbitrary v_m. We therefore conclude that

$$\int_0^{\pi}\int_0^{2\pi} \mathbf{E}_m(\theta',\phi) \cdot \mathbf{E}_n^*(\theta',\phi)\sin\theta' d\theta' d\phi = 0 \qquad (9.119)$$

i.e. *multiple independent beams can be formed by a lossless passive antenna only when the individual beam patterns are orthogonal over the unit sphere*. For one-dimensional arrays where the individual beams possess the same polarization the scalar product in (9.119) can be replaced by multiplication.

For a uniformly illuminated aperture as at the beginning of §9.21, the pattern is approximately

$$\frac{\sin(p-\theta_0)}{(p-\theta_0)} \qquad (p = \tfrac{1}{2}kd \sin \theta)$$

when it is sufficiently sharp. Also the integration may be extended to infinity without great loss. Then, since

$$\int_{-\infty}^{\infty} \frac{\sin x}{x} \frac{\sin(x+t)}{x + t}\, dx = \pi \frac{\sin t}{t} ,$$

the patterns $\sin(p-n_m\pi)/(p-n_m\pi)$ are orthogonal if the integers n_m and n_n are different. However, we have already seen that uniformly illuminated apertures are not optimal from the point of view of side lobes, so either resistance has to be deliberately introduced or the system made active instead of passive so as to render orthogonality unnecessary. Alternatively, the aperture may be split into independent apertures.

There are interesting facts to be gleaned from processing the data from independent apertures in an appropriate way. Consider the particular case of (9.111) with $\delta_n = (n-1)\delta$ and put $p = \frac{1}{2}(kd \sin \theta + \delta)$. If (9.111) is multiplied by $\exp\{-i(N-1)p\}$, the phase increases steadily from $\exp\{-i(N-1)p\}$ at the first element to $\exp\{i(N-1)p\}$ at the last so the point distant $\frac{1}{2}(N-1)d$ from the first may be regarded as the *phase centre* of the array. Also,

$$S_N(p) = \exp\{i(N-1)p\}\frac{\sin Np}{\sin p}$$

the subscript N indicating the number of elements under consideration. Suppose now that a further M elements are added at the right-hand end of the array. Their pattern will be $\exp(2iNp)S_M(p)$.

The *real* signal from the first aperture is $D_N(p)\cos\{\omega t+(N-1)p\}$ and from the second is $D_M(p)\cos\{\omega t+(2N+M-1)p\}$ where $D_N(p) = \sin Np/\sin p$. If these are multiplied together the result is

$$\tfrac{1}{2}D_N(p)D_M(p)\,[\cos\{2\omega t+(3N+M-2)p\}+\cos(M+N)p] \;.$$

If this is processed so that the sinusoidally varying term is eliminated the demodulated output

$$D(p) = D_N(p)D_M(p)\cos(M+N)p \tag{9.120}$$

may be generated.

The output (9.120) of a *multiplicative array* may be interpreted as the product of the directional responses D_N, D_M of the two individual parts of the aperture multiplied by the interference pattern of two elements situated at the phase centres of the two parts. If $M = N$, the first zero of $D(p)$ occurs at $p = \pi/4N$ whereas D_{2N} vanishes first at $p = \pi/2N$ so that the main beam obtained by multiplying the responses of two halves of an array has half the width of that of the whole aperture. Moreover, $D(p)$ is negative for $\pi/4N \leqslant p \leqslant 3\pi/4N$ so that, if the recorder accepts only positive D, the first side lobe can be suppressed. Thus signal processing can improve the directional qualities of an array.

Another form of processing is *rectification*. In linear rectification the output is the amplitude of the sinusoidally varying term. Therefore, if the signal from the whole $N + M$ array is rectified linearly, the output is

$$D_l(p) = |D_{N+M}(p)| \;.$$

In square-law rectification the output is the square of the amplitude so that

$$D_s(p) = \{D_{N+M}(p)\}^2 .$$

Clearly, rectification does not improve the main beam characteristics in the same way as multiplication. However, before deciding that multiplication is superior to rectification it is important to examine the performance when more than one source excites the array (Shaw and Davies 1964).

Let one wave be of unit amplitude from the direction indicated by p_1 and a second wave be of real amplitude A from the direction p_2. The two waves have the same frequency but differ in phase by ϕ. Then the real response of the N portion of the aperture can be expressed as

$$D_N(p_1)\cos \omega t + AD_N(p_2)\cos(\omega t+\phi) .$$

Consequently,

$$D_l = \{D^2_{N+M}(p_1)+A^2D^2_{N+M}(p_2)+2AD_{N+M}(p_1)D_{N+M}(p_2)\cos \phi\}^{\frac{1}{2}} ,$$

$$D_s = D_l^2 ,$$

$$D = \{D_N(p_1)D_M(p_1)+A^2D_N(p_2)D_M(p_2)\}\cos(N+M)p +$$

$$+ \quad AD_N(p_2)D_M(p_1)\cos\{(M+N)p-\phi\}+AD_N(p_1)D_M(p_2)\cos\{(M+N)p+\phi\} .$$

If there is sufficient incoherence between the waves for the average of ϕ to be zero, integration will lead to mean values

$$\overline{D}_s = D^2_{N+M}(p_1) + A^2D^2_{N+M}(p_2) ,$$

$$\overline{D} = \{D_N(p_1)D_M(p_1)+A^2D_N(p_2)D_M(p_2)\}\cos(N+M)p .$$

Thus the response for multiple targets can be evaluated by superposition (except for linear rectification) and the superiority of multiplicative processing is retained in these circumstances.

If ϕ remains relatively constant in the time interval of relevance, as is likely in radar applications, the performance depends upon the probability that ϕ assumes values which allow discrimination. It is found that multiplicative processing always gives better resolution than square-law rectification but the improvement falls off as the difference between source strengths increases unless the first sidelobe of the multiplicative system is suppressed.

Exercise

30. If

$$g(p) = \frac{\sin(p-\pi)}{p-\pi} + (2+4a)\frac{\sin p}{p} + \frac{\sin(p+\pi)}{p+\pi}$$

show that the patterns $g(p)$ and $g(p+q)$ are orthogonal for any value of a if $q = 3\pi$. Are there any other values of q for which this is true?

9.23. Time-varying arrays

In electronic beam scanning the shifting of the beam is accomplished by varying the properties of the array in time. This is one instance of altering array parameters with the objective of realizing desired radiation characteristics or improving the system's capacity to handle information. The parameters which might be varied are the weights to be attached to individual elements, the frequency of operation, and the physical dimensions of the array (including the physical location of the phase centre). The parts of an aperture can be sampled in time either by having elements placed there whose weights are zero except at certain time intervals or by having an antenna move physically over the aperture so that it is at a prescribed location at a set time. In the latter case we often speak of a *synthetic aperture* created by the movement of the antenna. Radio astronomers rely on the motion of the earth for a synthetic aperture of enormous proportions (Ryle 1952).

Consider the array of Fig. 9.11 with N odd and suppose that the phase δ_n varies with time according to the law

$$\delta_n = (n-\tfrac{1}{2}N-\tfrac{1}{2})\omega_0 t \ .$$

In effect, each element of the array is being sampled at intervals of $2\pi/\omega_0$. Then, if the phase of the incident wave is zero at the central element, the signal produced when the outputs of the elements are added is

$$S(\theta) = \cos \omega t \; D_N(\tfrac{1}{2}\omega_0 t+\tfrac{1}{2}kd \sin \theta) \ . \qquad (9.121)$$

For $|t| < \pi/\omega_0$ the maximum of D_N in (9.121) occurs at the angle θ satisfying

$$\sin \theta = -\omega_0 t/kd$$

provided, of course, that $\pi \leqslant kd$, i.e. the spacing between elements is at least half a wavelength. Thus as t increases the direction of the main beam moves from $\theta = \theta_m$ where $\theta_m = \sin^{-1}(\pi/kd)$ to $\theta = -\theta_m$. Owing to the periodic nature of D_N, a new main beam enters at $\theta = \theta_m$ as t passes through π/ω_0 while the old one leaves at $\theta = -\theta_m$. Consequently, the main beam is swept through the angle $2\theta_m$ at a repetition rate of $2\pi/\omega_0$. To extract all available information ω_0 must be equal to the signal bandwidth as pointed out at the commencement of §9.21. The system effectively provides N simultaneous information channels and can resolve signals from N different directions at a cost of a scanning bandwidth of $N\omega_0$.

Each channel has to cope with a different frequency and must be supplied with a narrow band filter to prevent the (wide-band) noise being N times worse on addition than in a single channel.

It is therefore evident that by varying the parameters of an array in time a multitude of radiation patterns can be generated (Shanks and Bickmore 1959, Shanks 1961, Milne 1964) which can be separated by proper processing. The information capacity of a solitary array is thereby improved substantially. Synthetic apertures may be subsumed under the same scheme because a moving antenna can be visualized as a series of fixed antennas which are sampled at the correct times. There is also no reason why the general principles cannot be applied to arrays which are not straight lines or planes (Johnson 1968).

REFERENCES: VOLUMES 1 AND 2

ABRAMOWITZ, M. and STEGUN, I.A. (1965). *Handbook of mathematical functions.* Dover, New York.
AHLBERG, J.H.E., NILSON, E.N., and WALSH, J.L. (1967). *The theory of splines and their applications.* Academic Press, New York.
AHLUWALIA, D.S., KELLER, J.B., and MATKOWSKY, B.J. (1974). *J. acoust. Soc. Am.* **55**, 7.
AHLUWALIA, D.S., LEWIS, R.M., and BOERSMA, J. (1968). *SIAM J. appl. Math.* **16**, 783.
ALBERTSEN, N.C., HANSEN, J.E., and EILSKOV JENSEN, N. (1972). Reps. 104, 108. Technical University of Denmark, Lyngby, Denmark.
ALLEN, J.L. (1961). *IRE Trans. Antennas Propag.* **AP-9**, 350.
ALTSHULER, E.E. (1961). *IEEE Trans. Antennas Propag.* **AP-9**, 324.
ARNBACK, J. (1969). *Electron. Lett.* **5**, 41.
ATKINSON, K.E. (1967). *SIAM J. numer. Anal.* **4**, 337.
BABICH, V.M. (1962). *Dokl. Akad. Nauk SSSR* **146**, 571.
BACH ANDERSEN, J. (1968). *Radio Sci.* **3**, 432.
BAḨAR, E. (1971). *J. math. Phys.* **12**, 186.
BAÑOS, A., Jr. and JOHNSTON, G.L. (1970). UCLA Rep. No. R-65.
BATES, R.H.T. (1968). *Proc. IEE* **115**, 1443.
— (1969). *Mon. Not. R. astr. Soc.* **142**, 413.
— (1971). *Int. J. eng. Sci.* **9**, 1107.
BATES, R.H.T., BOERNER, W.M., and DUNLOP, G.R. (1976). *Opt. Commun.* **18**, 421.
BATES, R.H.T. and GOUGH, P.T. (1975). *IEEE Trans. Comput.* **C-24**, 449.
BATES, R.H.T., GOUGH, P.T., and PETERS, T.M. (1974). *Int. Optical Computing Conf. Zurich*, p. 100. IEEE Catalog. No.74CH0862-3C.
BATES, R.H.T. and LEWITT, R.M. (1975). *Optik* **43**, 529.
BATES, R.H.T. and NAPIER, P.J. (1972). *Mon. Not. R. astr. Soc.* **158**, 405.
BATES, R.H.T., NAPIER, P.J., McKINNON, A.E., and McDONNELL, M.J. (1976). *Optik* **44**, 183, 253.
BATES, R.H.T. and WALL, D.J.N. (1976). *IEEE Trans. Antennas Propag.* **AP-24**, 251.
BATES, R.H.T. and WONG, C.T. (1974). *Appl. sci. Res.* **29**, 19.
BATORSKY, D.V. and FELSEN, L.B. (1973). *Radio Sci.* **8**, 547.
BAUM, C.E. (1971). *Interaction Notes* 63, 88.
BEAUBIEN, M.J. and WEXLER, A. (1968). *IEEE Trans.* **MTT-16**, 1007.
BECKER, K-D. and MEISTER, E. (1973). *Forschungsber. Landes Nordrhein-Westfalen* No. 1175.
BECKMANN, P. and FRANZ, W. (1957). *Z. Naturf.* **12a**, 533.
BENNETT, C.L. and WEEKS, W.L. (1968). Purdue University Tech. Rep. TR-EE68-11.
BIGGS, M.C. (1971). *J. Inst. Math. Appl.* **8**, 315.
BIRKHOFF, G. (1963). *Q. appl. Math.* **21**, 160.
— (1969). In *Approximation with special emphasis on spline functions* (ed. I.J. Schoenberg). Academic Press, New York.
BLADEL, J. van (1966). *Arch. elektr. Uebertr.* **20**, 447.
BLEISTEIN, N. (1976). *J. acoust. Soc. Am.* **59**, 1259.
— (1976*a*). *J. acoust. Soc. Am.* **60**, 1249.
BLEISTEIN, N. and BOJARSKI, N.N. (1975). University of Denver Rep. MS-R-7501.
BLEISTEIN, N. and COHEN, J.K. (1977). *J. math. Phys.* **18**, 194.
BLOOM, C.A. and MATKOWSKY, B.J. (1969). *Arch. ration. Mech. Anal.* **33**, 71.

BLOOM, C.O. and KAZARINOFF, N.D. (1976). *Short wave radiation problems in inhomogeneous media*. Springer, New York.
BOERSMA, J. (1974). *Proc. IEEE* **62**, 1475.
— (1975). *Q. J. Mech. Appl. Math.* **38**, 405.
— (1975*a*). *SIAM J. appl. Math.* **29**, 164.
BOJARSKI, N.N. (1967). Syracuse University Research Corporation Report.
— (1974). Naval Systems Command Rep. N00019-73-C-9312/F.
BOOR, C. de (1972). *J. Approx. Theory* **6**, 50.
— (1973). M.R.C. Tech. Rep. 1333, University of Wisconsin.
BORN, M. and WOLF, E. (1965). *Principles of optics* (3rd edn.). Pergamon Press, Oxford.
BOROVIKOV, V.A. (1962). *Dokl. Akad. Nauk SSSR* **144**, 527.
— (1973). *Inst. Prikl. Mat. Akad. Nauk SSSR*, Preprint 63.
BOUWKAMP, C.J. and De BRUIJN, N.G. (1946). *Philips res. Rep.* **1**, 135.
BOWMAN, J.J. (1970). *SIAM J. appl. Math.* **18**, 818.
BOYD, W.G.C. (1977). *Proc. R. Soc.* **356**, 315.
BROWN, A.L. and PAGE, A. (1970). *Elements of functional analysis*. Van Nostrand Rheinhold, New York.
BROWN, W.P., Jr. (1966). *J. math. Anal.* **15**, 355.
BROYDEN, C.G. (1970). *J. Inst. Math. Appl.* **6**, 79, 222.
BRYANT, T.G. and WEISS, J.A. (1968). *IEEE Trans.* **MTT-16**, 1021.
BUNEMAN, O. (1969). Rep. 294, Stanford University Institute for Plasma Research, Stanford, Calif.
BURKE, G.J.A. and SELDEN, E.S. (1973). *Antenna modelling program*. Rep. IS-R-72/10. Information Systems Division, MB Associates, Menlo Park, Calif.
BURTON, A.J. (1976). Natl. Phys. Lab. Rep. NPL OC5/535.
BUSLAEV, V.S. (1964). *Trudy matem. Inst. V.A. Steklov, Mat. Fiz.* **73**, 2, 14.
BUTLER, C.M. and WILTON, D.R. (1975). *IEEE Trans. Antennas Propag.* **AP-23**, 534.
BUTLER, G.F. (1972). RAE Tech. Rep. 72126. Royal Aircraft Establishment, Farnborough.
BUZBEE, B.L., DORR, F.W., GEORGE, J.A., and GOLUB, G.H. (1971). *SIAM J. numer. Anal.* **8**, 722.
BUZBEE, B.L., GOLUB, C.H., and NIELSON, C.W. (1970). *SIAM J. numer. Anal.* **7**, 627.
CARASSO, C. (1966). *Methodes numérique pour l'obtention de functions-spline*. Thèse de 3ème Cycle, Université de Grenoble.
CARRÉ, B.A. (1961). *Comput. J.* **4**, 73.
CERMAK, I.A. and SILVESTER, P. (1968). *Proc. IEE* **115**, 1341.
CHANG, S.K. and MEI, K.K. (1974). *URSI Symp. on Electromagnetic Wave Theory*, p. 160. Institution of Electrical Engineers, London.
CHESTER, W. (1950). *Phil. Trans. R. Soc.* **242A**, 527.
— (1950*a*). *Proc. R. Soc.* **203A**, 33.
CIARLET, P.G. (1975). *Seminaire de mathematiques supérieures*. Université de Montreal.
CLARRICOATS, P.J.B., OLVER, A.D., and CHONG, S.L. (1975). *Proc. Inst. Electr. Eng.* **122**, 1173.
CLARRICOATS, P.J.B., and SAHA, P.K. (1971). *Proc. Inst. electr. Eng.* **118**, 1167.
CLARRICOATS, P.J.B. and SALEMA, C.E.R.C. (1973). *Proc. Inst. electr. Eng.* **120**, 741.
CLEMMOW, P.C. and MUNFORD, C. (1952). *Phil. Trans. R. Soc. London* **245A**, 189.
COCHRAN, J.A. (1972). *The analysis of linear integral equations*. Mc Graw-Hill, New York.
COHEN, J.K. and BLEISTEIN, N. (1977). *SIAM J. appl. Math.* 32, 784.
COOLEY, J.W., LEWIS, P.A.W., and WELCH, P.D. (1967). *IEEE Trans.* **AU-15**, 79.
COOLEY, J.W. and TUKEY, J.W. (1965). *Math. Comput.* **19**, 297.

COONS, S.A. (1967). *MIT Project MAC*. Rep. No. MAC-TR-41, MIT, Cambridge, Mass.
CORR, D.G. and DAVIES, J.B. (1972). *IEEE Trans.* **MTT-20**, 669.
COX, M.G. (1972). *J. Inst. Math. Appl.* **10**, 134.
CURRY, H.B. and SCHOENBERG, I.J. (1947). *Bull. am. math. Soc.* **53**, 1114.
— (1966). *J. Analyse Math.* **17**, 71.
DAVIDON, W.C. (1959). AEC Res. Dev. Rep. ANL-5990.
DAVIES, J.B. and MUILWYK, C.A. (1966). *Proc. IEE* **113**, 277.
DESCHAMPS, G.A. (1971). *Electron Lett.* **7**, 684.
DESCHAMPS, G.A. and CABAYAN, H.S. (1972). *IEEE Trans. Antennas Propag.* **AP-20**, 268.
DINGLE, R.B. and MORGAN, G.J. (1967). *Appl. sci. Res.* **18**, 221.
DIXON, L.C.W. (1972). *Math. Prog.* **2**(3), 383.
— (1973). *Math. Prog.* **3**(3), 345.
DOVBYSH, L.N. (GAGEN-TORN) (1962). *Trudy matem. Inst. V.A. Steklov* **66**, 190.
— (1965). *Trudy matem. Inst. V.A. Steklov* **84**, 78.
— (1968), *Trudy matem. Inst. V.A. Steklov* **96**, 188.
DUFF, I.S. (1976). AERE Rep. No. CSS28, AERE, Harwell.
DuHAMEL, R.H. and ISBELL, D.E. (1957). *IRE natl. Conv. Rec.* Pt I, 119.
DUNCAN, J.W. (1967). *IEEE Trans.* **MTT-15**, 575.
DUNCAN, R.H. (1962). *J. Res. nat. Bur. Stand.* **66D**, 181.
FELSEN, L.B. (1976). *J. opt. Soc. Am.* **66**, 751.
FELSEN, L.B. and YEE, H.Y. (1968). *J. acoust. Soc. Am.* **44**, 1028.
FFOWCS WILLIAMS, J.E. and HAWKINGS, D.L. (1969). *Phil. Trans. Roy. Soc.* **264A**, 321.
FISHENDEN, R.M. and WIBLIN, E.R. (1949). *Proc. Inst. electr. Eng.* **96**, 5.
FLETCHER, R. (1970). *Comput. J.* **13**, 317.
— (1971). AERE Rep. TP 453. AERE, Harwell.
— (1972*a*). *Math. Prog.* **2**, 133.
— (1972*b*). AERE Rep. TP 478. AERE, Harwell.
— (1973). AERE Rep. CSS2. AERE, Harwell.
FLETCHER, R. and POWELL, M.J.D. (1963). *Comput. J.* **6**, 163.
FLETCHER, R. and REEVES, C.M. (1964). *Comput. J.* **7**, 149.
FOK, V.A. (1946). *J. Phys.* **10**, 399.
— (1965). *Electromagnetic diffraction and propagation problems.* Pergamon Press, Oxford.
FORGAN, D.H. (1974). RAE Tech. Rep. 74077. Royal Aircraft Establishment, Farnborough.
FORREST, A.R. (1968). Ph.D. Dissertation, University of Cambridge.
FORSYTHE, G.E. and MOLER, C. (1967). *Computer solution of linear algebraic equations.* Prentice-Hall, Englewood Cliffs, N.J.
FOX, L. (1962). *Numerical solution of ordinary and partial differential equations.* Pergamon Press, Oxford.
FOX, L., HENRICI, P., and MOLER, C. (1967). *SIAM J. Numer. Anal.* **4**, 89.
FRANZ, W. (1954). *Z. Naturf.* **9a**, 705.
FRANZ, W. and DEPPERMANN, K. (1952). *Annln Phys.* (6) **10**, 361.
FRANZ, W. and KLANTE, K. (1959). *IRE Trans. Antennas Propag.* **AP-7**, 568.
FREEMAN, T.L., DELVES, L.M., and REID, J.K. (1974). *J. Inst. Math. Appl.* **14**, 145.
FRIEDEN, B.R. and BURKE, J.J. (1972). *J. opt. Soc. Am.* **62**, 1202.
FRIEDLANDER, F.G. (1954). *Commun. pure appl. Math.* **7**, 705.
— (1973). *Proc. London math. Soc. 3*, **27**, 551.
FROST, O.L. (1972). *Proc. IEEE* **60**, 926.
GABOR, D. (1948). *Nature (London)* **161**, 777.
— (1949). *Proc. R. Soc.* **197A**, 454.
— (1951). *Proc. phys. Soc. B* **64**, 449.
— (1956). *Rev. mod. Phys.* **28**, 260.

GALINDO, V. (1964). *IEEE Trans. Antennas Propag.* **AP-12**, 403.
GANS, M.J. (1965). *Proc. IEEE* **53**, 1081.
GAZAZIAN, E.D. and KINBER, B. Ye. (1971). *Radiofizika* **14**, 1219.
GELDER, D. (1970). *Proc. IEE* **117**, 699.
GILL, P.E. and MURRAY, W. (1972). *J. Inst. Math. Appl.* **9**, 91.
— (1974). *Numerical methods for constrained optimisation.* Academic Press, New York.
GILL, P.E., MURRAY, W., and PICKEN, S.M. (1972). *Nat. phys. Lab.* N.A.C. 24.
GILL, P.E., MURRAY, W., and PITFIELD, R.A. (1972). *Nat. phys. Lab.* N.A.C. 11.
GOLDFARB, D. (1969). *SIAM J. appl. Math.* **17**, 739.
GOLDFELD, S.M., QUANDT, R.E., and TROTTER, H.F. (1966). *Econometrica* **34**, 541.
GORIAINOV, A.S. (1958). *Radio Eng. Electron. (USSR)* **3**, 23.
— (1961). *Radiotekh. Elektron.* **6**, 47.
GOUBAU, G. and SCHWERING, F. (1961). *IRE Trans. Antennas Propag.* **AP-9**, 248.
GOURLAY, A.R. and WATSON, G.A. (1973). *Computational methods for matrix eigenproblems.* Wiley, London.
GOURSAT, E. (1942). *Cours d'analyse mathématique,* Tomes II, III. Gauthier-Villars, Paris.
GREEN, A. (1976). Ph.D. Thesis, Polytechnic Institute, New York.
GREEN, H.E. (1965). *IEEE Trans. Microwave Theory Tech.* **MTT-13**, 676.
GREENSTADT, J. (1972). *Math. Comput.* **26**, 145.
GREISER, J.W. and MAYES, P.E. (1961). *Proc. natl. electron. Conf.* **17**, 193.
GRIMSHAW, R. (1966). *Commun. pure appl. Math.* **19**, 167.
HALLÉN, E. (1956). *IRE Trans. Antennas Propag.* **AP-4**, 479.
HARRINGTON, R.F. (1968). *Field computation by moment methods.* Macmillan, New York.
HART, J.F., CHENEY, E.W., LAWSON, C.L., MAEHLY, H.J., MESZTENYI, C.K., RICE, J.R., THATCHER, H.C., and WITZGALL, C. (1968). *Computer approximations.* Wiley, New York.
HAYES, W.P. (1970). *Proc. R. Soc.* **320A**, 209.
HEADING, J. (1975). *Proc. R. Soc. Edinburgh* **73A**, 51.
HEBDEN, M.D. (1973). AERE Rep. T.P. 515. AERE, Harwell.
HEINS, A.E. (1948). *Q. appl. Math.* **6**, 157, 215.
HESS, J.L. (1973). *Conf. Methods appl. Mech. Eng.* **2**, 1.
HESTENES, M.R. (1956). *Proc. Symp. appl. Maths* **6**, 83.
HESTENES, M. and STIEFEL, E. (1952). *J. Res. nat. Bur. Stand.* **49**, 409.
HIZAL, A. (1974). *J. Phys. D: Appl. Phys.* **7**, 248.
HOCKNEY, R.W. (1965). *J. Assoc. comput. Mach.* **12**, 95.
— (1970). *Methods comput. Phys.* **9**, 135.
HONG, S. (1967). *J. math. Phys.* **8**, 1223.
HOOP, A.T. de (1975). *Philips res. Rep.* **30**, 302.
HORNSBY, J.S. and GOPINATH, A. (1969). *IEEE Trans.* **MTT-17**, 684.
HOROWITZ, B.R. and TAMIR, T. (1971). *J. opt. Soc. Am.* **61**, 586.
HOWE, D. (1973). *J. Inst. Math. Appl.* **12**, 125.
HUANG, H.Y. (1970). *J. optim. Theory Appl.* **5**, 405.
IKEBE, Y. (1972). *SIAM Rev.* **14**, 465.
IMAI, I. (1954). *Z. Phys.* **137**, 31.
IMBRIALE, W.A. and MITTRA, R. (1970). *IEEE Trans. Antennas Propag.* **AP-18**, 633.
IRI, M. (1969). *Network flow, transportation and scheduling.* Academic Press, New York.
IRONS, B.M. and RAZZAQUE, A. (1972). In *The mathematical foundations of the finite element method with applications to partial differential equations* (ed. A.K. Aziz). Academic Press, New York.
ISBELL, D.E. (1960). *IRE Trans. Antennas Propag.* **AP-8**, 260.

JAMES, G.L. (1976). *The geometrical theory of diffraction.* Institution of Electrical Engineers, London.
JAMES, J.H. (1973). ARL Rep. ARL/R/R4 Admiralty Research Laboratory.
JOHNSON, M.A. (1968). *Proc. IEEE* **56**, 1801.
JONES, A.F. and MISELL, D.L. (1970). *J. Phys. A.* **3**, 462.
JONES, D.S. (1956). *IRE Trans.* **AP-4**, 297.
— (1963). *Phil. Trans. R. Soc.* **255A**, 341.
— (1964). *The theory of electromagnetism.* Pergamon Press, Oxford.
— (1966). *Generalised functions.* McGraw-Hill, New York.
— (1966a). *J. Inst. math. Appl.* **2**, 197.
— (1967). *Proc. Camb. phil. Soc.* **63**, 1145.
— (1969). *Phil. Trans. R. Soc.* **265A**, 1.
— (1972). *SIAM Rev.* **14**, 286.
— (1973). *J. Inst. Math. Appl.* **12**, 63.
— (1973a). *Proc. R. Soc. Edinburgh A* **71**, 263.
— (1973b). *Q. J. Mech. appl. Math.* **26**, 1.
— (1977). *Prog. aerospace Sci.* **17**, 149.
JONES, D.S. and JORDAN, D.W. (1969). *Introductory analysis,* Vol. I. Wiley, New York.
JONES, R.M. (1968). *Radio Sci.* **3**, 93.
KAMINETZKY, L. and KELLER, J.B. (1972). *SIAM J. appl. Math.* **22**, 109.
KAZARINOFF, N.D. and SENIOR, T.B.A. (1962). *IRE Trans. Antennas Propag.* **AP-10**, 634.
KELLER, J.B. (1957). *J. appl. Phys.* **28**, 426.
— (1962). *J. opt. Soc. Am.* **53**, 116.
KELLOGG, O.D. (1929). *Foundations of potential theory.* Ungar, New York.
KING, R. (1956). *The theory of linear antennas.* Harvard University Press, Cambridge, Mass.
KLEIN, C.A. and MITTRA, R. (1975). *IEEE Trans. Antennas Propag.* **AP-23**, 258.
KOGELNIK, H. and LI, T. (1966). *Proc. IEEE* **54**, 1312.
KRAVSTSOV, Yu. A. (1964). *Radiofizika* **7**, 664, 1049.
KRUSEMAN ARETZ, F.E.J. and ZONNEVELD, J.A. (1975). *Philips Res. Rep.* **30**, 288.
KULSRUD, H.E. (1961). *Commun. Assoc. comput. Mach.* **4**, 184.
KUNZ, K.S. (1957). *Numerical analysis.* McGraw-Hill, New York.
— (1963). *J. Res. nat. Bur. Stand.* **67D**, 417.
KUTT, H.R. (1973). Rep. WISK 132, National Research Institute of Mathematical Science, Pretoria.
LALESCO, T. (1912). *Introduction à la théorie des équations intégrales,* pt. III. Hermann & Fils, Paris.
LAMBERT, J.D. (1973). *Computational methods in ordinary differential equations.* Wiley, New York, 1973.
LAUWERIER, H.A. (1959). *K. ned. Akad. Wet. Amsterdam* **62A**, 476.
— (1960). *K. ned. Akad. Wet. Amsterdam* **63A**, 355.
— (1961). *K. ned. Akad. Wet. Amsterdam* **64A**, 123, 348.
LEE, S.W. (1972). *J. math. Phys.* **13**, 656.
LEE, S.W. and BOERSMA, J. (1975). *J. math. Phys.* **16**, 1746.
LEITH, E.N. and UPATNIEKS, J. (1963). *J. opt. Soc. Am.* **53**, 1377.
LEPPINGTON, F.G. (1968). *Proc. Camb. phil. Soc.* **64**, 1131.
— (1970). RAE Tech. Rep. 70183. Royal Aircraft Establishment, Farnborough.
LEVINSON, N. (1940). *Gap and density theorems.* Am. Math. Soc. Coll. Publ., Vol. 26.
LEWIS, R.M. (1965). *Arch. ration. Mech. Anal.* **20**, 191.
— (1969). *IEEE Trans. Antennas Propag.* **AP-17**, 308.
LEWIS, R.M., BLEISTEIN, N., and LUDWIG, D. (1967). *Commun. pure appl. Math.* **20**, 295.
LIEBECK, H. (1969). *Algebra for scientists and engineers.* Wiley, London.

LOGAN, N.A. (1969). Reps. LMSD-288087 and LMSD-288088, Missiles and Space Division, Lockhead Aircraft Corp.
LOOTSMA, F.A. (1972). *Numerical methods for non-linear optimisation.* Academic Press, New York.
LOWDON, T.A. (1970). *Q. J. Mech. appl. Math.* **23**, 315.
LUDWIG, A.C. (1969). *Jet Propulsion Lab. Pasadena SPS37-26* **4**, 200.
LUDWIG, D. (1966). *Commun. pure appl. Math.* **19**, 215.
— (1967). *Commun. pure appl. Math.* **20**, 103.
— (1975). *SIAM Rev.* **17**, 1.
LÜNEBURG, E. and WESTPFAHL, K. (1975). *Annln Phys.* **7**, 166.
McDONALD, B.H. and WEXLER, A. (1972). *IEEE Trans. Microwave Theory Tech.* **MTT-21**, 841.
MacDONALD, H.M. (1902). *Electric waves.* Cambridge University Press, London.
— (1913). *Phil. Trans. R. Soc.* **212A**, 299.
McLEOD, R. and MITCHELL, A.R. (1972). *J. Inst. Math. Applic.* **10**, 382.
MALYUZINEC, G.D. (1958). *Annln Phys.* **6**, 107.
MARIN, L. (1973). *IEEE Trans. Antennas Propag.* **AP-21**, 809.
— (1974). *IEEE Trans. Antennas Propag.* **AP-22**, 266.
MARIN, L. and LATHAM, R.W. (1972). *Proc. IEEE* **60**, 640.
MARSHALL, J.A. and MITCHELL, A.R. (1973). *J. Inst. Math. Appl.* **12**, 355.
MASTERMAN, P.H. and CLARRICOATS, P.J.B. (1971). *Proc. IEE* **118**, 51.
MATTHEWS, A. and DAVIES, D. (1971). *Comput. J.* **14**, 293.
MAUTZ, J.R. and HARRINGTON, R.F. (1969). *Appl. Sci. Res.* **20**, 405.
— (1973). Syracuse Univ. Tech. Rep. TR-73-9.
MEI, K.K. (1974). *IEEE Trans. Antennas Propag.* AP-22, 760.
MEINKE, H.H. and BAIER, W. (1966). *Nachrichtentech. Z.* **11**, 662.
MEINKE, H.H., LANGE, K.P., and RUGER, J.F. (1963). *Proc. IEEE* **51**, 1436.
MEYER, R.E. (1975). *Proc. IEEE* **63**, 1070.
— (1975*a*). *SIAM J. appl. Math.* **29**, 481.
— (1976). *J. math. Phys.* 17, 1039.
MIKHLIN, S.G. (1956). *Dokl. Akad. Nauk SSSR* **106** (3), 391.
MILLER, E.K. (1972). Lawrence Livermore Lab. Rep. UCRL-51276.
MILLER, E.K. and DEADRICK, F.J. (1973). Rep. UCRL 74818. Lawrence Livermore Laboratory, University of California.
MILLER, E.K. and MORTON, J.B. (1970). *IEEE Trans. Antennas Propag.* **AP-18**, 290.
MILNE, K. (1964). *Radio electron. Eng.* **28**, 89.
MILNE-THOMSON, L.M. (1960). *The calculus of finite differences.* Macmillan, London.
MITCHELL, A.R. (1969). *Computational methods in partial differential equations.* Wiley, London.
MITCHELL, A.R. and WAIT, R. (1977). *The finite element method in partial differential equations.* Wiley, New York.
MITTRA, R. (1963). *J. Res. Nat. Bur. Stand.* **67D**, 245.
MITTRA, R. and ITOH, T. (1971). *IEEE Trans.* **MTT-19**, 47.
MITTRA, R., ITOH, T., and LI, T. (1972). *IEEE Trans.* **MTT-20**, 96.
MORAWETZ, C.S. and LUDWIG, D. (1968). *Commun. pure appl. Math.* **21**, 187.
MORGAN, M.A. and MEI, K.K. (1974). *URSI Symp. on Electromagnetic Wave Theory,* p. 163. Institution of Electrical Engineers, London.
MÜLLER, C. and NIEMEYER, H. (1961). *Arch. ration. Mech. Anal.* **7**, 305.
MÜLLER, C. (1969). *Foundations of the mathematical theory of electromagnetic waves.* Springer, Berlin.
MUR, G., QUAK, D., and van DIJK, G.J. (1976). *ICCAD Conf. on Numerical Methods.* Santa Margherita Ligure.
NAPIER, P.J. and BATES, R.H.T. (1971). *Int. J. eng. Sci.* **9**, 1193.
— (1973). *Proc. Inst. electr. Eng.* **120**, 30.

NOBLE, B. (1960). *Proc. Symp. Int. Computer Center, Rome*, p. 540. Springer, New York.
— (1971). *Conf. on Application of Numerical Analysis, Dundee*, p. 137. Springer, New York.
NOBLE, B. and SEWELL, M.J. (1972). *J. Inst. Math. Appl.* **9**, 123.
NOKES, S.A. (1974). Ph.D. Thesis, University of London.
NOKES, S.A., BERNAL, M.J.M., and DAVIES, J.B. (1974). *1974 URSI Symp. on Electromagnetic Wave Theory, IEE Conf. Publ. No. 114*, p. 12, IEE, Stevenage.
OLVER, F.W.J. (1965). *J. Res. nat. Bur. Stand.* **69B**, 291.
— (1974). *Asymptotics and special functions*. Academic Press, New York.
OSHIRO, F.K., TORRES, F.P., and HEATH, H.C. (1966). U.S. Air Force Avionics Lab. Tech. Rep. AFAL-TR-66-162.
OTIS, G. (1974). *J. opt. Soc. Am.* **64**, 1545.
PALEY, R.E.A.C. and WIENER, N. (1934). *Fourier transforms in the complex domain*. Am. Math. Soc. Coll. Publ., Vol. 19.
PARINI, C., CLARRICOATS, P.J.B., and OLVER, A.D. (1975). *Electron. Lett.* **11**, 567.
PATHAK, P.H. and KOUYOUMJIAN, R.G. (1974). *Proc. IEEE* **62**, 1438.
PEARSON, L.W. and BUTLER, C.M. (1975). *IEEE Trans. Antennas Propag.* **AP-23**, 295.
PERRY, W.L. (1974). *IEEE Trans. Antennas Propag.* **AP-22**, 826.
PETERS, G. and WILKINSON, J.H. (1969). *Comput. J.* **12**, 398.
POCKLINGTON, H.C. (1897). *Proc. Camb. Phil. Soc.* **9**, 324.
POGGIO, A.J. and MILLER, E.K. (1970). MBA Rep. No. MB-TM-70/20.
POLAK, E. and RIBIERE, G. (1969). *Rev. Fr. Inform.* **16-R1**, 35.
PONTOPPIDAN, K. (1969). *European Microwave Conf., IEE Conf. Publ.* **58**, p. 99. IEE, Stevenage.
POTTER, P.D. (1967). *Microwave J.* **AP-15**, 727.
POULTON, G.T. (1975). *Proc. 5th European Microwave Conf.*, p. 61. IEE.
POWELL, M.J.D. (1966). In *Numerical analysis: an introduction* (ed. J. Walsh). Academic Press, London.
— (1971a). *J. Inst. Math. Appl.* **7**, 21.
— (1971b). U.K.A.E.A. Res. Rep. TP 459.
RA, J.W., BERTONI, H.L., and FELSEN, L.B. (1973). *SIAM J. appl. Math.* **24**, 396.
RALSTON, A. and WILF, H.S. (1967). *Mathematical methods for digital computers*, Vol. II. Wiley, New York.
RANSOM, P.L. and MITTRA, R. (1971). *Proc. IEEE* **59**, 1029.
RAWLINS, A.D. (1972). Ph.D. Thesis, University of Surrey.
REID, J.K. (1966). *Comput. J.* **9**, 200.
RICE, S.O. (1954). *Bell Syst. Tech. J.* **33**, 417.
RICHMOND, J.H. (1965). *Proc. IEEE* **53**, 796.
— (1966). *IEEE Trans. Antennas Propag.* **AP-14**, 782.
RICHTMYER, R.D. and MORTON, K.W. (1967). *Difference methods for initial-value problems*. Wiley-Interscience, New York.
ROBERTS, A. and RUNDLE, K. (1972). British Aircraft Corp. Rep. Aero MA 19.
ROBERTS, R. (1972). *Inst. Phys. Fibre Optical Communications*. Inst. of Physics.
— (1973). 8th Progress Rept, University of Dundee.
— (1975). *Electron Lett.* **11**, No.22.
ROZZI, T.E. (1973). *Int. J. circuit Theory Appl.* **1**, 161.
ROZZI, T.E. and MECKLENBRÄUKER, W.F.G. (1974). *Philips res. Rep.* M.S. 8433.
RUMSEY, V.H. (1957). *IRE natl. Conv. Rec.* Pt. I, 114.
RYAN, C.E., Jr. and PETERS, L. Jr. (1968). *IEEE Trans. Antennas Propag.* **AP-16**, 274.
RYLE, M. (1952). *Proc. R. Soc.* **211A**, 351.
SARD, A. and WEINTRAUB, S. (1971). *A book of splines*. Wiley, New York.

SAYRE, E.P. and HARRINGTON, R.F. (1968). *Proc. Int. Antennas Propagation Symp., Boston,* p. 160. IEEE.
SCHOENBERG, I.J. (1946). *Q. appl. Math.* **4**, 45, 112.
SEEGER, J.A. (1968). *Proc. IEEE* **56**, 1393.
SENIOR, T.B.A. (1969). *IEEE Trans. Antennas Propag.* **AP-17**, 378, 751.
— (1971). *Electron. Lett.* **7** (10), 87.
— (1971a). *Appl. sci. Res.* **23**, 459.
— (1972). *IEEE Trans. Antennas Propag.* **AP-20**, 326.
— (1975). *Radio Sci.* **10**, 645, 911.
SENIOR, T.B.A. and USLENGHI, P.L.E. (1971). *Radio Sci.* **6**, 393.
— (1973). *Radio Sci.* **8**, 247.
SHAMPINE, L.F., WATTS, H.A., and DAVENPORT, S.M. (1976). *SIAM Rev.* **18**, 376.
SHANKS, H.E. (1961). *IRE Trans. Antennas Propag.* **AP-9**, 162.
SHANKS, H.E. and BICKMORE, R.W. (1959). *Can. J. Phys.* **37**, 263.
SHANNO, D.F. (1970). *Math. Comput.* **24**, 647.
SHARPLES, A. (1962*a*). *Proc. Camb. Phil. Soc.* **58**, 662.
— (1962*b*). *Q. J. Mech. appl. Math.* **15**, 253.
SHAW, E. and DAVIES, D.E.N. (1964). *Radio electron. Eng.* **27**, 279.
SHAW, R.P. (1974). *J. acoust. Soc. Am.* **56**, 1354.
SHELTON, J.P. and KELLEHER, K.S. (1961). *IRE Trans. Antennas Propag.* **AP-9**, 154.
SILVESTER, P. (1968). *Proc. IEE* **115**, 43.
— (1970). *IEEE Trans.* **MTT-18**, 63.
SILVESTER, P. and HSIEH, M.S. (1971). *Proc. Inst. electr. Eng.* **118**, 1743.
SNYDER, A.W. and LOVE, J.D. (1975). *IEEE Trans. Microwave Theory Tech.* **MTT-23**, 134.
SOLOGUB, V.G. (1972). *USSR comput. Math. math. Phys.* **12**, 135.
SONDHI, M.M. (1972). *Proc. IEEE* **60**, 842.
STINEHELFER, H.E., Sr. (1968). *IEEE Trans.* **MTT-16**, 439.
STRANG, G. and FIX, G.J. (1973). *An analysis of the finite element method.* Prentice-Hall, Englewood Cliffs, N.J.
TAKAHASI, H. and MORI, M. (1973). *Num. Math.* **21**, 206.
TAYLOR, C.D. and WILTON, D.R. (1972). *IEEE Trans. Antennas Propag.* **AP-20**, 772.
TAYLOR, T.T. (1955). *IRE Trans. Antennas Propag.* **AP-3**, 16.
TEE, G.J. (1963). *Computer J.* **6**, 177.
TESCHE, F.M. (1973). *IEEE Trans. Antennas Propag.* **AP-21**, 53.
TESCHE, F.M. and NEUREUTHER, A.R. (1970). *IEEE Trans. Antennas Propag.* **AP-18**, 692.
THIELE, G.A., TRAVIESO-DAVIES, M., and JONES, H.S. (1969). *Proc. Conf. on Environmental Effects on Antenna Performance.* Pergamon Press.
THONG, T. and LIU, B. (1977). *IEEE Trans.* **CAS-24**. 132.
TIHONOV, A.N. (1944). *Dokl. Akad. Nauk SSSR* **39**, 332.
— (1964). *Sov. Math. Dokl.* **5**, 835.
TITCHMARSH, E.C. (1939). *The theory of functions.* Oxford University Press, London.
TOKATLY, V.I. and B. Ye. KINBER (1971). *Radiofizika* **14**, 761.
URSELL, F.J. (1957). *Proc. Camb. phil. Soc.* **53**, 115.
— (1968). *Proc. Camb. phil. Soc.* **64**, 171.
VAN BLARICUM, M.L. and MITTRA, R. (1975). *IEEE Trans. Antennas Propag.* **AP-23**, 777.
VAN BUREN, A.L. (1970), NRL Rep. 7160.U.S. Govt., Washington.
VAN DANTZIG, D. (1958). *K. ned. Akad. Wet. Amsterdam* **61A**, 384.
VAINIKKO, G.M. (1968). *Dokl. Akad. Nauk SSSR* **179**, 1029.
— (1965). *Uch. Zap. Tartusk. Inst.* **73**, 182.
VARGA, R.S. (1962). *Matrix iterative analysis.* Prentice-Hall, Englewood Cliffs, N.J.
WACHSPRESS, E.L. (1973). *J. Inst. Math. Applic.* **11**, 83.

WAIT, J.R. and CONDA, A.M. (1958). *Trans. Inst. radio Eng.* **AP-6**, 1957.
— (1959). *J. Res. nat. Bur. Stand.* **63D**, 181.
WAIT, R. (1973). Ministry of Defence Res. Rep. University of Dundee.
WALL, D.J.N. and BATES, R.H.T. (1975). *IEEE Trans. Microwave Theory Tech.* **MTT-23**, 605.
WATERMAN, P.C. (1965). *Proc. IEEE* **53**, 805.
— (1971). *Phys. Rev. D* **4**, 825.
WEINSTEIN, L.A. (1948). *Izv. Akad. Nauk Ser. Fiz.* **12**, 144, 166.
— (1969). *The theory of diffraction and the factorization method.* Golem Press. Boulder, Colorado.
WESTON, V.H. (1962). *IRE Trans. Antennas Propag.* **AP-10**, 775.
— (1965). *Radio Sci.* **69D**, 1257.
WHEELER, H.A. (1965). *IEEE Trans.* **MTT-13**, 172.
WHITE, F.P. (1922). *Proc. R. Soc.* **100A**, 505.
WHITE, W.D. (1962). *IRE Trans. Antennas Propag.* **AP-10**, 430.
WHITING, K.B. (1968). *IEEE Trans.* **MTT-16**, 889.
WHITTAKER, E.T. and ROBINSON, G. (1952). *The calculus of observations.* Blackie, London.
WILKINSON, J.H. (1965). *The algebraic eigenvalue problem.* Clarendon Press, Oxford.
WILLIAMS, W.E. (1959). *Proc. R. Soc.* **252A**, 376.
— (1961). *Appl. Sci. Res.* **9B**, 21.
WILLIAMS, W.F. (1965). *Microwave J.* **8**, 79.
WILTON, D.T. (1973). Ministry of Defence Res. Rep., Pts 1, 2, University of Dundee.
WOLF, E. (1969). *Opt. Commun.* **1**, 153.
— (1970). *J. opt. Soc. Am.* **60**, 18.
WOLF, E. and SHEWELL, J.R. (1970). *J. math. Phys.* **11**, 2254.
WOLFE, P. (1971). *SIAM Rev.* **13**, 185.
WOOD, P.J. (1970). *Electron Lett.* **6**, 326.
— (1971). *Marconi Rev.* **34**, 149.
— (1972). *Marconi Rev.* **35**, 121.
WOODWARD, P.M. and LAWSON, J.D. (1948). *J. Inst. electr. Eng.* **95**, 363.
WU, T.T. (1956). *Phys. Rev.* **104**, 1201.
WU, T.T. and RUBINOW, S.I. (1956). *J. appl. Phys.* **27**, 1032.
YASKOVA, G.N. and YAKOVLEV, M. (1962). *Trudy matem. Inst. V.A. Steklov* **66**, 182.
YEE, H.Y. and FELSEN, L.B. (1969). *IEEE Trans. Microwave Theory Tech.* **MTT 17**, 73, 671.
YEE, H.Y., FELSEN, L.B., and KELLER, J.B. (1969). *SIAM J. appl. Math.* **16**, 268.
YOUNG, D.M. (1971). *Iterative solution of large linear systems.* Academic Press, London.
ZOUTENDIJK, G. (1970). In *Integer and non-linear programming* (ed. J. Abadie). North-Holland, Amsterdam.

INDEX: VOLUMES 1 AND 2

[Volume 1 comprises pp. 1-410; Volume 2 comprises pp. 411-864.]